FORTSCHRITTE
DER CHEMIE ORGANISCHER
NATURSTOFFE

PROGRESS IN THE CHEMISTRY
OF ORGANIC NATURAL PRODUCTS

PROGRÈS DANS LA CHIMIE
DES SUBSTANCES ORGANIQUES
NATURELLES

HERAUSGEGEBEN VON   EDITED BY   RÉDIGÉ PAR

L. ZECHMEISTER

CALIFORNIA INSTITUTE OF TECHNOLOGY, PASADENA

VIERUNDZWANZIGSTER BAND
TWENTY-FOURTH VOLUME   VINGT-QUATRIÈME VOLUME

VERFASSER   AUTHORS   AUTEURS
K. BIEMANN · H. ERDTMAN · H. FRAENKEL-CONRAT
O. HOFFMANN-OSTENHOF · H. KINDL · T. NORIN · R. TSCHESCHE
A. B. TURNER · F. L. WARREN

MIT 25 ABBILDUNGEN WITH 25 FIGURES AVEC 25 ILLUSTRATIONS

1966

SPRINGER SCIENCE+BUSINESS MEDIA, LLC

ISBN 978-3-7091-8145-4   ISBN 978-3-7091-8143-0 (eBook)
DOI 10.1007/978-3-7091-8143-0

© 1966 SPRINGER SCIENCE+BUSINESS MEDIA NEW YORK
ORIGINALLY PUBLISHED BY SPRINGER-VERLAG / WIEN · NEW YORK 1966
SOFTCOVER REPRINT OF THE HARDCOVER 1ST EDITION 1966
LIBRARY OF CONGRESS CATALOG CARD NUMBER AC 39-1015

Titel Nr. 8231

Inhaltsverzeichnis.
Contents. — Table des matières.

Cyclite: Biosynthese, Stoffwechsel und Vorkommen. Von H. KINDL

und O. HOFFMANN-OSTENHOF, Organisch-chemisches Institut der

The Chemistry of the Order Cupressales. By H. ERDTMAN, Department of Organic Chemistry, Royal Institute of Technology and T. NORIN, Swedish Forest Products Research Laboratory, Stockholm

Mass Spectrometry
of Selected Natural Products.

By **Klaus Biemann**, Cambridge, Massachusetts.

With 12 Figures.

Contents.

I. Introduction.

During the past three decades the organic chemist has become in-creasingly used to take advantage of more and more complex instrumenta-tion and physical measurements in lieu of laborious, time-consuming and often ambiguous chemical transformations. Mass spectrometry is perhaps the most recent, most complex and most expensive addition to this field. In view of the astonishingly quick acceptance of nuclear magnetic reso-nance by the organic chemist it is, in retrospect, surprising that he has neglected mass spectrometry for such a long time. This can be explained, in part, by the complexity of the instrumentation and some technical shortcomings of the earlier commercially available instruments but, to an even greater extent, it reflects also the prejudices against a technique that was originally mainly used for quantitative gas analysis. The usefulness of mass spectrometry as a qualitative technique in organic chemistry rather than a tool for quantitative analysis was more and more recognized towards the end of the last decade. A rather spectacular development followed during the intervening few years to the point that now any reasonably well equipped modern organic laboratory is supplied with, or at least has access to, one or more mass spectrometers suitable for work on organic compounds.

Within the realm of organic chemistry the technique has become much more important, if not indispensable, for the natural products chemist while its application to synthetic problems is much less pro-nounced. And even within the field of natural products chemistry application and success vary from alkaloids, in particular indole alkaloids on the one hand, in which hardly any work is done nowadays without involving a mass spectrum, to polycyclic aromatic compounds on the other hand, where the applicability is much more restricted.

While over the past few years a number of reviews (*18, 178, 197*) and books (*17, 56, 57, 58*) covering the entire area of the application of mass spectrometry to natural products chemistry or specific aspects thereof have appeared, it may now be appropriate to summarize the place which mass spectrometry presently occupies in this field.

Let us first consider the attributes which made mass spectrometry so attractive for the work on the determination of the structure of natural products. Since the spectrum can be obtained on a very small amount of not even pure material, it is an excellent method for following the isolation of a particular substance or detecting a new compound present in a mixture being separated. The mass spectrum of the isolated pure substance permits one to draw certain conclusions concerning its structure which may range from the deduction of the entire structure in very

References, pp. 86—98.

advantageous cases to more information about the molecular size or some particular structural detail in less favorable situations.

A consequence of the sensitivity and the tolerance of even large amounts of impurities is the information that can be gained from simple chemical reactions performed on a milligram or sub-milligram level. Comparison of the mass spectrum of the crude product, without any purification, with that of the starting material often yields information concerning the presence (or absence) of certain reactive groupings within the molecule and sometimes even pinpoints the area in which this group is located. Of particular value are reactions which lead to the specific incorporation of heavy isotopes (most commonly deuterium) and comparison of the mass spectrum of the product with the one obtained with unlabeled reagent. These reactions have been summarized (*17, 57*) and the more important ones are mentioned in later Sections.

·While discussing the unique advantages of mass spectrometry one should not fail to point out the one major handicap of the technique, namely the requirement of a certain vapor pressure at a temperature below the thermal decomposition point. Considerable improvements have been made over the last five years in instrumental design and development of experimental techniques, and it can now be said that the vast majority of compounds up to a molecular weight of 1000 may be expected to give usable mass spectra (unless the compound were particularly heat sensitive). This is precisely the range into which most of the non-polymeric natural products fall.

Considerable progress can be expected in an area that is not a part of conventional natural products chemistry, namely the detection and characterization of relatively small and often simple molecules produced from certain sources in extremely small quantities, and thus defying isolation in visible and weighable amounts. Pheromes and sex attractants produced by insects fall in this category, and their identification is of considerable interest as they are one of the means for active (or passive) communication between insects over very long distances; furthermore, they may be used for a chemical correlation of various species (*141*). Due to the nature of these substances (relatively small molecules of considerable volatility) they lend themselves particularly well to characterization by mass spectrometry which is sufficiently sensitive to detection, while their low concentration virtually rules out all other techniques of identification unless one works up large amounts of source material – a time-consuming and expensive approach.

In contrast to the interest mass spectrometry has aroused among natural product chemists, the technique has had a somewhat less startling influence on synthetic chemistry. This is mainly due to the fact that neither the sensitivity nor the particular information which mass spectrometry offers is of such a vital consequence in synthetic work. There are always the few milligrams of quite pure material available to make good and immediate use of more conventional approaches such as infrared and NMR spectroscopy and to obtain routine combustion analyses. Unless a product is the result of a very obscure and unexpected side reaction, sufficient information about the system is available and the product can be identified in a conventional manner. For this reason mass spectrometry plays as yet much more the role of an auxiliary technique in synthetic chemistry as compared to the work on natural products.

But even within the latter field there developed a remarkable gradient in the usefulness and success of mass spectrometry. While there is hardly a single new recent alkaloid structure in whose determination mass spectrometry did not play an important role – and there are, in fact, a considerable number of structures which were deduced exclusively on mass spectrometric evidence – there are very few carbohydrates, for example, whose structures were determined mass spectrometrically although the number of papers concerned with the mass spectra of carbohydrates is considerable. Various other types of compounds fall in between. The mass spectra of fatty acids, amino acids and peptides, for example, have been investigated in considerable detail and this work has led to a satisfactory number of structures determined by this technique. On the other hand, the ratio of comparative work on known substances versus structures determined is much less favorable in the case of steroids and terpenes. It might be instructive to shed some light on the reasons for these variations, since they represent an excellent illustration for the somewhat selective applicability of mass spectrometry which should be taken into account whenever its use for the solution of a structural problem is considered.

The remarkable success of mass spectrometry in the elucidation of the structure of alkaloids, particularly indole alkaloids during the past few years is undoubtedly due to a number of factors, not necessarily all related to the technique itself.

Firstly, indole alkaloids represent a system of a nitrogen-containing heteroaromatic nucleus which gives the molecule a certain degree of stability after electron impact, leading to abundant, well recognizable molecular ions as well as preventing easy cleavage of the bonds present in the aromatic part. Secondly, the alicyclic nitrogen-containing system provides some bonds causing the specific fragmentation of that part of the molecule and the ensuing pattern is thus characteristic of a particular ring system. By comparison, steroids, for example, lack the aromatic system (except the estrogens) as well as the strongly fragmentation directing nitrogen, while they possess many carbon-carbon bonds of similar tendency to cleave upon electron impact (58). This makes a mass

References, pp. 86—98.

spectrum much more susceptible to the influence of minor structural changes which can be so manyfold as to be difficult to assess and predict. Furthermore, we know from experience that the functional groups present in indole alkaloids are very limited. For example, carbomethoxy groups are very common while carbethoxy groups are virtually unknown. So are aromatic ethoxyls in contrast to the abundant methoxyl groups. Biogenetic considerations, if not experimentally proven at least empirically established, virtually exclude various positions from bearing certain substituents. All these pieces of information often lead quickly to a working hypothesis which can be subjected to rather direct proof by relating it to one of the ever growing number of known alkaloids; and even in this comparison mass spectrometry can be used again to advantage.

Last, but not least, it seems that an important contributor to the numerical success of mass spectrometry in the elucidation of the structure of indole alkaloids was the fact that many of them had been isolated during the latter part of the last decade as a result of what one may call the "post-reserpine rush". The screening of plants of the *Apocynaceae* family has produced a large number of alkaloids which, however, did not lend themselves to quick structure determination by conventional techniques until the power of mass spectrometry was demonstrated in this area (*21*).

One or more of these factors are missing in most other areas to which mass spectrometry has been applied. Most often the lack of the last one, viz. the availability of a number of already isolated substances awaiting structural elucidation by a more suitable technique has prevented equally spectacular success in quantity. This might be true for amino acids which lend themselves rather well to mass spectrometry (*34*), because of their relatively linear structure and the presence of the fragmentation-directing amino group as well as other functionalities.

The mass spectrometric behavior of fatty acids and related compounds has been studied extensively (*179*) and while their reasonable volatility and linear structure aids in both the determination and interpretation of the spectra, the lack of strongly fragmentation-directing functional groups as well as the co-occurrence of fatty acids of unknown structure with large amounts of known related compounds hinders work in this field. The growing interest of the biochemist in mass spectrometry and the more sophisticated use of gas chromatography in conjunction with the mass spectrometer (*177, 206, 207*) might, however, change the situation rapidly in the near future.

On the other side of the scale there are the polycyclic aromatic substances which either have no substituents that are susceptible to characteristic fragmentation (such as certain aromatic oxygen heterocycles) or which differ mainly in the position of substituents on an aromatic system (such as in prophyrines). In these cases the mass spectrum may only permit recognition of the molecular size of the compound and the nature of some of the substituents – information which may or may not be vital.

Before entering into a discussion of specific examples of structural problems that have been solved with the aid of mass spectrometry, it may be worthwhile to briefly summarize the present state of *instrumen-*

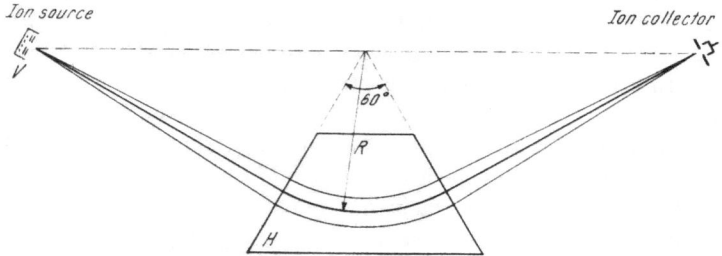

Fig. 1. Principle of a sector (60°) instrument.

tation from the point of view of the natural product chemist. The well established magnetic deflection instrument has held the dominant place while the more novel systems, such as time-of-flight, quadrupole, monopole and synchrotron-type instruments did not survive the rigorous reliability tests in the hands of the organic chemist. Reliable performance over a wide mass range is most important in this field, while high speed and ultra-high sensitivity, characteristics which are important for the physical or analytical chemist, are of secondary significance for the organic chemist.

It also turned out that the lower cost of construction of some of these resolving systems vanishes in the final product because the auxiliary components such as inlet systems, vacuum systems, recording systems, etc. contribute most to the final price of the instrument.

In a magnetic-deflection instrument, the positive ions produced in the ion source by bombardment of the molecules present (at a pressure of 10^{-4}–10^{-6} mm. Hg) with electrons of 10–70 eV energy are first accelerated by an electric field of a few thousand volts. Upon entrance into a magnetic field they are deflected and refocused at a collector slit. Ion source and flight path are held at a pressure of 10^{-6} mm. or lower. Commonly used angles of deflection are 60°, 90° *(Fig. 1)* or 180°. The relationship between the mass of an ion (m), its charge (e), the accelerating potential (V), the magnetic field (H) and the radius (R) of the path which the ion describes is related as follows (C is a constant):

$$\frac{m}{e} = \frac{H^2 R^2}{2 V} C.$$

References, pp. 86—98.

Thus, using a fixed position (R) for the collector slit, the mass of a unipositively charged ion passing through the slit is determined by the magnetic field (H) as well as the accelerating potential (V). To obtain a mass spectrum (i. e. a recording of the abundance of ions versus their mass-to-charge ratio) one can vary either H or V. The ion-current passing through the collector slit is measured with the aid of a Faraday collector or an electron-multiplier followed by amplification. Because of the great variations of abundance of significant ions as well as the high signal-to-noise ratio of a mass spectrometer, 3- or 5-channel recording oscillographs are commonly used to display the spectrum.

An integral part of any mass spectrometer used to produce spectra of a stationary sample is the so-called inlet system which facilitates vaporizing the sample into a fixed volume from which it streams in well controlled fashion into the ion source and from where it can be removed later by a separate vacuum system.

A mass spectrometer, consisting of inlet system; ion source with flight path; collector with amplifier and recording system; and vacuum system for sample handling and spectrometer tube, represents, therefore, a reasonably complex instrument combining high-vacuum technique, high-voltage problems and electronics. This, combined with the fact that, in contrast to other spectroscopic techniques used by the organic chemist, the sample is placed into the instrument (rather than a separate cell or container) is the reason why the successful and trouble-free operation of a mass spectrometer requires a certain extent of experimental skill and experience.

A more detailed discussion of the physical principle of all these mass resolving systems and their design would be beyond the scope of this review and the reader is referred to various sources in which this subject is discussed in detail (*15, 146*).

The de-emphasis of the quantitative aspects and the desire to obtain usable qualitative mass spectra of compounds of rather low volatility or thermal instability led to the development of techniques for the introduction of the sample directly into the ion source near the electron beam, permitting the determination of spectra of compounds with a vapor pressure as low as 10^{-6} mm. Hg at 200°. This is generally accomplished by the use of vacuum locks which range from simple homemade designs to elaborate automatic versions; and such devices are now commonly available with commercial instruments.

The ion sources themselves are still almost exclusively of the electron beam type, the only novel development being the field-ionization principle (*10*) in which a very high, locally concentrated potential is used to extrude an electron from the molecule. The resulting mass spectrum is characterized by a much more abundant molecular ion and the virtual absence of fragment ions, a characteristic which might be very useful in combination with a more conventional form of ionization that produces the structurally characteristic fragments. The methods of ionization generally used may be most in need of a revolutionary change, and field-ionization or photo-ionization with light beams of rather short wavelengths (*16, 50*) might become much more widely used in the near future.

Presently, a second generation of mass spectrometers is coming more and more in use and these are so-called double focusing or high resolution instruments. Their focusing properties are orders of magnitudes better than those of single focusing conventional instruments and thus permit an accuracy in the mass measurement in the order of a millimass unit*, making possible the determination of the elemental composition of the ions directly from their mass (*15*). Obviously, such data are of great value in the interpretation of the spectrum of an unknown compound as well as in the determination of the elemental composition of a new natural product long before it can be isolated in a highly purified state and in quantities sufficient for conventional elemental analysis.

The approach to the interpretation of complete high resolution mass spectra differs somewhat from that used in conventional mass spectrometry and it will be discussed in more detail in a later Section (p. 77).

Having outlined the present state of mass spectrometry in relation to natural product chemistry in a comparative and general manner, the various aspects shall be illustrated and documented on the basis of a detailed discussion of specific examples.

Before doing so the basic *principles of the interpretation of mass spectra* shall be briefly reviewed. It is generally acknowledged that upon electron impact the molecule becomes ionized (by loss of one or more electrons to form singly or multiply charged ions) and then may or may not decompose into fragments. Which of the many bonds present in a molecule of reasonable size will be cleaved depends mainly on the bond strength (the majority of bonds broken are C—C bonds, however) and the energy content (inductive or resonance stabilization) of the neutral fragment (molecule or radical) and of the positive ion. The latter is the most important aspect, and the formation of a well stabilized ion is thus an important requirement of any fragmentation process proposed in the course of the prediction of the mass spectrum of a compound of known structure or in the rationalization of a mass spectrum for the purpose of supporting a suggested structure of a new compound. In the course of multistep fragmentation processes it is important to choose a path during which n-1 bonds are made for n bonds broken, to assure that the energy requirement of the proposed process is a reasonable one.

During the past few years it has become commonplace to depict sometimes very detailed fragmentation processes which ionized organic compounds are assumed to undergo. This is vitally important in some situations but a waste of time or even confusing in others. A few examples will bear this out.

In the course of the determination of a structure, mass spectrometric evidence has to be accompanied by a rational explanation why a compound should, upon electron impact, decompose into positively

---

* 0,001 atomic mass unit.

References, pp. 86—98.

charged particles whose mass-to-charge ratio corresponds to abundant signals in the spectrum. Such cleavage processes have to be of low energy requirement, i. e. cleavage of as few bonds as possible, avoidance of unpaired electrons, production of positive charges which are delocalized by inductive or resonance effects. Analogies with known systems which give the same (or analogous) peaks in the mass spectrum provide welcome and convincing support.

One of the best ways to generate such analogies for immediate or future reference is the determination of mass spectra of authentic compounds and correlation of the data obtained with the known structure of these substances. If carried out on a series of related compounds (rather than on a single one), the general applicability of the method is tested at the same time, which is very important because a process that is peculiar only to a single compound is rather useless as a general analogy. A large amount of work has been done in this direction and it is, in fact, the basis – directly or indirectly – for the actual success over the past few years of mass spectrometry in the application to structural problems.

In some instances this desirability to accumulate knowledge about generally applicable systems and fragmentation processes has led to the presentation of highly speculative, involved fragmentation processes supposed to rationalize more or less abundant (sometimes even minute) peaks in a mass spectrum of a known compound. While any well supported fact or theory is a contribution to the field, unsupported speculations create only confusion and are even prone to discredit the technique. Some of the reasons for presenting such speculations seem to be the discrepancy between the ease with which a mass spectrum can be obtained and rationalized on paper and the great amount of experimental work that is required to put these rationalizations on a firm basis for example through extensive isotope labeling and high resolution spectra.

With more and more organic laboratories becoming equipped with mass spectrometers, the tendency becomes greater and greater to publish interpretations of the mass spectra of often rather trivial known compounds, putting forward mechanistic proposals which are as detailed as they are unsupported by experiments. Most of them would not survive rigorous experimental testing, but since this is a difficult and laborious task (precisely the reason why it was not done in the first place), such speculations will probably remain a part of the literature. One can hope that this fad will eventually subside and make space for a more cautious approach.

One of the reasons for this situation is the wide range of significance of the various pieces of information and conclusions provided by mass spectrometry.

The original data, namely the mass of the ion produced, is perhaps the most fundamental fact concerning a compound which one can obtain in an experiment and which may give a mass spectrometric argument whose value is based on undisputable ground. It should not be overlooked that the mass of the ion indicates only the number and kind of elements present* in the ion and that the conclusions that follow are not such fundamental facts. For example, which individual atoms (of those elements occurring more than once) are present can be decided unequivocally only by labeling each one with a stable isotope and this is already prohibitive. Fortunately, often only one or a few arrangements are plausible.

Least can be said about which atoms are connected with each other in the ion formed upon electron impact followed by cleavage of one or more bonds; and extreme caution should be observed when discussing the "structure" of an ion. The organic chemist associates something very definite with this term, namely the location and direction of bonding of all the atoms in a species. What we know (and for the practical use of the data need to know) is only that a certain molecule can form an ion which contains a specified number of atoms arranged in such a way that a positive charge can be well stabilized (if the ion is an abundant one). Furthermore, we can assume that this arrangement can be formed from the original molecular ion without requiring large amounts of energy (i. e. no more than the appearance potential** of the particular ion would indicate, after accounting for the energy content of the neutral fragment(s) produced).

Thus, the representations depicted for these ions, while by necessity drawn in the manner used for real chemical structures, are the least certain part of the interpretation of a mass spectrum.

Indeed, many different representations of a given ion can often be drawn with the same degree of justification (or lack thereof). Fortunately, the structural information that can be derived from a mass spectrum is based on the appearance of ions indicating the relationship of various atoms and groups *within the original molecule* (the item of interest) while their interconnection *within the fragment-ion* is not part of the argument***.

---

* The number of possible combinations is considerable for data obtained with a conventional spectrometer but is greatly reduced, sometimes to a unique one, if accurate mass measurements with a reliable high resolution instrument are performed (see Table 10, p. 79).

** The minimum electron energy of the bombarding beam at which this ion is formed.

*** The recent finding that the C_6H_5-ion that appears in the mass spectra of many compounds, e. g. benzophenone, is best represented by $HC{\equiv}C{-}CH{=}CH{-}$ $-CH{=}CH{\oplus}$ rather than by

Finally, an important aspect of the description of fragmentation processes is that of "electron bookkeeping". It permits one to outline (or predict) consecutive bond cleavages in a logical manner and in accordance with the principles governing the stability of ions and radicals. Electron bookkeeping also eliminates the inadvertent postulation of intermediates or product ions involving more electrons than atomic theory permits. Arrows have been used to outline the flow of electrons during the bond breaking (and bond making) processes but again one has to keep in mind that the meaning of this notation differs greatly from the one it has in chemical reactions, where an arrow indicates the shift of an electron or pair of electrons, whichever is indicated, in the direction of the arrow.

This specific meaning should by no means be carried over into mass spectrometric arguments because much too little information is available at present either on the electron densities in the molecular ion of a compound sufficiently complex to be of interest to the organic chemist, or on the transition state during decomposition. It was suggested, again mainly for bookkeeping purposes (58), to assume localization of the positive charge at a particular area of the molecule, preferably at heteroatoms or π-system of aromatic rings. This is followed by homolytic or heterolytic fission of bonds (indicated by arrows ⌢⃗ or fishhooks ⌢, respectively). One should, however, keep in mind that we have no data at the present time permitting direct determination of the location of the positive charge within a molecular ion (if one can speak at all of such a discrete localization), nor can one experimentally distinguish between moving single electrons versus electron pairs.

As an example, we shall follow the various symbolisms *(Chart 1)* which can be drawn to depict the ionization of a molecule of aspidospermidine (14, see Table 1, p. 18) and its fragmentation to ions of m/e 124 and 130, respectively. The resulting ions are the same regardless whether one assumes ionization at the alicyclic nitrogen or at the aromatic system, whether one assumes homolytic (⌢⃗) or heterolytic (⌢⃗) cleavage of ring C and whether this is followed or preceeded (not shown) by cleavage

is an excellent illustration *(155)*. The appearance of a C_6H_5-ion is still indicative of the presence of a phenyl group and in a structural argument we will still write it as a phenyl ion rather than the linear isomer, in order to retain the purpose of the mass spectral data, namely to support the presence of a phenyl group in the original molecule. There is no objection to this practice, as long as one keeps in mind that the pictorial expression does not necessarily represent the actual structure of the ion while it travels down the flight path of the spectrometer.

of the $C_{(10)}$–$C_{(11)}$ bond. Thus, the type of pictorial expression chosen is relatively irrelevant and serves mainly a bookkeeping purpose unless one wishes to specifically express the differences between two proposals for the purpose of discussion.

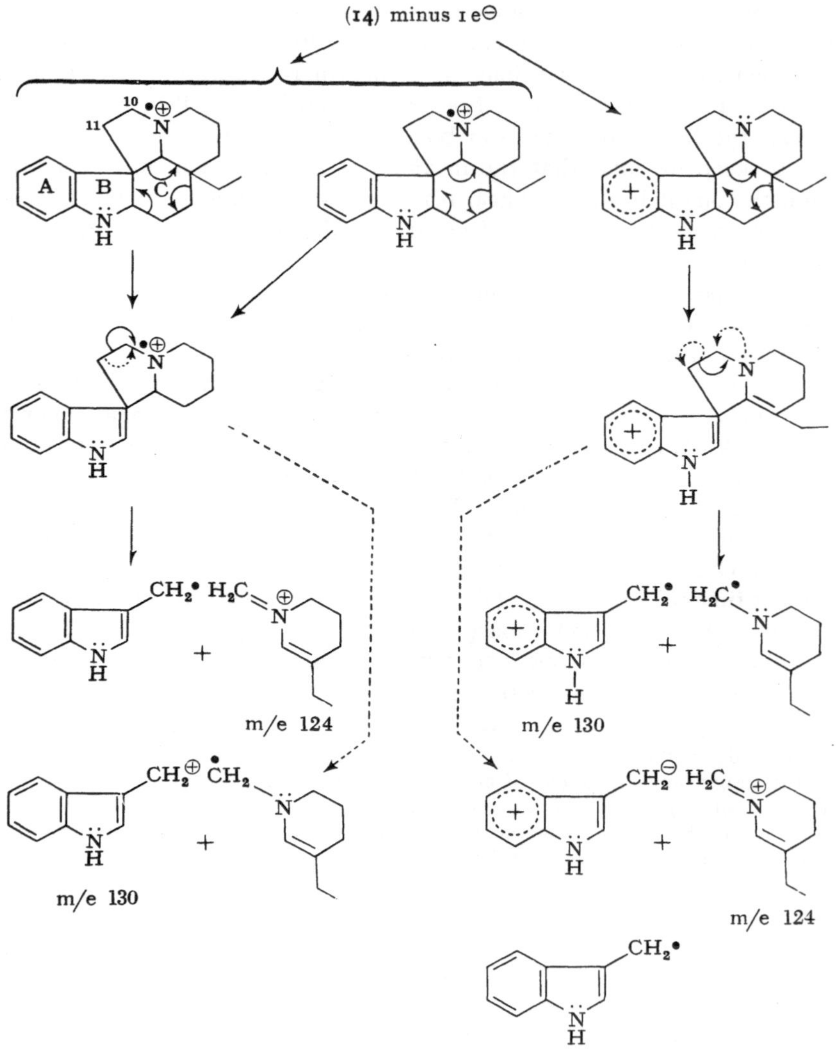

$(\mathbf{14})$ minus $1 e^{\ominus}$

m/e 124

m/e 130

m/e 130

m/e 124

Chart 1. Various Notations and Possibilities to Depict the Fragmentation of the Aspidospermidine Skeleton.

In the later Sections of this review the notation is presented which was employed by the author of the paper under discussion, to avoid any misrepresentations of his personal views.

References, pp. 86—98.

The "structures" drawn in the fragmentations outlined above are to be taken with the potential tations that confuses that follow form the are one might propose that the ion of m/e 130 undergoes ring expansion to a quinolinium ion (*191*).

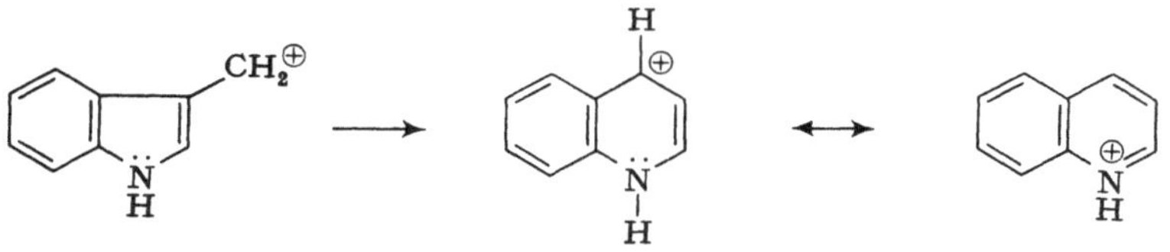

The only justification for writing this ring expansion is the prettiness of the resulting ion. It is, however, completely unjustified in the absence of appearance potential measurements, and, at the least, confuses the structural argument.

II. Indole Alkaloids.

1. Correlation of Closely Related Alkaloids.

As was briefly indicated above, indole alkaloids and related substances lend themselves, for various reasons, especially well to mass spectrometric structure determination and thus the work in this field represents an excellent example for the various aspects of the applicability of mass spectrometry to structure determination. The first instance in which mass spectrometry was employed to settle the structure of an alkaloid (*21*) and thus triggered the rapid development and remarkable success of this approach, has demonstrated three basic principles, namely (a) the so-called "mass spectrometric shift" technique, (b) the value of chemical conversion of the original molecule by a simple chemical reaction on a very small scale, and (c) the marking of specific areas of the molecule by the introduction of heavy isotopes for the purpose of obtaining structural information from the mass spectrum of the resulting labeled derivative.

This example involved the *correlation of the carbon skeleton* proposed for sarpagine (**1**) with a known compound (**4**) which has the same carbon skeleton derived from the dihydroindole alkaloid ajmaline (**3**)* *(Chart 2)*. While at that time compounds of such complexity had not been subjected to mass spectrometry, it was predicted that the fragmentation of such a system will be virtually independent of the presence or absence of a substituent in the indole moiety, such as a methoxyl or an N-methyl group, and that therefore the mass spectra of the two compounds (**2** and **4**) should be virtually identical with the exception that all peaks that are due to ions containing the indole ring would appear 16 mass units higher in (**2**) derived from sarpagine than in (**4**), obtained from ajmaline.

---

* References to established structures of known alkaloids (*45, 144*).

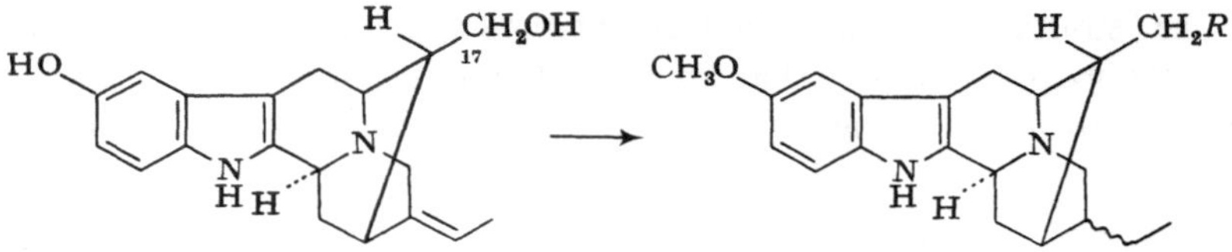

Chart 2. Interconversion of the Sarpagine and Ajmaline Skeletons.

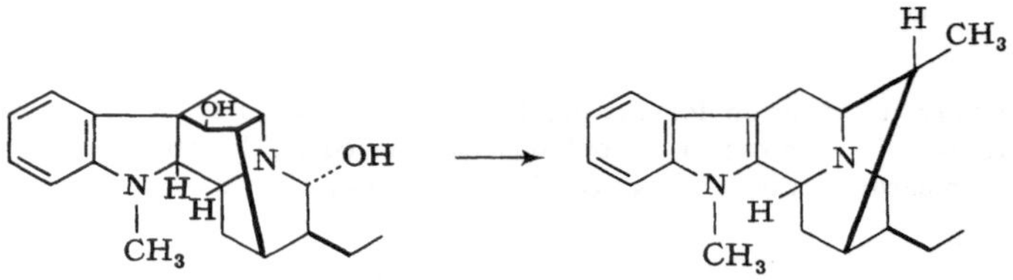

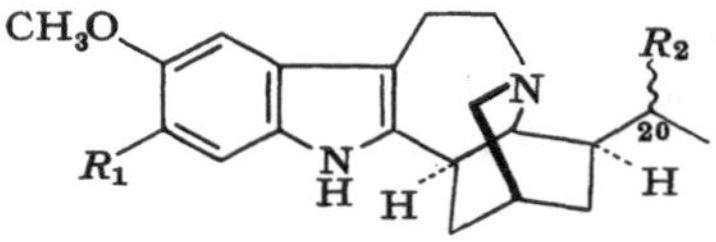

(6) $R_1 = R_2 = H$. Ibogaine.
(7) $R_1 = CH_3O, R_2 = H$. Ibogaline.
(8) $R_1 = H, R_2 = OH$. Iboxygaine.
(9) $R_1 = H, R_2 = D$.

The basis for this prediction was the assumption that the bonds that are cleaved during the fragmentation of such a molecule will be those of the alicyclic part and thus uninfluenced by the distant substituent. The mass spectra *(Fig. 2)* of these two compounds did, in fact, exhibit this relationship, and even without any further interpretation of the spectra one could conclude that sarpagine had, indeed, the proposed carbon skeleton. Since the preparation of (2) from (1) involved the removal of the primary hydroxy group, it was necessary to mark its original position in the molecule, which was achieved by its replacement by deuterium. The mass spectrum of the resulting monodeutero derivative (5) revealed the presence of a —CH_2D group confirming the attachment of the hydroxyl to $C_{(17)}$.

At that stage it was, of course, necessary to show that the fragmentation pattern is, indeed, characteristic for a given carbon skeleton, and the mass spectrum of ibogaine (6), a compound isomeric to (2), was measured to demonstrate that a different system gives, in fact, rise to a very different mass spectrum.

Thus, the ground work was laid to facile and speedy correlation of indole alkaloids that are of the same skeletal type but differ by the

References, pp. 86—98

number and kind of small substituents in the aromatic moiety – a pheno-
menon that leads to the wide variety of alkaloids in plants. In this respect
the mass spectrometric technique became an extremely convenient and
highly sensitive replacement of the classical approach to structure determination, namely chemical degradation to smaller fragments which lend
themselves to chemical identification. In the mass spectrometer the
degradation is achieved by the electron beam and the degradation products are identified by their mass, after separation in the magnetic field.

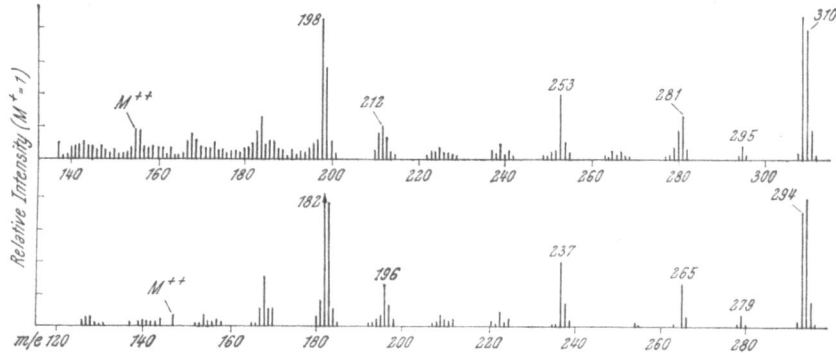

Fig. 2. **Mass spectrum of a) compound (2) and b)** compound (4). According to BIEMANN [From: J. Amer.
Chem. Soc. **83**, 4801 (1961).]

This approach was immediately used (28, 29) for the confirmation
of the previously suggested structures of two derivatives of ibogaine,
namely ibogaline (7) and iboxygaine (8). The mass spectrum of the former
exhibited, in comparison with that of (6), a shift of 30 mass units (H versus
CH_3O) for all those peaks containing the aromatic portion, while the
pattern of the remaining ones was exactly as in the spectrum of (6). To
conclusively correlate iboxygaine (8) with ibogaine and confirm the
position of the hydroxyl group, the H in the latter was replaced by
deuterium. The resulting compound (9) was physically identical with
ibogaine, except that the mass spectrum revealed the presence of a
—CHD—CH_3 group in place of the ethyl group of (6), indicating that the
hydroxyl group had been at $C_{(20)}$.

Another early use of this "mass spectrometric shift technique" is
the proof of the structure of quebrachamine (10), containing a ring
system novel at that time, for which no analogous compound was available. A methoxy analogue (11) could, however, be synthesized from
aspidospermine (12) by a series of unequivocal reactions, and again the
mass spectra have revealed that quebrachamine differs from (11) only

by the absence of the methoxy group and must thus have structure (10) (*38*, *39*).

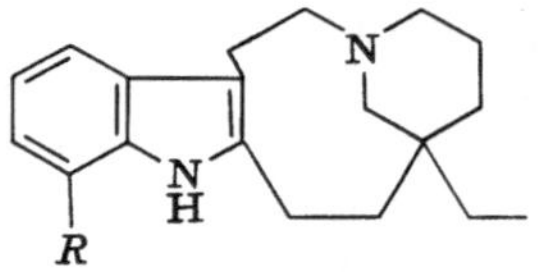

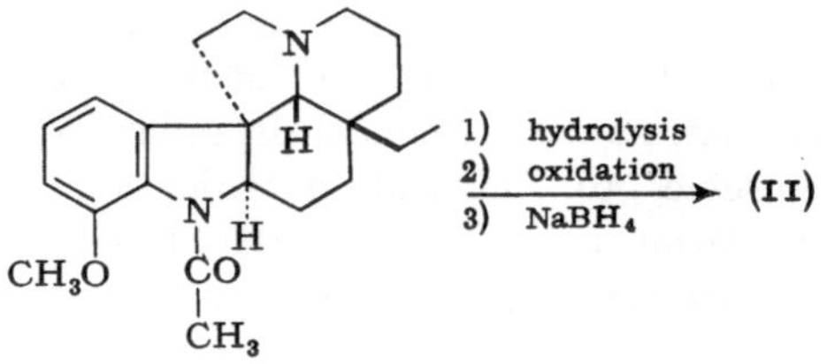

(10) R = H. Quebrachamine.
(11) R = CH₃O.

(12) Aspidospermine.

2. Dihydroindole Derivatives.

Shortly before, the structure of aspidospermine (12) had been finally settled by X-ray crystallography (*149*) and the interpretation of the mass spectrum of its hydrolysis product, deacetylaspidospermine (13) has made it possible to assign structures to a number of minor alkaloids (14–18) of *Aspidosperma quebracho blanco* (*30*, *40*), most of them isolated in quantities permitting only the determination of mass and UV spectra but insufficient for a classical structure proof.

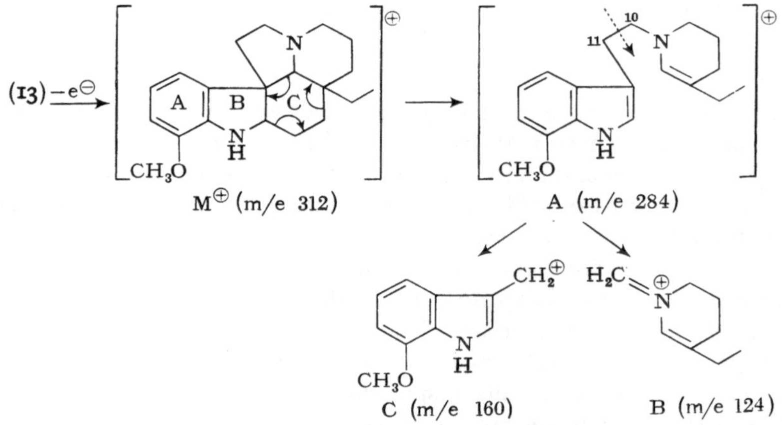

Chart 3. Fragmentation of Deacetylaspidospermine (13).

This system is characterized by the elimination of carbon atoms 2 and 3 of ring C as ethylene, followed by cleavage of the "tryptamine-bond" ($C_{(10)}$—$C_{(11)}$) and retention of the charge either at $C_{(10)}$ (to give the 124-ion) or at $C_{(11)}$ to give the ion of m/e 160 *(Chart 3)*. These are principal fragmentation modes leading to most of the significant peaks in the spectrum. Further characteristic fragments are those at m/e 174, a higher homologue (including $C_{(10)}$) of ion C as well as that at m/e 152 which corresponds to m/e 124 (A) plus C_2H_4, i. e. the $C_{(3)}$—$C_{(4)}$ bridge *(Fig. 3)*.

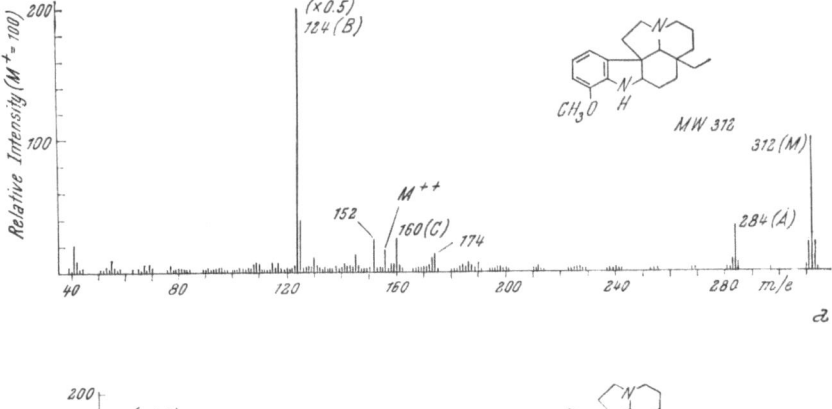

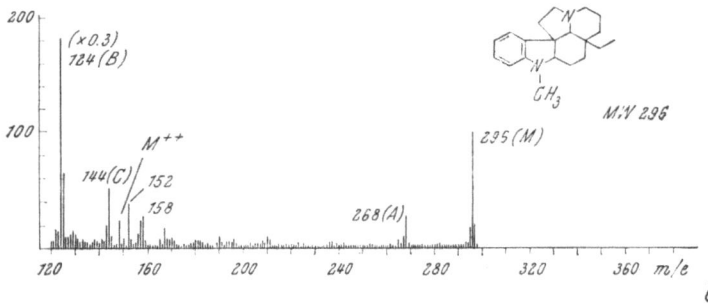

Fig. 3. Mass spectrum of a) deacetylaspidospermine (13) and b) N$_a$-methylaspidospermidine (15, Table 1, p. 18). According to BIEMANN, SPITELLER-FRIEDMANN and SPITELLER. [From: J. Amer. Chem. Soc. 85, 631 (1963).]

Thus, all derivatives which differ in substitution patterns on the aromatic moiety (the dihydroindole system) of these alkaloids reveal this difference in the mass of both the molecular ion which, of course, must contain all substituents as well as in ions A and C which contain the entire dihydroindole system. This then is the simplest situation interpreted most easily except that it is impossible on the basis of a mass spectrum to deduce the exact position of a substituent in the benzene ring. That is best accomplished by a consideration of the aromatic region of the NMR spectrum of the alkaloid or by comparison of its ultraviolet spectrum with that of authentic systems representing the various possible substitution patterns. This is one of the many examples where the combination of mass spectrometry with one or more other techniques is required to arrive at a unique solution of a problem.

Table 1. Alkaloids Related to Aspidospermine (12, p. 19).

	R_1	R_2	R_3	R_4	References
Aspidospermidine (14)	H	H	H	H	(30, 40)
N-Methylaspidospermidine (15)	CH_3	H	H	H	
N-Methyldeacetylaspidospermine (16)	CH_3	CH_3O	H	H	
Deacetylpyrifolidine (17)	H	CH_3O	CH_3O	H	(90)
Pyrifolidine (18)	CH_3CO	CH_3O	CH_3O	H	
Demethylaspidospermine (19)	CH_3CO	HO	H	H	(100)
Demethoxyvallesine (20)	HCO	H	H	H	
Demethoxyaspidospermine (21)	CH_3CO	H	H	H	
Demethoxypalosine (22)	C_2H_5CO	H	H	H	
Spegazzinine (23)	CH_3CO	HO	H	3-OH	(86)
Spegazzinidine (24)	CH_3CO	HO	HO	3-OH	
Aspidolimine (25)	C_2H_5CO	HO	CH_3O	21-OH	(104, 181)
Limaspermine (26)	C_2H_5CO	HO	H	21-OH	(170)
16-Methoxylimaspermine (27)	C_2H_5CO	HO	CH_3O	3-OH	
Vindoline (28)	CH_3	H	CH_3O	3-OH, 3-CO_2CH_3, 4-$OCOCH_3$, Δ^6	(109)
Vindorosine (29)	CH_3	H	H	3-OH	(153)
Aspidoalbine (30)	C_2H_5CO	HO	CH_3O	CH_3O, 3-CO_2CH_3, 4-$OCOCH_3$, Δ^6	(85)
Aspidolimidine (31)	CH_3CO	HO	CH_3O	19,21-ether	(104)
N-Acetylaspidoalbidine (32)	CH_3CO	H	H	19,21-ether	(52)
17-Methoxy-N-formylaspidoalbidine (33)	HCO	CH_3O	H	19,21-ether	(205)
17-Methoxy-N-acetylaspidoalbidine (34)	CH_3CO	CH_3O	H	19,21-ether	(205)
Dichotamine (35)	HCO	CH_3O	H	19,21-lactone	(52)

These few examples have sufficed to convince the workers in this field of the unique potentialities of mass spectrometry in the structure determination of indole alkaloids, and consequently of alkaloids with other aromatic systems; and it was thus employed in the majority of

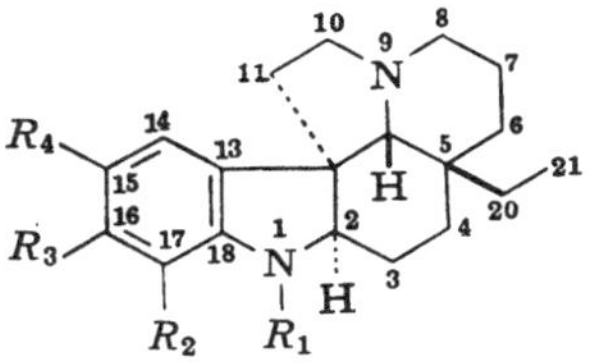

(12)

$R_1 = CH_3CO$, $R_2 = CH_3O$, $R_3 = R_4 = H$. Aspidospermine.

structure determinations of indole alkaloids published since 1962. *Table 1* summarizes the naturally occurring alkaloids of the *aspidospermine skeleton*, whose structures were totally or in part deduced by mass spectrometry. Similar lists can be compiled for the closely related aspidofractinine (38) and vincadifformine (39) types as well as others which will be mentioned later (Tables 2–7, pp. 25–37).

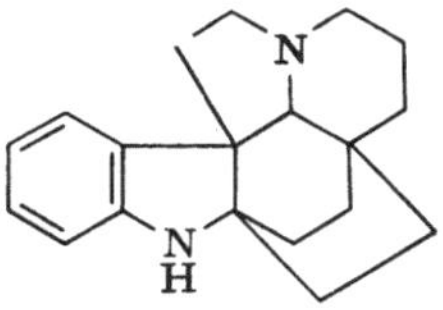

(38)
Aspidofractinine.

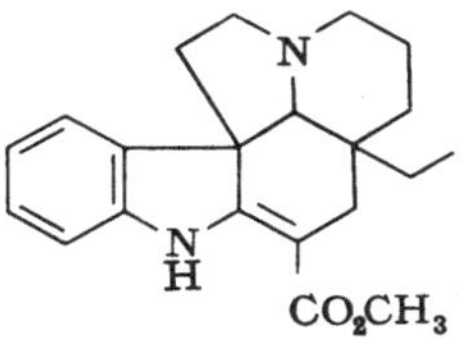

(39)
Vincadifformine.

Once it had been established that the mass spectrometric shift technique can be relied upon in those cases where the difference in substitution is located at a site remote from the bonds whose cleavage gives rise to the more abundant fragments, it became of interest to apply this technique to new compounds which differ from a known one by the presence of additional functional groups or substituents in the alicyclic region, that is in that part of the molecule that does, in fact, undergo fragmentation.

Evaluation of the many examples studied in recent years has lead to the conclusion that the influence of such substituents on the fragmentation process differs widely. In general it can be predicted or rationalized on the basis of the nature and location of the structural change. If it neither prevents a typical fragmentation process nor triggers a new one, the spectrum will resemble that of the unsubstituted molecule, except that some of the peaks will be shifted by the appropriate mass.

2*

Thus the mass spectrum of spegazzinidine (**24**) dimethyl ether (*86*) exhibits a peak at m/e 354 (M-44), corresponding to fragment A (see p. 16, Chart 3), in this case formed by the loss of vinyl alcohol instead of ethylene, because of the oxygen at $C_{(3)}$. Further cleavage of the $C_{(10)}$—$C_{(11)}$ bond leads to fragment B at m/e 124 as in all derivatives with unsubstituted piperidine ring, and to fragment C (m/e 232) which contains R_1, R_2 and R_3.

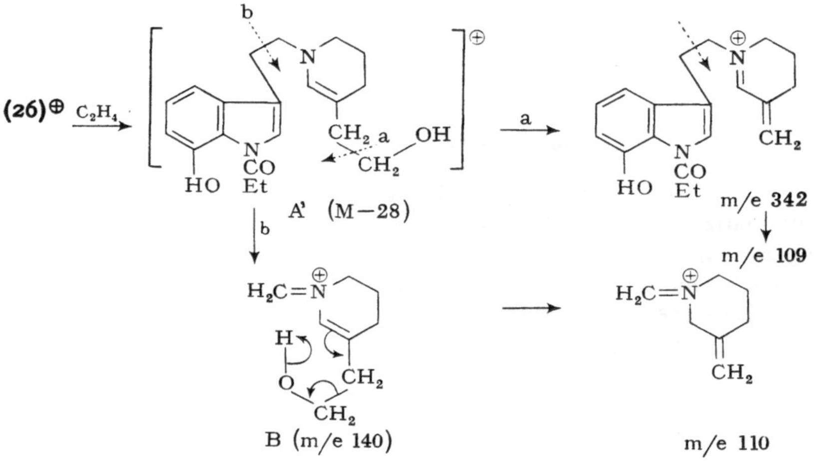

Chart 4. Fragmentation of Limaspermine (**26**).

Similarly, a hydroxyl group at $C_{(21)}$, as in limaspermine (**26**), does not alter or prevent the processes outlined above for (**13**, Chart 3) but merely introduces an additional possibility for fragmentation, namely the loss of $C_{(21)}$, either from fragment A′ or B (*170*) *(Chart 4)*. These processes lead merely to three additional peaks at m/e 342, 110 and 109 without obscuring the "aspidospermine pattern" of peaks A, B and C (m/e 146, after elimination of CH_3CHCO).

Neither does the ether bridge in aspidoalbidine (**36**) derivatives (**30—32**) alter the fragmentation of the molecules (fragment B shifts from m/e 124 to m/e 138) except that there is a tendency to eliminate the elements of ethylene oxide (44 m. u.) from the tetrahydrofurane ring as well as ethylene ($C_{(3)}$—$C_{(4)}$) (*52*) *(Chart 5)*.

A small, seemingly inconsequential change, namely the replacement of CH_2 at $C_{(20)}$ by CO, as in dichotamine (**35**) (p. 18) considerably supresses the formation of the fragment of type B (the very abundant ion of m/e 124 in all compounds with unmodified piperidine ring and ethyl group) (*52*). Since the aromatic ion of type C is quite abundant, one might rationalize this by assuming that ions B′ and B″ are energetically

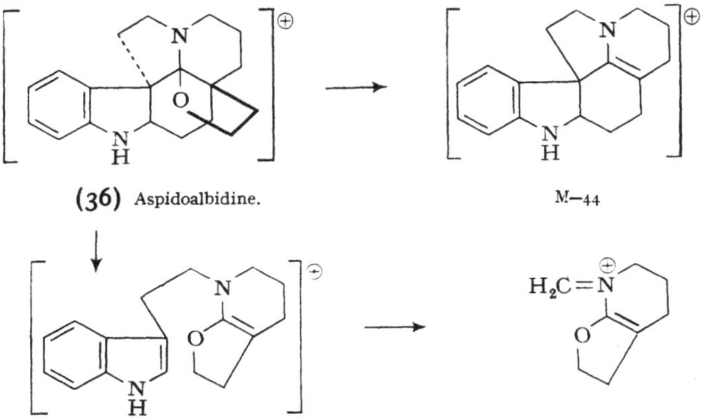

(36) Aspidoalbidine.   M-44

Chart 5. Fragmentation of Aspidoalbidine (**36**).

less favorable and thus cannot compete well for the positive charge with the aromatic ion formed by cleavage of the $C_{(10)}$—$C_{(11)}$ bond.

The cyclopropene-ion B″ would arise via loss of CO_2 from the molecular ion (which leads to an abundant fragment) and further cleavage of the product ion in the usual pattern outlined above for (**13**, Chart 3). In such instances, where the presence of a substituent obscures the typical fragmentation pattern of a particular system, it is often impossible to immediately recognize the nature of a new alkaloid from its mass spectrum, even if it is related to a mass spectrometrically well known type. Not infrequently, a characteristic fragmentation pattern is suddenly revealed when subjecting the compound to a simple chemical reaction, which happens to remove this "blocking group" or converts it to one that does not interfere with the "normal" fragmentation.

For example, lithium aluminium hydride reduction of the lactone ring leads to $C_{(21)}$-alcohols which behave "normally" as was discussed above for limaspermine (**26**, p. 18) and permitted recognition of the aspidospermine skeleton as well as placement of the lactone ring (*52*).

Similarly, the presence of the aspidospermine skeleton in the *Vinca* alkaloid vindoline (**28**) became apparent only when the mass spectrum of an accidentally obtained degradation product (**37**) was determined

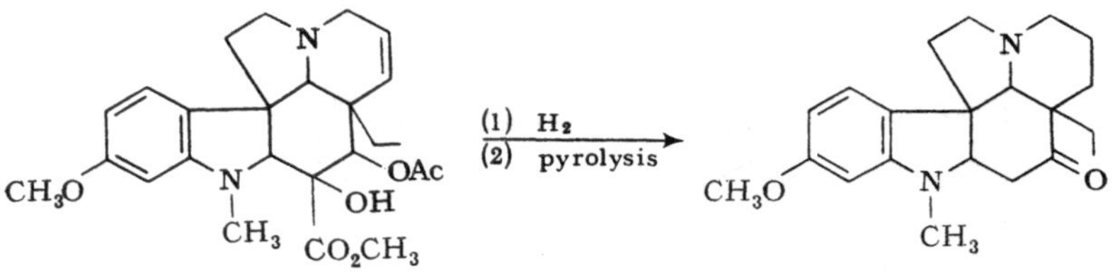

(28)
Vindoline.

(37)

(*109*). This ketone (37) gives a typical aspidospermine pattern with the elimination of $C_{(3)}$ and $C_{(4)}$ as ketene (instead of ethylene) in the first step. The great number of substituents in vindoline itself made it impossible to recognize this ring system.

A rather large group of indole alkaloids contain a 2,3-double bond and a 3-carbomethoxy group, such as in vincadifformine (39) and akuammicine (40). The former contains the aspidospermine carbon skeleton but the presence of the double bond prevents the elimination of $C_{(3)}$ and $C_{(4)}$ (as ethylene or substituted ethylene), while at the same time it facilitates the cleavage of ring C (in retro-Diels-Alder fashion) to yield ultimately the typical fragment of m/e 124 in abundance (*88*).

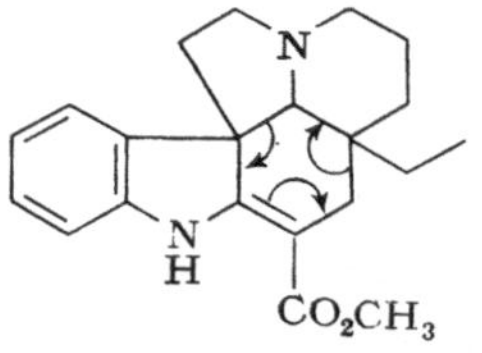

(39)
Vincadifformine.

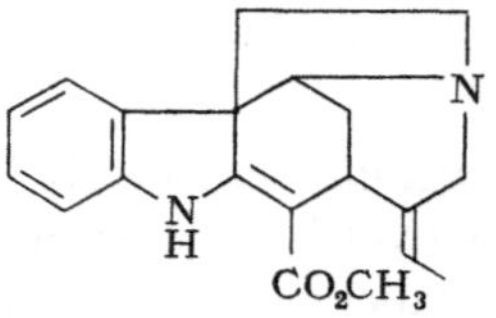

(40)
Akuammicine.

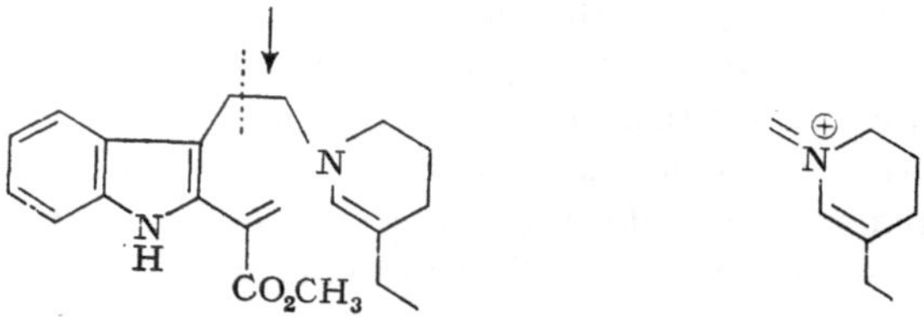

Removal of the double bond by reduction with zinc and acid to yield 3-carbomethoxy-aspidospermidine (42) *(Chart 6)* or, after acid-catalyzed decarboxylation and reduction of the intermediary indolenine (41) catalytically or with lithium aluminium hydride to aspidospermidine (14), results in compounds which both undergo the fragmentation typical of the aspidospermine class. These two reactions, therefore, are generally

References, pp. 86—98.

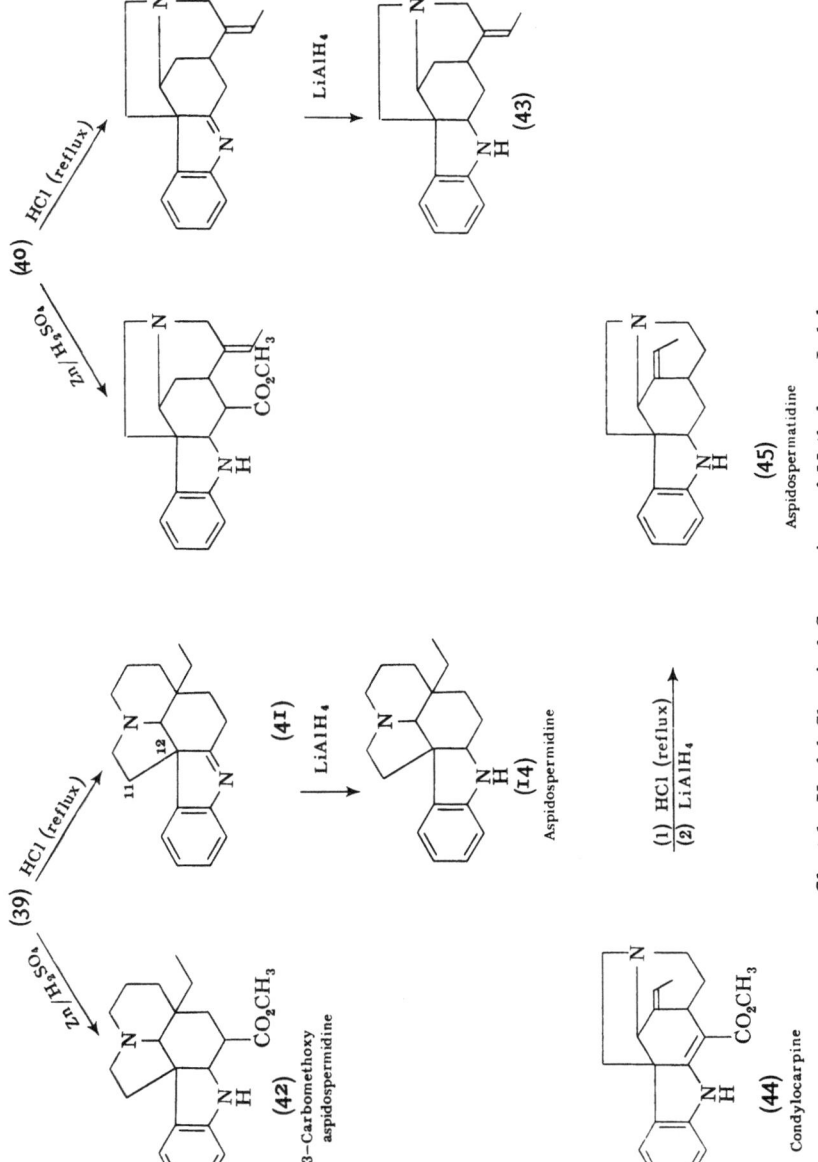

Chart 6. Useful Chemical Conversions of Methylene Indoles.

performed whenever an alkaloid exhibits the typical ultraviolet spectrum of α-methyleneindolines, (λ_{max} 320–330, $\varepsilon = 10000$–20000) because the mass spectrum of the product will show whether it belongs to the vincadifformine (39), akuammicine (40) or condylocarpine (44) class, considering the typical differences in the mass spectrometric behavior of (14), (44) and (45) as well as of the corresponding 3-carbomethoxy derivatives which shall be discussed later.

The 1,2-double bond present in the indolenine (41), which, incidentally, is a minor alkaloid (1,2-dehydroaspidospermidine) of *A. quebracho blanco*, completely changes the fragmentation when compared with the saturated

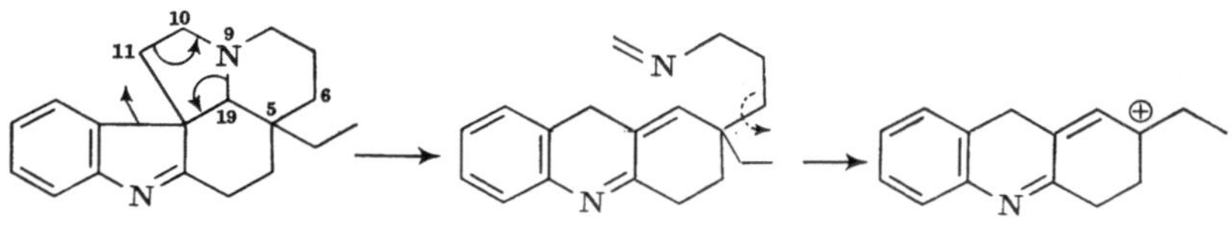

(41)

Chart 7. Fragmentation of an Indolenine.

system (e. g. 13). In contrast to the 2,3-double bond of (39), which facilitates the cleavage of ring C, the 1,2-double bond prevents it. Cleavage of the 10,11; 9,19 and 5,6-bonds accompanied by expansion of ring C was postulated for the process leading to a major fragment (m/e 210) from (41), in addition to the mere loss of the ethyl group (39) *(Chart 7).*

Lithium aluminium hydride reduction restores the aspidospermine behavior and it should be noted that sodium borohydride reduces (41) to quebrachamine (10) which exhibits a mass spectrum which is, of course, again very different from that of the aspidospermine type. Thus, if a substance is suspected to have a structure like (41) [whether it originates from a natural source or was obtained chemically from a compound related to (39)], reduction with LiAlH₄ and NaBH₄, respectively, yields two products, whose mass spectra will solve the problem, including the presence and location of any substituents in the derivative of (41).

This approach or variations thereof, was used in the elucidation of the structure of *vincadifformine* and several derivatives listed in *Table 2.* An exception of the chemical behavior discussed above is represented by the acid decarboxylation of the 20-ketoderivative minovincine (50) which does not yield the 20-keto analogue of (41) but cyclizes (21 → 2) to 20-oxo-aspidofractinine (38) with C₍₂₀₎-carbonyl. This unexpected reaction was obvious from the mass spectrum of the product and led to the discovery of this facile conversion of the pentacyclic aspidospermine skeleton to the hexacyclic aspidofractinine system equally abundant in nature (*182*).

References, pp. 86—98.

Table 2. Some Alkaloids Related to Vincadifformine (39)

	R_1	R_2	R_3		References
Vincadifformine (39)	H	H	H		(88)
Tabersonine (46)	H	H	H	Δ^6	(173)
Minovine (47)	CH$_3$	H	H		(150)
Minovincinine (48)	H	H	OH		(172)
Echitovenine (49)	H	H	OAc		(70)
Minovincine (50)	H	H	=O		(172)
16-Methoxyminovincine (51)	H	CH$_3$O	H		(172)

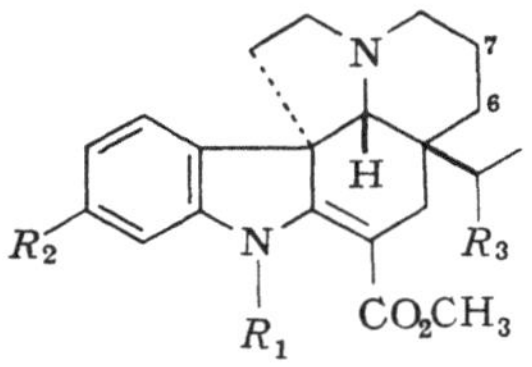

(39)

This ring system was originally proposed independently for pleiocarpine (63) (137) and refractine (62) (89). In the latter case, mass spectrometry played a major role. Refractine was suspected to be related to aspidospermine and exhibited indeed the fragment due to loss of the $C_{(2)}$—$C_{(3)}$ bridge, but it showed an intense peak at m/e 109 rather than at m/e 122 as would be expected if the two hydrogens lacking were responsible for a double bond or additional ring involving the carbons 5–10, 19, 20 and 21. A linkage between the piperidine system and the dihydroindole moiety was thus indicated, and the most plausible one is a $C_{(2)}$—$C_{(21)}$ bond as also suggested by the NMR spectrum (no ethyl group).

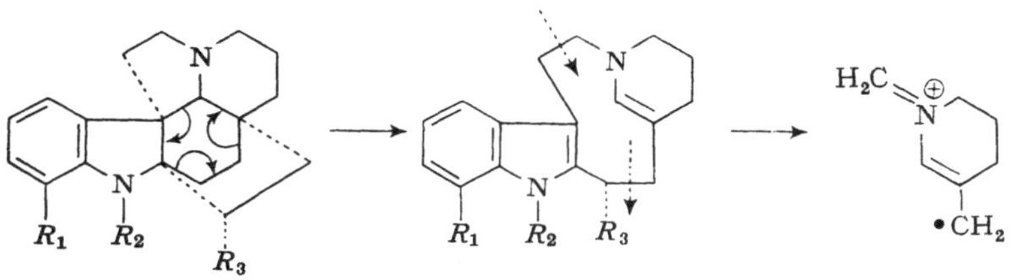

(62)
$R_1 = CH_3O$, $R_2 = HCO$, $R_3 = CO_2CH_3$. Refractine.

(65)
$R_1 = H$, $R_2 = R_3 = CO_2CH_3$. Pleiocarpine.

D (M-28)   m/e 109

While one would not expect this system to yield a peak at m/e 124, there is still one observed (although by far not as exceptionally abundant as in the aspidospermine class). It is rationalized by a transfer of the hydrogen from $C_{(3)}$ to $C_{(21)}$ or from a hydrogen bearing substituent (CH_3 or OH) from $C_{(3)}$ to $C_{(21)}$ (81).

The typical fragmentation pattern (Fig. 4) led to, or aided in, the determination of a large number of indole alkaloids (Table 3) since it is not only indicative of the ring system but also reveals the presence

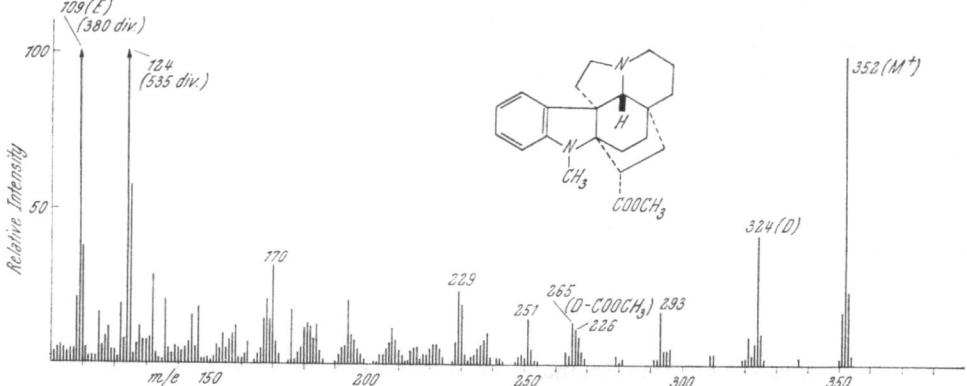

Fig. 4. Mass spectrum of pleiocarpinine (63). According to THOMAS, ACHENBACH and BIEMANN. [From: J. Amer. Chem. Soc. 88, 1537 (1966).]

and location of substituents on the basis of the masses of fragments D, E and F (m/e 124 or its analogue).

As mentioned earlier, aside from the aspidospermine and aspidofractinine group which both have 19 carbon atoms in the basic skeleton, there exist two other large groups of dihydroindole alkaloids, namely the akuammicine (40) and condylocarpine (44) types (cf. p. 26). They differ only in the position of the C_2-side chain (biogenetically due to condensation of the β-position of the indole with one or the other of the carbons adjacent to N_b) and contain a total of 18 carbon atoms in the ring system.

The structure of akuammicine itself had been determined without recourse to mass spectrometry, but in the course of the determination of the aspidospermatine structure (86, p. 31) the mass spectrum of 2,14-dihydro-decarbomethoxyakuammicine [43, later found in nature and named tubifolidine (138)] was measured and interpreted as follows (30) (Chart 8).

Thus it has become rather routine to determine whether an alkaloid with the typical carbomethoxy-α-methylene indoline chromophore does indeed belong to the akuammicine group and what the substitution pattern is. For placing the substituents in the alicyclic moiety it is

References, pp. 86—98.

Table 3. Alkaloids Related to Aspidofractinine (38).

	R_1	R_2	R_3	R_4	References
Aspidofractinine (38)	H	H	H	H	(87)
17-Methoxyaspidofractinine (52)	CH₃O	H	H	H	(105)
16,17-Dimethoxyaspidofractinine (53)	CH₃O	H	H	16-CH₃O	(105)
N-Formyl-52 (54)	CH₃O	HCO	H	H	(93)
N-Formyl-53 (55)	CH₃O	HCO	H	16-CH₃O	(93)
Aspidofiline (56)	HO	CH₃CO	H	H	(106)
Refractidine (57)	H	HCO	H	CH₃O	(106)
Pyrifoline (58)	CH₃O	CH₃CO	H	CH₃O	(137)
Kopsinine (59)	H	H	CO₂CH₃	H	(70)
Venalstoline (60)	H	—	CO₂CH₃	Δ⁶	(89)
Aspidofractine (61)	H	HCO	CO₂CH₃	H	(89)
Refractine (62)	CH₃O	HCO	CO₂CH₃	H	(137)
Pleiocarpinine (63)	H	CH₃	CO₂CH₃	H	(137)
Pleiocarpinilam (64)	H	CH₃	CO₂CH₃	10-oxo	(87)
Pleiocarpine (65)	H	CO₂CH₃	CO₂CH₃	H	(137)
Pleiocarpine lactam (66)	H	CO₂CH₃	CO₂CH₃	10-oxo	(87)
Dioxopleiocarpine (67)	H	H	CO₂CH₃	10,11-dioxo	(1)
Kopsinoline (68) (= 59-Nb-oxide)	H	H	CO₂CH₃	9-oxide	(200)
Pleiocarpoline (69) (= 63-Nb-oxide)	H	CH₃	CO₂CH₃	9-oxide	(200)
Pleiocarpoline (70) (= 65-Nb-oxide)	H	CO₂CH₃	CO₂CH₃	9-oxide	(200)

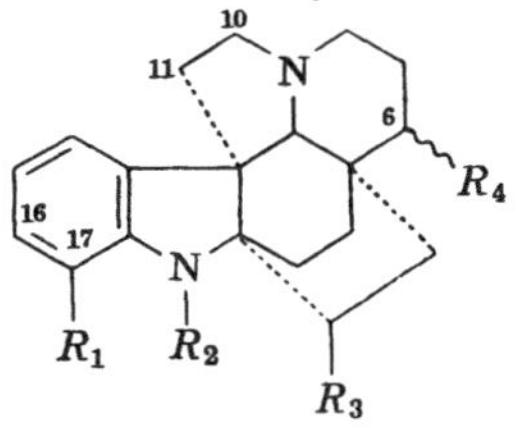

(38) R_1 — R_4 = H. Aspidofractinine.

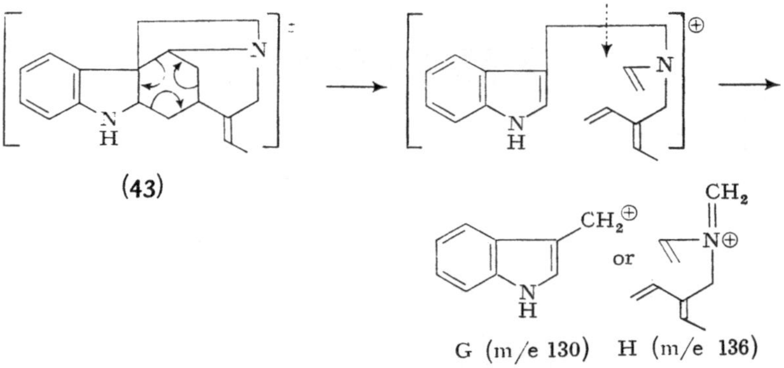

(43)

G (m/e 130)   H (m/e 136)

Chart 8. Fragmentation of Decarbomethoxydihydroakuammicine (43).

necessary to reduce the 19,20-double bond (if it is present) because this permits a further fragmentation mode, namely cleavage of the 17,20-bond which results in an ion of mass 199 (*30*) *(Chart 9)*.

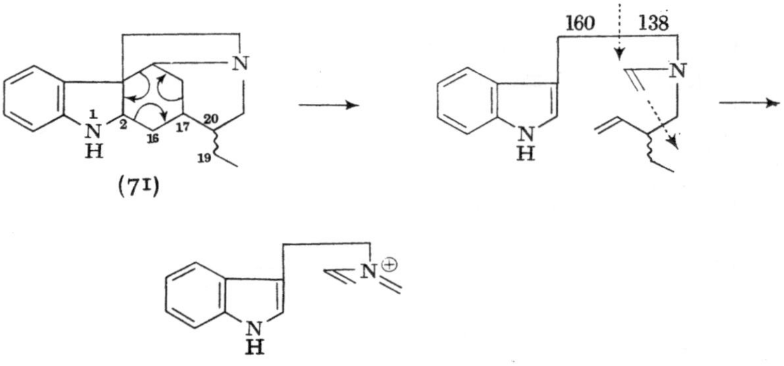

(71)

I (m/e 199)

Chart 9. Fragmentation of Decarbomethoxytetrahydroakuammicine (71).

The mass spectrometric behavior typical of (71) was observed upon catalytic hydrogenation of two other alkaloids, henningsamine (after deacetylation) and henningsoline. The spectra revealed that the products were the hydroxymethylene derivatives of (71) (74 and 75); and since the reduction involved the loss of one oxygen atom and four hydrogens (this can be seen from the difference between the spectra of starting material and product), one could write structures (72) and (73) for henningsamine and henningsoline, respectively (*32, 33*).

References, pp. 86—98.

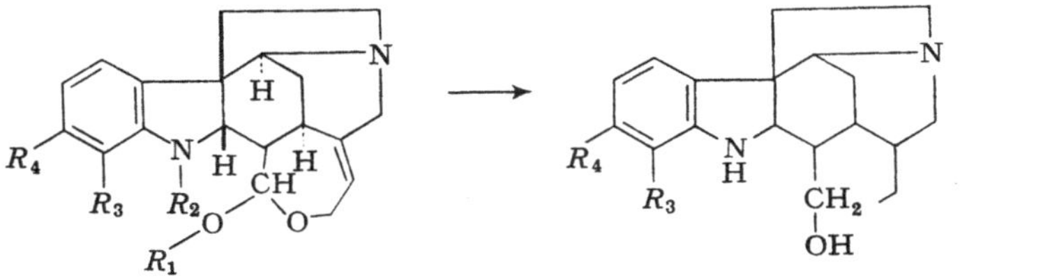

(72) $R_1 = R_2 = CH_3CO$, $R_3 = R_4 = H$. Henningsamine.
(73) $R_1 = H$, $R_2 = CH_3CO$, $R_3 = OH$, $R_4 = CH_3O$. Henningsoline.

(74) $R_3 = R_4 = H$.
(75) $R_3 = OH$, $R_4 = CH_3O$.

The structures of a number of alkaloids *(Table 4)* related to akuammicine have been determined in part or solely on the basis of the mass spectra of the products in which the carbomethoxy group and the 2,16-double bond (and 19,20-double bond, if present) had been removed *(61)*.

Table 4. Alkaloids Related to Akuammicine (40).

	R_1	R_2	R_3	R_4	2,16-double bond	References
Tubifoline (76)	H	H	H	H	no, Δ^1	
Tubifolidine (77)	H	H	H	H	no	*(139)*
19,20-Dihydro-akuammicine (78)	CO_2CH_3	H	H	H	yes	
Lochneridine (79).......	CO_2CH_3	H	OH	H	yes	*(154)*
Echitamidine (80)	CO_2CH_3	OH	H	H	yes	*(91)*
Compactinervine (81) ...	CO_2CH_3	OH	OH	H	yes	*(92)*
Mossambine (82)	CO_2CH_3	$\Delta^{19,20}$		OH	yes	*(151)*

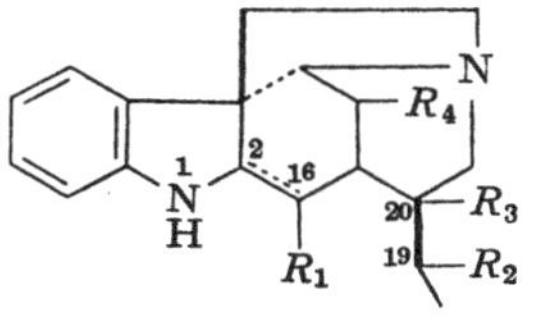

(40)

The process analogous to the cleavage of the 17,20-bond in (71) revealed the structure of aspidospermatidine (45, p. 31) because its dihydro derivative (79) gave, in addition to the very intense ion at m/e 138 (like 71), a peak at m/e 227, which was best rationalized by placing the ethyl group at $C_{(14)}$ *(Chart 10)*.

On the basis of this interpretation the structures of various aspidospermatidine-aspidospermatine alkaloids were derived which, incidentally, represented the first new alkaloid carbon skeleton deduced entirely by mass spectrometry *(30)*. Later the carbomethoxy-α-methylene indoline derivative of aspidospermatidine was found to be represented by the alkaloid condylocarpine (**44**, p. 31) on the basis of reductive decarboxylation experiments (**44** → **45**) as well as by reduction to tetrahydrocondylocarpine. The spectrum of the latter exhibited an intense peak (fragment H) at m/e 186 ($138 + CO_2CH_3-H$), due to the presence of the carbomethoxy group at $C_{(16)}$ (*26*).

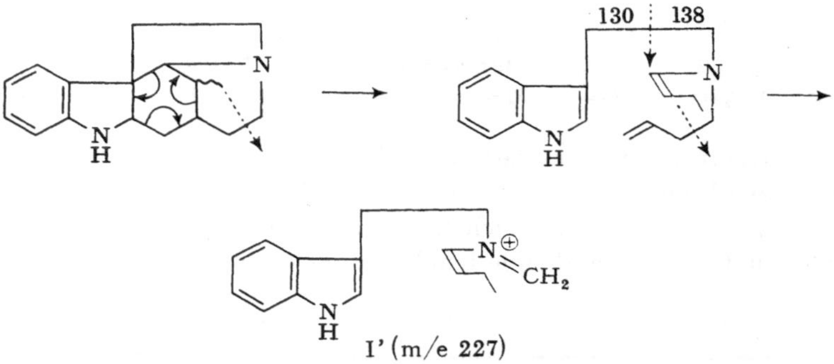

Chart 10. Fragmentation of Dihydroaspidospermatidine (**79**).

Table 5 summarizes these alkaloids of the aspidospermatidine-condylocarpine group.

An interesting variation of the akuammicine skeleton is one occurring in various *Picralima* alkaloids and related substances. The two systems e. g. 1,2-dihydrostrictamine (**92**) (*183*) and 2,16-dihydroakuammicine (**93**) differ in fact only by the interchange of the attachment of two bonds to the α- and β-carbon atoms of the dihydroindole system (*163*) (*Chart 11*).

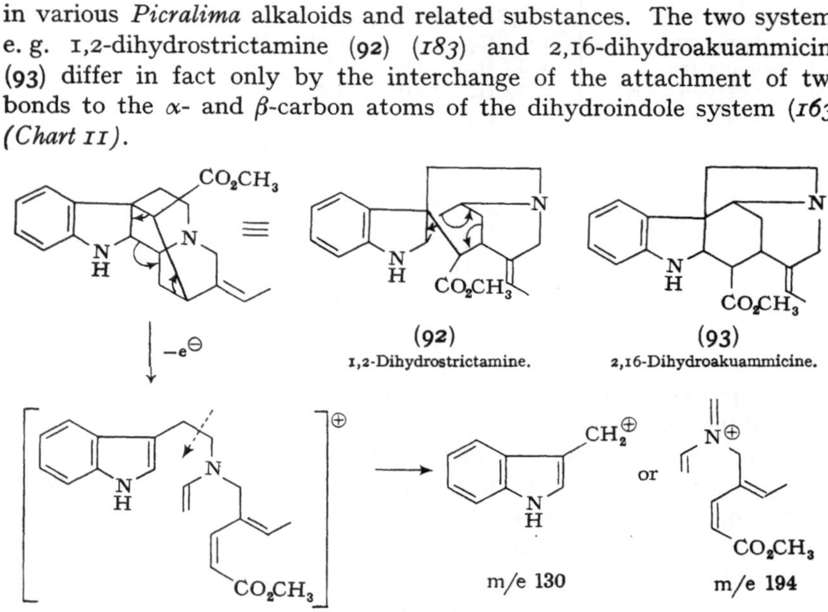

Chart 11. Fragmentation of 1,2-Dihydrostrictamine (**92**) and its Relationship to that of 2,16-Dihydroakuammicine (**93**).

References, pp. 86—98.

Table 5. Alkaloids Related to Aspidospermatidine (45) and Condylocarpine (44).

	R_1	R_2	R_3	R_4	Double bond 2,16	Double bond 14,19	References
Condylocarpine (44)	CO_2CH_3	H	H	H	yes	yes	(26)
Aspidospermatidine (45)	H	H	H	H	no	yes	
N-Methylaspidospermatidine (83)	H	CH_3	H	H	no	yes	
N-Acetylaspidospermatidine (84)	H	CH_3CO	H	H	no	yes	(30, 40)
Deacetylaspidospermatine (85)	H	H	CH_3O	H	no	yes	
Aspidospermatine (86)	H	CH_3CO	CH_3O	H	no	yes	
Dihydroaspidospermatine (87)	H	CH_3CO	CH_3O	H	no	no	
11-Hydroxy-N-acetylaspidospermatidine (88)	H	CH_3CO	H	OH	no	yes	(105)
Tubotaiwine (89)	CO_2CH_3	H	H	H	yes	no	(205)
11-Methoxy-14,19-dihydrocondylocarpine (90)	CO_2CH_3	$\Delta^{1,2}$	H	CH_3O	yes	no	(105)
Precondylocarpine (91)	$\left[\begin{array}{c}CO_2CH_3\\ CH_2OH\end{array}\right]$	$\Delta^{1,2}$	H	H	no	yes	(205)

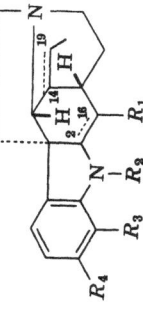

This relationship carries over into the mass spectrometric behavior inasmuch as cyclic opening of one ring can occur in a fashion very similar to the fragmentation discussed earlier for the akuammicine degradation product (43, p. 28). The resulting fragments of m/e 130 and 194 obtained from (92) correspond exactly to ions G and H (see p. 28) formed from (93).

It may be noted at this point that the formation of ions of m/e 194 and 130 can be visualized in a slightly different fashion involving transfer of a hydrogen and avoiding rupture of the piperidine ring (163) *(Chart 12).*

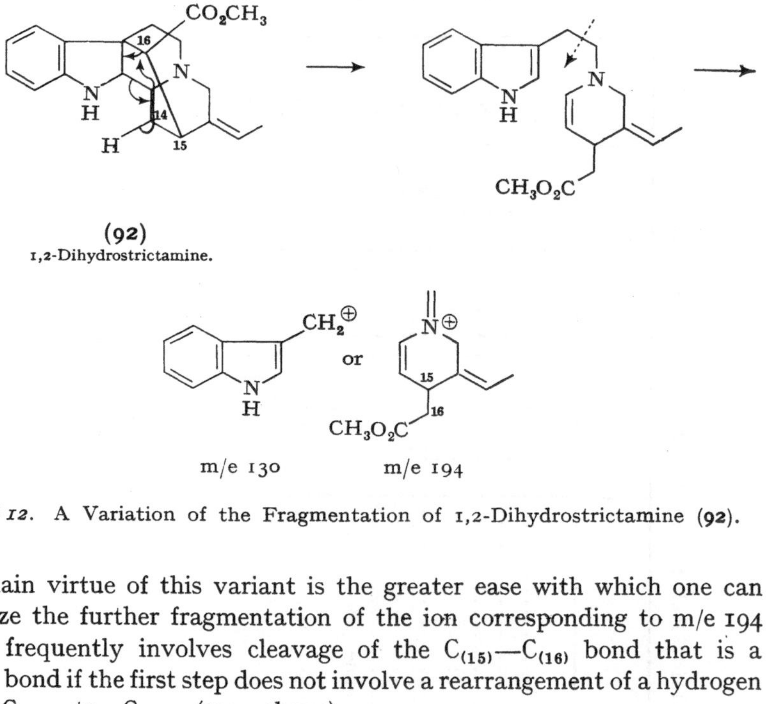

(92)
1,2-Dihydrostrictamine.

m/e 130   m/e 194

Chart 12. A Variation of the Fragmentation of 1,2-Dihydrostrictamine (**92**).

The main virtue of this variant is the greater ease with which one can visualize the further fragmentation of the ion corresponding to m/e 194 which frequently involves cleavage of the $C_{(15)}$—$C_{(16)}$ bond that is a double bond if the first step does not involve a rearrangement of a hydrogen from $C_{(14)}$ to $C_{(16)}$ (see above).

The two structural types (92) and (93) are thus, in a sense, "mass spectrometrically equivalent". The possibility has always to be kept in mind that two different compound types can give very similar mass spectra, if the first step in the fragmentation of each is a very efficient one and leads to identical species from each one of the starting structures; this will then, of course, give rise to identical further fragmentation. Since the rate of decomposition in the first step will hardly be the same in both instances, competing fragmentation processes will be more pronounced in one than in the other, giving rise to differences in intensity ratios as well as to additional peaks. The mass spectra of two such compounds (e. g. **92** and **93**) are thus quite similar, but not identical. The differentiation of compounds of type (**92**) and (**93**) always requires further chemical and spectroscopic evidence.

Table 6. Alkaloids Related to Picraline (**94**)

	R_1	R_2		References
Picraline (**94**)	CO_2CH_3	CH_2OCOCH_3	2,5-oxide	(*51, 163*)
Akuammiline (**95**)	CO_2CH_3	CH_2OCOCH_3	1,2-double bond	(*162*)
Strictamine (**96**)	CO_2CH_3	H	1,2-double bond	(*183*)
ψ-Akuammigine (**97**)	CO_2CH_3	CH_2O— to $C_{(2)}$	N_1-methyl	(*162*)
Akuammine (**98**) = 10-hydroxy-ψ-aku- ammigine				(*162*)

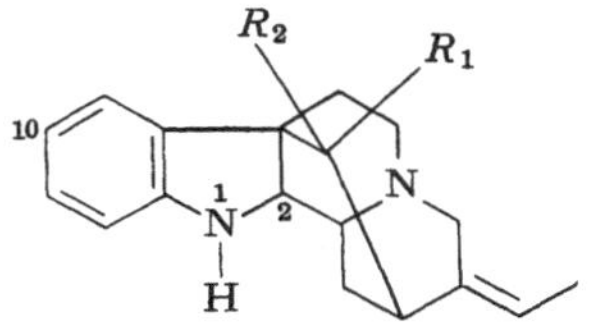

(**94**)

Table 6 reveals that all naturally occurring alkaloids possessing the carbon skeleton represented by (**92**) have a functional group (or double bond) at $C_{(2)}$. Under certain reducing conditions a rearrangement to a quite different system takes place:

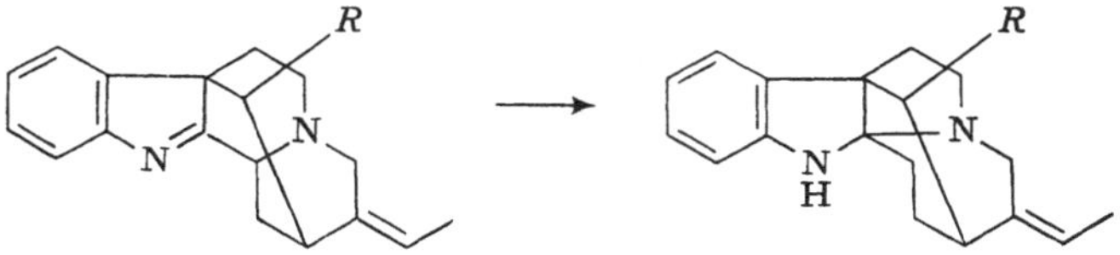

The latter compound has a very typical UV spectrum (*123*) and gives rise to a characteristic fragment (m/e 157, if unsubstituted).

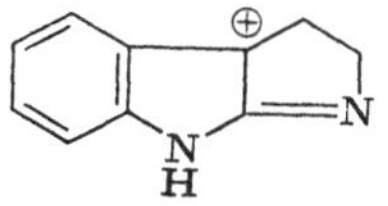

(m/e 157)

Arguments of this type, in addition to other chemical and spectroscopic results, have played an important role in the determination of the alkaloids listed in Table 6.

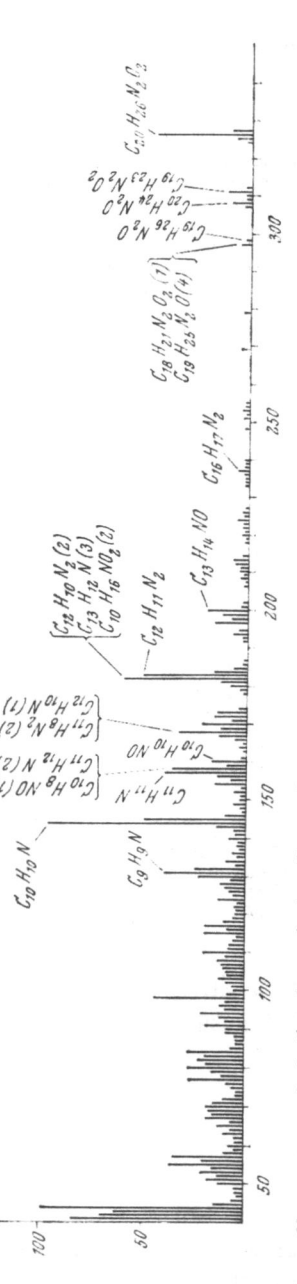

Fig. 5. Mass spectrum of ajmaline (3, p. 14). According to BIEMANN, BOMMER, BURLINGAME and McMURRAY. [From: J. Amer. Chem. Soc. 86, 4624 (1964).]

There is one more type of dihydroindole alkaloids which has been investigated in greater detail by mass spectrometry, namely the *ajmaline* group (3, p. 35) (*23, 24, 191*). In contrast to the aspidospermine, aspidofractinine, akuammicine and aspidospermatine groups discussed earlier, the structures of most of the alkaloids related to ajmaline were known and the mass spectrometric work was thus directed towards a study of the fragmentation of this ring system, with the exception of quebrachidine (**102**, p. 36) the structure of which was deduced by mass spectrometry (*108*).

The fragmentation of the ajmaline skeleton represents, as shall be seen, a bridge between the dihydroindole alkaloids and the more fully aromatic indole derivatives discussed earlier. The former undergo mainly fission of the carbocyclic ring adjacent to the dihydroindole system and thus lead to fragments representing either the alicyclic or the aromatic portion while the latter have the tendency to retain the second nitrogen with the indole part to give β-carboline ions.

The mass spectrum of (3) *(Fig. 5)* *(Chart 13)* is thus characterized by the peaks at m/e 144 and 158, representing part of the "tryptamine" portion of the molecule, as well as N_a-methyl-β-carboline ions at m/e 182 and 183. The situation is, however, complicated by the fact that the peaks at m/e 158 and 182 have been shown to be multiplets by high resolution mass spectrometry (see later, p. 77) and contain also ions of the composition $C_{10}H_8NO$ (m/e 158) and $C_{13}H_{12}N$ as well as $C_{10}H_{16}NO_2$ (both of mass 182) (*24*).

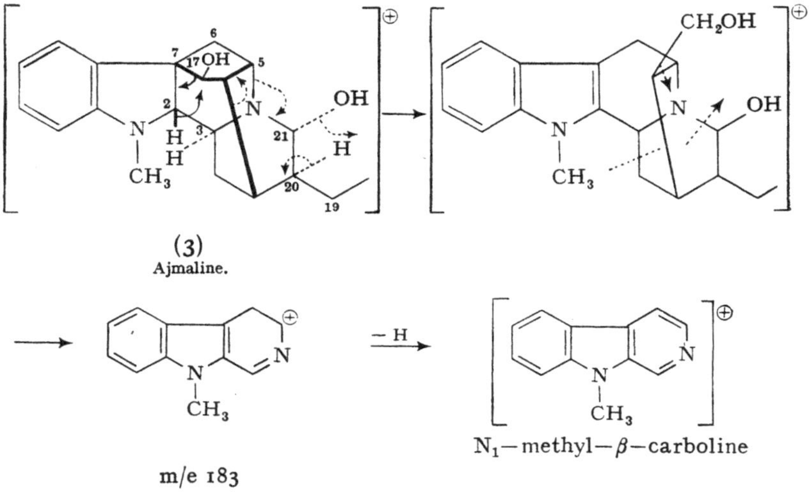

(3)
Ajmaline.

N_1—methyl—β—carboline

m/e 183

Chart 13. Fragmentation of Ajmaline (3).

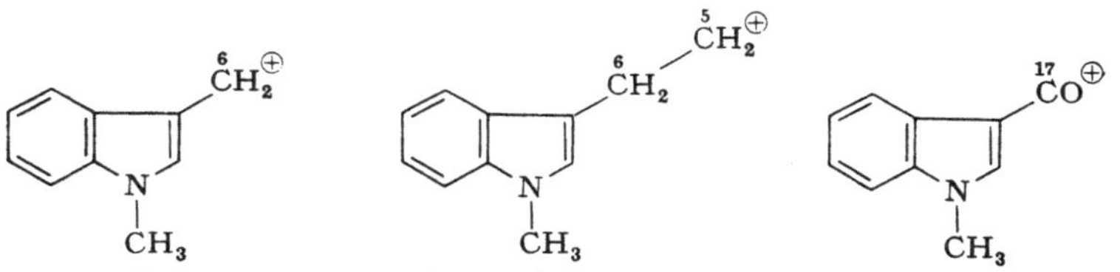

m/e 144 m/e 158 ($C_{11}H_{12}N$) m/e 158 ($C_{10}H_8N_2$)

Cleavage along the dotted arrows in (3), followed by rupture of the $C_{(2)}$—$C_{(3)}$ bond and migration of $C_{(17)}$ from $C_{(7)}$ to $C_{(2)}$ is proposed to lead upon dehydration to the $C_{13}H_{12}N$-ion of mass 182 *(Chart 14)*.

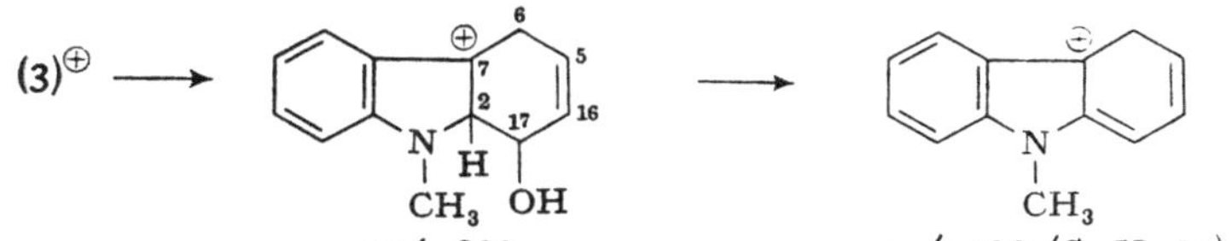

$(3)^{\oplus}$ →

m/e 200 m/e 182 ($C_{13}H_{12}N$)

Chart 14. Formation of Ion m/e 182 from Ajmaline (3).

The last step is the reason for the unexpected difference between the mass spectrum of ajmaline (3) and its 17-keto derivative ajmalidine (99). Lacking a 17-hydroxyl group, the ion of m/e 198 from the latter cannot eliminate water and thus a strikingly intense peak remains while only a very small peak appears at m/e 182 (23).

3*

This behavior may be taken as another example of a slight change in molecular structure that can give rise to a considerable change in the mass spectrum, here because a certain mode of fragmentation becomes impossible.

All of the alkaloids related to ajmaline and ajmalidine (e. g. 100–102) fit into this pattern because the structural variations generally encountered in this system (19,20-double bonds, presence or absence of the 21-hydroxyl as well as substituents in the aromatic ring) do not much influence the mass spectra.

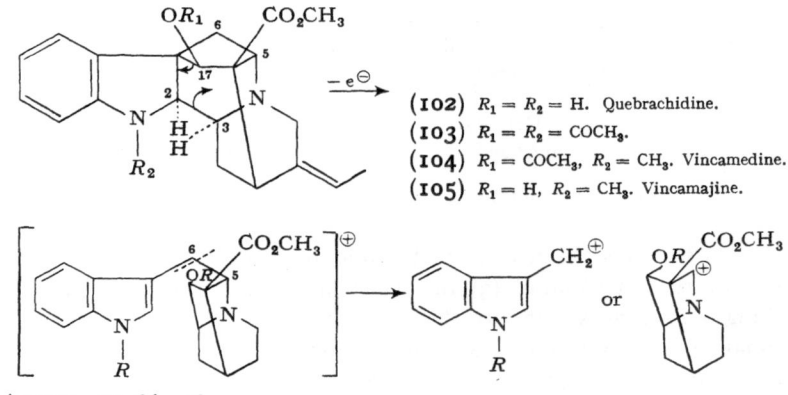

(3) R_1 = OH; R_2 = H, OH; R_3 = H. Ajmaline.
(99) R_1 = OH, R_2 = =O, R_3 = H. Ajmalidine.
(100) R_1 = OH, R_2 = =O, R_3 = CH$_3$O. Vomalidine.
(101) R_1 = H; R_2 = H, OH; R_3 = H; $\Delta^{19,20}$. Tetraphyllicine.

One of the most striking influences of a stereochemical detail upon mass spectra was revealed in the course of the determination of the structure of quebrachidine (102) (108). Its spectrum exhibited both the usual indole peaks and peaks due to the alicyclic part of the molecule – features more indicative of the aspidospermine-aspidospermatine type. Such a relationship could also be rationalized on the basis of the occurrence of (102) in a *Vinca* species until it was observed that the mass spectrum of the O,N-diacetyl derivative (103) clearly relates it to vincamedine (104) and its deacetyl derivative vincamajine (105), an alkaloid possessing the tetraphyllicine skeleton (102) (but at that time of unspecified stereochemistry) (126). It was then realized that the steric arrangement of the hydrogen at C$_{(2)}$ is vital for the first step in the fragmentation processes outlined above for ajmaline which leads to the indolization of the system (i. e. to a conversion to the sarpagine (1)-like skeleton).

(102) R_1 = R_2 = H. Quebrachidine.
(103) R_1 = R_2 = COCH$_3$.
(104) R_1 = COCH$_3$, R_2 = CH$_3$. Vincamedine.
(105) R_1 = H, R_2 = CH$_3$. Vincamajine.

In the 2-epi(β) configuration the hydrogen atom cannot be transferred to $C_{(17)}$, but the $C_{(2)}$—$C_{(3)}$ bond migrates instead. Cleavage of the 5,6-bond easily leads then to ions containing either the indole portion or the piperidine part and fragmentation of the latter, if highly substituted, as in (103–105). On the basis of this observation the 2-epi configuration was assigned to these alkaloids (103–105), which was corroborated by measuring the spectrum of authentic 2-epi-21-deoxyajmaline (*108*).

3. Indole Derivatives.

Chemically, biogenetically and even mass spectrometrically closely related to the ajmaline (3) group is *sarpagine* (1) and a number of other alkaloids based on the same carbon skeleton *(Table 7)*. The two types differ in the presence or absence of a bond linking $C_{(17)}$ to $C_{(7)}$. The rupture of this bond and transfer of a hydrogen from $C_{(2)}$ to $C_{(17)}$ converts the skeleton of (3) to that of (1); and since this process is suggested to be a favored fragmentation for (3), the mass spectrum of (1) and related substances (e. g. 106–111) exhibit some of the peaks discussed above with the ajmaline group. The most characteristic aspect is the formation of the β-carboline ions (*20, 188*), which appear as two peaks at m/e 168 and 167 for alkaloids unsubstituted in the indole portion or $C_{(3)}$ (e. g. 106–108), at m/e 182, 183 for 1-methyl derivatives (e. g. 4, p. 14), at

Table 7. Alkaloids Related to Sarpagine (1).

	R_1	R_2	R_3	R_4	R_5	References
Sarpagine (1)	OH	H	H	CH_2OH	H	(*20, 21, 188*)
Polyneuridine (106)....	H	H	CH_2OH	CO_2CH_3	H	(*5*)
Akuammidine (107) ...	H	H	CO_2CH_3	CH_2OH	H	(*159*)
Polyneuridine aldehyde (108)	H	H	CHO	CO_2CH_3	H	(*129*)
Perivine (109)					OH	(*110*)
Voacarpine (110)	H	H	CO_2CH_3	CH_2OH	OH	(*81*)
Voacoline (111 a) $R_1 = H$; $R_2 = CH_3$..						(*143*)
Voamonine (111 b) $R_1 = OH$; $R_2 = H$ ..						(*143*)

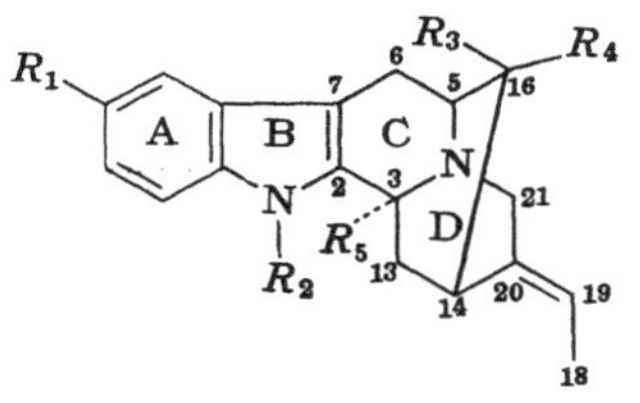

(1)

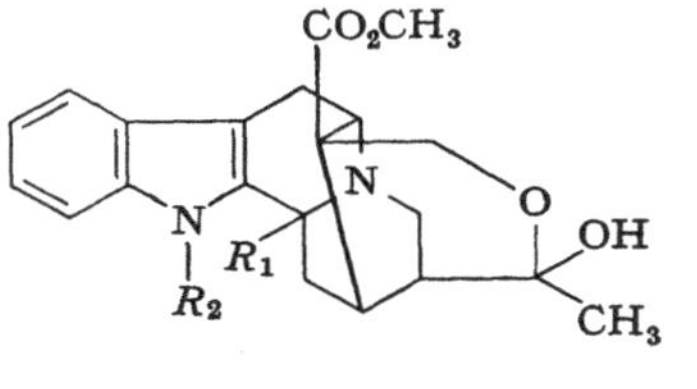

(111)

m/e 184, 185 for phenols (e. g. 1) or 3-hydroxy derivatives (109–111) and at m/e 198, 199 for aromatic methoxyl derivatives (e. g. 2). Its genesis is best explained by a rupture of three (3,13; 4,21; and 5,16) and making of two (3,4 and 13,21) bonds accompanied by transfer of hydrogens (for detailed suggestions see 20):

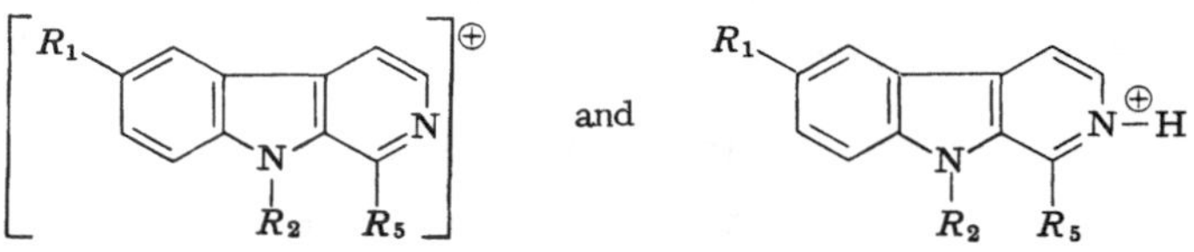

These two peaks, in addition to the loss of one hydrogen (at $C_{(3)}$) from the molecular ion are typical of the sarpagine skeleton. The absence of the "indole-peaks" at m/e 130, 143 and 144 (or their analogues in substituted derivatives) distinguish the sarpagine type from the dihydro-indoles capable of carboline ion formation such as the ajmaline derivatives (see discussion above). Most other peaks in the mass spectra of sarpagine derivatives are due to the loss of parts of the periphery, such as the substituents R_3 and R_4 without or with $C_{(16)}$ (5, 20, 81, 143, 159).

An interesting effect of the stereochemical difference between (106) and (107) was observed, namely the preferred loss of water from the former. This was first explained (5) as an elimination of H_2O in the form of the 17-hydroxyl* with the hydrogen at $C_{(6)}$, because these are close in (106) but far apart in the isomer (107). When it was, however, found that the 1-methyl derivative (e. g. degradation product of vincamedine, 104) does not eliminate water to a comparable extent, it was concluded that this difference in the spectra of (106) and (107) is mainly due to a pre-ferred dehydration of (106) in the heated inlet system, involving the loss of the 17-hydroxyl group together with the hydrogen at $N_{(1)}$ and forma-tion of a 1,17-bond (159).

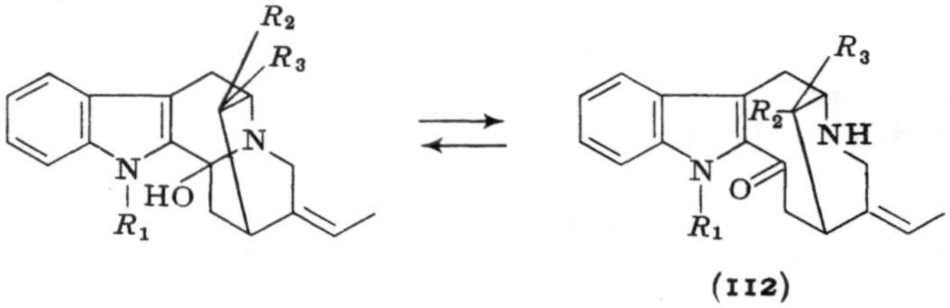

(112)
(113)
($R_1 = R_3 = $ H, $R_2 = CO_2CH_3$; N—CH$_3$ instead of NH).

Chart 15. Tautomerism of 2-Acylindole Alkaloids with sec. N_b.

* The 17-position is the carbon atom in R_3, if R = H.
References, pp. 86—98.

The 3-hydroxy derivatives (109, 110) (81, 110) represent a special system as these compounds are in equilibrium with medium-ring 2-acylindoles (112) of the vobasine type (113) *(Chart 15)*.

The latter gives rise to a very intense peak (e. g. m/e 180 in 112) representing the piperidine ring with the substituent at $C_{(16)}$ (if either R_2 or $R_3 = H$) as well as the ethylene group (55, 176).

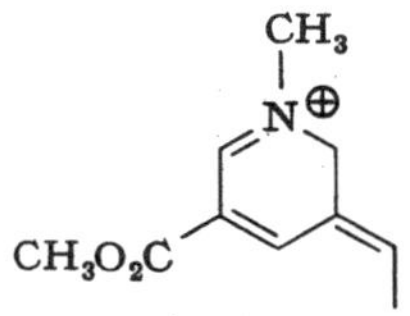

Chemical interconversion (81, 110) of one form into the other provides valuable information when carried out in conjunction with mass spectrometry as it provides support for the presence of such an acylindole-3-hydroxytetrahydrocarboline system and also permits to locate the substituents.

CH$_3$

CH_3O_2C

m/e 182

The hexacyclic alkaloids voacoline and voamonine gave mass spectra with characteristic β-carboline peaks (see p. 37), and opening the hemiketal ring by chemical means and removal of the hydroxymethylene group yielded the pentacyclic sarpagineskeleton (i. e. 114) (143) which exhibited the mass spectrum typical of this system.

(111 a) →

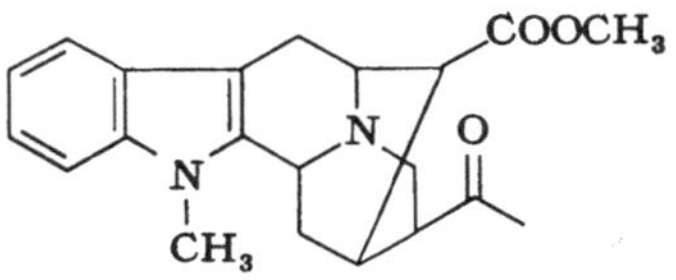

COOCH$_3$

N

CH$_3$

(114)

Sarpagine.

The foregoing discussion, leading in some detail from aspidospermine to sarpagine may suffice to illustrate the usefulness of mass spectrometry in this particular field, especially if used in conjunction with not only other physical methods (which have been mentioned only occasionally, because they are not related to the subject of this Review) but also with chemical transformations that sometimes convert one system into another.

It would be too far-reaching to outline the many other types of alkaloids that have been investigated mass spectrometrically. Since these papers concern mainly the mass spectra of largely known compounds with only occasional contributions to the elucidation of a new structure it suffices to summarize this work briefly as follows.

The mass spectrum of uleine (115) has been interpreted (127) and used to establish the structure of related alkaloids such as des-N-methyl-uleine (116), dasycarpidone (117), des-N-methyl-dasycarpidone (118), dasycarpidol (119), and 1,13-dihydro-13-hydroxy-uleine (120) (129, 160).

		R_1	R_2
(115)	Uleine	CH_2	CH_3
(116)	Des-N-methyl-uleine	CH_2	H
(117)	Dasycarpidone	O	CH_3
(118)	Des-N-methyl-dasycarpidone	O	H
(119)	Dasycarpidol	H, OH	CH_3
(120)	1,3-Dihydro-13-hydroxy-uleine	H, CH_2OH	CH_3

The alkaloid apparicine, giving a mass spectrum rather similar to that of uleine, turned out to have a related but definitely different carbon skeleton and was assigned structure (121) (128).

(121) R = CH_2. Apparicine.
(122) R = CH_2OH, CO_2CH_3. Vallesamine.
(123) R = CH_2OCOCH_3, CO_2CH_3.

Vallesamine (122) and O-acetyl-vallesamine (123) were found to be related to apparicine (204).

The mass spectra of the *eburnamine* (124) class were found to be very characteristic (171, 184, 190), particularly due to the preferential loss of either the ethyl group or atoms 5, 14, 17, 18 and 19 via a retro-Diels-

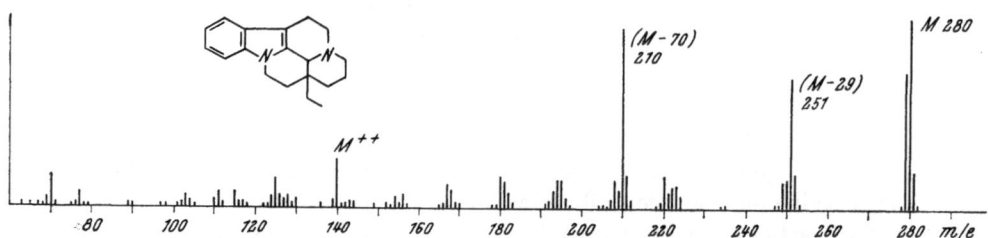

Fig. 6. Mass spectrum of dihydroeburnamenine. According to SCHNOES, BURLINGAME and BIEMANN. [From: Tetrahedron Letters 1962, 993.]

Alder cleavage of ring C. Vincamine (125) follows this pattern and was the basis for the characterization of a new alkaloid as 11-methoxyvinca-mine (126) (171). The mass spectrum of dihydroeburnamenine is shown in *Fig. 6*.

Other tetrahydrocarboline derivatives such as the *yohimbines* (127), ajmalicine (128), and related alkaloids (6, 193, 194) undergo an analogous fragmentation process which seems to be favored for (128) in comparison with (127) since the resulting ion of m/e 156 is much more abundant in the former. This may be due to the allylic nature of the $C_{(14)}$—$C_{(15)}$ bond in (128) resulting in an allylic radical for the neutral fragment *(Chart 16)*.

References, pp. 86—98.

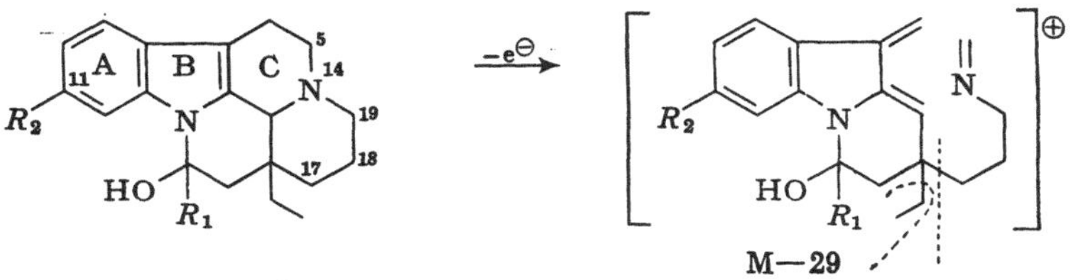

(124) $R_1 = R_2 =$ H. Eburnamine.
(125) $R_1 = CO_2CH_3$, $R_2 =$ H. Vincamine.
(126) $R_1 = CO_2CH_3$, $R_2 = OCH_3$. 11-Methoxyvincamine.

The β-carboline fragments (rings A, B and C) formed by loss of rings D and E are present but much less abundant than in the sarpagine-type (see p. 37). The loss of one hydrogen (at $C_{(3)}$) is however very pronounced

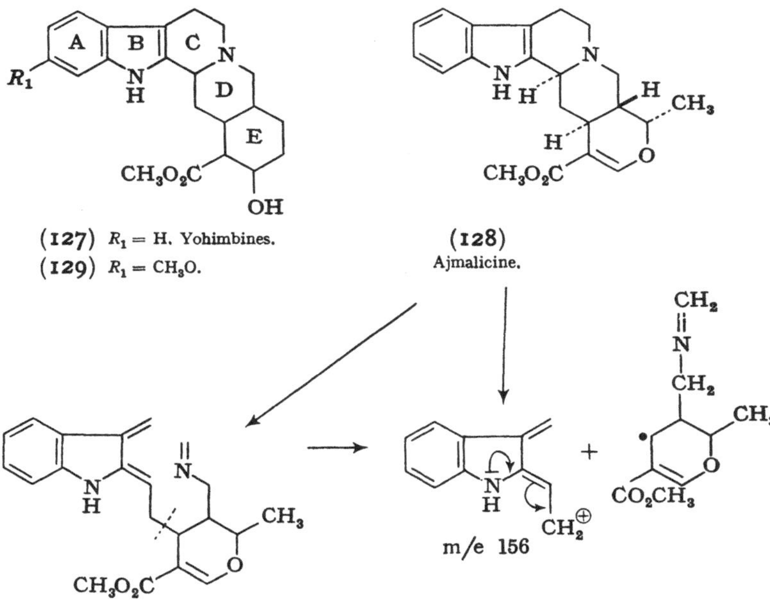

(127) $R_1 =$ H. Yohimbines.
(129) $R_1 = CH_3O$.

(128)
Ajmalicine.

m/e 156

Chart 16. Fragmentation of the Yohimbine Skeleton.

making this a characteristic of all tetrahydro-β-carboline alkaloids (e. g. 1, 124, 127 and 128) in which the resulting ion represents a fully conjugated ammonium ion*:

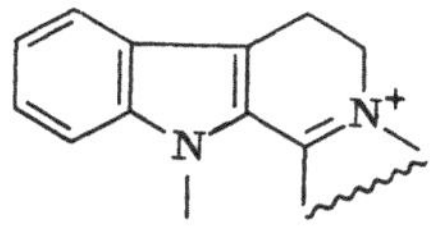

* The published (189) mass spectrum of kopsine shows a very intense peak for loss of hydrogen which was attributed to an (unspecified) hydrogen atom whose

Naturally occurring methoxy derivatives of yohimbine (e. g. **129**) (*193*) and ajmalicine as well as unsaturated and/or methoxylated derivatives of dihydrocoreantheol (**130**) and tetrahydrogeissoschizine (**131**) have been characterized by mass spectrometry (*194*).

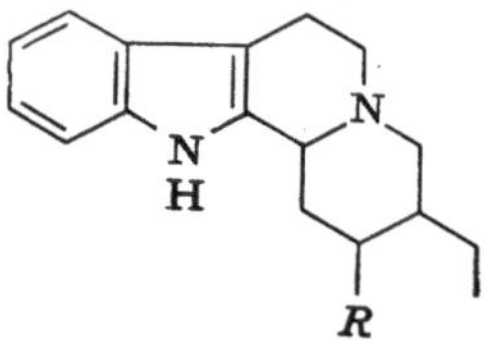

(**130**) R = CH$_2$CH$_2$OH. Dihydrocoreantheol.
(**131**) R = CH(CO$_2$CH$_3$)CH$_2$OH. Tetrahydrogeissoschizine.

The *oxindoles* (*103*) and *pseudoindoxyls* (*101*) corresponding to yohimbine and ajmalicine exhibit mass spectra which are, of course, very different from the corresponding indoles (**127** and **128**, respectively). They are characterized by a very intense peak due to cleavage of the spiropyrrolidine ring with charge retention favored on the piperidine moiety. In the case of the oxindoles the resulting species is a radical ion, e. g. m/e 223 from mitraphylline (**132**).

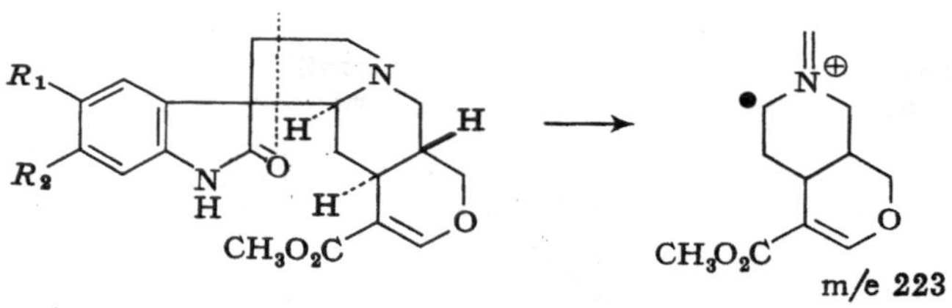

(**132**) R$_1$ = R$_2$ = H. Mitraphylline.
(**133**) R$_1$ = R$_2$ = OCH$_3$, 20-Hα. Carapanaubine.

The structure of carapanaubine was proposed mainly on the basis of its mass spectrum which showed this typical peak but had a molecular weight 60 mass units heavier than that of mitraphylline (**132**). Since the difference corresponds to the presence of two methoxyl groups, which must be in the aromatic part, structure (**133**) was indicated and finally established by partial synthesis (*103*).

The pseudoindoxyls behave similarly with the exception that the hydrogen at C$_{(14)}$ is transferred from the alicyclic moiety to the indole system, resulting in a species of m/e 222 from the pseudoindoxyl (**134**) of ajmalicine, for example (*101*):

loss is a favorable process, although kopsine (*112*, *189*), which is related to pleiocarpine (**65**, p. 27) does not contain such a hydrogen. The mass spectrum of kopsine does, in fact, not exhibit such a peak. The published data must have been due to a misidentification of the mass of the molecular ion peak.

References, pp. 86—98.

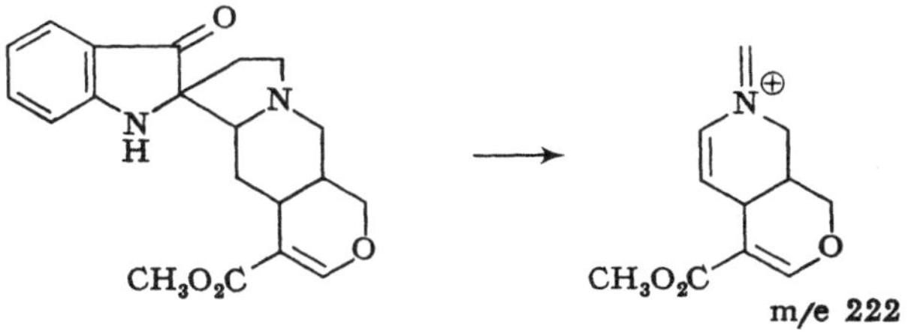

(**134**) → m/e 222

It is clear that the mass of this conspicuous peak, taken in conjunction with that of the molecular ion, reveals much about the structure of an oxindole or pseudoindoxyl alkaloid; and since they are generally related to naturally occurring indole alkaloids, their complete structure can be easily deduced and corroborated by partial synthesis.

The mass spectra of norfluorocurine (**135**) (*116*) and fluorocarpamine (**136**) (*131*), the pseudoindoxyl derivatives of pleiocarpaminol (**137**) and pleiocarpamine (**138**) (*116*) are also characterized by a single, very intense peak (at m/e 121) representing the piperidine ring *(Chart 17)*.

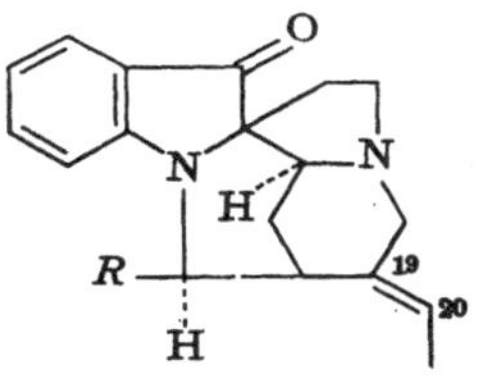

(**135**) R = CH$_2$OH. Norfluorocurine.
(**136**) R = CO$_2$CH$_3$. Fluorocarpamine.
(**139**) R = CH$_2$OH. 19,20-Dihydro.

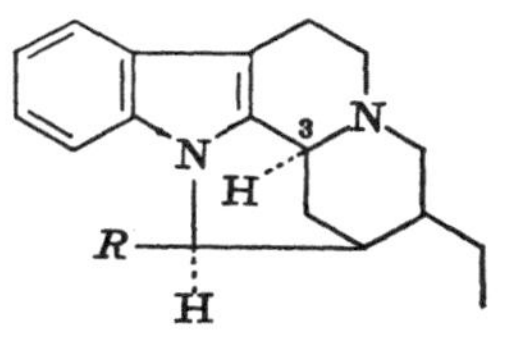

(**137**) R = CH$_2$OH. Pleiocarpaminol.
(**138**) R = CO$_2$CH$_3$. Pleiocarpamine.

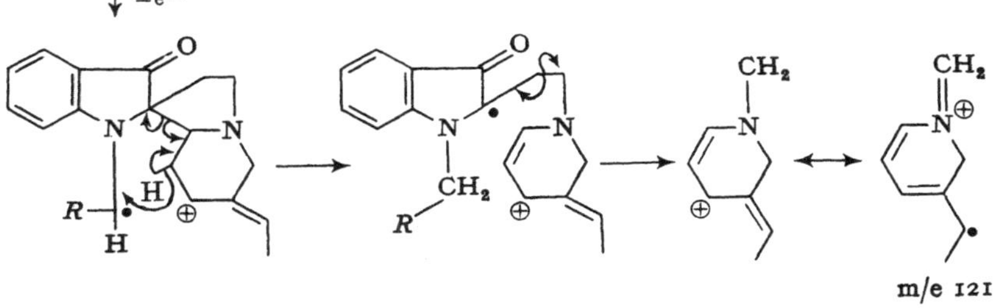

m/e 121

Chart 17. Fragmentation of Some Pseudoindoxyl Derivatives.

While the process depicted above must not necessarily be correct in all details, there is hardly any doubt that the ion of mass 121 represents the atoms shown. If it appears in abundance in a mass spectrum of a pseudoindoxyl alkaloid it may thus be taken as indicative of the system present in (**135**). Surprisingly, the 19,20-dihydroderivative (**134**) does not exhibit a peak at m/e 123 of any significant intensity, a result which reduces the diagnostic value of this fragmentation and casts some doubt on the proposed mechanism which would have to depend entirely on the allylic carbonium ion formed in the first step. The most abundant peak in the spectrum of (**139**) is due to the loss of R (*116*).

The mass spectra of (**137**) and (**138**) are quite uninteresting, consisting of a fragment which results from loss of R as well as of a peak at m/e 180. The following representation was given for this ion without any explanation (*116*):

$$(\text{137}) \text{ or } (\text{138}) \xrightarrow{-e^{\ominus}}$$

m/e 180

It may be noted that the spectra of (**137**) and (**138**) do not contain an intense M-1 peak although loss of the hydrogen at $C_{(3)}$ would seem to give rise to the conjugated ammonium ion discussed on p. 41. In this particular case such an ion would probably be too strained to be energetically favorable.

The more aromatic and the less substituted an indole alkaloid becomes, the less characteristic its mass spectrum will be. Its intense molecular ion peaks provide a clear indication of the molecular weight, but the often only weak fragment peaks are due to the loss of small substituents and thus are not very informative. Fortunately, such systems have generally very characteristic UV spectra and the few hydrogens present, mostly in specific environments, give rise to specific NMR signals. A combination of these techniques frequently permits the elucidation of the structure of such compounds with little effort.

For example, the UV spectrum of isotuboflavine was very similar to that of tuboflavine (known to have structure **140**) suggesting that the two compounds have the same aromatic system. Their mass spectra, which indicated that they are isomers, were similar, in general, but differed distinctly. For example, while tuboflavine has the tendency to lose the carbonyl and the ethyl group (as C_2H_4) *(Chart 18)*, isotuboflavine lost, in addition, two carbon atoms. For isomers of the same chromophore these data permit only a different position of the ethyl group in the same ring. Since the differences in the mass spectra cited above could best be rationalized by placing the ethyl group closer to the benzene ring, structure (**141**) was assigned to isotuboflavine, a conclusion supported by the differences in the NMR signals due to the ethyl group and the vinylic proton (*2*).

References, pp. 86—98.

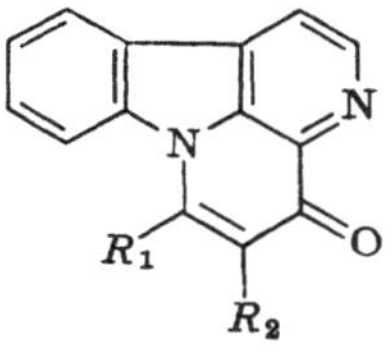

(140) $R_1 = H$, $R_2 = C_2H_5$. Tuboflavine.
(141) $R_1 = C_2H_5$, $R_2 = H$. Isotuboflavine.
(142) $R_1 = CH_3$, $R_2 = H$. Norisotuboflavine.

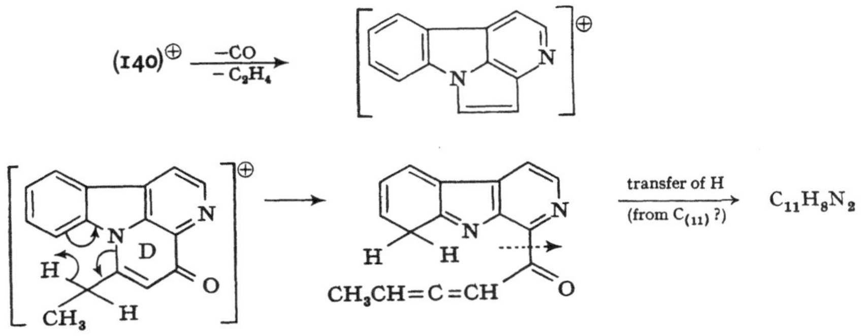

Chart 18. Structures and Fragmentation of Some Aromatic *Pleiocarpa* Alkaloids.

The mass spectrum of a congener of these alkaloids clearly showed it to be a lower homologue of (141) (but not of 140) since the molecular ion and the peak due to loss of CO were found 14 mass units lower while those arising by the loss of ring D remained the same. It is therefore represented by structure (142) and was called norisotuboflavine (2).

The extreme of aromaticity and lack of substituents (as far as indole alkaloids are concerned) is represented by indolopyridocoline (143) which was isolated along with its dihydroproduct (144) from *Gonioma kamassi* E. MEY (131).

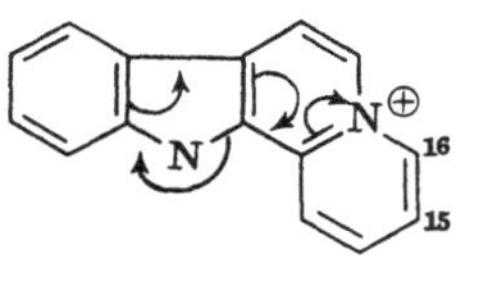

(143) Indolopyridocoline.
(144) (15,16-dihydro).

The mass spectrum of (143) (which in the gas phase probably is the uncharged resonance form indicated by the arrows) consists almost exclusively of the molecular ion (m/e 218) and the corresponding doubly charged species (m/e 109). The even molecular weight requires the presence of an even number of nitrogens and this information, in combination with the characteristic ultraviolet spectrum, permitted the identification of the natural product as (143) which was confirmed by comparison with an authentic sample.

It should be noted that saturation of the 15,16-double bond introduces a hydrogen atom that can easily be lost. The spectrum of (**144**) is, therefore, not an exact analogue of that of (**143**) shifted by two mass units but exhibits a very intense peak at M-1:

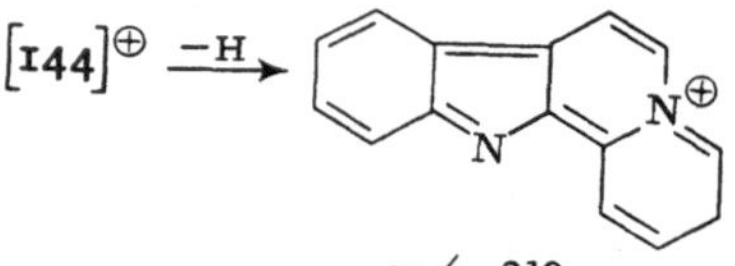

$$[\mathbf{144}]^{\oplus} \xrightarrow{-H}$$

m/e 219

4. "Dimeric" Indole Alkaloids.

While the above discussion has by no means encompassed all types of indole alkaloids that have been investigated mass spectrometrically, it will have sufficed to outline the general approach and some specific results.

It may be worth-while to mention briefly the role that mass spectrometry has played in the elucidation of the structure of some more complex molecules, namely naturally occurring compounds that consist of two indole alkaloid rests (generally one indole and one dihydroindole), joined together via a bond between the alicyclic moiety of one and the aromatic nucleus of the other. These systems are therefore not true dimeric alkaloids (such as the curare-type); and the problem is to deduce the structure of the two non-identical components and the two points of attachment.

The first alkaloid of this type whose structure was elucidated was *voacamine* (**145**) (*53*) which consists of a molecule of voacangine (**146**) bound to a derivative of vobasine (**147**). Its structure was deduced mainly on the basis of chemical evidence (*53*) but the interpretation of the mass spectrum was also quite informative, since it showed peaks characteristic for the iboga-skeleton present in voacangine (e. g. m/e 136) as well as for the vobasine type (e. g. m/e 180) (*53, 55*). While the spectrum of voacamine is not as ideally suited for the recognition of the two moieties as in some other cases (e. g. pleiomutine and villalstonine, p. 51, 52), it lent considerable support to the structure of (**145**) once the nature of the two parts had been determined (or at least suspected) by chemical means.

There arose one problem which caused a certain amount of confusion and, at least for some time, cast doubt on the reliability of mass spectrometry in this field. The molecular weight of voacamine as determined by mass spectrometry was 718 rather than 704, if the peak of highest mass and reasonable intensity was considered due to the molecular ion of the compound, as is common practice. This could have been neglected as due to contamination with a higher homologue or a keto-analogue of (**145**) (*55*); it was found, however, (*53*) that the higher homologue was not an impurity but was formed by a thermal reaction during the heating of the sample to vaporize it into the ion source of the spectrometer. A change of the ratio of the intensities of the peaks at m/e 704 and 718 during the process of heating was a clear indication of such a reaction. It was suggested that this was due to an intermolecular transmethylation process involving a carbomethoxy group and a basic nitrogen, followed by a Hofmann elimination; thus the quarternary base was converted into a tertiary one which then volatilized (*53*). This proposal was analogous to the one made much earlier (*111*) to explain the formation of trimethyl-amine during the pyrolysis of voacamine.

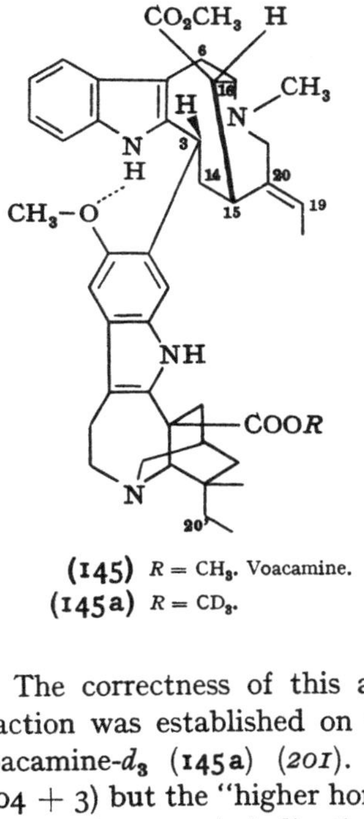

(**145**) *R* = CH₃. Voacamine.
(**145a**) *R* = CD₃.

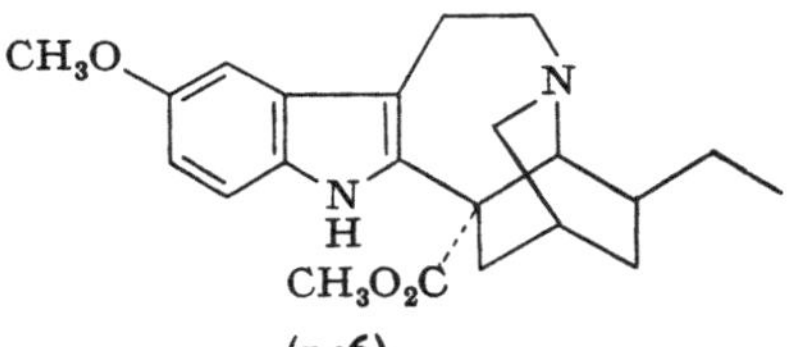

(**146**)
Voacangine.

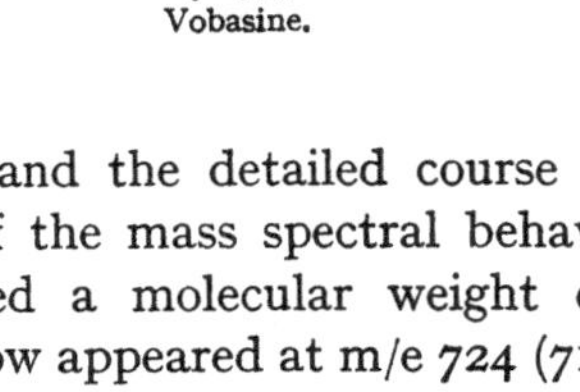

(**147**)
Vobasine.

The correctness of this assumption and the detailed course of the reaction was established on the basis of the mass spectral behavior of voacamine-d_3 (**145a**) (*201*). It exhibited a molecular weight of 707 (704 + 3) but the "higher homologue" now appeared at m/e 724 (718 + 6) (with little at 721), indicating that the voacangine carbomethoxy group of one molecule of voacamine methylates the amino-nitrogen of the vobasine moiety of another molecule. (The distinction between the four nitrogens present in (**145**) could be made on the basis of the fragment ions which appear in the spectrum while pyrolysis proceeds.)

These results are significant because they clearly establish the requirements for the occurrence of such thermal "homologizations" and thus restore confidence in the reliability of mass spectrometry in the determination of true molecular weights. For such a thermal transmethylation to occur at a reasonable rate (and thus give rise to artificial peaks of reasonable intensity) there must be present an alkylable nitrogen atom and a group that can alkylate a carbomethoxy group, for example. The latter should be a reactive one, such as present in 3-carbomethoxy indole alkaloids which are known to undergo decarbomethoxylation upon heating with hydrazine, in contrast to "normal" carbomethoxy groups that merely form hydrazides (see the example of vinblastine below). In agreement with this conclusion is the fact that pleiomutine (150, p. 51) and villalstonine (151, p. 52) which both lack a carbomethoxy group on the indole moiety, do not exhibit thermal transmethylation. A further requirement for this thermal reaction to occur is low volatility of the sample which should, of course, not vaporize below the temperature at which the transmethylation sets in. The low pressure in the ion source precludes this bimolecular reaction from taking place after volatilization.

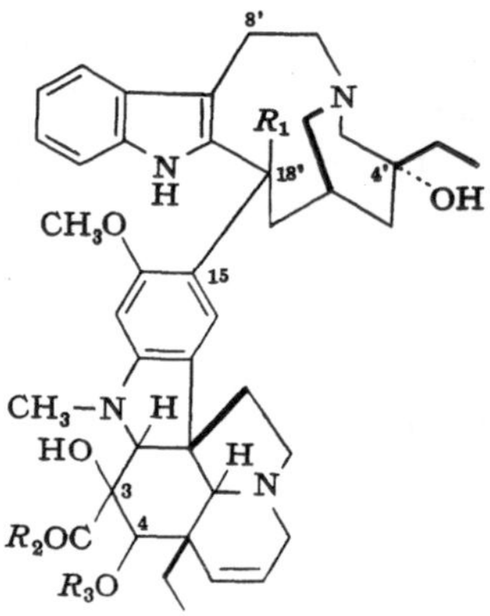

(148) R_1 = COOCH$_3$, R_2 = OCH$_3$, R_3 = COCH$_3$. Vinblastine.
(149) R_1 = R_3 = H, R_2 = NHNH$_2$.

The problem of defining the molecular weight was aggravated in the case of *vinblastine* (158), the mass spectrum of which showed that the heaviest ion is at m/e 838. Only after studying its thermal reactivity in the mass spectrometer and accurate mass measurements on the ions of m/e 810, 824 and 838 (which were found to correspond, respectively, to $C_{46}H_{58}N_4O_9$, $C_{47}H_{60}N_4O_9$ and $C_{48}H_{62}N_4O_9$) it became clear that the molecular weight of vinblastine is 810 and that the homologous ions are formed by thermal transmethylation which in this molecule can even occur twice (47). Support of this explanation was derived from the mass spectrum of the product formed by treatment of vinblastine with hydrazine, which then had a molecular weight of 710 and was thermally stable. Decarboxylation of the voacangine moiety and hydrazinolysis of the two ester groups in the vindoline part had taken place and produced a molecule lacking alkylating groups. The structure of vindoline was finally established on the basis of chemical and spectral evidence to be (**148**) and the hydrazinolysis product corresponds to (**149**) (*158*).

Especially useful for the determination of the structure of such "pseudo-dimeric" alkaloids is a mass spectrum if it shows most or all of the characteristic peaks of both units. In this case the spectrum eliminates the need for chemical cleavage followed by isolation and characterization of the cleavage products which often is difficult and time-consuming. While one will always wish to support the mass spectrometric evidence by such a chemical conversion and characterization, this is much easier if one has a clear notion which compounds are to be looked for and thus can design the cleavage and isolation procedures accordingly.

An excellent example is the determination of the *pleiomutine* structure (*199*). Its mass spectrum indicated a molecular weight of 630 and precise mass measurement revealed the elemental composition $C_{41}H_{50}N_4O_2$. The details of the spectrum indicated that one of the components may be of the pleiocarpine type (intense peaks at m/e 109 and 124). The loss of CH_3O and CO_2CH_3, respectively, from the molecular ion required the presence of a carbomethoxy group. Finally, comparison of the spectrum of pleiomutine *(Fig. 7)* with those of various pleiocarpine derivatives revealed that almost all peaks of the spectrum of pleiocarpinine (Fig. 4, p. 26) are present, either as such (m/e 109, 124) or shifted by 278 mass units (m/e 507, 519, 544, 571, 599, 602 and 630, corresponding to 229, 251, 266, 293, 321, 324 and 352 in Fig. 4). Subtraction of the UV spectrum and molecular weight (minus one hydrogen) of pleiocarpinine from the UV spectrum and molecular weight of pleiomutine has revealed that the other part of the molecule should be an indole derivative of mass 280 minus one hydrogen, more specifically a species $C_{19}H_{24}N_2$. This is precisely the formula and chromophore of dihydroeburnamenine; and an inspection of the high mass section of Fig. 7 reveals that the two important peaks that are not compatible with a fragmentation of the pleiocarpinine moiety are those at M-29 and M-70, ions that are indeed prominent in the spectra of eburnamine derivatives (see Fig. 6, p. 40, and p. 41). Furthermore, the spectrum has made it possible to narrow down the areas of the two molecules which cannot be involved in the junction of the two in the "dimer". The presence of the peaks at M-29, M-70, 124 and 109 eliminate the carbon atoms 7', 17', 18', 19', 20' and 21' of the eburnamine part and the carbon atoms 3, 4, 5, 6, 7, 8, 10, 19 and 20 of the pleiocarpinine moiety. Furthermore, the peak at m/e 378 ($C_{24}H_{30}N_2O_2$) corresponds to pleiocarpinine plus two aliphatic carbon atoms which leaves virtually only $C_{(14')}$ or $C_{(15')}$ for the attachment on the eburnamin part.

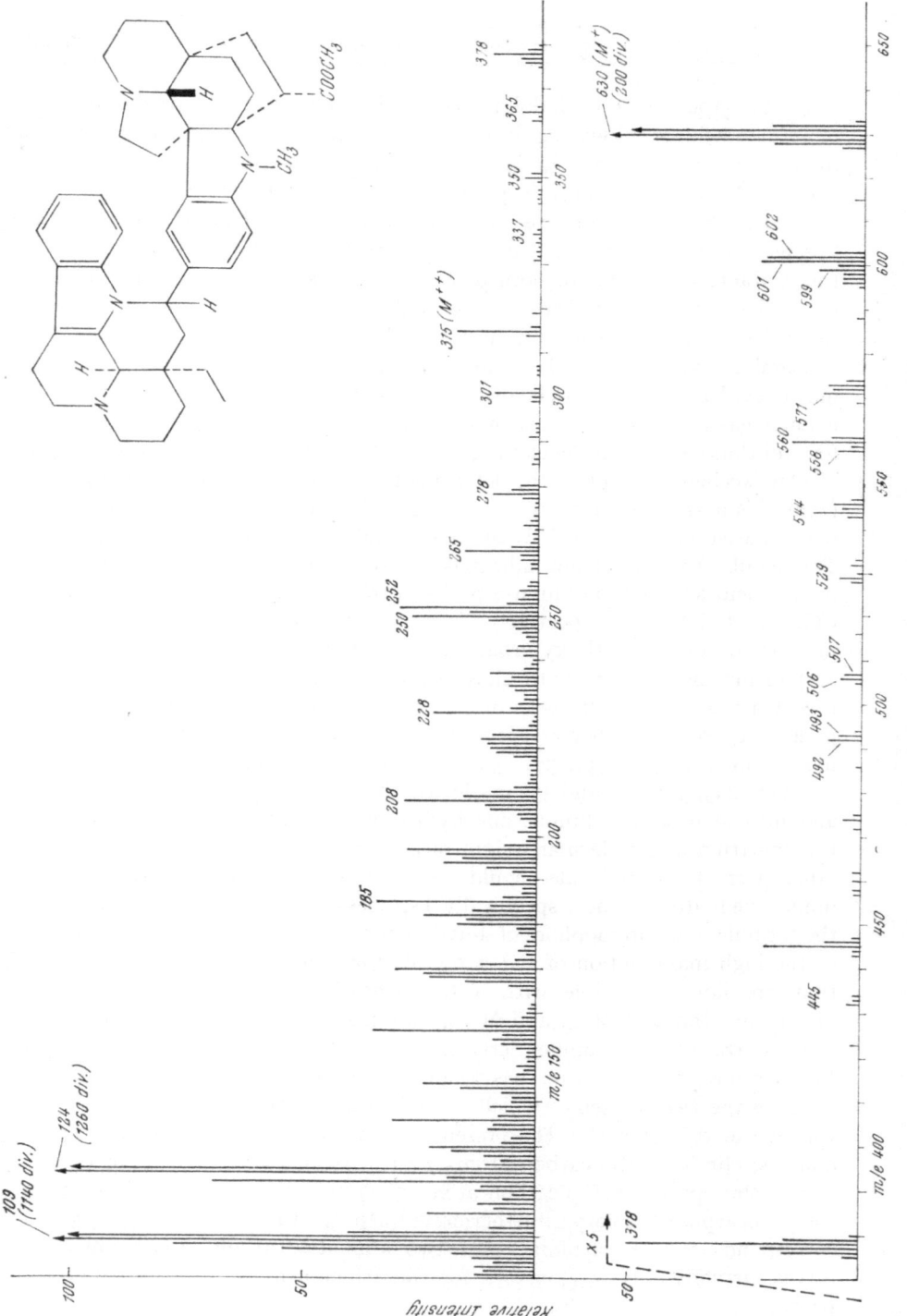

Fig. 7. Mass spectrum of pleiomutine. According to Thomas, Achenbach and Biemann. [From: J. Amer. Chem. Soc. **88**, 1537 (1966).]

That the pleiocarpine segment is attached via its benzene ring to $C_{(14')}$ was also shown by mass spectrometry. Cleavage of pleiomutine with deuterophosphoric acid produced pleiocarpinine containing only two deuterium atoms in the benzene ring and in the same position (15 and 17, determined by NMR) as obtained when pleiocarpine itself was treated with D_3PO_4. A detailed analysis of the mass spectrum of the dihydro-eburnamine, produced by cleavage of pleiomutine with tin and DCl, revealed that it contained deuterium at $C_{(14')}$ (in addition to 3', 9', 10', 11', 12' and 15') which can be rationalized only by the original presence of a 14'-substituent. Final assignment of the 14',15-bond and thus of structure (**150**) rests on NMR evidence.

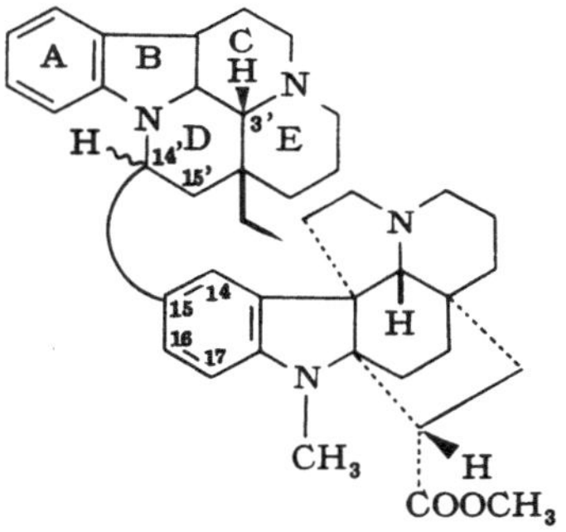

(**150**)
Pleiomutine.

The interpretation of the mass spectrum of pleiomutine may be regarded as an extreme and sophisticated example of that basic principle, the "mass spectrometric shift" technique, discussed on p. 13, if one considers pleiomutine to be merely a derivative of pleiocarpinine bearing a substituent (14-eburnamyl) at position 15.

Another "dimeric" indole alkaloid in whose structure determination mass spectrometry played a major role is *villalstonine* (**151**) (*114*). Most important was the observation that the mass spectrum of an isomerization product, villamine (**152**), which showed a molecular weight of 660, contained the peaks (m/e 322, 263 and 180) of pleiocarpamine (**138**) which had also been obtained from villalstonine by chemical cleavage. It thus seemed that electron impact cleaves the alkaloid into the two parts and pyrolysis of villamine yielded, in fact, (**138**) and another product (**153**, called macroline) whose mass spectrum exhibited all the peaks (m/e 338, 307, 320, 251, 197, 181 and 170) which were prominent, in addition to those of (**138**), in the spectrum of villamine. Unfortunately, macroline was not a known substance and its structure had to be clarified first. A methyl-vinyl ketone moiety as well as two N-methyl groups were deduced from the NMR spectrum. The UV spectrum was that of an indole and the most intense peak (m/e 197) suggested a N_a-N_b-dimethyl-β-carboline moiety. All this evidence, combined with the fragments at m/e 320 (M—H_2O) and 307 (M—CH_2OH) led to structure (153) as a working hypothesis; this was confirmed by the analogies between the mass spectrum of macroline and that of a substance [derived from ajmaline (3), p. 14], which differed from (153) only by the presence of a sec.-butyl group in place of the methyl-vinyl ketone group.

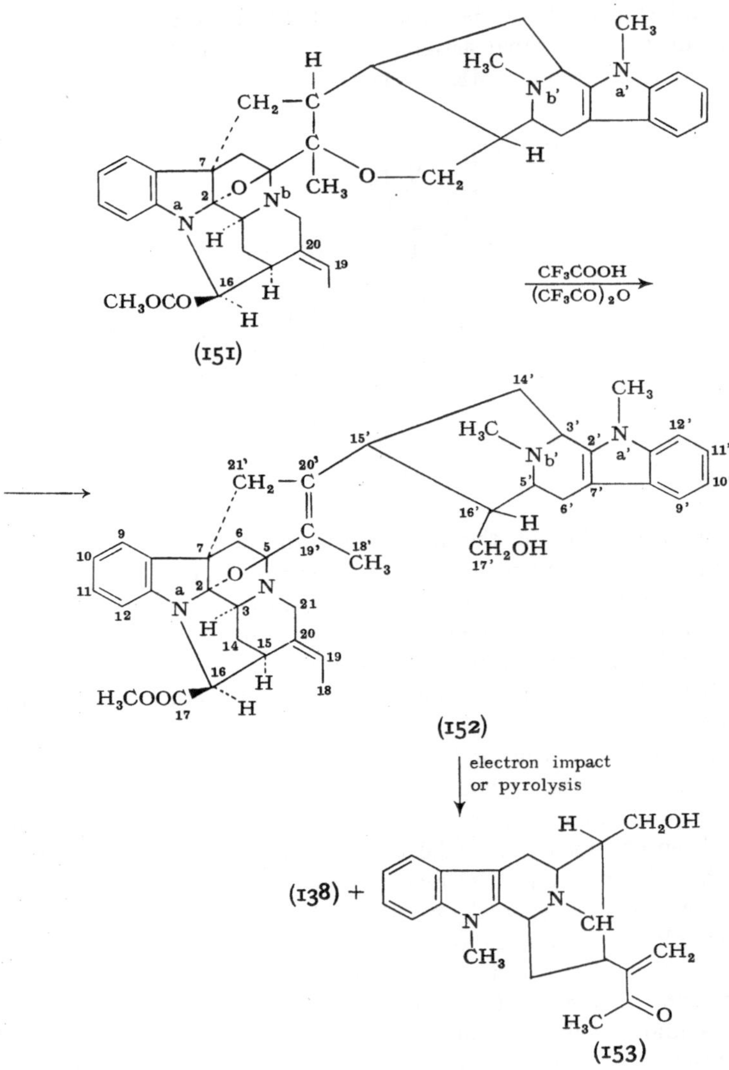

(151)

$$\xrightarrow[\text{(CF}_3\text{CO)}_2\text{O}]{\text{CF}_3\text{COOH}}$$

(152)

electron impact or pyrolysis

(138) +

(153)

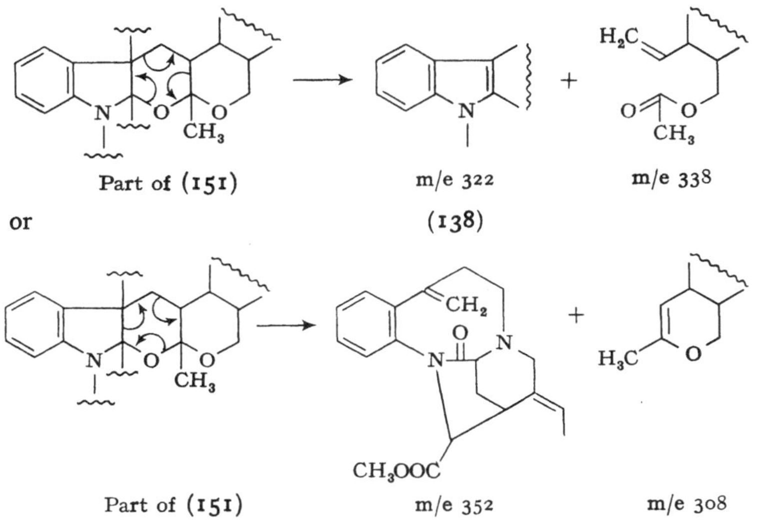

Part of (151)   m/e 322   m/e 338

or   (138)

Part of (151)   m/e 352   m/e 308

Chart 19. Cleavage of Villalstonine (151).

A detailed analysis of the NMR spectrum of villamine and comparison with the spectra of (138) and (153) has led to the conclusion that the pleiocarpamine part is present in the dihydroindole state and that the vinyl-ketone is absent. In the light of the facile pyrolysis of the "dimer" into the two indole alkaloids, a Diels-Alder type linkage, as shown in structure (152), was proposed for villamine.

Finally, it was possible to reconvert villamine to villalstonine, a reaction interpreted to involve addition of the macroline-hydroxyl group to the dihydropyrane system. The mass spectrum of villalstonine is very similar to that of villamine but contains in addition intense peaks at m/e 107, 121, 135 and 352. This spectrum was taken as a clue for the structural difference between the two compounds. The elemental composition of the m/e 352 ion was found to be $C_{21}H_{24}N_2O_3$, i. e. CH_2O more than pleiocarpamine (138). The ketal structure would permit cleavage along either of two pathways and thus account for the peaks at m/e 322, 338 and 352 *(Chart 19).*

The acyl-aniline (m/e 352) contains a piperidine ring which conceivably may give rise to the peaks at m/e 107, 121 and 135 (pyridine ions with one, two or three carbon atoms). It may be recalled that oxindoles and pseudoindoxyls (p. 42 ff.) also have a tendency to give very intense peaks due to the alicyclic part of the molecule.

The elucidation of structure (151) for villalstonine (*114*) represents another excellent example for the power of the combination of NMR and mass spectral data.

In this Section an attempt was made to illustrate the contributions of mass spectrometry over the past five years to an admittedly very narrow, but in itself highly diversified and interwoven field of natural product chemistry. Since it is impossible to devote similar detail to all other areas, the following will concern a selection of recent work to outline the suitability of mass spectrometric techniques to subjects other than indole alkaloids.

III. Tetrahydroisoquinoline Alkaloids.

For the purpose of discussion this Section will go beyond the group commonly known as tetrahydroisoquinoline alkaloids since the mass spectrometric characteristics of some compounds not actually containing an intact isoquinoline system is related to those which do contain it.

A fitting link between this and the previous Section is the alkaloid tubulosine (**154**) in whose structure determination mass spectrometry played a major role (*48*). Its molecular weight of 475 and combustion data required the elemental composition $C_{29}H_{37}N_3O_3$. Chemical information and NMR data indicated the presence of a phenolic hydroxyl, two aromatic methoxyls, two NH groups and one ethyl group.

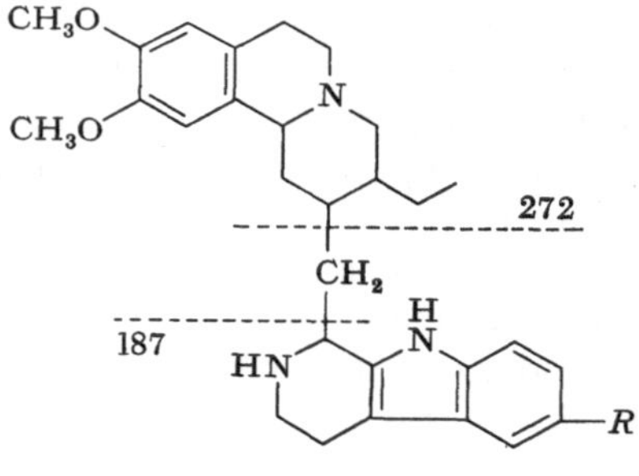

(**154**) $R = OH$. Tubulosine.
(**157**) $R = H$.

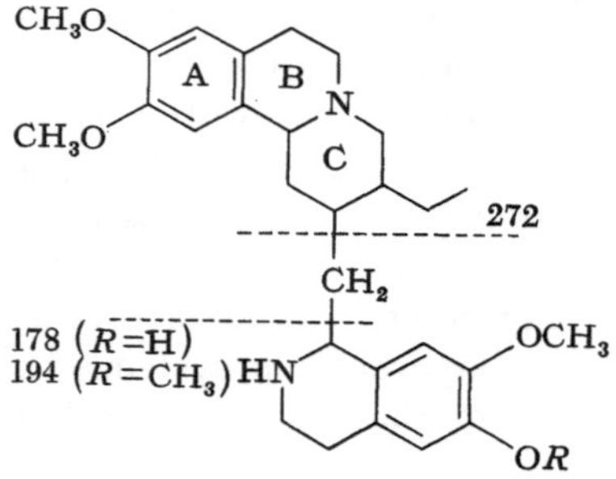

(**155**) $R = H$. Cephaeline.
(**156**) $R = CH_3$. Emetine.

The mass spectrum had many peaks in common with that of cephaeline (**155**) (*59*) and emetine (**156**) (*59, 192*), namely all those which represent the "upper" tetrahydroisoquinoline portion, including ring C (e. g. m/e 272). On the other hand, the peak that occurs at m/e 178 in (**155**) and is due to the "lower" tetrahydroisoquinoline, was replaced by one at m/e 187 which indicated the absence of rings D and E of (**155**). In their place there should be a species of 187 mass units containing two nitrogens and the phenolic hydroxyl group. If one is familiar with the mass spectra of indole alkaloids it needs little intuition to suspect the presence of a hydroxylated tetrahydro-β-carboline system and to write

References, pp. 86—98.

structure (154) for tubulosine in which only the exact position of the hydroxyl group is uncertain (so far as one relies on the biogenetic precedence of the location of the two methoxyls as in all the Ipecacuanha alkaloids).

The correctness of this assignment was supported by the mass spectrum of the synthetic base (157) which was analogous to the one of tubulosine except that the molecular weight and the mass of all peaks due to the carboline moiety were found 16 mass units lower. Finally, deoxygenation led to (157) and the mass spectrum of natural and synthetic (racemic) base were then identical (48). The position of the phenolic hydroxyl was assigned on the basis of the UV spectrum of a product in which the tetrahydrocarboline system had been selectively aromatized to a β-carboline (152).

The high tendency to cleavage of the bond attached to the "benzylamine"-carbon in tetrahydroisoquinoline derivatives governs many of the mass spectral characteristics of these compounds. It is most pronounced in those where this bond is the only link between the two parts of the molecule, such as in hydrastine (158) which exhibits a very intense peak at m/e 190 but no observable molecular ion (62, 161).

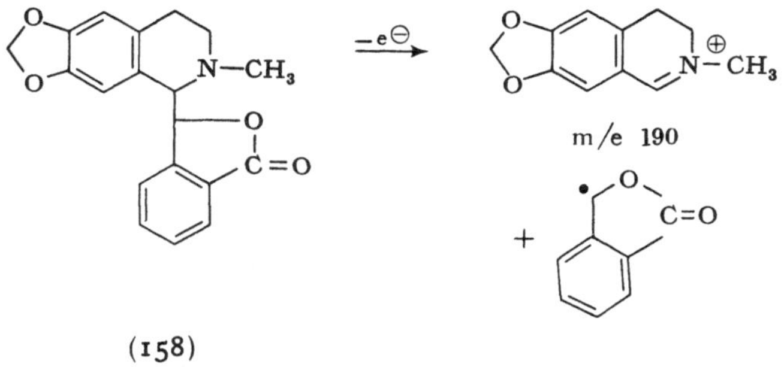

(158)
Hydrastine.

The mass of the resulting well stabilized ammonium ion is indicative of the substitution pattern of the tetrahydroisoquinoline moiety which may occur in nature not only with a methylenedioxy group but also in various stages of hydroxylation and methoxylation as well as with or without an N-methyl group.

The presence of a bond between the two phenyl groups leads to the *aporphine alkaloids* of which nornuciferine (159) is one of the simplest examples (161). Cleavage of the benzylic C—C bond does not lead directly to a fragment but rupture of the benzylic C—H bond gives rise to a sometimes very intense M-1 peak. Rupture of the benzylic C—N bond initiates a retro-Diels-Alder fragmentation of ring B with expulsion of $CH_2=NH$ (29 mass units). This cleavage is particularly useful for establishing the extent of substitution at the nitrogen (e. g. loss of 43 mass units indicates N-methyl).

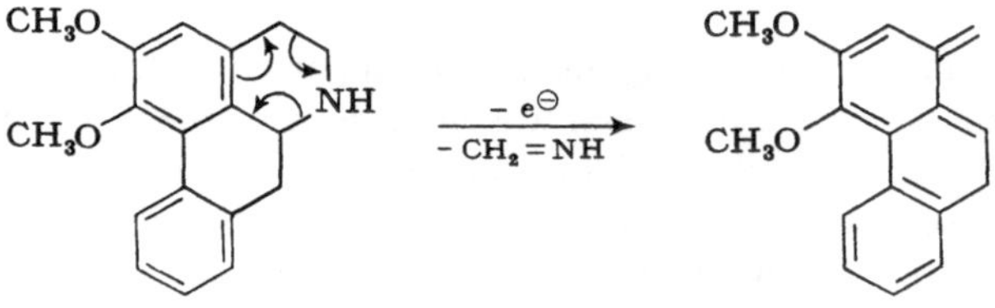

(159)
Nornuciferine.

M-29

Otherwise these spectra are not very informative, because the only peaks of reasonable intensity correspond to the loss of CH_3 and CH_3O from the molecular ion as well from the retro-Diels-Alder product (see above). The resulting ions are, of course, the more abundant the more methyl groups are present. Peaks at m/e 152 and 165 are always present to a significant extent and have been suggested to serve as an indication for the presence of an aporphine system (*161*). High resolution spectra showed them to be due to $C_{12}H_8$ (biphenylene?) and $C_{13}H_9$ (fluorenyl ion?) which indicates that they are structurally not very specific and are, in fact, found in the spectra of many compounds that can conceivably produce this highly aromatic system (morphine alkaloids, for example) (*202*).

At present no new alkaloid structure rests on mass spectrometric evidence of this type but the technique has been used in the characterization of degradation products of larger (dimeric) alkaloids.

A further variant is represented by the *xylopine* (160) *group*, and again the rupture of the benzylic C—C bond leads to the major peaks in the spectrum (m/e 164, 149 and 190 in the case of 160) (*62*) *(Chart 20)*.

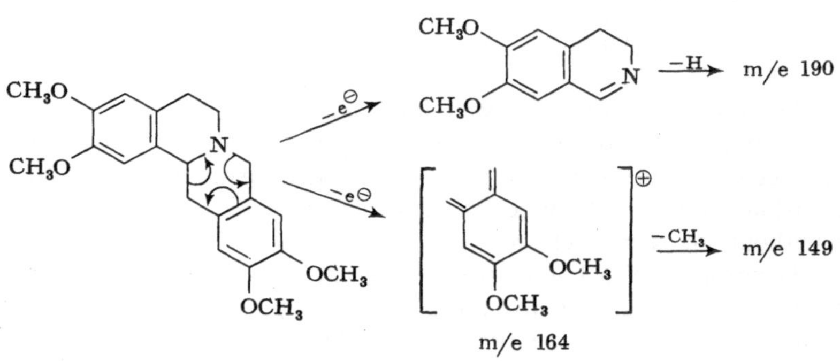

(160) Xylopine.

Chart 20. Fragmentation of Xylopine (160).

References, pp. 86—98.

Even compounds which are not exactly tetrahydroisoquinolines such as the *protopine* (161) *group* undergo a very similar fragmentation during which the missing bond is actually formed (62, 66) *(Chart 21)*.

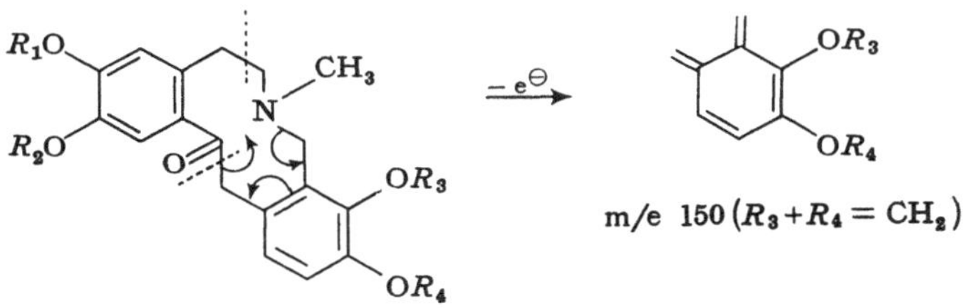

(161) $R_1 + R_2 = CH_2$; $R_3 + R_4 = CH_2$. Protopine.
(162) $R_1 = R_2 = CH_3$; $R_3 + R_4 = CH_2$. Cryptopine.
(163) $R_1 + R_2 = CH_2$; $R_3 = R_4 = CH_3$. Allocryptopine.
(164) $R_1 = R_2 = R_3 = R_4 = CH_3$. Muramine.

Chart 21. Fragmentation of Protopine Derivatives (161–164).

Less abundant, but quite significant ions arise by cleavage along the dotted lines in (161) with shift of a hydrogen from the nitrogen-containing part to the benzoyl moiety. For protopine these peaks are found at m/e 163 and 190. The substitution pattern of certain isomers is clearly apparent from only these three peaks as can be seen by comparison of the spectrum of cryptopine (162) and allocryptopine (163).

Recently, the proposed structure (164) (66) of muramine was verified using this technique, since the mass spectrum of (164) exhibited indeed by far the most intense peak at m/e 164 (requiring R_3 and R_4 to be methyl), while the appearance of relatively intense peaks at m/e 179 and 206 (as well as the molecular weight of 385) requires all substituents R to be methyl. The exact location of the methoxyl groups within each of the two benzene rings is, of course, not evident from the mass spectrum but biogenetic reasoning as well as the NMR spectrum point to the pattern shown in (164).

While all these structural types give mass spectra that are not as rich in information as those of most indole alkaloids, they easily permit assignment of the type to which a new alkaloid (or a newly isolated known one) may belong. Even if the botanical source of the material does not itself indicate that one deals with a true or modified tetrahydroisoquinoline, this can be deduced from the aromatic chromophore and the presence of one nitrogen atom (odd molecular weight). The general appearance of the mass spectrum and the mass of the most intense peaks lead easily to a working hypothesis which then can be tested in detail.

As a reverse example the structure of the *papaver alkaloid* rhoeadine might be cited. It had been assigned structure (165) but the mass spectrum clearly did not agree with this proposal since such a compound should, in analogy to hydrastine (158, p. 55) exhibit an intense peak at m/e 190 and no signal for the molecular ion (see p. 55). The spectrum of rhoeadine did, however, show a peak for the molecular ion (m/e 383) that was one half as high as the most intense one, which was found at m/e 177. The mass spectra of various Hofmann degradation products led to the conclusion that the nitrogen is not attached to the benzylic carbon atom, i. e. that rhoeadine cannot be a true tetrahydroquinoline and thus must have a seven-membered ring *B* such as in (165).

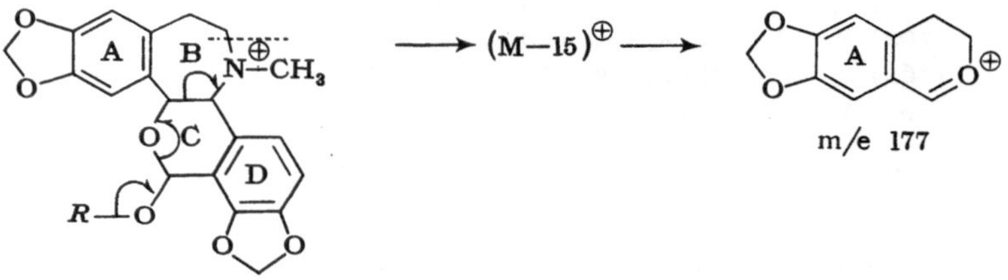

$$\longrightarrow (M-15)^{\oplus} \longrightarrow$$

m/e 177

(165) $R = CH_3$. Rhoeadine.
(166) $R = H$.
(167) $R = C_2H_5$.
(168) $R = H$ (dimethoxy instead of methylenedioxy in ring *A*). Glaucamine.
(169) $R = CH_3$ (dimethoxy instead of methylenedioxy in ring *A*). Glaudine.

The mass spectra of various derivatives particularly of the lower and higher homologues (166) and (167), of rhoeadine have led to the postulation of the fragmentation process depicted above. This evidence alone would, of course, be insufficient as a proof for structure (165), but the spectra of a series of degradation products as well as analogues such as glaucamine (168) and its methyl ether, glaudine (169), are all in support of this novel ring system. For the detailed arguments the original papers should be consulted (*168, 169, 180, 186, 189*).

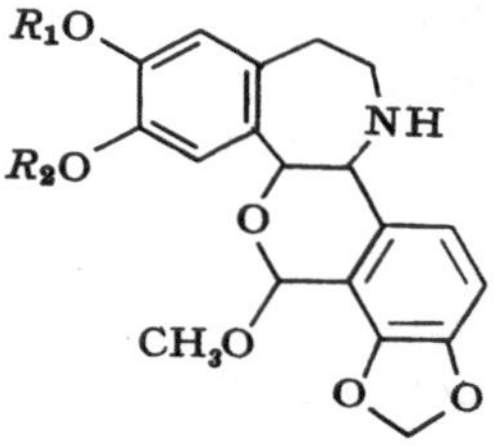

(170) $R_1 + R_2 = CH_2$. Papaverrubines A and E.
(171) $R_1 = R_2 = CH_3$. Papaverrubine B.
(172) $R_1 + R_2 = H + CH_3$. Papaverrubine D.

References, pp. 86—98.

This revision of the rhoeadine structure has led to the reassignment of a series of related alkaloids which exhibited the same type of mass spectrum as rhoeadine. These were the papaverrubines A and E (170), B (171) and D (172).

IV. Bisbenzyl-tetrahydroisoquinoline Alkaloids.

Some of the fragmentation processes discussed above for the simple tetrahydroisoquinoline alkaloids are also reflected in the mass spectra of bisbenzyl-tetrahydroisoquinoline alkaloids. Due to the more complex nature of these substances a few interesting variations of these processes are observed.

Dauricine (173) is a "dimer" (here also the two parts are hardly ever exactly identical) joined by a single ether bond, and its mass spectrum is thus similar to that of the "monomers" (80). The spectrum exhibits a very small molecular ion peak (at m/e 624) and consists of almost a single peak only (at m/e 206) which is due to the cleavage of the doubly benzylic bond with charge retention on the dihydroquinolinium ion. Since both tetrahydroquinoline moieties are identical only one peak of this type results. Charge retention on the benzyl group proper is not favored and the ion at mass 418 is thus small but still perceptible.

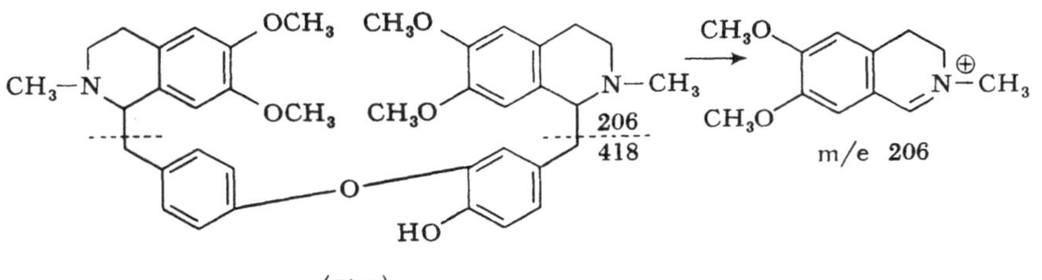

(173)
Dauricine.

A very interesting change takes place if the two tetrahydroisoquinoline groups are also linked by an ether bond as in berbamine (174) (71) and related systems in which merely one or the other ether bond is attached to different carbon atoms (of the same rings, as in 174) such as oxy-canthine (a'—b bond) or thaliciberine (a—c' bond) (80). In all these types the mass spectrum is dominated by a very intense doubly charged peak (as evidenced by the isotope peak appearing half a mass unit higher) due to a fragment arising from the cleavage of both isoquinoline-benzyl bonds in a doubly charged molecular ion (Fig. 8).

Additional proof for the fact that the ion of mass 198 is indeed a doubly charged ion comes also from accurate mass measurements which in this case gave a value of 198.1015 that is incompatible with the mass of combinations of integral numbers of C, H, N and O (with no more than two nitrogen atoms) but fits well with $C_{11.5}H_{14}NO_2$, i. e. it must be due to $C_{23}H_{28}N_2O_4{}^{\oplus\oplus}$ (*71*). Such evidence is desirable because the peak at m/e 198.5 could be also due to a small amount of $C_{23}H_{29}N_2O_4{}^{\oplus\oplus}$ in which case the first mentioned argument for the doubly charged nature of m/e 198 would no longer hold. Accurate mass measurement is of use only if one of the elements is present as an odd number, e. g., in the case of obegamine (*175*) the mass of the corresponding ion at m/e 191 (382/2) fits $C_{11}H_{13}NO_2{}^{\oplus}$ just as well as $C_{22}H_{26}N_2O_4{}^{\oplus\oplus}$.

The singly charged "top-half" leads, after loss of a hydrogen, also to a peak of reasonable intensity and these data settle the sum of all substituents on the two tetrahydroisoquinoline systems taken together. The molecular weight can be deduced from the corresponding peak (which is here much more abundant than in *173*) and gives, after subtraction of the "top-half" the combined mass of the two benzyl moieties.

It is obvious that the change of the position of the ether linkage between the two quinoline parts or the two benzyl groups does not materially change the fragmentation process, and thus the spectra of oxycanthine and thaliciferine are of the same type as that of berbamine (*174*) and obegamine (*175*).

A drastic change takes place if the ether linkages connect different rings, i. e. a "head-to-tail" arrangement such as in cycleanine (*176*) or chondrodendrine (*177*) rather than the "head-to-head"-type represented by berbamine (*174*).

In contrast to the "head-to-head"-type, these compounds do not exhibit such an unusually intense doubly charged ion although one should expect a stable benzyl-dihydroisoquinolinum ion in this case *(Chart 22)*. Instead, the corresponding singly charged ion, involving

References, pp. 86—98.

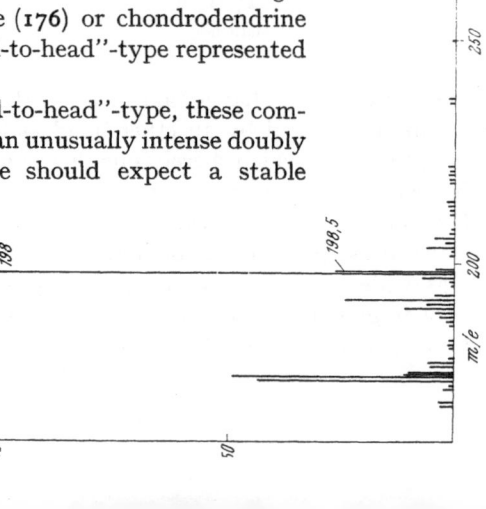

Fig. 8. Mass spectrum of berbamine (174). According to Das, Sangster and Biemann (unpublished).

transfer of a hydrogen from the rest of the molecule, causes the most intense peak in the spectrum. The reason for this different behavior may be the simple fact that in this case the formation of the doubly charged ion would involve cleavage of one Ar—C—¦—C—N— bond with an overall energetically less favorable charge and electron distribution.

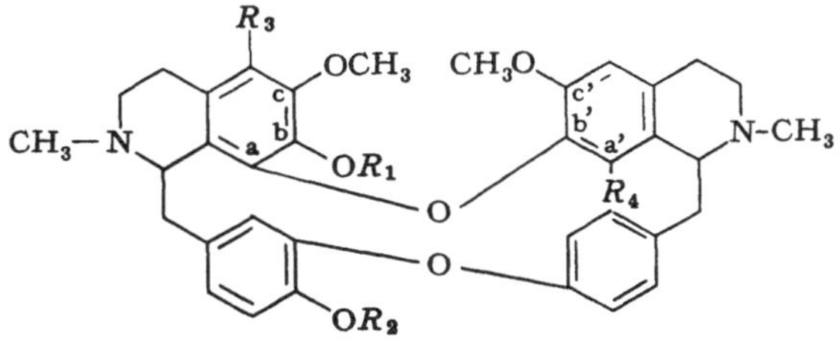

(174) $R_1 = CH_3$, $R_2 = R_3 = R_4 = H$. Berbamine.
(175) $R_1 = R_2 = R_3 = R_4 = H$. Obegamine.
(178) $R_1 = R_2 = CH_3$, $R_3 = H$, $R_4 = CH_3O$.
(179) $R_1 = R_2 = CH_3$, $R_3 = CH_3O$, $R_4 = H$.

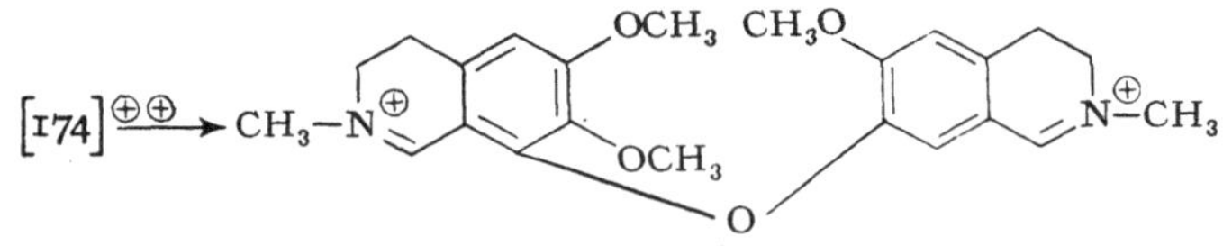

m/e 198

Retention of the positive charge rather than an electron on the Ar—C—N carbon atom is much more favorable while there is not that much difference in resonance stabilization between a benzyl radical and a benzyl ion which do not bear nitrogen *(Chart 23)*.

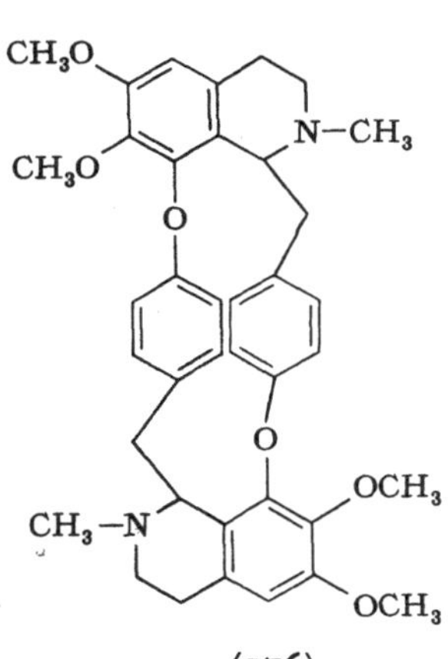

(176)
Cycleanine.

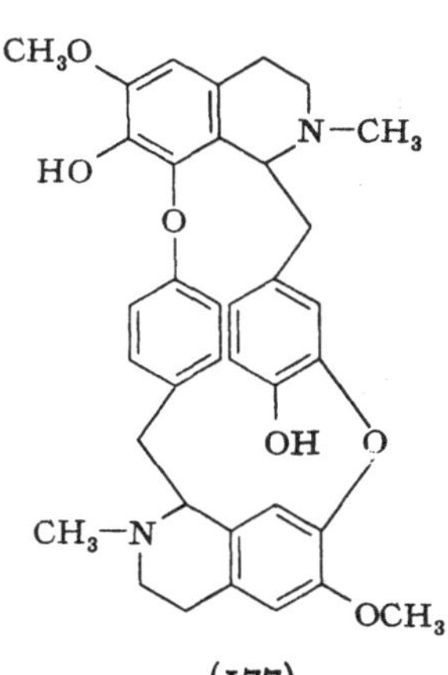

(177)
Chondrodendrine.

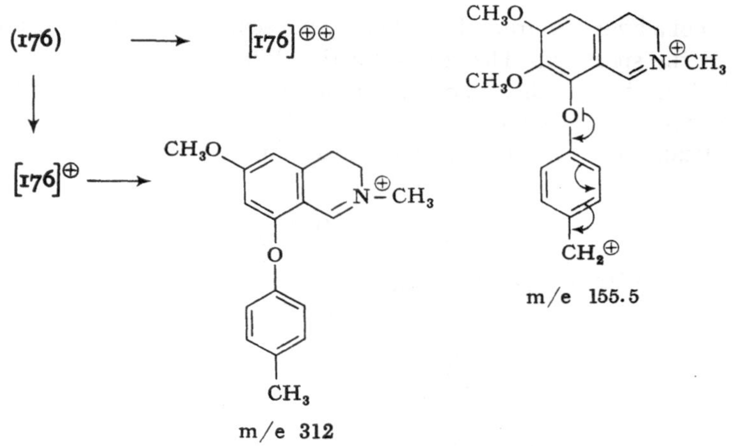

(176) $\longrightarrow$ $[176]^{\oplus\oplus}$

$[176]^{\oplus}$ $\longrightarrow$

m/e 312

m/e 155.5

Chart 22. "Head-to-tail" Bisbenzyltetrahydroisoquinoline Alkaloids.

better than

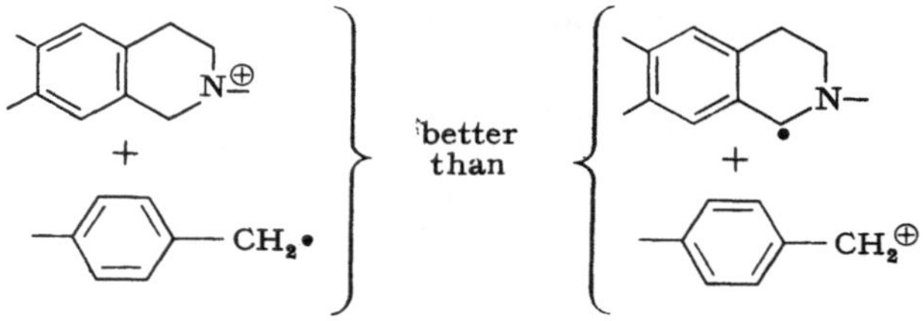

Chart 23. Fragmentation of Cyclamine (176).

Thus, mass spectrometry can be used to distinguish easily the three types of bisbenzyl-tetrahydroisoquinoline alkaloids (173, 174 and 177) as well as to reveal the sum of the molecule. More subtle distinctions can be made when small peaks are considered which are due to either one of the isoquinoline groups (or their loss from the molecule), but this is more difficult and the result is not necessarily unambiguous. Similar considerations were, however, the reason for suspecting that the previously proposed (164) structure (179, p. 61) for the alkaloid hernandezine might be incorrect. The mass spectrum was more compatible with formula (180) as it showed a peak at M-191 which is interpreted as the loss of a mono-methoxylated N-methyl tetrahydroisoquinoline ring (185). Chemical degradations proved this assumption to be correct.

V. Polycyclic Tetrahydroisoquinoline Alkaloids.

Chemically a number of *Amaryllidaceae alkaloids*, namely the buphanisine (180) group could also be considered as tetrahydroisoquinoline derivatives. Their mass spectra lack intense peaks at the lower mass range which are often so useful for the characterization of various areas in the molecule, as contrasted with the peaks at high mass, which reflect merely the loss of small substituents or a few carbon atoms of a ring system (*63, 64, 97*). This behavior is due to the polycyclic nature of the ring system and the high degree of unsaturation generated upon loss of some of these small parts.

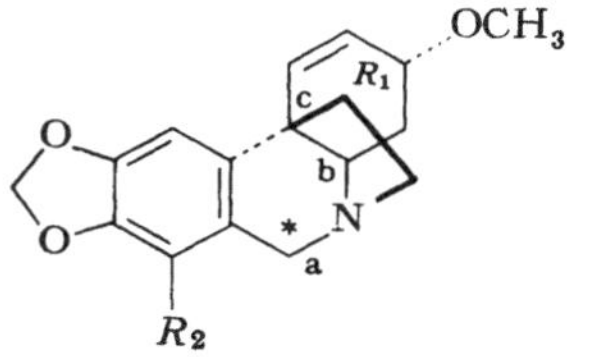

(**180**) $R_1 = R_2 =$ H. Buphanisine.
(**181**) $R_1 =$ OH, $R_2 =$ CH$_2$O. Ambelline.

(**184**) $R_1 =$ H, $R_2 =$ CH$_2$O;
HO + CH$_2$O instead of O—CH$_2$—O. Amaryllisine.
(**185**) $R_1 =$ H, $R_2 =$ CH$_2$O. Buphanidrine.

The absence of a substituent at the benzylamine carbon (* in **180**), the loss of which gives rise to the most characteristic peaks in the spectra of other tetrahydroisoquinoline alkaloids, is mainly responsible for this difference and forces the molecule to undergo fission of one of the other benzylic bonds (*a, b* or *c* in **180**) which in itself does not lead to fragments but to further bond cleavages. One of the more unusual features of the spectra of (**180**) and other compounds containing the same carbon skeleton is the fact that many of the fragment ions do not contain nitrogen (as determined by high resolution mass spectrometry). While this is contrary to most alkaloids discussed in the previous sections, it is understood if one considers that the benzylamine bond *a* is easily broken (just as in benzylamine itself) with retention of the positive charge at the aromatic portion. The elimination of azabutadiene is one of the more interesting features. It can be rationalized in the following way (*97*) *(Chart 24)*, which is only one of several possible processes differing mainly in the sequence of the bond cleavages; this can, of course, not be distinguished.

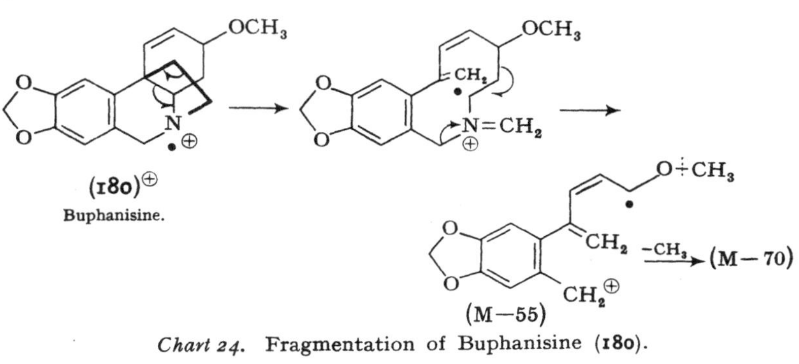

Chart 24. Fragmentation of Buphanisine (**180**).

The two resulting fragments (M-55 and M-70) give rise to the two most intense peaks (aside from the very high molecular ion peak) in the spectrum of buphanisine (180). These processes are, however, of limited value even among closely related substances. Ambelline (181), for example, exhibits neither a peak at M-55 nor one at M-70. A peak of reasonable intensity at M-71 is rationalized by elimination of $-CH_2-CH=N-CH_3$ (i. e. one hydrogen atom more than azabutadiene) from the ion that arises by loss of methyl from the molecular ion. In the spectrum of powellane (182) the peak corresponding to M-55 has become very small and it is absent in that of deacetylbowdensine (183) which is quite barren of peaks (except that corresponding to the molecular ion), although it contains two functionalities more than (182).

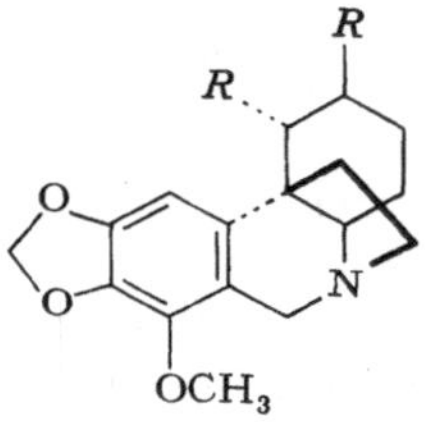

(182) R = H. Powellane.
(183) R = OH. Deacetylbowdensine.

These examples illustrate the variations in the applicability of mass spectrometry which is so useful in the work with indole alkaloids. Much less detailed information can be deduced from the spectra discussed above, although the Amaryllidaceae alkaloids also occur to a considerable extent in nature and have related structures. At present, the determination of their mass spectra remains an exercise in the rationalization of possible fragmentation processes. An exception is amaryllisine (184, p. 63) which was shown by the "shift technique" to correspond to buphanidrine (185) but to differ in aromatic substitution (64).

Morphine (186) is also a derivative of tetrahydroisoquinoline; its mass spectral behavior was the subject of a paper, along with a series of related compounds (7). Unfortunately, the lengthy discussion referred exclusively to rather speculative proposals concerning the fragmentation of this ring system. It lost sight of the question whether or not a generally meaningful interpretation in terms of the structure of these molecules is feasible to the point at which a mass spectrum can be used to indicate the structure of a new alkaloid of this type. Since no spectra are presented, the reader might be inclined to assume that the peaks discussed are all due to abundant ions, which does not seem to be the case (202). One general aspect seems to be certain, namely that all morphine derivatives give rise to a peak (m/e 162 in 186) *(Chart 25)* that can be attributed to the tetrahydroisoquinoline system and thus allows conclusions concerning the substitution pattern of rings C and D, and by inference (i. e. by the mass of the fragment lost) also tells us something about the substitution in rings A and B.

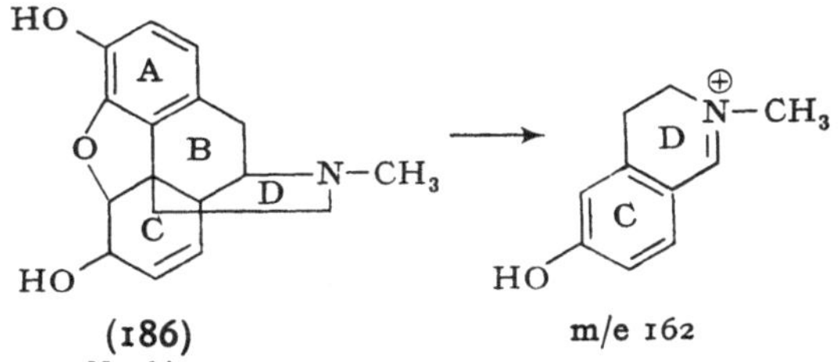

Chart 25. Major Fragmentation Mode of the Morphine (186) Type.

High resolution mass spectra have shown that the majority of the ions have lost the nitrogen and some of them also part or all of the oxygen (202). The reason seems to be the presence of a preformed diphenyl system and of the nitrogen atom in the alicyclic periphery, as well as the well known tendency of phenols and aromatic ethers to eliminate the oxygen function (as CO or CHO).

VI. Miscellaneous Alkaloids and Other Nitrogen Containing Natural Products.

Having dwelt in some detail with large groups of alkaloids it is impossible, for lack of space, to present a discussion of the considerable amount of work that has been done in other areas. For the purpose of providing a lead to supplemental reading, the major groups whose mass spectra have been studied in greater detail are listed in Table 8, without claiming completeness.

Table 8. Literature for the Mass Spectra of Some Alkaloids.

Group	References
Physostigmine	(65, 191)
Colchicine	(67, 210)
Cinchonine	(192)
Piperidine alkaloids	(196)
Tobacco alkaloids	(98)
Tropane alkaloids	(44, 166)
Pyrrolizidine alkaloids	(68, 156)
Lupine alkaloids	(62, 157)
Lycopodium alkaloids	(148)
Porphyrines	(124, 125)
Steroidal alkaloids and aminosteroids	(60, 96, 167, 203)

These investigations were mainly concerned with the determination
of the mass spectra of known substances and their rationalization in
terms of their structure. During this work one has sometimes occasion
to uncover a previous incorrect assignment of a structure. A noteworthy
example is the alkaloid carpaine for which structure (187) seemed well
established (175) until the mass spectrum revealed that it has a molecular
weight twice that of (187) and must, therefore, represent the macrocyclic
dimer (188) (195).

(187)   (188)
                                      Carpaine.

At this point a few comments concerning steroidal alkaloids and
synthetic amino steroids should be made. If one considers the complexity
of the mass spectra of steroids, one is at a first glance surprised by the
paucity of peaks in the spectra of most of the saturated, nitrogen-

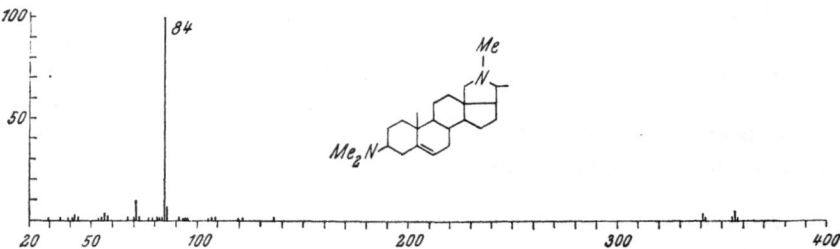

Fig. 9. Mass spectrum of conessine (189). According to DOLEJŠ, HANUŠ, CERNÝ and ŠORM. [From: Coll. Czech. Chem. Commun. **28**, 1584 (1963).]

containing steroid derivatives. The mass spectrum (96) of conessine (189)
(Fig. 9) consists almost exclusively of one peak (at m/e 84) which is
due to the dimethylamino group and carbon atoms 1, 2 and 3 – a fragment
prominent for all dimethylamino steroids, particularly if associated
with three consecutive ring carbons (i. e. in rings *A* or *D* of the steroid
skeleton) (167, 203). This is the only information that can be deduced
with some degree of certainty, in addition to the molecular weight.
References, pp. 86—98.

That it is the dimethylamino group which leads to the virtual absence of peaks at the medium or high mass range, is borne out by the mass spectrum of 5 α-conanine (190) which exhibits in addition to the molecular ion peak a very intense M-15 fragment (loss of the methyl group α to the nitrogen) as well as a peak at m/e 71, attributed to the tertiary nitrogen and the four carbon atoms associated with it (203).

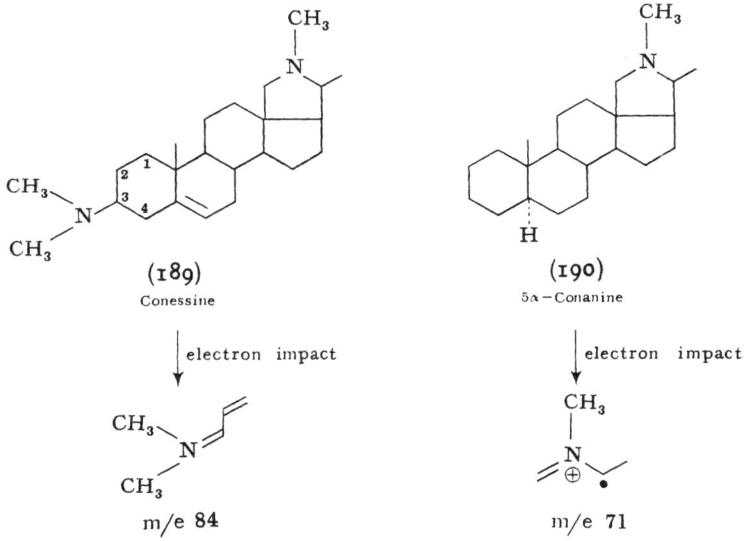

(189)
Conessine

(190)
5α−Conanine

electron impact

electron impact

CH_3
N
CH_3

m/e 84

CH_3
N⊕
•

m/e 71

Consideration of the available spectra of other nitrogen substituted steroids (96, 167) [except the "aza-steroids" in which a skeletal carbon is replaced by nitrogen (165)] indicates that the most useful information that can be gained is the environment of the nitrogen atom but little can be said about the carbon skeleton itself. Deamination might be a useful approach if it becomes necessary to obtain mass spectrometric information on the skeleton.

The great difference in the fragmentation between amino-substituted steroids and the nitrogen-free analogues reflects the fragmentation-directing effect of nitrogen (with its free electron pair) and its tendency to retain and stabilize the positive charge.

VII. Quarternary Bases.

The previous discussions were concerned with the behavior of substances which have a reasonable volatility without change and thus the mass spectrum obtained is that of the compound itself. Quarternary bases represent, however, instances in which the thermal decomposition of non-volatile materials is reproducible and, to a certain extent, predictable. The mass spectra of the products are, therefore, indicative of the structure of the original compounds (*115, 132*).

This thermal decomposition can proceed in three ways: (a) Dealkylation, e. g. formation of methyl iodide and the tertiary base from methiodides; (b) Hofmann degradation, which may lead to HX and up to three different tertiary bases; and (c) cleavage of one of the four bonds by attack of the neighboring carbon atom by the anion, i. e. substitution *(Chart 26)*.

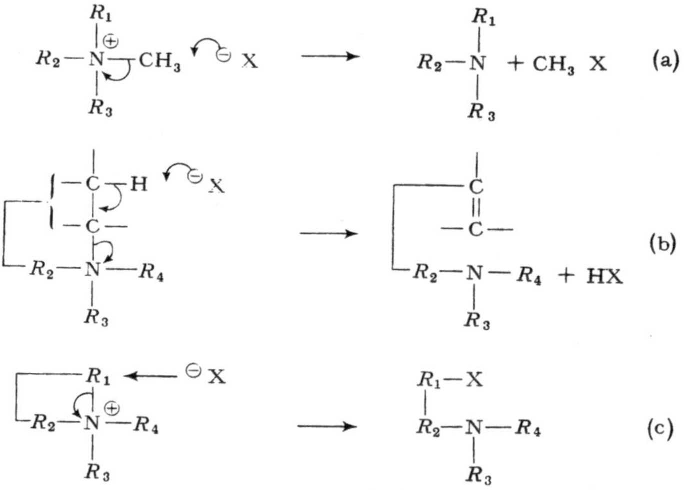

Chart 26. Thermal Decomposition of Quaternary Salts.

The determination of the structure of the quarternary salt is very simple if it behaves according to (a) and if the mass spectrum of the corresponding tertiary base is known. The two spectra will be identical, except that the spectrum of the alkyl halide is superimposed (e. g. m/e 142 and 127 for methyl iodide). Methiodides of the ajmalicine type (**128**, p. 41) fall into this class (*115*).

Compounds undergoing Hofmann elimination (case b) yield mass spectra which are quite different from that of the tertiary base because opening of one bond has led to a different carbon skeleton. The methochlorides of coryantheine (skeleton **130**, p. 42) belong to this class and seem to react mainly by cleavage of the bond between rings C and D (*115*).

Substitution was observed in few cases only, e. g. with the methiodide of pleiocarpamine (**138**, p. 43).

Similar observations have been made with quarternary salts of tetrahydroisoquinolines (*102*). The dimethyl derivative (**191**) reacts exclusively by demethylation and one obtains a spectrum of N-methyl-tetrahydroisoquinoline + CH$_3$I. The benzyl substituted derivative (**192**) gives, however, rise to the Hofmann product (**193**) indicating the increased facility for elimination of a benzylic hydrogen as HX *(Chart 27)*.

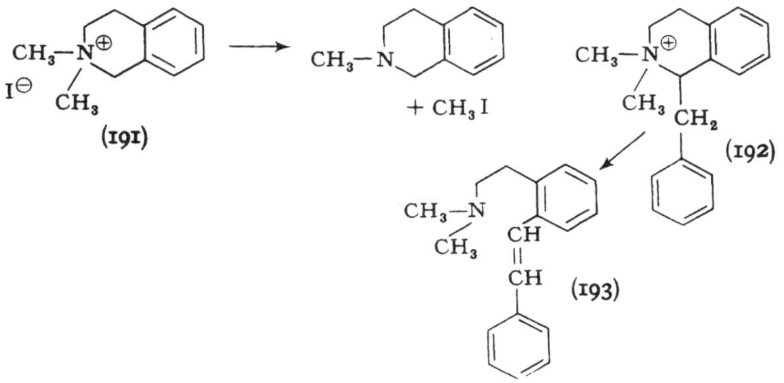

Chart 27. Thermal Decomposition of Quaternary Tetrahydroisoquinolines.

VIII. Natural Products Other than Alkaloids.

The mass spectra of fatty acids and related substances, steroids and triterpenes, carbohydrates, amino acids, peptides, nucleosides and, to a lesser extent, many other types have been investigated rather thoroughly. Some of this work has been summarized earlier and needs not to be reiterated in detail. Fatty acids, their esters and other long-chain *lipids* are particularly suitable for mass spectrometric structure determination because their linear nature minimizes complex fragmentation and re-arrangement processes and most (but not all) fragments are thus directly and simply related to the structure (*179*). A noteworthy, more recent contribution concerned the structure determination of the prostaglandines (e. g. **194**) which would have been impossible without the help of mass spectrometry because of the scarcity of the material and the difficulties in separating and purifying very closely related substances.

Oxidative degradation to acidic products and esterification have provided chemical cleavage products whose mass spectra permitted identification or determination of their structure. For example, upon treatment with alkali to cleave the β-hydroxycyclopentanone ring (known to be present from chemical and spectroscopic studies) followed by oxidative ozonolysis, acetyl-prostaglandine-E$_1$ methyl ester gave a mixture of acids, which after separation and conversion to the methyl esters were identified as the esters of succinic acid (**195**), suberic acid (**196**), 2-acetoxyheptanoic acid (**197**) and 1,2,8-octanetricarboxylic acid (**198**). The formation of these products suggested structure (**194**) for prostaglandine E_1 (*12, 13*).

HO$_2$C CH$_2$ CH$_2$ CO$_2$H
(**195**)

HO$_2$C(CH$_2$)$_6$CO$_2$H
(**196**)

CH$_3$(CH$_2$)$_4$CH—CO$_2$H
|
OAc

(**197**)

HO$_2$C—CH$_2$—CH–(CH$_2$)$_6$CO$_2$H
|
COOH

(**198**)

CH$_3$(CH$_2$)$_4$CH—CH=CH—CH—CH–(CH$_2$)$_6$CO$_2$H
|
OAc HO   O

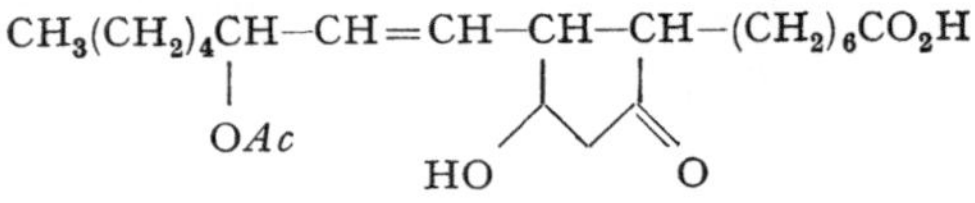

(**194**)
Prostaglandine E$_1$.

The esters of (**195**) and (**196**) were recognized as known compounds while the structures of (**197**) and (**198**) were deduced on the basis of previous mass spectrometric work with substituted carboxylic esters and the conclusions were corroborated by synthesis of authentic specimens.

Steroids and related compounds, such as di- and triterpenes, represent perhaps the group of compounds most thoroughly investigated (*58*). An enormous amount of work has been invested in determining and interpreting the mass spectra of known compounds with comparatively little application to structural problems. There are two reasons for this, namely the relative scarcity of new steroids of unknown structure (in comparison to the large number of known ones) and the relative un-specificity of mass spectrometry in the type of structural variations occurring in steroids (except the amino-substituted analogues mentioned) which require the availability of the spectra of closely related substances for their interpretation.

The valuable contribution which this work, undertaken mainly by the Stanford group under DJERASSI (*58*), has provided is the information on the complexity of electron impact induced fragmentation reactions of organic molecules. The polycyclic nature of these substances practically excludes simple fragmentation processes and requires cleavage of a number of bonds for the production of a fragment (except the simple loss of the $C_{(17)}$ side-chain or methyl substituents). Elaborate isotope labeling has shown that even processes that appear to be simple and straightforward (*54*) may be much more involved (*94*). One of the most valuable general conclusions drawn from the mass spectra of a series of steroids with a keto group in various positions concerned the so-called McLafferty rearrangement (*147*) that operates in carbonyl compounds with a hydrogen in the γ-position *(Chart 28)*.

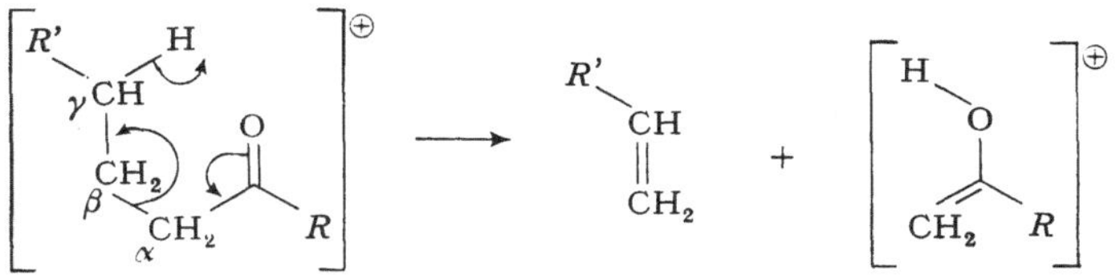

Chart 28. Principle of the "McLafferty-rearrangement".

The polycyclic system of steroidal ketones restricts the approach of the nearest γ-hydrogen and the carbonyl oxygen. Therefore, the minimum distance between the two varies greatly. The hydrogen rearrangement was indeed found to depend on the Hγ—O interatomic distance. The spectra of appropriately deuterated ketones indicate that the rearrangement occurs, for example, in 16-keto steroids involving a hydrogen atom (at $C_{(21)}$ or $C_{(22)}$) which is only 1.5 Å away, but it does not occur in 11-keto or 15-keto steroids because the γ-hydrogens at $C_{(1)}$ and $C_{(7)}$, respectively, are separated by 1.8 Å and 2.3 Å [see expression (199) which combines these positions to save space, while the work was done, of course, on the individual monoketones] (*84*).

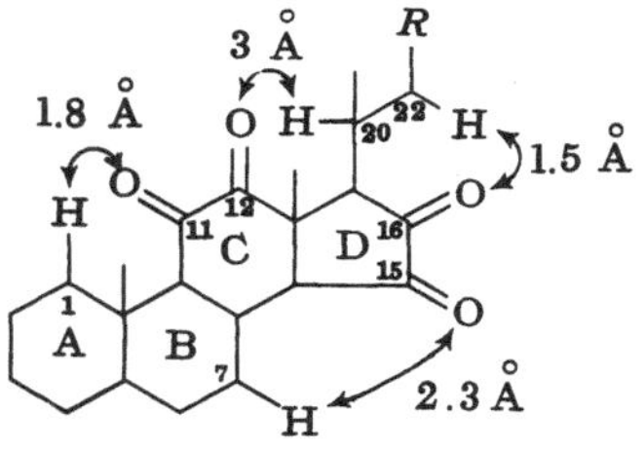

(199)

Considering the large distance (3 Å) between the 12-keto oxygen and the γ-hydrogen (at $C_{(20)}$) one should not expect such rearrangement. The mass spectrum of 20,20-α_2-5 α-pregnan-12-one does, however, show that ring D is lost with transfer of deuterium to the oxygen-containing ion (rings A, B and C). This result is rationalized by assuming prior cleavage of the $C_{(13)}$—$C_{(17)}$ bond which then permits a much closer approach of the carbonyl oxygen and the 20-hydrogen (*95*). This finding obviously restricts the usefulness of any such distance-consideration as a structural argument for molecules in which cleavage of a bond can decrease an otherwise too large distance (> 1.8 Å).

The mass spectrometric behavior of *carbohydrate derivatives*, such as O-acetates (*27, 75, 135*), O-methyl ethers (*119–122, 134, 136, 137*), isopropylidene ethers (*76*) and thioacetals (*72–74, 77–79*) have been the subject of a considerable number of investigations. For reasons similar to that cited in the case of the steroids, the work was almost exclusively restricted to the spectra of known substances (mostly monosaccharides). The results may be significant not only because of their potential applicability for the determination of the structure of new carbohydrates but because they open the way for identification of carbohydrates of known structure isolated in submilligram amounts, made possible by recent advances in the gas chromatography of carbohydrate derivatives. Mass spectrometry permits the determination of the size of the molecule, for which thioacetals or thioketals are most suitable because of the presence of the sulfur atom which has, in general, the tendency to stabilize the molecular ion. The ring size is best obtained in peracetates or permethyl ethers: Loss of the substituent α to the ether linkage is a favorable process, particularly in furanosides where it gives rise to intense peaks due to the two particles, e. g. m/e 245 and m/e 145 in the spectra of pentaacetylhexofuranoses, such as (200) *(Chart 29)*. Similar loss of CH_2OAc in hexopyranoses is observed, but is much less abundant.

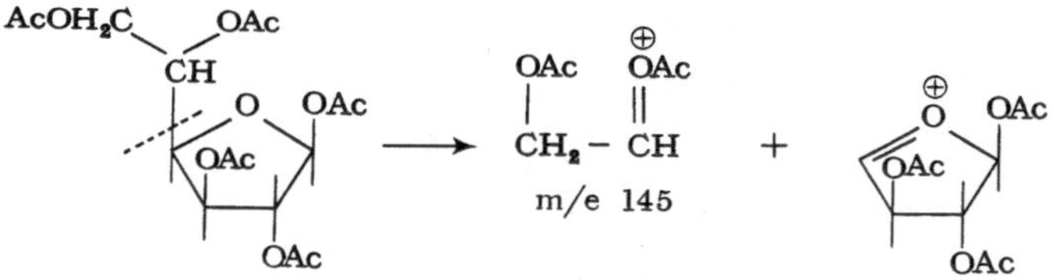

Chart 29. Fragmentation of a Hexafuranose.

While the position of substituents (or the lack thereof, as in deoxysugars) can be deduced from the spectra of cyclic forms, it is preferable to convert the substance to the linear isomer by opening the acetal (or ketal) bond and blocking the free carbonyl group. This is best accomplished by conversion to the corresponding thioacetals or ketals. The products of simple cleavage of single bonds are much easier to interpret than the more involved fragmentation processes of the cyclic derivatives. The position of methylene groups in deoxysugars *(73, 74, 79)* as well as aminosugars *(77, 78)* are quite easily deduced from the mass spectra of their thioacetals.

An extension of the work on carbohydrates has led to the exploration of the feasibility of mass spectrometry for the characterization of *nucleosides*. The advances in the work on the structure of nucleic acids has not only uncovered the presence of small amounts of "odd" nucleosides which differ from the common ones mostly in the presence of a modified base-unit, but also requires a method to identify these in the minute amounts of material available.

References, pp. 86—98.

Mass spectra are indeed useful for this purpose (*35, 41*), because they reveal the molecular weight of the nucleosides, the mass of the sugar moiety (a small peak at m/e 117 for deoxypentoses, 133 for pentoses) and of the base by cleavage of the N—C bond at *a* and transfer of a hydrogen from the sugar to the base part whenever the latter retains the charge. Thus 2-deoxyadenosine (**201**) exhibits a peak at m/e 251 for the molecular ion, the most intense peak at m/e 135 (134 + 1) and a small one at m/e 117. There is always present a peak due to the loss of the part encompassing carbons 3, 4 and 5 with the associated atoms except the hydrogen of the 5-hydroxyl group. From these few fragments, considerable information can be obtained concerning the nature of both the base and the sugar.

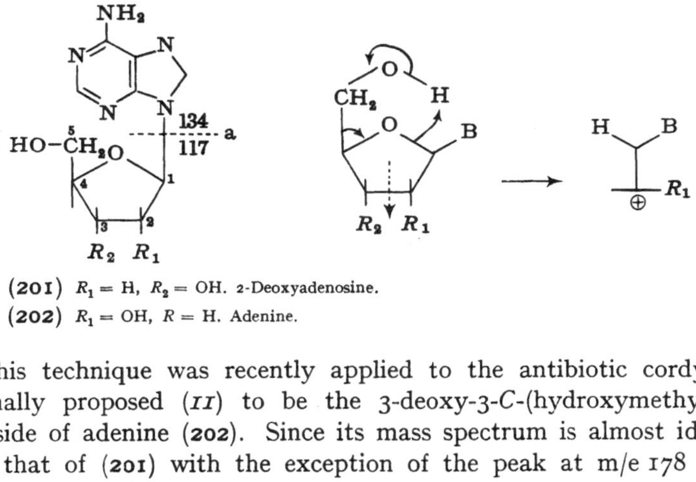

(**201**) R_1 = H, R_2 = OH. 2-Deoxyadenosine.
(**202**) R_1 = OH, R = H. Adenine.

This technique was recently applied to the antibiotic cordycepin, originally proposed (*11*) to be the 3-deoxy-3-*C*-(hydroxymethyl)aldo-tetroside of adenine (**202**). Since its mass spectrum is almost identical with that of (**201**) with the exception of the peak at m/e 178 (M-73) instead of the one at M-89, its structure must be that of a 3-deoxypentoside (such as **202**) (*113*).

The particular suitability of mass spectrometry for the determination of the structure of linear molecules and especially of the position and size of substituents along the chain has made it attractive as a technique for the determination of the *amino acid sequence in oligopeptides*. A fast and reliable method would be a welcome relief from the tedious chemical and chromatographic techniques presently required, and it was mainly for this reason that the sequence determination was explored already in the beginning of the application of mass spectrometry when working with relatively complex molecules.

The major difficulty was, of course, the low volatility of free peptides, and several approaches had been followed to overcome it. It was reported briefly that the volatility of the N-trifluoroacetyl methyl esters (**203**, $R = CF_3CO$) of very simple di- and tripeptides is sufficient to permit their introduction into a mass spectrometer through a conventional inlet system and that the spectrum contains features that are indicative of the amino acid sequence (*3*, *198*). At about the same time it was suggested to simplify the volatility problem as well as the spectra of peptides by conversion of the $-N-CO-CH(R)-N-$ backbone [in which the carbonyl group gives rise to additional cleavages and rearrangement processes (*17*)] to the $-N-CH_2-CH(R)-N-$ grouping. The products of the lithium aluminum hydride reduction of peptides are polyamino alcohols (**204**) *(Chart 30)* in which this unit is repeated

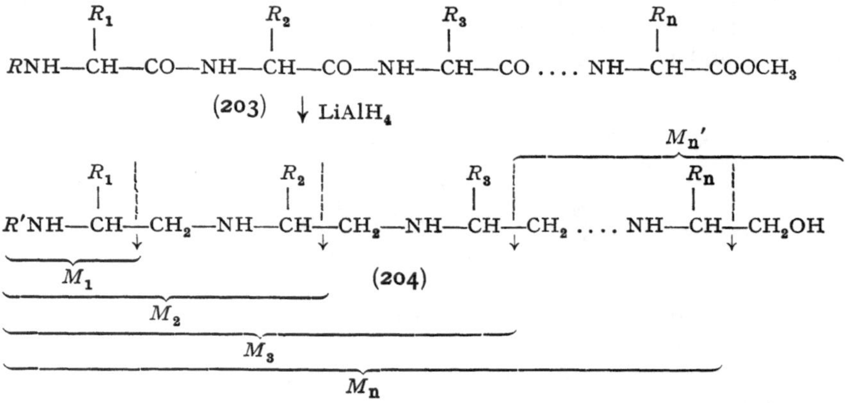

$$
\begin{array}{cccc}
R_1 & R_2 & R_3 & R_n \\
| & | & | & | \\
RNH{-}CH{-}CO{-}NH{-}CH{-}CO{-}NH{-}CH{-}CO \ldots\ldots NH{-}CH{-}COOCH_3
\end{array}
$$

(**203**) $\downarrow$ LiAlH$_4$

$$
\begin{array}{cccc}
 & & & M_n{'} \\
R_1 & R_2 & R_3 & R_n \\
| & | & | & | \\
R'NH{-}CH{-}CH_2{-}NH{-}CH{-}CH_2{-}NH{-}CH{-}CH_2 \ldots\ldots NH{-}CH{-}CH_2OH
\end{array}
$$

(**204**)

M_1

M_2

M_3

M_n

Chart 30. Principal Fragmentation of Polyamino Alcohols Derived from Peptides.

uniformly for each additional amino acid. The spectrum is dominated by fragments due to cleavage of the C—C bond between consecutive nitrogen atoms (which are well known to facilitate rupture of the adjacent C—C bond) and by preferential charge retention at the more substituted carbon atom, i. e. that representing the N-terminal side of the peptide (*19*, *31*, *42*).

Subtraction of the mass of RNHCH ($R = C_2H_5$ if derived from an N-acetyl peptide) from M_1 gives the mass of R_1 and thus reveals the N-terminal amino acid. Subtraction of ($M_1 + CH_2NHCH$) from M_2 reveals the side-chain (R_2) of the second amino acid, and so forth to the C-terminal amino acid. The same can be done to follow the sequence from the C-terminal to the N-terminal, using the set of generally somewhat less abundant peaks, $M_n{'}$ etc. This has been accomplished with molecules of the size of pentapeptides that contain aliphatic amino acids as well as lower peptides of polyfunctional amino acids (*31*).

With the advent of techniques that permit vaporizing the sample directly into the ion source of the spectrometer, the possibility of using N- and C-terminally protected peptides (without removal of the peptide-carbonyl) has reappeared (17, 117, 118, 130, 145). It was found accidentally in the work on the peptidolipid fortuitine (205) that polypeptide esters acylated at the amino nitrogen with a long-chain fatty acid are rather volatile and give spectra which can be interpreted relatively simply in terms of the amino acid sequence (8). Successive subtraction of the repeating units $+ R_1 \ldots _n$, is analogous to the principle outlined above for polyamino alcohols, except that in the peptide, cleavage may take part either at the $-CH(R)-CO-$ bond or at the $-CO-N-$ bond, with charge retention at the $-CHR$ and $-CO$ group, respectively.

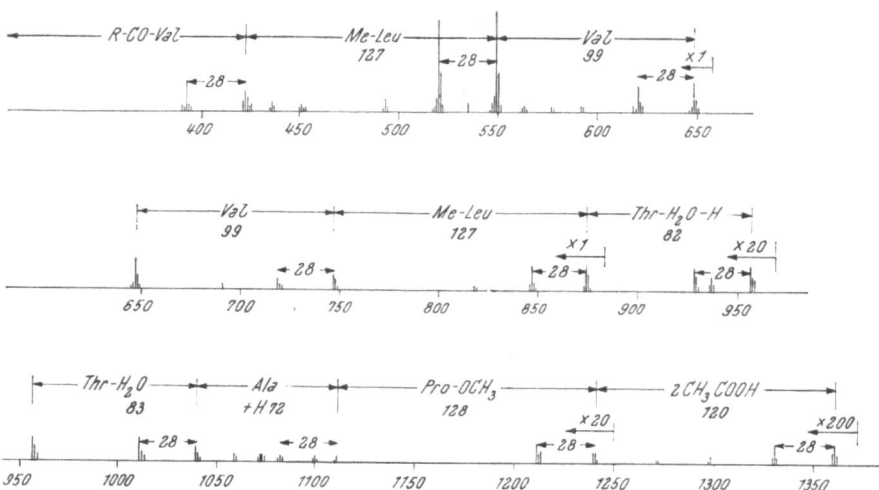

Fig. 10. Mass spectrum of fortuitine (**205**). According to BARBER, JOLLÈS, VILKAS and LEDERER. [From: Biochem. Biophys. Res. Commun. 18, 469 (1965).]

The presence of the fragments containing the fatty acid (and, therefore, containing all the amino acids from the N-terminal end to the point of cleavage) moves the corresponding peaks into the region of relatively high mass. The many fragments formed from oligopeptides by various rearrangement processes which are triggered by the amide carbonyl and represent parts of the center of the chain (17, 117, 198), are of much lower mass and do, therefore, not interfere, as long as the aliphatic chain (R in **203**) contains a large number of carbon atoms, making R a heavy substituent.

The first example of this approach was the determination of the fortuitine structure, a naturally occurring mixture of two nonapeptides containing a C_{20} and a C_{22} acyl group, whose mass spectrum exhibited, in the region of m/e 390 to m/e 1359, intense peaks due mainly to the cleavage of the $-CO-N$ bonds *(Fig. 10)*. The presence of a mixture of different acyl groups was an advantage rather than a disadvantage because it provided additional confirmation that these pairs of peaks of the same intensity ratio and occurring 28 mass units apart, represent, in fact, fragments retaining the N-terminal acyl groups; thus they represent the N-terminal sequence. The peaks agreed with the C-terminal sequence thr(ac)—thr(ac)—ala—pro—OCH$_3$ known from chemical studies (142), but in addition revealed the remaining sequence, as well as the presence of two previously overlooked N-methyl-leucine units in the molecule, shown in (205) (8).

$$CH_3(CH_2)_{18-20}CO\text{-val—Meleu—val—val—Meleu—thr}(Ac)\text{—thr}(Ac)\text{—ala—pro—}$$
$$—OCH_3$$

(205)

Fortuitine.

The same approach has been used successfully to complete the amino acid sequence of peptidolipin NA, a cyclic heptapeptide (9) and of an N-acylpentapeptide ester which occurs in *Mycobacterium johnei* (140). A detailed investigation of the spectra of N-acyl peptide esters with the aim of using them for the determination of the amino acid sequence in protein-derived oligopeptides was undertaken (49, 211). Methods for the N-acylation of peptides with fatty acids have been described (49, 133).

The use of N-trifluoroacetylpeptide esters has been revived recently (208, 209) and results were reported on the correlation of their mass spectra with the structure known for these synthetic samples. An interpretation scheme essentially identical with that described above for amino alcohols from peptides (but starting at the C-terminal end) and with that used in the case of fortuitine and similar peptidolipids, was proposed (208). It was outlined on the examples N—TFA—Phe—ile—gly—leu—val—OCH$_3$ and N—TFA—leu—phe—gly—leu—met—OCH$_3$*, "peeling" the amino acids off from the C-terminal end of the peptide. It is important for the success of the technique to be familiar with the fragmentations which amino acids undergo in their side-chain [these are well known from a study of amino acid esters (4, 37)] and with the particular fragmentation behavior of N—TFA peptides which have the tendency to give fragments that have lost either the TFA group or part of the N-terminal sequence (198, 208). Most limiting is the circumstance that the correct interpretation of such a spectrum necessitates identification of the molecular ion; this requirement is quite limiting in the case of such complex molecules, which, if derived from a large peptide by partial hydrolysis, will often contain impurities or another peptide. The examples of the successful application of this and the lipido-peptide technique involved only the

---

* TFA = trifluoroacetyl.

References, pp. 86—98.

mass spectrometrically simple amino acids, such as the aliphatic ones, including anhydrothreonine and methionine as well as phenylalanine. These do not cause complications due to functional groups in the side-chain and thus give rise to spectra in which peaks due to the cleavage at the amide linkages indeed predominate. All the approaches discussed above have always given good results in these cases. But the final test is the application to more complex peptides (in terms of amino acids present) of the type obtained upon partial hydrolysates of proteins.

A radically different approach has been outlined more recently (*43*), making use of the unique elemental composition of the fragments due to cleavage of the $-CH(R)-CO-$ and $-CO-NH-$ bonds, respectively, rather than hoping for their exceptional abundance (which is certainly not always justified, particularly in the presence of polyfunctional amino acids). The method permits objective, automatic interpretation of the amino acid sequence if the vast number of the necessary comparisons are made by a computer. As this technique involves the principles of high resolution mass spectrometry, it will be discussed in detail in the next Section.

IX. High Resolution Mass Spectrometry.

The material reviewed in the previous Section concerned mainly spectra and results obtained with conventional mass spectrometers producing spectra which permit unequivocal assignment of the integral mass of the various ions formed. From the integral mass one must guess at the elements present in such ions as outlined above. The more is known about the compound in question, the easier is this estimation. An extreme example would be a substance known to be a hydrocarbon lacking strong absorption in the ultraviolet and found to have a molecular weight of 210 on the basis of its mass spectrum. Its composition must be $C_{15}H_{30}$ unless it were a highly cyclic $C_{16}H_{18}$-system. Even these two possibilities can easily be distinguished by consideration of the character of the spectrum which is unique for an alkene or cycloalkane as compared with a polycyclic system containing eight rings or double bonds. Similarly, two alkaloids whose mass spectra are identical with the exception of a shift by 30 mass units will differ by a CH_3O group versus H. Their elemental composition must thus differ by CH_2O which settles the question of the elemental composition if one of the two compounds is known.

In many instances the situation is not that simple and the primary data (namely the integral mass) permit many possibilities for the elemental composition, particularly for fragment ions for which, of course, most of the bonding requirements of molecules do not hold. It is very helpful in such instances to determine the mass accurately enough to calculate the elemental composition, an approach introduced to chemical mass spectrometry by Beynon (14). *Table 9* lists the nuclidic masses of the elements commonly found in organic molecules, while in *Table 10* all ions containing up to five nitrogen and eight oxygen atoms are listed that can possibly occur in the region between masses 210.075 and 210.175, i. e. over 0.1 mass unit. (There are 85 such species of nominal mass 210, ranging from mass 209.978 to mass 210.235. In order to decrease the size of the Table, a narrower region has been chosen for this example.) It will be noted that an accuracy of $\pm$ 0.0006 mass units is necessary to uniquely determine the composition of an ion at that mass. Resolving multiplets would sometimes be impossible, since a resolving power of 1 part in 155.000 would be required to resolve mass 210.1031 ($C_{13}H_{12}N_3$) from mass 210.1045 ($C_{15}H_{14}O$); but it is fortunate that their elemental compositions differ considerably and such a doublet does not frequently occur in the spectrum of a pure compound or a mixture of related substances. Thus, mass measurement is the most important aspect of the technique and the term "high resolution mass spectrometry" is somewhat misleading as far as its major application is concerned, although resolving capability is a necessary consequence of mass measurement.

Table 9. Nuclidic Masses of Elements Commonly Occurring in Organic Compounds.

Isotope	Atomic weight ($C^{12} = $ 12.000000)
H^1	1.007825
H^2	2.014102
C^{12}	12.000000
C^{13}	13.003354
N^{14}	14.003074
N^{15}	15.000108
O^{16}	15.994915
O^{17}	16.999133
O^{18}	17.999160
F^{19}	18.998405
Si^{28}	27.976927
Si^{29}	28.976491
Si^{30}	29.973761
P^{31}	30.973763
S^{32}	31.972074
S^{33}	32.971461
S^{34}	33.967865
Cl^{35}	34.968855
Cl^{37}	36.965896
Br^{79}	78.918348
Br^{81}	80.916344
J^{127}	126.904352

Obviously, the determination of the elemental composition of an organic molecule is an important and valuable aspect of high resolution mass spectrometry. On the other hand, it will certainly not replace the conventional combustion technique which will always be faster and cheaper if a sufficient amount of highly purified material is available. One has also to keep in mind that mass spectrometry is not generally applicable and requires a substance that can be vaporized without decomposition and gives a measurable molecular ion. It has unique advantages, however, in cases where the substrate is available only in submilligram amounts thus precluding extensive purification and when it would be very time-consuming and costly to prepare more. This situation is frequently encountered in natural products research and biochemistry; mass spectrometric composition determination will, therefore, be most important in these fields. The determination of the

References, pp. 86—98.

Table 10. Mass of Combinations of C, H, N and O
Falling in the Range of Mass 210.0750 to 210.1750.

Mass	C	H	N	O
210.1408	16	18	0	0
210.1283	15	16	1	0
210.1157	14	14	2	0
210.1031	13	12	3	0
210.0905	12	10	4	0
210.1719	10	20	5	0
210.0780	11	8	5	0
210.1045	15	14	0	1
210.0919	14	12	1	1
210.1732	12	22	2	1
210.0793	13	10	2	1
210.1606	11	20	3	1
210.1481	10	18	4	1
210.1355	9	16	5	1
210.1620	13	22	0	2
210.1494	12	20	1	2
210.1368	11	18	2	2
210.1242	10	16	3	2
210.1117	9	14	4	2
210.0991	8	12	5	2
210.1256	12	18	0	3
210.1130	11	16	1	3
210.1004	10	14	2	3
210.0879	9	12	3	3
210.0753	8	10	4	3
210.1566	6	20	5	3
210.0892	11	14	0	4
210.0766	10	12	1	4
210.1454	7	20	3	4
210.1328	6	18	4	4
210.1202	5	16	5	4
210.1341	8	20	1	5
210.1216	7	18	2	5
210.1090	6	16	3	5
210.0964	5	14	4	5
210.0838	4	12	5	5
210.1103	8	18	0	6
210.0978	7	16	1	6
210.0852	6	14	2	6

elemental composition ($C_{46}H_{58}N_4O_9$) of vinblastine was mentioned on p. 48. Other examples of biologically highly active substances are streptonigrin [$C_{31}H_{36}N_4O_8$ found for the hexamethyldihydro derivative (174)] and the frog venom batrachotoxin the elemental composition of which, $C_{24}H_{33}NO_4$, was determined using only 50 μg of incompletely purified material (69). The number of such instances, in which the elemental composition had been thus derived long before a sufficient amount was available for combustion analyses, increases fast as the capability of the technique is becoming more widely known.

The determination of the elemental composition of a molecule is, however, only one small contribution of high resolution mass spectrometry. It permits equally well direct determination of the composition of fragment ions and thus greatly aids in the correct interpretation of a mass spectrum, particularly of an unknown compound or of a feature that is difficult to explain at first glance. One of the many examples of this use is the final explanation of the considerable differences between the mass spectrum (Fig. 5, p. 34) of ajmaline (3) containing a secondary hydroxyl group and the corresponding ketone, ajmalidine (99), which exhibited an intense peak at m/e 198, lacking in Fig. 5 (see p. 34). Equally important is the verification of the elemental composition of an ion whenever its genesis is discussed for the purpose of "determining" a fragmentation mechanism or its "structure". Much of the speculation based on integral masses obtained from conventional mass spectra would be more convincing if accurate mass determinations were available.

These examples involve what is essentially low resolution mass spectrometry supported by individual mass measurements on peaks found (or expected) to be important in the interpretation. This approach has developed because of the tediousness of carrying out these highly accurate measurements and the calculations involved. Further examples involve some final points concerning the structure of the macrolide filipin (107); and a mass spectrometric investigation of mycolic acids (oxygenated, branched carboxylic acids which contain about 80 carbon atoms) has been reported (99). In these cases the size of the molecules (molecular weights 1000–1500) was the more important problem, taxing the instrument more than the accuracy of the mass measurement, which does not have to be very precise to distinguish between one oxygen more or less in nitrogen-free substances.

More recently, some techniques were developed (25, 46, 82, 83) that provide a complete mass spectrum in terms of accurate masses, and thus elemental compositions of the ions formed, almost as easily as one can record a conventional low resolution spectrum. It is now possible by using a reasonably large and fast computer to process the data, which consist of a few hundred seven-digit numbers (the precise masses).

To obtain such data automatically the spectrum is recorded simultaneously (by admixture) with a compound whose mass spectrum is completely known and differs in elemental composition (e. g. a fluorocarbon) from that of the compound in question. A parameter related to the mass of the ions (e. g. the distance along a photographic plate placed in the focal plane of a Mattauch-Herzog type mass spectrometer, or time, if scanned using an electron multiplier at the focal point of a Nier-Johnson type instrument) is recorded along with the intensity information in computer compatible input form (punched cards or paper tape, or magnetic tape). The computer accomplishes then the following operations (*46, 82, 83*): (a) identification of the exact center of peak; (b) calculation of all masses using a set of selected ions of the added compound as mass standards; (c) identification of all ions to the added compound and their removal from the data; (d) calculation of all possible elemental compositions of all ions due to the compound under investigation; and (e) arranging these data in a form suitable for interpretation.

	CH	CHO	CHO2	CHN	CHNO	CHNO2
43		2/ 3 0*				
51	4/ 3 0*					
63	5/ 3 0*					
65	5/ 5 0***					
72					3/ 6 0****	
74	6/ 2 1*					
77	6/ 5 0***					
78	6/ 6 0***					
79	6/ 7 0*					
90	7/ 6 0**					
91	7/ 7 1******					
92	7/ 8 0*****					
102	8/ 6 0*					
103	8/ 7 0****					
104	8/ 8 1*******					
105	8/ 9 1*****					
117				8/ 7 0*		
118				8/ 8 0*		
119				8/ 9 0**		
120				8/10 0**		
163					10/13 1*****	
	CH	CHO	CHO2	CHN	CHNO	CHNO2

Fig. 11. "Element map" of phenylethyl acetamide.

This approach not only removes the aspect of subjectivity (selection of certain peaks thought to be important) but also opens the way to a more direct interpretation of the mass spectrum in terms of the structure of the compound. In the past one had as the only experimentally determined fact the integral mass and thus became, by necessity, accustomed to the use of the mass as parameter, projecting into it the possible nature of the corresponding fragment. Since we have means for direct determination of the composition, mass becomes unimportant and is replaced by consideration of the elemental composition. For this reason, it is most desirable to group the ions first according to their heteroatom content and then by increasing carbon and hydrogen content which gives a much clearer picture of the distribution of the elements within the molecule than if they were listed by increasing mass only.

As an example, such data are shown *(Fig. 11)* for a simple molecule in the form of an *"element map"*, a term chosen to emphasize the fact that such a presentation is related to the distribution of heteroelements within the molecule (25).

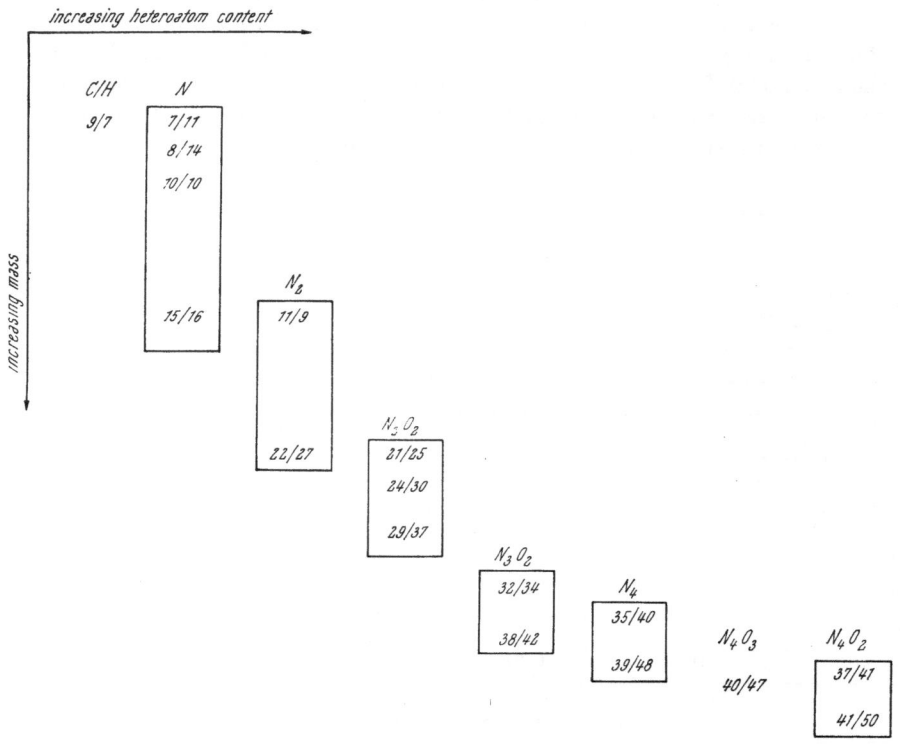

Fig. 12. Schematic element map of pleiomutine (150, p. 51).

Starting from the left, the vertical columns list the nominal mass of the ions found, followed by all those ions containing only carbon and hydrogen, then those containing C, H and O, C, H and N, etc., to C, H, N, O_2. To designate their composition it is then only necessary to list the number of carbons and hydrogens; e. g. an entry 8/10 in the C, H, N column refers to an ion $C_8H_{10}N$. The next numerical value is the discrepancy between the found mass and that corresponding to this composition (zero implies that this discrepancy was less than ± 1 millimass unit). The asterisks represent a logarithmic indication of the relative intensity (8 asterisks correspond to the most intense peak, 1 asterisk corresponds to one peak 0.3% as intense).

References, pp. 86—98.

The data show that in this compound there are up to eight carbon atoms which may appear as fragments not containing nitrogen or oxygen. Their C/H ratio is typically aromatic. The next atom must be nitrogen, because there are a number of ions of the $C_8H_{7-10}N$ type. The entire molecule has the composition $C_{10}H_{13}NO$ (entry with the largest number of carbon and heteroatoms). The oxygen and nitrogen atoms must, therefore, be associated with the ninth and tenth carbon atoms. This is supported by the appearance of a C_3-fragment which in turn contains both the nitrogen and the oxygen, as well as by a C_2-fragment containing oxygen only. This distribution of elements can best be explained with the assumption that the compound in question is phenyl ethyl acetamide, $C_6H_5CH_2CH_2NHCOCH_3$.

A much more complex example is represented by the "element map" of pleiomutine *(Fig. 12)* (150, p. 51), in a schematic representation, to save space. The entries shown indicate the lowest and highest number of carbons and hydrogens in each heteroatom group and intermediate entries are listed only where they are particularly abundant. The entry with the highest number of C, H, N and O atoms is 41/50 in the N_4O_2 group, indicating $C_{41}H_{50}N_4O_2$ to be the elemental composition of the molecule. While other combinations such as $C_{40}H_{54}O_6$ (630.3920) and $C_{38}H_{52}N_2O_5$ (630.3907) would also fit the experimentally determined mass (630.3915) very closely, they are excluded because the low mass region requires the presence of nitrogen (eliminating $C_{40}H_{54}O_6$) and the other one cannot correspond to an intact molecule.

The absence of hydrocarbon ions (except C_9H_7) excludes compounds with contiguous carbon skeletons or those which are able to eliminate or expel all six heteroatoms. The ions containing only one nitrogen have an alicyclic C/H ratio up to C_8, but an aromatic one beyond C_{10}, indicating the presence of an alicyclic nitrogen as well as an N-heteroaromatic system that contains up to 15 carbon atoms. The N_2-group appears in the range of C_{11}–C_{22} while it requires at least 32 carbon atoms to retain three nitrogens and 35 carbons to retain four nitrogens. One can thus conclude that the nitrogen atoms are rather equally distributed over the entire molecule but not bunched together in one segment.

Some information concerning the distribution of the oxygen atoms can also be gleaned from two independent aspects of Fig. 12. While there are ions that contain only N_2, and others containing N_2O_2, there are none with two nitrogens and only one oxygen; this implies that the two oxygens are connected to one of two N_2-moieties present in the compound. Similar information is obtained from a consideration of the N_4O_2, N_4O_3 and N_4 ions which indicate that one oxygen can be lost only with one carbon atom (as CH_3O) and two can be lost with two carbon atoms, as CH_3OCO or $CH_3O + CO$; the former is more probable (since loss of CO as such is not observed as the first step as indicated by the absence of an ion $C_{40}H_{50}N_2O$) and makes the presence of a carbomethoxy group highly plausible. The carbon range of the N_2 species which centers around half of that of the entire molecule makes the presence of a "dimeric" indole alkaloid quite likely. From there on the interpretation follows more conventional lines as outlined on p. 8, except that the elemental compositions rather than masses are used.

Obviously, this approach is most useful if very little is known concerning the compound. It is much less useful when the problem involves mainly intensity considerations, for example when mass spectra of labeled and unlabeled compounds are compared or when the "shift technique" is used in its strictest sense (namely when the spectra are exactly analogous due to simple substituents in the aromatic portion). Intensity relationships are not so obvious from the element maps because of the logarithmic representation and the fact that the presence of an additional methoxyl group, for example, shifts an ion in two directions, viz. to the right in the column containing one more oxygen, as well as 30 mass units down. Thus, for this purpose one will preferably compare the conventional spectra (which, incidentally can be produced by the computer automatically from the high resolution data), although a verification of the coincidence of the expected elemental compositions provide a welcome additional support for the conclusions reached from the conventional representation.

Since the elemental composition of the ions produced from a compound represents the most fundamental data concerning its structure that can be determined directly by experiment and because of their "quantized" character*, high resolution mass spectra are potentially most suitable to independent interpretation by a computer. In principle, the conclusions drawn from the element maps of phenyl ethyl acetamide and pleiomutine could also be arrived at automatically, since they were based mainly on the differences in the elemental composition of various ions.

It is, for example, possible to identify the most probable molecular ion (even in the presence of impurity peaks in that mass region or if its abundance is too low for observation) by having the computer select that particular ion which is related to most of the other ions by mass differences corresponding to simple combinations of elements easily lost as fragments (36).

---

* The exact mass of an ion is, of course, a parameter independent of the structure or experimental parameters in contrast to, for example, the wavelength at which a group absorbs infrared radiation.

References, pp. 86—98.

Spectra of molecules of a type which exhibit specific structural restrictions which can conveniently be incorporated into a computer program, are already amenable to *computer interpretation*. For example, peptides are restricted to the general formula (203, p. 74) ($R = H$, COOH instead of COOCH$_3$), and the groups $R_1 \ldots R_n$ are limited to those of the amino acids present. Thus, the mass spectrum of a peptide containing a well recognizable group (i. e. one which differs greatly from the elemental composition of an amino acid) at the N-terminal (206, $R =$ trifluoroacetyl, carbobenzoxy or phthaloyl) can be interpreted automatically (43, 184a) if the computer is given the mass of R and of all possible side-chains to be considered (i. e. all those in naturally occurring amino acids). By checking the presence of ions corresponding to the elemental compositions of fragments A_1 through A_n and B_1 through B_n, as well as the molecular ion for all possible combinations of amino acids, the sequence can be found regardless of the abundance of the ions, a parameter which is considered only if two or more sequences are possible on the basis of composition data. In the absence of a recognizable molecular ion the molecular weight can be deduced as outlined above (36), considering that the high mass fragments must bear a definite relationship to the molecule (loss of H$_2$O, CO$_2$ or any R_1–R_n groups present) (206).

$$
\begin{array}{cccc}
R_1 & R_2 & R_3 & R_n \\
| & | & | & | \\
R\text{—NH—CH—CO—NH—CH—CO—NH—CH—CO} & \ldots & \text{NH—CH—CO} & \text{OH} \\
A_1 & A_2 & A_3 & A_n \\
B_1 & B_2 & B_3 & B_n
\end{array}
$$

(206)

It will probably be possible in the near future to go beyond the present stage of merely producing the mass spectrum of a compound, leaving the interpretation entirely to the experience, intuition and ingenuity of the investigator. Especially the part which experience plays is to a large extent a matter of deriving the arithmetic difference between the masses of various ions that comprise the spectrum and having at one's fingertips the most frequently occurring masses and mass differences, as well as their structural significance. This aspect of the interpretation can most efficiently be handled by a computer thus opening the way for the presentation of a "pre-digested" spectrum, whose interpretation is then less time-consuming. How far the completely computerized interpretation of high resolution mass spectra can be successfully applied in general (as contrasted to the much more specialized case of peptides discussed above) remains to be seen.

The vast majority of mass spectrometric work in the near future will, of course, still be undertaken with conventional instruments and non-automated interpretation techniques – an approach whose potential power has been pointed out and reiterated in this Review.

References.

1. ACHENBACH, H. and K. BIEMANN: 10,11-Dioxopleiocarpine, a New Alkaloid from *Pleiocarpa mutica* BENTH. Tetrahedron Letters **1965**, 3239.

2. — — Isotuboflavine and Norisotuboflavine, Two New Alkaloids Isolated from *Pleiocarpa mutica* BENTH. J. Amer. Chem. Soc. **87**, 4177 (1965).

3. ANDERSSON, C.-O.: Mass Spectrometric Studies on Amino Acid and Peptide Derivatives. Acta Chem. Scand. **12**, 1353 (1958).

4. ANDERSSON, C.-O., R. RYHAGE, S. STÄLLBERG-STENHAGEN and E. STENHAGEN: Mass Spectrometric Studies. Ark. Kemi **19**, 405 (1962).

5. ANTONACCIO, L. D., N. A. PEREIRA, B. GILBERT, H. VORBRUEGGEN, H. BUDZIKIEWICZ, J. M. WILSON, L. J. DURHAM and C. DJERASSI: Alkaloid Studies. XXXIII. Mass Spectrometry in Structural and Stereochemical Problems. VI. Polyneuridine, A New Alkaloid from *Aspidosperma polyneuron* and Some Observations on Mass Spectra of Indole Alkaloids. J. Amer. Chem. Soc. **84**, 2161 (1962).

6. ARNDT, R. R. and C. DJERASSI: Alkaloid Studies. LV. 19-Dehydroyohimbine, an Alkaloid from *Aspidosperma pyricollum.* Experientia **21**, 566 (1965).

7. AUDIER, H., M. FETIZON, D. GINSBURG, A. MANDELBAUM and TH. RÜLL: Mass Spectrometry of the Morphine Alkaloids. Tetrahedron Letters **1965**, 13.

8. BARBER, M., P. JOLLÈS, E. VILKAS and E. LEDERER: Determination of Amino Acid Sequences in Oligopeptides by Mass Spectrometry. I. Structure of Fortuitine, an Acyl Nonapeptide Methyl Ester. Biochem. Biophys. Res. Commun. **18**, 469 (1965).

9. BARBER, M., W. A. WOLSTENHOLME, M. GUINAND, G. MICHEL, B. C. DAS and E. LEDERER: Determination of Amino Acid Sequences in Oligopeptides by Mass Spectrometry. II. Structure of Peptidolipin NA. Tetrahedron Letters **1965**, 1331.

10. BECKEY, H. D., H. KNOPPEL, G. METZINGER and P. SCHULZE: Advances in Experimental Techniques, Applications, and Theory of Field Ion Mass Spectrometry. Adv. Mass Spectrometry **3**, 225 (1966).

11. BENTLEY, H. R., K. G. CUNNINGHAM and F. S. SPRING: Cordycepin, a Metabolic Product from Cultures of *Cordyceps militaris* (LINN.) LINK. Part II. The Structure of Cordycepin. J. Chem. Soc. (London) **1951**, 2301.

12. BERGSTRÖM, S., R. RYHAGE, B. SAMUELSSON and J. SJÖVALL: Degradation Studies on Prostaglandins. Prostaglandins and Related Factors. 10. Acta Chem. Scand. **17**, 2271 (1963).

13. — — — — Prostaglandins and Related Factors. 15. The Structures of Prostaglandin E_1, $F_{1\alpha}$, and $F_{1\beta}$. J. Biol. Chem. **238**, 3555 (1963).

14. BEYNON, J. H.: High Resolution Mass Spectrometry of Organic Materials. Adv. Mass Spectrometry **1**, 328 (1959).

15. — Mass Spectrometry and Its Applications to Organic Chemistry. London: Elsevier. 1960.

16. BEYNON, J. H.: New Ionization Methods. EUCHEM Symposium on Mass Spectrometry. Sarlat (France), Sept. 1965.

17. BIEMANN, K.: Mass Spectrometry: Applications to Organic Chemistry. New York: McGraw-Hill. 1962.

18. — The Application of Mass Spectrometry in Organic Chemistry: Determination of the Structure of Natural Products. Angew. Chem. **74**, 102 (1962); Intern. Ed. **1**, 98 (1962).

19. — The Application of Mass Spectrometry in Amino Acid and Peptide Chemistry. Chimia **14**, 393 (1960).

20. — Application of Mass Spectrometry to Structure Problems. IV. The Carbon Skeleton of Sarpagine. J. Amer. Chem. Soc. **83**, 4801 (1961).

21. — The Determination of the Carbon Skeleton of Sarpagine by Mass Spectrometry. Tetrahedron Letters **1960**. No. 15, 9.

22. BIEMANN, K. et al.: to be published.

23. BIEMANN, K., P. BOMMER, A. L. BURLINGAME and W. J. McMURRAY: The High Resolution Mass Spectra of Ajmalidine and Related Substances. Tetrahedron Letters **1963**, 1969.

24. — — — — High Resolution Mass Spectra of Ajmaline and Related Alkaloids. J. Amer. Chem. Soc. **86**, 4624 (1964).

25. BIEMANN, K., P. BOMMER and D. M. DESIDERIO: Element-Mapping, a New Approach to the Interpretation of High Resolution Mass Spectra. Tetrahedron Letters **1964**, 1725.

26. BIEMANN, K., A. L. BURLINGAME and D. STAUFFACHER: Application of Mass Spectrometry to Structure Problems. VII. Condylocarpine. Tetrahedron Letters **1962**, 527.

27. BIEMANN, K., D. C. DE JONGH and H. K. SCHNOES: Application of Mass Spectrometry to Structure Problems. XIII. Acetates of Pentoses and Hexoses. J. Amer. Chem. Soc. **85**, 1763 (1963).

28. BIEMANN, K. and M. FRIEDMANN-SPITELLER: Application of Mass Spectrometry to Structure Problems. V. Iboga Alkaloids. J. Amer. Chem. Soc. **83**, 4805 (1961).

29. — — Mass Spectrometric Evidence for the Structure of Iboxygaine and its Tosylate. Tetrahedron Letters **1961**, 68.

30. BIEMANN, K., M. FRIEDMANN-SPITELLER and G. SPITELLER: An Investigation by Mass Spectrometry of the Alkaloids of *Aspidosperma quebracho blanco*. Tetrahedron Letters **1961**, 485.

31. BIEMANN, K., F. GAPP and J. SEIBL: Application of Mass Spectrometry to Structure Problems. I. Amino Acid Sequence in Peptides. J. Amer. Chem. Soc. **81**, 2274 (1959).

32. BIEMANN, K., J. S. GROSSERT, J. M. HUGO, J. OCCOLOWITZ and F. L. WARREN: The Indole Alkaloids. Part IV. The Structure of Henningsamine. J. Chem. Soc. (London) **1965**, 2814.

33. BIEMANN, K., J. S. GROSSERT, J. OCCOLOWITZ and F. L. WARREN: The Indole Alkaloids. Part V. The Structure of Henningsoline. J. Chem. Soc. (London) **1965**, 2818.

34. BIEMANN, K., C. LIORET, J. ASSELINEAU, E. LEDERER and J. POLONSKY: On the Structure of Lysopine, a New Amino Acid Isolated from Crown Gall Tissue. Biochim. Biophys. Acta **40**, 369 (1960).

35. Biemann, K. and J. A. McCloskey: Application of Mass Spectrometry to Structure Problems. VI. Nucleosides. J. Amer. Chem. Soc. 84, 2005 (1962).

36. Biemann, K. and W. J. McMurray: Computer-Aided Interpretation of High Resolution Mass Spectra. Tetrahedron Letters 1965, 647.

37. Biemann, K., J. Seibl and F. Gapp: Mass Spectra of Organic Molecules. I. Ethyl Esters of Amino Acids. J. Amer. Chem. Soc. 83, 3795 (1961).

38. Biemann, K. and G. Spiteller: Application of Mass Spectrometry to Structure Problems. VIII. Quebrachamine. J. Amer. Chem. Soc. 84, 4578 (1962).

39. — — The Structure of Quebrachamine. Tetrahedron Letters 1961, 299.

40. Biemann, K., M. Spiteller-Friedmann and G. Spiteller: Application of Mass Spectrometry to Structure Problems. X. Alkaloids of the Bark of Aspidosperma quebracho blanco. J. Amer. Chem. Soc. 85, 631 (1963).

41. Biemann, K., S. Tsunakawa and J. A. McCloskey: to be published.

42. Biemann, K. and W. Vetter: Separation of Peptide Derivatives by Gas Chromatography Combined with the Mass Spectrometric Determination of the Amino Acid Sequence. Biochem. Biophys. Res. Commun. 3, 578 (1960).

43. Biemann, K., B. R. Webster and C. Cone: Computer-Aided Interpretation of High Resolution Mass Spectra. II. Amino Acid Sequence of Peptides. J. Amer. Chem. Soc. 88, 2597 (1966).

44. Blossey, E. C., H. Budzikiewicz, M. Ohashi, G. Fodor and C. Djerassi: Mass Spectrometry in Structural and Stereochemical Problems. XXXIX. Tropane Alkaloids. Tetrahedron 20, 585 (1964).

45. Boit, H.-G.: Ergebnisse der Alkaloid-Chemie bis 1960. Berlin: Akademie-Verl. 1961.

46. Bommer, P., W. J. McMurray and K. Biemann: Techniques in the High Resolution Mass Spectrometry of Complex, Polyfunctional Organic Molecules. 12th Annu. Conf. Mass Spectrometry and Allied Topics, Montreal, June 1964.

47. — — High Resolution Mass Spectra of Natural Products. Vinblastine and Derivatives. J. Amer. Chem. Soc. 86, 1439 (1964).

48. Brauchli, P., V. Deulofeu, H. Budzikiewicz and C. Djerassi: The Structure of Tubulosine, A Novel Alkaloid from Pogonopus tubulosus (DC.) Schumann. J. Amer. Chem. Soc. 86, 1895 (1964).

49. Bricas, E., J. van Heijenoort, M. Barber, W. A. Wolstenholme, B. C. Das and E. Lederer: Determination of Amino Acid Sequences in Oligopeptides by Mass Spectrometry. IV. Synthetic N-Acyl Oligopeptide Methyl Esters. Biochemistry 4, 2254 (1965).

50. Brion, C. E.: Windowless Photoionization Source for High Resolution Mass Spectrometers. Analyt. Chemistry 37, 1706 (1965).

51. Britten, A. Z., G. F. Smith and G. Spiteller: The Structure of Picraline. Chem. and Ind. 1963, 1492.

52. Brown, K. S., H. Budzikiewicz and C. Djerassi: Alkaloid Studies. XLII. The Structures of Dichotamine, 1-Acetylaspidoalbidine and 1-Acetyl-17-hydroxy-aspidoalbidine: Three New Alkaloids from Vallesia dichotoma Ruiz et Pav. Tetrahedron Letters 1963, 1731.

53. Büchi, G., R. E. Manning and S. A. Monti: Voacamine and Voacorine. J. Amer. Chem. Soc. 86, 4631 (1964).

54. Budzikiewicz, H. and C. Djerassi: Mass Spectrometry in Structural and Stereochemical Problems. I. Steroid Ketones. J. Amer. Chem. Soc. 84, 1430 (1962).

55. BUDZIKIEWICZ, H., C. DJERASSI, F. PUISIEUX, F. PERCHERON et J. POISSON: Alcaloïdes des Voacanga. Contribution à la structure de la voacamine et de la voacorine, observations sur les spectres de masse de la vobasine et de ses dérivés. Application de la spectrométrie de masse aux problèmes de détermination de structure et de stéréochimie, XXXVIII. Bull. soc. chim. France 1963, 1899.

56. BUDZIKIEWICZ, H., C. DJERASSI and D. H. WILLIAMS: Interpretation of Mass Spectra of Organic Compounds. San Francisco: Holden-Day. 1964.

57. — — — Structure Elucidation of Natural Products by Mass Spectrometry. Vol. 1, Alkaloids. San Francisco: Holden-Day. 1964.

58. — — — Structure Elucidation of Natural Products by Mass Spectrometry. Vol. 2, Steroids, Terpenoids, Sugars, and Miscellaneous Classes. San Francisco: Holden-Day. 1964.

59. BUDZIKIEWICZ, H., S. C. PAKRASHI und H. VORBRÜGGEN: Die Isolierung von Emetin, Cephaelin und Psychotrin aus *Alangium lamarckii* und die Identifizierung von Almarckine mit N-Methylcephaelin. Tetrahedron 20, 399 (1964).

60. BUDZIKIEWICZ, H., J. M. WILSON und C. DJERASSI: Massenspektroskopie und ihre Anwendung auf strukturelle und stereochemische Probleme. 15. Mitt. Steroidsapogenine. Monatsh. Chem. 93, 1033 (1962).

61. BUDZIKIEWICZ, H., J. M. WILSON, C. DJERASSI, J. LÉVY, J. LE MEN and M.-M. JANOT: Mass Spectrometry in Structural and Stereochemical Problems. XIX. Akuammicine and Related Alkaloids. Tetrahedron 19, 1265 (1963).

62. BURLINGAME, A. L.: Mass Spectra of Complex Organic Molecules. Ph. D. thesis, Massachusetts Inst. Technol., 1962.

63. — Application of High Resolution Mass Spectrometry in Molecular Structure Studies. Adv. Mass Spectrometry 3, 744 (1966).

64. BURLINGAME, A. L., H. M. FALES and R. J. HIGHET: The Structure of Amaryllisine. J. Amer. Chem. Soc. 86, 4976 (1964).

65. CLAYTON, E. and R. I. REED: Electron Impact and Molecular Dissociation. XV. Mass Spectra of Physostigmine and Some Related Compounds. Tetrahedron 19, 1345 (1963).

66. CROSS, A. D., L. DOLEJŠ, V. HANUŠ, M. MATUROVÁ and F. ŠANTAVÝ: Structure of the Alkaloid Muramine. Coll. Czech. Chem. Commun. 30, 1335 (1965).

67. CROSS, A. D., F. ŠANTAVÝ and B. TRIVEDI: Substances Isolated from Plants of the Subfamily Wurmbaeoideae and Their Derivatives. LIV. Constitution of Oxycolchicine. Coll. Czech. Chem. Commun. 28, 3402 (1963).

68. CULVENOR, C. C. J., J. D. MORRISON, A. J. C. NICHOLSON and L. W. SMITH: Alkaloids of *Crotalaria trifoliastrum* WILLD. Austral. J. Chem. 16, 131 (1963).

69. DALY, J. W., B. WITKOP, P. BOMMER and K. BIEMANN: Batrachotoxin. The Active Principle of the Colombian Arrow Poison Frog. J. Amer. Chem. Soc. 87, 124 (1965).

70. DAS, B., K. BIEMANN, A. CHATTERJEE, A. B. RAY and P. L. MAJUMDER: The Alkaloids of the Bark of *Alstonia venenata* R. BR. Tetrahedron Letters 1965, 2239.

71. DAS, B., A. SANGSTER and K. BIEMANN: unpublished.

72. DE JONGH, D. C.: Mass Spectrometry in Carbohydrate Chemistry. Diethyl Dithioacetal and Dithioketal Peracetates. J. Amer. Chem. Soc. 86, 3149 (1964).

73. — Mass Spectrometry in Carbohydrate Chemistry. Ethylene Dithioacetal Peracetates. J. Amer. Chem. Soc. 86, 4027 (1964).

74. DeJongh, D. C.: Mass Spectrometry in Carbohydrate Chemistry. Dithio-acetals of Common Monosaccharides. J. Organ. Chem. (USA) **30**, 1563 (1965).

75. DeJongh, D. C. and K. Biemann: Application of Mass Spectrometry to Structure Problems. XIV. Acetates of Partially Methylated Pentoses and Hexoses. J. Amer. Chem. Soc. **85**, 2289 (1963).

76. — — Mass Spectra of O-Isopropylidene Derivatives of Pentoses and Hexoses. J. Amer. Chem. Soc. **86**, 67 (1964).

77. DeJongh, D. C. and S. Hanessian: Assignment of the Position of Amino Groups in Amino Sugars by Mass Spectrometry. J. Amer. Chem. Soc. **87**, 1408 (1965).

78. — — Characterization of Amino Sugars by Mass Spectrometry. J. Amer. Chem. Soc. **87**, 3744 (1965).

79. — — Characterization of Deoxy Sugars by Mass Spectrometry. J. Amer. Chem. Soc. **88**, 3114 (1966).

80. DeJongh, D. C., S. R. Shrader and M. P. Cava: The Mass Spectrometry of Some Bisbenzyltetrahydroisoquinoline Alkaloids. J. Amer. Chem. Soc. **88**, 1052 (1966).

81. Denayer-Tournay, M., J. Pecher, R. H. Martin, M. Friedmann-Spiteller et G. Spiteller: Alcaloïdes indoliques. V. Structure de la voacarpine. Bull. soc. chim. Belges **74**, 170 (1965).

82. Desiderio, D. M.: Computer Techniques in High Resolution Mass Spectro-metry. Ph. D. thesis, Massachusetts Inst. Technol., 1965.

83. Desiderio, D. M. and K. Biemann: Computer Techniques for the Fast and Facile Conversion of Line Positions to Elemental Compositions of Ions Recorded on Photographic Plates. 12th Annu. Conf. Mass Spectrometry and Allied Topics, Montreal, June 1964.

84. Djerassi, C.: Isotope Labelling and Mass Spectrometry of Natural Products. In: The Chemistry of Natural Products, Vol. 3, p. 159. London: Butterworths. 1964.

85. Djerassi, C., L. D. Antonaccio, H. Budzikiewicz, J. M. Wilson and B. Gil-bert: Mass Spectrometry in Structural and Stereochemical Problems. XVI. Structure of the *Aspidosperma* Alkaloid Aspidoalbine. Tetrahedron Letters **1962**, 1001.

86. Djerassi, C., H. W. Brewer, H. Budzikiewicz, O. O. Orazi and R. A. Corral: Mass Spectrometry in Structural and Stereochemical Problems. III. Spegazzinine and Spegazzinidine. Experientia **18**, 113 (1962).

87. Djerassi, C., H. Budzikiewicz, R. J. Owellen, J. M. Wilson, W. G. Kump, D. J. Le Count, A. R. Battersby und H. Schmid: Die Massenspektren von Alkaloiden der Refractin-Pleiocarpin-Klasse und die Struktur von Aspido-fractinin, einem Nebenalkaloid aus *Aspidosperma refractum* Mart. 26. Mitt. über Massenspektroskopie und ihre Anwendung auf strukturelle und stereo-chemische Probleme. Helv. Chim. Acta **46**, 742 (1963).

88. Djerassi, C., H. Budzikiewicz, J. M. Wilson, J. Gosset, J. Le Men and M.-M. Janot: Mass Spectrometry in Structural and Stereochemical Problems. VII. Vincadifformine. (Alcaloïdes des Pervenches. XXI.) Tetrahedron Letters **1962**, 235.

89. Djerassi, C., T. George, N. Finch, H. F. Lodish, H. Budzikiewicz and B. Gilbert: Mass Spectrometry in Structural and Stereochemical Problems. V. Refractine and Aspidofractine. J. Amer. Chem. Soc. **84**, 1499 (1962).

90. DJERASSI, C., B. GILBERT, J. N. SHOOLERY, L. F. JOHNSON and K. BIEMANN: The Constitution of Pyrifolidine. Experientia 17, 162 (1961).

91. DJERASSI, C., Y. NAKAGAWA, H. BUDZIKIEWICZ, J. M. WILSON, J. LE MEN, J. POISSON and M.-M. JANOT: Mass Spectrometry in Structural and Stereochemical Problems. XIII. Echitamidine. Tetrahedron Letters 1962, 653.

92. DJERASSI, C., Y. NAKAGAWA, J. M. WILSON, H. BUDZIKIEWICZ, B. GILBERT and L. D. ANTONACCIO: Alkaloid Studies. XLI. Structure of the Aspidosperma Alkaloid Compactinervine. Experientia 19, 467 (1963).

93. DJERASSI, C., R. J. OWELLEN, J. M. FERREIRA and L. D. ANTONACCIO: Alkaloid Studies. XXXVII. The Structure of Aspidofiline. Experientia 18, 397 (1962).

94. DJERASSI, C., R. H. SHAPIRO and M. VANDEWALLE: Mass Spectrometry in Structural and Stereochemical Problems. LXXXI. Stereospecificity in a Hydrogen-Transfer Reaction Characteristic of 6-Keto Steroids. J. Amer. Chem. Soc. 87, 4892 (1965).

95. DJERASSI, C. and L. TÖKÉS: Mass Spectrometry in Structural and Stereochemical Problems. XCIII. Further Observations on the Importance of Interatomic Distance in the McLafferty Rearrangement. Synthesis and Fragmentation Behavior of Deuterium-Labeled 12-Keto Steroids. J. Amer. Chem. Soc. 88, 536 (1966).

96. DOLEJŠ, L., V. HANUŠ, V. ČERNÝ and F. ŠORM: On Steroids. LXXVIII. Mass Spectra of Holarrhena Alkaloids. Coll. Czech. Chem. Commun. 28, 1584 (1963).

97. DUFFIELD, A. M., R. T. APLIN, H. BUDZIKIEWICZ, C. DJERASSI, C. F. MURPHY and W. C. WILDMAN: Mass Spectrometry in Structural and Stereochemical Problems. LXXXII. Fragmentation of Some Amaryllidaceae Alkaloids. J. Amer. Chem. Soc. 87, 4902 (1965).

98. DUFFIELD, A. M., H. BUDZIKIEWICZ and C. DJERASSI: Mass Spectrometry in Structural and Stereochemical Problems. LXXII. Fragmentation Processes of Some Tobacco Alkaloids. J. Amer. Chem. Soc. 87, 2926 (1965).

99. ETEMADI, A. H., A.-M. MIQUEL, E. LEDERER et M. BARBER: Sur la structure des acides α-mycoliques de Mycobacterium kansasii. Spectrometrie de masse à haute résolution pour des masses de 750 à 1200. Bull. soc. chim. France 1964, 3274.

100. FERREIRA, J. M., B. GILBERT, R. J. OWELLEN and C. DJERASSI: The Alkaloids of Aspidosperma discolor A. DC. Experientia 19, 585 (1963).

101. FINCH, N., I. H. C. HSU, W. I. TAYLOR, H. BUDZIKIEWICZ and C. DJERASSI: Mass Spectrometry in Structural and Stereochemical Problems. XLVII. Some Observations on Mass Spectra of Pseudoindoxyl Alkaloids. J. Amer. Chem. Soc. 86, 2620 (1964).

102. FRANCK, B.: unpublished results from the author's laboratory.

103. GILBERT, B., J. A. BRISSOLESE, N. FINCH, W. I. TAYLOR, H. BUDZIKIEWICZ, J. M. WILSON and C. DJERASSI: Mass Spectrometry in Structural and Stereochemical Problems. XX. Carapanaubine, a New Alkaloid from Aspidosperma carapanauba and Some Observations on Mass Spectra of Oxindole Alkaloids. J. Amer. Chem. Soc. 85, 1523 (1963).

104. GILBERT, B., J. A. BRISSOLESE, J. M. WILSON, H. BUDZIKIEWICZ, L. J. DURHAM and C. DJERASSI: The Alkaloids of Aspidosperma limae WOODSON: Aspidolimidine, Aspidolimine, Demethoxypalosine and Aspidocarpine. Chem. and Ind. 1962, 1949.

105. Gilbert, B., A. P. Duarte, Y. Nakagawa, J. A. Joule, S. E. Flores, J. A. Brissolese, J. Campello, E. P. Carrazzoni, R. J. Owellen, E. C. Blossey, K. S. Brown, Jr. and C. Djerassi: Alkaloid Studies. L. Alkaloids of Twelve *Aspidosperma* Species. Tetrahedron **21**, 1141 (1965).

106. Gilbert, B., J. M. Ferreira, R. J. Owellen, C. E. Swanholm, H. Budzikie-wicz, L. J. Durham and C. Djerassi: Mass Spectrometry in Structural and Stereochemical Problems. II. Pyrifoline and Refractidine. Tetrahedron Letters **1962**, 59.

107. Golding, B. T., R. W. Rickards and M. Barber: Determination of Molecular Formulae of Polyols by Mass Spectrometry of their Trimethylsilyl Ethers. The Structure of the Macrolide Antibiotic Filipin. Tetrahedron Letters **1964**, 2615.

108. Gorman, M., A. L. Burlingame and K. Biemann: Application of Mass Spectrometry to Structure Problems. XI. The Structure of Quebrachidine. Tetrahedron Letters **1963**, 39.

109. Gorman, M., N. Neuss and K. Biemann: Vinca Alkaloids. X. The Structure of Vindoline. J. Amer. Chem. Soc. **84**, 1058 (1962).

110. Gorman, M. and J. Sweeny: Vinca Alkaloids. XXII. Perivine. Tetrahedron Letters **1964**, 3105.

111. Goutarel, R., F. Percheron et M.-M. Janot: Alcaloïdes des *Voacanga*: structure de la voacamine. C. R. hebd. Séances Acad. Sci. **243**, 1670 (1956).

112. Govindachari, T. R., B. R. Pai, S. Rajappa, N. Viswanathan, W. G. Kump, K. Nagarajan und H. Schmid: Über die Struktur des Kopsins. Helv. Chim. Acta **45**, 1146 (1962).

113. Hanessian, S., D. C. DeJongh and J. A. McCloskey: Further Evidence on the Structure of Cordycepin. Biochim. Biophys. Acta **117**, 480 (1966).

114. Hesse, M., H. Hürzeler, C. W. Gemenden, B. S. Joshi, W. I. Taylor und H. Schmid: Die Struktur des *Alstonia*-Alkaloides Villalstonin. Helv. Chim. Acta **48**, 689 (1965).

115. Hesse, M., W. Vetter und H. Schmid: Das massenspektrometrische Verhalten quartärer Stickstoffverbindungen. Helv. Chim. Acta **48**, 675 (1965).

116. Hesse, M., W. v. Philipsborn, D. Schumann, G. Spiteller, M. Spiteller-Friedmann, W. I. Taylor, H. Schmid und P. Karrer: Die Strukturen von C-Fluorocurin, C-Mavacurin und Pleiocarpamin. 57. Mitt. über Curare-Alkaloide. Helv. Chim. Acta **47**, 878 (1964).

117. Heyns, K. und H. F. Grützmacher: Massenspektrometrische Untersuchungen. IV. Massenspektren von *N*-Acetyl-peptiden einfacher Monoaminocarbonsäuren. Liebigs Ann. Chem. **669**, 189 (1963).

118. — — Massenspektrometrische Untersuchungen von acetylierten Peptiden. Tetrahedron Letters **1963**, 1761.

119. Heyns, K. und D. Müller: Massenspektrometrische Untersuchungen. VI. Massenspektrometrische Untersuchungen deuteriummarkierter Methyl-2,3,4-tri-O-methyl-β-arabopyranoside. Tetrahedron **21**, 55 (1965).

120. — — Massenspektrometrische Untersuchungen. VIII. Die Massenspektren permethylierter N-Acetyl-aminozucker. Tetrahedron **21**, 3151 (1965).

121. Heyns, K. und H. Scharmann: Massenspektrometrische Untersuchungen. II. Massenspektren von Derivaten der Monosaccharide und Aminozucker. Liebigs Ann. Chem. **667**, 183 (1963).

122. — — Massenspektrometrische Untersuchungen. VII. Der Einfluß von Ring-größe und Stereochemie in den Massenspektren der permethylierten Pentosen. Tetrahedron **21**, 507 (1965).

123. HODSON, H. F. and G. F. SMITH: The Structure of Folicanthine. Part. II. J. Chem. Soc. (London) **1957**, 1877.

124. HOFFMANN, D. R.: Mass Spectra of Porphyrins and Chlorins. J. Organ. Chem. (USA) **30**, 3512 (1965).

125. JACKSON, A. H., G. W. KENNER, K. M. SMITH, R. T. APLIN, H. BUDZIKIEWICZ and C. DJERASSI: Pyrroles and Related Compounds. VIII. Mass Spectrometry in Structural and Stereochemical Problems. LXXVI. Mass Spectra of Porphyrins. Tetrahedron **21**, 2913 (1965).

126. JANOT, M.-M., J. LE MEN, J. GOSSET et J. LÉVY: Dégradation de la vincamédine et configuration absolue des alcaloïdes apparentés: vincamajine, akuammidine, polyneuridine, voachalotine et macusine A. Alcaloïdes des pervenches, 23e mém. Bull. soc. chim. France **1962**, 1079.

127. JOULE, J. A. and C. DJERASSI: Alkaloid Studies. Part XLV. Mass Spectrometry in Structural and Stereochemical Problems. Part XLII. Some Aspects of the Chemistry and Mass Spectrometry of Uleine. J. Chem. Soc. (London) **1964**, 2777.

128. JOULE, J. A., H. MONTEIRO, L. J. DURHAM, B. GILBERT and C. DJERASSI: Alkaloid Studies. Part XLVIII. Structure of Apparicine, an Aspidosperma Alkaloid. J. Chem. Soc. (London) **1965**, 4773.

129. JOULE, J. A., M. OHASHI, B. GILBERT and C. DJERASSI: Alkaloid Studies. LIII. Structures of Nine Alkaloids from *Aspidosperma dasycarpon* A. Dc. Tetrahedron **21**, 1717 (1965).

130. JUNK, G. A. and H. J. SVEC: Mass Spectrometric Identification of Dipeptides. Analyt. Biochem. **6**, 199 (1963).

131. KASCHNITZ, R. und G. SPITELLER: Anwendung der Massenspektrometrie zur Strukturaufklärung von Alkaloiden. 7. Mitt. Neue Alkaloide aus *Gonioma Kamassi* E. MEY. Monatsh. Chem. **96**, 909 (1965).

132. KHAN, Z. M., M. HESSE und H. SCHMID: Quartäre Alkaloide aus *Pleiocarpa mutica* BENTH. 2. Mitt. über das massenspektrometrische Verhalten quartärer Stickstoffverbindungen. Helv. Chim. Acta **48**, 1957 (1965).

133. KIRYUSHKIN, A. A., YU. A. OVCHINNIKOV, M. M. SHEMYAKIN, V. N. BOCHKAREV, B. V. ROZINOV and N. S. WULFSON: Mass Spectrometric Determination of Amino Acid Sequence in Peptides. II. A Convenient Method of Converting Peptides to Acylpeptide Esters. Tetrahedron Letters **1966**, 33.

134. KOCHETKOV, N. K., O. S. CHIZHOV and B. M. ZOLOTAREV: Mass Spectrometer Study of Carbohydrates. Methyl Ethers of Certain Methyl Deoxy Hexosides. Doklady Akad. Nauk (USSR) **165**, 569 (1965) [Chem. Abstr. **64**, 6738 (1966)].

135. KOCHETKOV, N. K., N. S. WULFSON, O. S. CHIZHOV and B. M. ZOLOTAREV: Mass Spectrometric Study of Carbohydrates. Methyl Ethers and Acetates of Glycosides. Doklady Akad. Nauk (USSR) **151**, 336 (1963) [engl. transl. **1963**, 543].

136. — — — — Mass Spectrometric Investigation of Carbohydrates. Fragmentation Pattern of 2,3,4,6-Tetramethyl-α-methyl-D-glucoside. Izv. Akad. Nauk (SSSR), Ser. Khim. **1965**, 776 [Chem. Abstr. **64**, 799 (1966)].

137. — — — — Mass Spectrometry of Carbohydrate Derivatives. Tetrahedron **19**, 2209 (1963).

138. KUMP, W. G., D. J. LE COUNT, A. R. BATTERSBY und H. SCHMID: Die Struktur von Pleiocarpin, Pleiocarpinin und Kopsinin. Helv. Chim. Acta **45**, 854 (1962).

139. KUMP, W. G., M. B. PATEL, J. M. ROWSON und H. SCHMID: Indoalkaloide aus den Blättern von *Pleiocarpa pycnantha* (K. SCHUM.) STAPF *var. tubicina* (STAPF) PICHON. 7. Mitt. über Pleiocarpa-Alkaloide. Helv. Chim. Acta **47**, 1497 (1964).

140. LANEELLE, G., J. ASSELINEAU, W. A. WOLSTENHOLME et E. LEDERER: Détermination de séquences d'acides aminés dans les oligopeptides par la spectrométrie de masse. III. Structure d'un peptidolipide de *Mycobacterium johnei*. Bull. soc. chim. France **1965**, 2133.

141. LAW, J. H., E. O. WILSON and J. A. McCLOSKEY: Biochemical Polymorphism in Ants. Science **149**, 544 (1965).

142. LEDERER, E.: Biosynthesis, Structure and Biological Action of the Lipids of the Tubercle Bacillus. Angew. Chem. **76**, 241 (1964); Intern. Ed. **3**, 393 (1964).

143. LHOEST, G., R. DE NEYS, N. DEFAY, J. SEIBL, J. PECHER et R. H. MARTIN: Alcaloïdes indoliques. VII. Structure de la voacoline. Bull. soc. chim. Belges **74**, 534 (1965).

144. MANSKE, R. H. F.: The Alkaloids. Chemistry and Physiology. Vol. I—VIII. New York: Academic Press. 1950—65.

145. MANUSADZHYAN, V. G., A. M. ZYAKOON, A. V. CHUVILIN and YA. M. VARSHAVSKIĬ: Application of Mass Spectrometric Methods to the Study of Amino Acid Derivatives and Short Peptides. II. Mass Spectrometric Analysis of Ethyl Esters of N-Acyl Peptides. Izv. Akad. Nauk Arm. SSR, Khim. **17**, 143 (1964) [Chem. Abstr. **61**, 9019 (1964)].

146. McDOWELL, C. A.: Mass Spectrometry. McGraw-Hill Series in Advanced Chemistry. New York: McGraw-Hill. 1963.

147. McLAFFERTY, F. W.: Mass Spectrometric Analysis. Molecular Rearrangements. Analyt. Chem. **31**, 82 (1959).

148. MacLEAN, D. B.: Lycopodium Alkaloids. XIII. Mass Spectra of Representative Alkaloids. Canad. J. Chem. **41**, 2654 (1963).

149. MILLS, J. F. D. and S. C. NYBURG: The Molecular Structure of Aspidospermine. Tetrahedron Letters **1959**, No. 11, 1.

150. MOKRÝ, J., L. DÚBRAVKOVÁ and P. ŠEFČOVIČ: Alkaloide aus *Vinca minor* L. X. Vincadin, Minovin und Vincorin. Experientia **18**, 564 (1962).

151. MONSEUR, X., R. GOUTAREL, J. LE MEN, J. M. WILSON, H. BUDZIKIEWICZ et C. DJERASSI: Structure de la mossambine (= Diplorrhyncine). Alcaloïdes du *Diplorrhyncus condylocarpon* ssp. *mossambicensis* BENTH DUVIGN. (Apocynacées), 2e note. Application de la spectrographie de masse aux problèmes de structure et de stéréochimie, X. Bull. soc. chim. France **1962**, 1088.

152. MONTEIRO, H., H. BUDZIKIEWICZ, C. DJERASSI, R. R. ARNDT and W. H. BAARSCHERS: Alkaloid Studies. LIV. Structure of Deoxytubulosine and Interconversion with Tubolosine. Chem. Commun. **1965**, 317.

153. MOZA, B. K., J. TROJÁNEK, V. HANUŠ and L. DOLEJŠ: On Alkaloids. XIII. On the Mass Spectra of Vindorosine. Coll. Czech. Chem. Commun. **29**, 1913 (1964).

154. NAKAGAWA, Y., J. M. WILSON, H. BUDZIKIEWICZ and C. DJERASSI: Alkaloid Studies. XLI. The Constitution of Lochneridine. Chem. and Ind. **1962**, 1986.

155. NATALIS, P. and J. L. FRANKLIN: Ionization and Dissociation of Diphenyl and Condensed Ring Aromatics by Electron Impact. II. Diphenylcarbonyls and Ethers. J. Physic. Chem. **69**, 2943 (1965).

156. NEUNER-JEHLE, N., H. NESVADBA und G. SPITELLER: Anwendung der Massenspektrometrie zur Strukturaufklärung von Alkaloiden. 6. Mitt. Pyrrolizidinalkaloide aus dem Goldregen. Monatsh. Chem. **96**, 321 (1965).

157. — — — Schlüsselbruchstücke in den Massenspektren von Alkaloiden. 3. Mitt. Monatsh. Chem. **95**, 687 (1964).

158. NEUSS, N., M. GORMAN, W. HARGROVE, N. J. CONE, K. BIEMANN, G. BÜCHI and R. E. MANNING: Vinca Alkaloids. XXI. The Structure of the Oncolytic Alkaloids Vinblastine (VLB) and Vincristine (VCR). J. Amer. Chem. Soc. **86**, 1440 (1964).

159. OHASHI, M., H. BUDZIKIEWICZ, J. M. WILSON, C. DJERASSI, J. LÉVY, J. GOSSET, J. LE MEN and M.-M. JANOT: Mass Spectrometry in Structural and Stereochemical Problems. XXXVI. Alkaloids of Periwinkles. 27. The Mass Spectra of Stereoisomers of the Sarpagine-Akuammidine Group. Tetrahedron **19**, 2241 (1963).

160. OHASHI, M., J. A. JOULE, B. GILBERT and C. DJERASSI: Structures of Five New *Aspidosperma* Alkaloids Related to Uleine. Experientia **20**, 363 (1964).

161. OHASHI, M., J. M. WILSON, H. BUDZIKIEWICZ, M. SHAMMA, W. A. SLUSARCHYK and C. DJERASSI: Mass Spectrometry in Structural and Stereochemical Problems. XXXI. Aporphines and Related Alkaloids. J. Amer. Chem. Soc. **85**, 2807 (1963).

162. OLIVIER, L., J. LÉVY, J. LE MEN, M.-M. JANOT, H. BUDZIKIEWICZ et C. DJERASSI: Structure et configuration absolue de la pseudo-akuammigine, de l'akuammine et de l'akuammiline. (Alcaloïdes du *Picralima nitida* STAPF. Apocynacées) (10e mém.) Bull. soc. chim. France **1965**, 868.

163. OLIVIER, L., J. LÉVY, J. LE MEN, M.-M. JANOT, C. DJERASSI, H. BUDZIKIEWICZ, J. M. WILSON et L. J. DURHAM: Structure de la ψ-akuammigine et de la picraline. Alcaloïdes du *Picralima nitida* STAPF (Apocynacées), 6e mém. Application de la spectrographie de masse aux problèmes de détermination de structure et de stéréochimie, XXVIII. Bull. soc. chim. France **1963**, 646.

164. PADILLA, J. and J. HERRÁN: Hernandezine. A New Alkaloid of the Bisbenzylisoquinoline Series. Tetrahedron **18**, 427 (1962).

165. PANDIT, U. K., W. N. SPECKAMP and H. O. HUISMAN: Heterocyclic Steroids. III. Mass Spectra of 6-Aza-steroids with Two Aromatic Rings. Tetrahedron **21**, 1767 (1965).

166. PARELLO, J., P. LONGEVIALLE, W. VETTER et J. A. McCLOSKEY: Structure de la philalbine. Application de la résonance magnétique nucléaire et de la spectrométrie de masse à l'étude des dérivés du tropane. Bull. soc. chim. France **1963**, 2787.

167. PELAH, Z., D. H. WILLIAMS, H. BUDZIKIEWICZ and C. DJERASSI: Mass Spectrometry in Structural and Stereochemical Problems. LX. Electron Impact Induced Fragmentation of Steroidal Dimethylamines. J. Amer. Chem. Soc. **87**, 574 (1965).

168. PFEIFER, S., S. K. BANERJEE, L. DOLEJŠ und V. HANUŠ: Zur Struktur der Papaverrubine. Pharmazie **20**, 45 B (1965).

169. PFEIFER, S., I. MANN, L. DOLEJŠ und V. HANUŠ: Zur Struktur von Oreodin und Oreogenin. Pharmazie **20**, 585 (1965).

170. PINAR, M., W. v. PHILIPSBORN, W. VETTER und H. SCHMID: Limaspermin. Helv. Chim. Acta **45**, 2260 (1962).

171. Plat, M., D. Dohkac Manh, J. Le Men, M.-M. Janot, H. Budzikiewicz, J. M. Wilson, L. J. Durham et C. Djerassi: Structure de la vincamine et de la méthoxy-11 vincamine. Alcaloïdes des pervenches, 24e mém. Application de la spectrographie de masse aux problèmes de détermination de structure et de stéréochimie, XII. Bull. soc. chim. France **1962**, 1082.

172. Plat, M., J. Le Men, M.-M. Janot, H. Budzikiewicz, J. M. Wilson, L. J. Durham et C. Djerassi: Structure de quatre alcaloïdes de la Petite Pervenche (*Vinca minor* L.). (—)-Vincadifformine, minovincine, minovincinine et méthoxy-16 minovincine. Alcaloïdes des pervenches, 26e mém. Application de la spectrographie de masse aux problèmes de détermination de structure et de stéréochimie, XVII. Bull. soc. chim. France **1962**, 2237.

173. Plat, M., J. Le Men, M.-M. Janot, J. M. Wilson, H. Budzikiewicz, L. J. Durham, Y. Nakagawa and C. Djerassi: Mass Spectrometry in Structural and Stereochemical Problems. IX. Tabersonine (Alkaloids of *Amsonia tabernaemontana* Walt. III.). Tetrahedron Letters **1962**, 271.

174. Rao, K. V., K. Biemann and R. B. Woodward: The Structure of Streptonigrin. J. Amer. Chem. Soc. **85**, 2532 (1963).

175. Rapoport, H., H. D. Baldridge, Jr., and E. J. Volchek, Jr.: The Lactone Ring of Carpaine. J. Amer. Chem. Soc. **75**, 5290 (1953).

176. Renner, U., D. A. Prins, A. L. Burlingame und K. Biemann: Die Struktur der 2-Acylindol-Alkaloide Vobasin, Dregamin und Tabernaemontanin. Helv. Chim. Acta **46**, 2186 (1963).

177. Ryhage, R.: Use of a Mass Spectrometer as a Detector and Analyzer for Effluent Emerging from High Temperature Gas Liquid Chromatography Columns. Analyt. Chemistry **36**, 759 (1964).

178. Ryhage, R. and E. Stenhagen: Mass Spectrometry in Lipid Research. J. Lipid Res. **1**, 361 (1960).

179. — — Mass Spectrometry of Long-Chain Esters. In: F. W. McLafferty: Mass Spectrometry of Organic Ions, p. 399. New York: Academic Press. 1963.

180. Šantavý, F., J. L. Kaul, L. Hruban, L. Dolejš, V. Hanuš, K. Bláha and A. D. Cross: Constitution of Rhoeadine and Isorhoeadine. Coll. Czech. Chem. Commun. **30**, 335 (1965).

181. Schmid, H. und M. Pinar: Aspidolimin. Helv. Chim. Acta **45**, 1283 (1962).

182. Schnoes, H. K. and K. Biemann: Interconversion of the Aspidospermine Skeleton to the Refractine Type. J. Amer. Chem. Soc. **86**, 5693 (1964).

183. Schnoes H. K., K. Biemann, J. Mokrý, I. Kompiš, A. Chatterjee and G. Ganguli: Strictamine. J. Organ. Chem. (USA) **31**, 1641 (1966).

184. Schnoes, H. K., A. L. Burlingame and K. Biemann: Application of Mass Spectrometry to Structure Problems. IX. The Occurrence of Eburnamenine and Related Alkaloids in *Rhazya stricta* and *Aspidosperma quebracho blanco*. Tetrahedron Letters **1962**, 993.

184a. Senn, M. and F. W. McLafferty: Biochem. Biophys. Res. Comm. **23**, 381 (1966).

185. Shamma, M., B. S. Dudock, M. P. Cava, K. V. Rao, D. R. Dalton, D. C. DeJongh and S. R. Shrader: Revised Structures of Hernandezine and Thalsimine. Mass Spectrometry of a Bisbenzylisoquinoline Alkaloid. Chem. Commun. **1966**, 7.

186. Slavík, J., L. Dolejš, K. Vokáč and V. Hanuš: The Structure of Glaucamine, an Alkaloid from *Papaver glaucum* Boiss. et Hauskn. Mass Spectrometric Study of Fragmentation of Rhoeadine Alkaloids. Coll. Czech. Chem. Commun. **30**, 2864 (1965).

187. SLAVÍK, J., V. HANUŠ, K. VOKÁČ and L. DOLEJŠ: Mass Spectrometric Study of Homologues of Rhoeadine and Structure of Dubirheine. Coll. Czech. Chem. Commun. **30**, 2464 (1965).

188. SPITELLER, G., C. BRUNNÉ, K. HEYNS und H. F. GRÜTZMACHER: Die massenspektrometrische Strukturuntersuchung thermisch labiler und schwer flüchtiger organischer Verbindungen. Z. Naturforsch. **17 b**, 856 (1962).

189. SPITELLER, G., A. CHATTERJEE, A. BATTACHARYA und A. DEB: Anwendung der Massenspektrometrie zur Strukturaufklärung von Alkaloiden, 2. Mitt. Monatsh. Chem. **93**, 1220 (1962).

190. SPITELLER, G. und M. SPITELLER-FRIEDMANN: Zur massenspektrometrischen Untersuchung labiler und schwerflüchtiger organischer Verbindungen. Monatsh. Chem. **94**, 742 (1963).

191. — — Schlüsselbruchstücke in den Massenspektren von Alkaloiden, 1. Mitt. Tetrahedron Letters **1963**, 147.

192. — — Schlüsselbruchstücke in den Massenspektren von Alkaloiden, 2. Mitt. Tetrahedron Letters **1963**, 153.

193. — — Anwendung der Massenspektrometrie zur Strukturaufklärung von Alkaloiden, 1. Mitt. Die basischen Inhaltsstoffe der Rinde von *Aspidosperma oblongum* A. DC. Monatsh. Chem. **93**, 795 (1962).

194. — — Anwendung der Massenspektrometrie zur Strukturaufklärung von Alkaloiden, 3. Mitt. Die Nebenalkaloide aus der Rinde von *Aspidosperma oblongum* A. DC. Monatsh. Chem. **94**, 779 (1963).

195. SPITELLER-FRIEDMANN, M. und G. SPITELLER: Anwendung der Massenspektrometrie zur Strukturaufklärung von Alkaloiden, 5. Mitt. Die Struktur des Carpains. Monatsh. Chem. **95**, 1234 (1964).

196. — — Schlüsselbruchstücke in den Massenspektren von Alkaloiden, 4. Mitt. Piperidin-Alkaloide. Monatsh. Chem. **96**, 104 (1965).

197. STENHAGEN, E.: Massenspektrometrie als Hilfsmittel bei der Strukturbestimmung organischer Verbindungen, besonders bei Lipiden und Peptiden. Z. analyt. Chem. **181**, 462 (1961).

198. — Jetziger Stand der Massenspektrometrie in der organischen Analyse. Z. analyt. Chem. **205**, 109 (1964).

199. THOMAS, D. W., H. ACHENBACH and K. BIEMANN: 15-(14'-Eburnamyl)-pleiocarpinine (Pleiomutine), a New Dimeric Indole Alkaloid from *Pleiocarpa mutica* BENTH. J. Amer. Chem. Soc. **88**, 1537 (1966).

200. — — Revised Structures of the Pleiocarpa Alkaloids Pleiocarpoline (Pleiocarpine N_b-oxide), Pleiocarpolinine (Pleiocarpinine N_b-oxide), and Kopsinoline (Kopsinine N_b-oxide). J. Amer. Chem. Soc. **88**, 3423 (1966).

201. THOMAS, D. W. and K. BIEMANN: Thermal Methyl Transfer. Mass Spectrum of Voacamine-d_3. J. Amer. Chem. Soc. **87**, 5447 (1965).

202. TSUNAKAWA, S.: unpublished results from the author's laboratory.

203. VETTER, W., P. LONGEVIALLE, F. KHUONG-HUU-LAINE, Q. KHUONG-HUU et R. GOUTAREL: Alcaloïdes stéroïdiques, XVIII. La spectrométrie de masse dans la détermination des structures des stéroïdes aminés. Bull. soc. chim. France **1963**, 1324.

204. WALSER, A. und C. DJERASSI: Alkaloidstudien, XLIX. Die Strukturen von Vallesamin und O-Acetyl-vallesamin. Helv. Chim. Acta **47**, 2072 (1964).

205. — — Alkaloidstudien. LII. Die Alkaloide aus *Vallesia dichotoma* RUIZ et PAV. Helv. Chim. Acta **48**, 391 (1965).

206. WATSON, J. T. and K. BIEMANN: High Resolution Mass Spectra of Compounds Emerging from a Gas Chromatograph. Analyt. Chemistry 36, 1135 (1964).
207. — — Direct Recording of High Resolution Mass Spectra of Gas Chromatographic Effluents. Analyt. Chemistry 37, 844 (1965).
208. WEYGAND, F., A. PROX, H. H. FESSEL und K. K. SUN: Massenspektrometrische Sequenzanalyse von Peptiden als N-Trifluoracetyl-peptid-ester. Z. Naturforsch. 20 b, 1169 (1965).
209. WEYGAND, F., A. PROX, W. KÖNIG und H. H. FESSEL: Massenspektrometrische und gaschromatographische Sequenzanalyse von Peptiden. Angew. Chem. 75, 724 (1963).
210. WILSON, J. M., M. OHASHI, H. BUDZIKIEWICZ, F. ŠANTAVÝ and C. DJERASSI: Mass Spectrometry in Structural and Stereochemical Problems. XXXIII. Colchicine Alkaloids. Tetrahedron 19, 2225 (1963).
211. WULFSON, N. S., V. N. BOCHKAREV, B. V. ROZINOV, M. M. SHEMYAKIN, YU. A. OVCHINNIKOV, A. A. KIRYUSHKIN and A. I. MIROSHNIKOV: Mass Spectrometric Determination of Amino Acid Sequence in Peptides. III. Peptides Containing Aspartic and Glutamic Acid Residues. Tetrahedron Letters 1966, 39.

(Received, March 17, 1966.)

Pflanzliche Steroide mit 21 Kohlenstoffatomen.

Von RUDOLF TSCHESCHE, Bonn.

Mit 4 Abbildungen.

Inhaltsübersicht.

aus *P. terminalis*, O-Desacyl-pachysandrin B 120. — h. An $C_{(3)}$ und
$C_{(20)}$ aminierte Derivate und Δ^5: Irehdiamine A und B, Kurchessin
(= Saracodinin?), Saracoccin (= Saracocin), 3 α,20 S·Bisdimethyl-
amino-Δ^5-pregnen 122. — i. An $C_{(3)}$ und $C_{(20)}$ aminierte Derivate mit
einer OH-Gruppe an $C_{(18)}$, Δ^5: Holarrhimin, Monomethylholarrhimine I
und II, Tetramethylholarrhimin, Holarrhidin 122. — j. Conessin-
derivate mit O-Funktionen an $C_{(3)}$ bzw. $C_{(7)}$, Δ^5: Latifolin, Norlatifolin,
Latifolinin, Funtulin, Funtudienin, Holonamin 124. — k. Conanin-
derivate mit N-Funktion an $C_{(3)}$, 5 α: Malouphyllamin, Dihydro-conessin,
Dihydro-conessimin, Dihydro-conkuressin, Funtessin 126. — l. Conenin-
derivate mit N-Funktion an $C_{(3)}$, Δ^5: Conarrhimin, Conamin, Conessimin,
Conimin, Isoconessimin, Conessin, 7 α-Hydroxyconessin, 12 β-Hydroxy-
conessin, Holafrin, Holarrhenin, Holarrhetin 127. — m. Conanin-derivate
mit Doppelbindungen $\Delta^{5,18}$: Conkurchin (= Irehlin), Conessidin 132. —
n. Kurcholessin 133. — o. D-Homo-androstan-Alkaloide: Dictyolucidin,
Dictyolucidamin 134.

I. Einleitung.

Im Tierreich sind Pregnanderivate in Form des Progesterons und der
Nebennierenrinden-Hormone Cortison, Aldosteron usw. seit den dreißiger
Jahren bekannt und in der Struktur gesichert; später wurde für sie
Δ^5-Pregnen-3β-ol-20-on als biogenetische Vorstufe erkannt. Im
Pflanzenreich wurde als erstes C_{21}-Steroidderivat 1936 von W. Karrer (79)
Diginin aus den Blättern von *Digitalis purpurea* L. isoliert und als D-Digi-
nose-glykosid erkannt. In der Folgezeit haben dann der Arbeitskreis
um Satoh sowie der meine aus verschiedenen Digitalis-Varietäten
weitere Glykoside dieses Typs isolieren können. Ihre Auffindung ge-
schah im Verlauf der Bemühungen, die Cardenolid-nebenglykoside dieser
Pflanzen zu isolieren. Da sie weder durch charakteristische Farbreaktionen, noch durch besondere physiologische Effekte auffallen, war ihre
Erfassung mehr oder weniger dem Zufall überlassen. Die Mengen in den
Blättern liegen mit Ausnahme des Digipurpurins zwei Zehnerpotenzen
unterhalb derjenigen der Cardenolidglykoside, so daß große Quantitäten
von Ausgangsmaterial aufgearbeitet werden müssen, um die C_{21}-Steroid-
glykoside zu gewinnen.

Einfacher liegen in bezug auf die vorhandenen Mengen vielfach die
Verhältnisse in der Familie der Asclepiadaceen. Hier haben Cornforth
und Earl (30) 1939 als erstes Glykosid ein Derivat des Sarcostins aus
Sarcostemma australe R. Br. gewonnen. Die Glykoside dieser Familie liegen
vorwiegend als Ester vor, in denen Essigsäure, Benzoesäure, Zimtsäure und
andere Säuren an sekundäre Hydroxyle im Aglykon gebunden sind.

Weitere Vertreter dieser Esterglykoside wurden durch REICHSTEIN und Mitarb., durch MITSUHASHI und Mitarb. sowie durch die Bonner Arbeitsgruppe isoliert. Trotz der besseren Zugänglichkeit verursachen hier die vielfach geringe Kristallisationsneigung und die leichte Isomerisierbarkeit durch Säuren wie Basen erhebliche Schwierigkeiten. Die Isolierung von Δ^5-Pregnen-3β-ol-20-on und seines 5α,6-Dihydroderivates [TSCHESCHE und SNATZKE (202)] in Form von Glucosiden aus *Xysmalobium undulatum* R. Br. ist besonders bemerkenswert. Sie wurden damit zum ersten Mal im Pflanzenreich gefunden; dieser Befund legte die Vermutung nahe, daß die erstgenannte Verbindung biogenetischer Vorläufer für die C_{21}-Steroide des Pflanzenreiches ebenso wie des Tierreiches sein könnte. Im allgemeinen scheinen sich größere Mengen an C_{21}-Steroidglykosiden und Cardenolidglykosiden nebeneinander in einer Pflanze auszuschließen; meist findet man nur geringe oder sehr geringe Mengen des anderen Typs vor, so daß eine gemeinsame direkte Vorstufe für beide nicht ausgeschlossen schien.

Aus der Familie der Apocynaceen sind bisher nur wenige N-freie Vertreter bekannt geworden, die bei der Suche nach Vorläufern der entsprechenden Steroidalkaloide aufgefunden werden konnten. Die interessanteste Entdeckung ist hierbei von LEBOEUF, CAVÉ und GOUTAREL (103) gemacht worden, die in *Holarrhena floribunda* Progesteron auffanden, das bis dahin nur aus dem Tierreich bekannt war. Diese Familie liefert jedoch die große Gruppe der Alkaloide vom Pregnantyp, die man sich durch Aminierung in der 3- und in der 20-Stellung entstanden denken kann. Diese Überlegung wird durch die Auffindung N-freier Vertreter gestützt, deren Struktur in auffallender Weise mit derjenigen verschiedener Alkaloide dieses Typs korrespondiert. Sie werden fast alle in freier, nicht glykosidierter Form gefunden. Conessin als Hauptalkaloid dieser Gruppe ist seit langem bekannt [WARNECKE (208), 1886, erstes kristallisiertes Präparat] und wurde erstmals aus der Rinde von *Holarrhena antidysenterica* isoliert. Später konnten durch die Arbeitskreise um SIDDIQUI, BERTHO, GOUTAREL, ŠORM, TSCHESCHE u. a. zahlreiche weitere Alkaloide aus dieser und anderen Arten der gleichen Familie isoliert werden.

Unter Verwendung von mit ^{14}C an $C_{(21)}$ markiertem Δ^5-Pregnen-3β-ol-20-on-glucosid ließ sich zeigen, daß diese Verbindung als Vorstufe für die Biogenese der Cardenolid- und Bufadienolidglykoside [TSCHESCHE und Mitarb. (184, 193)] geeignet ist. Man muß daraus schließen, daß dieses Steroid im Pflanzenreich eine ebenso wichtige Rolle als Relaissubstanz wie im Tierreich spielt und daß zum mindesten in allen solchen Pflanzen mit dem Auftreten von C_{21}-Steroidderivaten zu rechnen ist, in denen solche Glykoside nachgewiesen wurden.

Für die N-freien C_{21}-Steroide des Pflanzenreiches wurde von uns die Bezeichnung Digitanol- bzw. beim Vorliegen einer Δ^5-Doppelbindung Digitenolderivate vorgeschlagen, nach der ersten Auffindung einer derartigen Verbindung (Diginin) in *Digitalis purpurea* L. Diese Bezeichnung wurde später auch auf die Aglykone der Esterglykoside der Familie der Asclepiadaceen und die N-freien Vorstufen der C_{21}-Steroidalkaloide übertragen.

II. N-freie Pregnanderivate aus Scrophulariaceen.

Tabelle 1 (S. 103) gibt die einzelnen Glykoside in ihrer Zusammensetzung, die vorhandenen Zucker, die Herkunft, die Aglykone und deren Summenformeln wieder.

Die Struktur der einzelnen Aglykone.

a) Diginigenin (1): Dieses Aglykon findet sich sowohl in dem schon 1936 in den Blättern von *Digitalis purpurea* L. aufgefundenen Diginin (Zucker ist die *D*-Diginose) (*79*) wie im später von Satoh und Mitarb. (*156*) isolierten Digitalonin (Zucker ist die *D*-Digitalose). Seine Strukturformel ist mehrfach variiert worden (*161, 188*) und konnte erst neuerdings mittels moderner physikalischer Methoden durch Shoppee, Lack und Robertson (*164*) verbessert und vermutlich geklärt werden. Die ersten Arbeiten über dieses Aglykon stammen von Shoppee und Reichstein (*167*) sowie später von Shoppee allein (*160, 161*) und haben die Natur der funktionellen Gruppen geklärt. Die Verknüpfung mit einem bekannten Steroidderivat gelang Press und Reichstein (*146*), die einen schon von Shoppee mit Hilfe der Wolff-Kishner-Reaktion dargestellten Kohlenwasserstoff als $5\alpha,14\beta,17\beta$ H-Pregnan erkannten. Eine Vorstufe dieses Kohlenwasserstoffes könnte vermutlich $5\alpha,14\beta,17\beta$H-Pregnan-3.20-dion gewesen sein. Die Schwierigkeiten der weiteren Strukturaufklärung bestanden vornehmlich darin, wie der im Diginigenin aufgefundene Sauerstoffring unterzubringen sei. Massenspektrometrie, magnetische Kernresonanz und optische Rotationsdispersionsmessungen haben zu einer befriedigenden Formulierung nach (1) geführt (*164, 165, 185*).

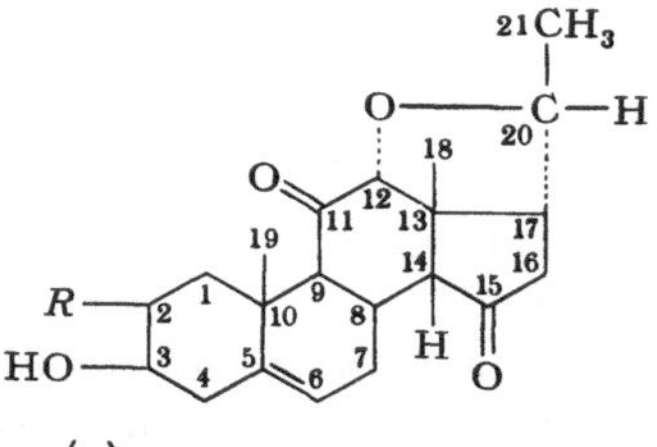

(1) Diginigenin: $R =$ H.
(2) Digifologenin: $R =$ OH.

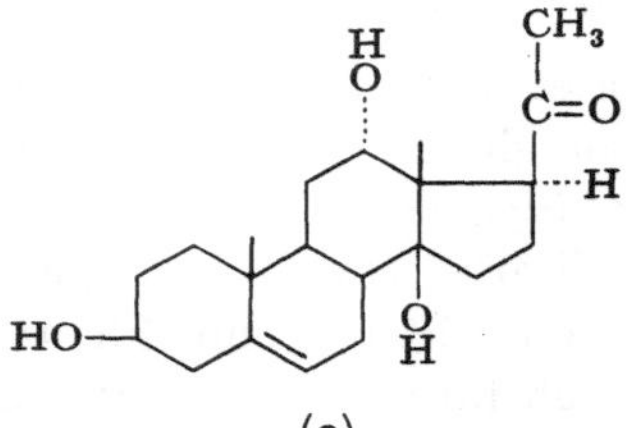

(3)
Digipurpurogenin I.

Literaturverzeichnis: SS. 138—148.

Tabelle 1. Pflanzliche Pregnanderivate.

Glykosid	Zusammensetzung	Zucker*	Vorkommen	Aglykone	Zusammensetzung
Diginin (79)	$C_{28}H_{40}O_7$	1 Dn	Digitalis purpurea und D. lanata	Diginigenin	$C_{21}H_{28}O_4$
Digitalonin (156)	$C_{28}H_{40}O_8$	1 Dig	D. sp.	Diginigenin	$C_{21}H_{28}O_4$
Digifolein (188)	$C_{28}H_{40}O_8$	1 Dn	D. purpurea und D. lanata	Digifologenin	$C_{21}H_{28}O_5$
Lanafolein (188)	$C_{28}H_{40}O_8$	1 Ol	D. lanata	Digifologenin	$C_{21}H_{28}O_5$
Digipronin (156, 195)	$C_{28}H_{40}O_9$	1 Dig	D. sp.	Digiprogenin	$C_{21}H_{28}O_5$
Digipurpurin (189)	$C_{39}H_{62}O_{14}$	3 Dx	D. purpurea	Digipurpurogenin	$C_{21}H_{32}O_4$
Purpnin (156, 155)	$C_{39}H_{62}O_{13}$	3 Dx	D. sp.	Purpnigenin	$C_{21}H_{32}O_4$
Purpronin (155)	$C_{39}H_{60}O_{14}$	3 Dx	D. sp.	Purprogenin	$C_{21}H_{30}O_5$
Digacetinin (190)	$C_{43}H_{64}O_{16}$	3 Dx 1 Essigsäure	D. sp.	Digacetigenin	$C_{21}H_{29}O_5$ + 1 Acetyl

* Dn = D-Diginose, Ol = D-Oleandrose, Dx = D-Digitoxose, Dig = D-Digitalose.

b) *Digifologenin* (2): Sehr nahe verwandt mit Diginigenin ist Digifologenin, das in Form des Glykosids Digifolein als ein steter Begleiter des Diginins in den Blättern von *D. purpurea* L. vorkommt und früher vermutlich nicht abgetrennt worden ist (*188*). In *D. lanata* findet sich an seiner Stelle Lanafolein (*188*), in dem der Zucker D-Oleandrose die D-Diginose ersetzt und der hier zum ersten Mal in der Natur beobachtet wurde. Die Ähnlichkeit des Digifologenins in den Farbreaktionen, der optischen Drehung, in der UV- und IR-Absorption, zusammen mit der Möglichkeit der Herstellung eines Acetonids und der Angreifbarkeit mit Perjodsäure, ließen auf ein Hydroxydiginigenin schließen. Es wird ein 2β-Hydroxyderivat angenommen (*185, 188, 194*), eine Struktur, die von SHOPPEE (*166*) auf Grund physikalischer Messungen ebenfalls bevorzugt wird. Partialsynthetische Versuche sind bisher weder am Diginigenin noch am Digifologenin ausgeführt worden.

c) *Digipurpurogenin I* (3): Das Glykosid Digipurpurin (*189*) befindet sich in den Blättern von *D. purpurea* L. in vergleichsweise großer Menge; es kann bis 10% der Cardenolidglykoside ausmachen, in den Blättern von *D. lanata* Ehrh. kommt es nicht vor. Es enthält 3 Mole *D*-Digitoxose als Zuckerkomponente. Bei der vorsichtigen sauren Hydrolyse entsteht Digipurpurogenin I, das als Δ^5-14β-Pregnen-3β,12α, 14-triol-20-on erkannt worden ist (*186*). Bei Einwirkung stärkerer Säure wird es in Digipurpurogenin II umgelagert, hierbei tritt Umklappen der OH-Gruppe an $C_{(12)}$ in die β-Konfiguration ein. Der Mechanismus dieser Umlagerung ist nicht bekannt. Bei der Behandlung mit Alkali und anschließend mit Säure liefert Digipurpurogenin I Isodigipurpurogenin I, das einen Cyclohalbacetalring zwischen $C_{(12)}$ und $C_{(20)}$ enthält. Durch das Alkali ist eine Isomerisierung der Seitenkette an $C_{(17)}$ nach α erfolgt, womit die Ausbildung eines Halbacetalringes möglich wird. Auch Digipurpurogenin II erleidet unter den gleichen Bedingungen dieselbe Isomerisierung an $C_{(17)}$, ein Halbacetalring ist aber wegen der 12β-Stellung der OH-Gruppe jetzt nicht mehr möglich (*187*). Mitsuhashi und Nomura (*114*) haben aus *Metaplexis japonica* Benzoylramanon (s. S. 108) isoliert. Bei der alkalischen Verseifung des Benzoesäureesters an $C_{(12)}$ entstanden Ramanon und Isoramanon, die mit Isodigipurpurogenin II bzw. Digipurpurogenin II identisch sind.

Tschesche und Mitarb. (*186*) haben durch Marker-Abbau von Rockogenin und 12-epi-Rockogenin und anschließende Reduktion der Δ^{16}-Doppelbindung 5α-Pregnan-3β,12β-diol-20-on und 5α-Pregnan-3β,12α-diol-20-on hergestellt. Beide konnten aus Digipurpurogenin I und II durch Abspaltung der OH-Gruppe an $C_{(14)}$ mit Thionylchlorid/Pyridin und katalytische Hydrierung der Doppelbindungen Δ^5 und Δ^{14} hergestellt werden. Damit wurde sowohl die Verknüpfung mit einem bekannten Steroidderivat erreicht als auch die sterische Stellung der Hydroxylgruppe an $C_{(12)}$ in beiden Verbindungen gesichert. Später haben Mitsuhashi und Nomura (*115*) Dihydro-isodigipurpurogenin II-diacetat ausgehend von Hecogenin hergestellt, wobei sie über das 14,16-Dien und über das 16-En-14,15-epoxid vorgingen. Dessen Hydrierung lieferte das gesuchte Dihydro-isodigipurpurogenin II-diacetat. Diese Reaktionsfolge sichert noch einmal die 14β-Stellung der Hydroxylgruppe.

d) Digiprogenin (4): Digiprogenin ist das Aglykon des Digipronins (*156*, *195*), der Zucker dieses Glykosids ist *D*-Digitalose. Digipronin (= „γ"-Digipronin) geht mit Alkali in „α"-Digipronin über, Bezeichnungen, die von Satoh (*153*, *154*, *156*) gewählt wurden, aber im Hinblick auf die nunmehr geklärten stereochemischen Verhältnisse in diesem Molekül nicht glücklich sind, denn γ ist in bezug auf die Stellung des H an $C_{(14)}$ in Wirklichkeit α, während α die β-Konfiguration an diesem C-Atom darstellt. Energischere Hydrolyse mit Säure entfernte außer dem Zucker auch die OH-Gruppe an $C_{(17)}$, die wie Shoppee und Lack (*163*) mittels Kernresonanz und optischer Rotationsdispersionsmessungen zeigen konnten, die β-Konfiguration einnimmt. Die Seitenkette ist daher im

Literaturverzeichnis: SS. 138—148.

Digiprogenin wie im Diginigenin und Digifologenin α-orientiert. Die neue Doppelbindung Δ^{16} war mit Zink und Essigsäure auf Grund ihrer Lage zwischen zwei Ketogruppen hydrierbar. Die Oppenauer-Oxydation lieferte nunmehr Δ^4-Pregnen-3,11,15,20-tetron, wobei die Seitenkette an $C_{(17)}$ nunmehr β-orientiert ist. Diese Verbindung konnte aus 11α-Hydroxyprogesteron partialsynthetisch gewonnen werden. Durch mikrobiologische Oxydation mit Hilfe von *Calonectria decora* ließ sich eine 15α-Hydroxylgruppe einführen. Oxydation mit Chromsäure ergab hieraus das gleiche Δ^4-ungesättigte Pregnen-3,11,15,20-tetron. Damit ist die Verknüpfung des Digiprogenins mit einem bekannten Steroid erreicht (*153*).

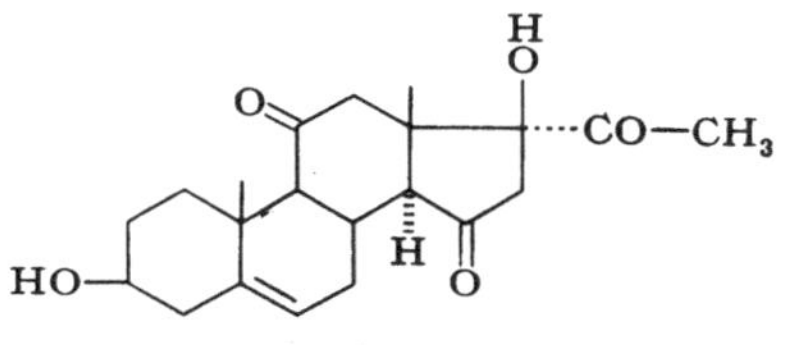

(4) Digiprogenin.

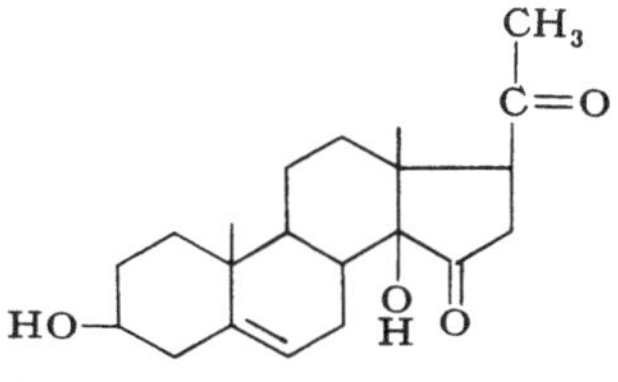

(5) Purpnigenin.

(6) Purprogenin = x-Ketopurpnigenin.

e) Purpnigenin (5): Dieses Aglykon entsteht bei der milden Säure-hydrolyse aus Purpnin unter Abspaltung von 3 Molen *D*-Digitoxose; es wurde erstmals von SATOH (*155*) isoliert. Die Konstitution des Aglykons konnte ISHII (*53–55*) in folgender Weise sichern: Aus $\Delta^{4,14}$-Pregnadien-3,20-dion wurde mit Perbenzoesäure das 14α,15α-Epoxid hergestellt, das mit Perchlorsäure unter Öffnung des O-Ringes die entsprechende 14β,15α-Dihydroxyverbindung lieferte. Oxydierte man Purpnigenin nach Oppenauer am $C_{(3)}$, so konnte die gleiche Verbindung erhalten werden. Purpnigenin ist daher vermutlich Δ^5-14β-Pregnen-3β,14,15α-triol-20-on.

f) Purprogenin (6): Das von SATOH (*155*) isolierte Glykosid Purpronin spaltete bei der vorsichtigen Säurehydrolyse in Purprogenin und 3 Mole *D*-Digitoxose. Purprogenin ergab bei der Wolff-Kishner-Reduktion ein Bis-desketoderivat, das auf die gleiche Weise auch aus Purpnigenin durch Entfernung nur einer Ketogruppe erhalten werden konnte. Purprogenin muß daher ein Ketopurpnigenin sein. Wegen der leichten Bildung eines Dioxims werden die Stellungen 1 und 12 für die zweite Ketogruppe im Purprogenin in Betracht gezogen.

g) Digacetigenin (7): Digacetinin ist ein Glykosid, das ebenfalls 3 Mole *D*-Digitoxose enthält, zusätzlich aber noch 2 Mole Essigsäure, von denen eines im Aglykon, das andere im Zuckerteil gebunden ist. Digacetinin wurde von TSCHESCHE und Mitarb. (*190*) erstmals aus einem Digitalismuster unbekannter Provenienz isoliert und später von REICHSTEIN (*150*) auch als Begleiter des Handelsdigitoxins aufgefunden. Die zunächst in Erwägung gezogene Konstitution (*191*) wurde durch SHOPPEE und LACK (*162*) auf Grund physikalischer Messungen geändert. Danach ist auch hier die Seitenkette an $C_{(17)}$ α-orientiert. Der Essigsäurerest dürfte an ein OH an $C_{(1)}$ gebunden sein. Die Verknüpfung dieses Genins mit einem bekannten Steroid konnte bisher nicht erreicht werden.

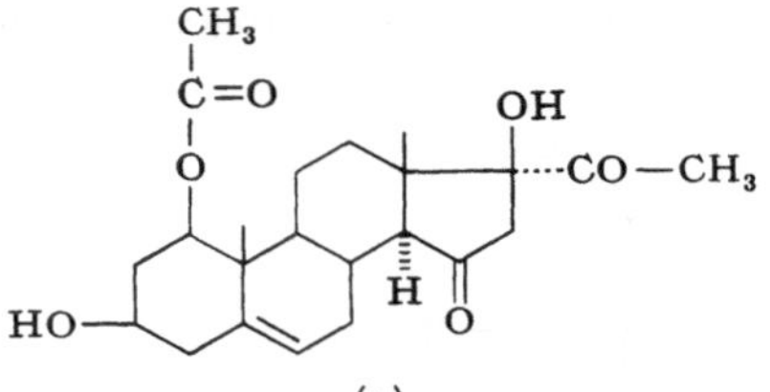

(7)
Digacetigenin.

III. N-freie Pregnanderivate der Asclepiadaceen.

In dieser Gruppe sind bisher fast ausschließlich nur die Aglykone der natürlich vorkommenden Glykoside (nicht diese selbst) bekannt. Die Ursache ist darin zu suchen, daß die Glykoside stets als Mischung nahe verwandter Verbindungen vorkommen, die nur sehr schwierig zu

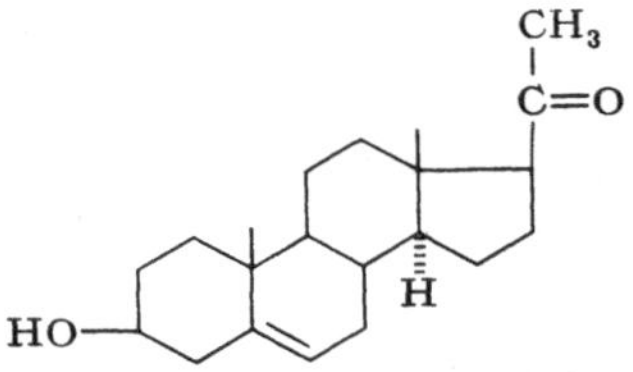

(8)
Δ^5-Pregnen-3β-ol-20-on.

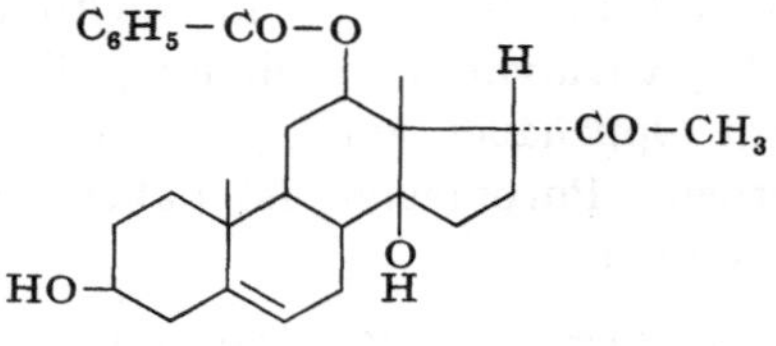

(9)
Benzoylramanon.

trennen sind. Vielfach kristallisieren auch die Aglykone schlecht, da in ihnen meist ein oder mehrere Säurereste gebunden sind, die dazu variieren können. Ferner erleiden die Esterglykoside bei saurem oder alkalischem pH verhältnismäßig leicht Umlagerungen oder Wasserabspaltungen im Aglykon, die ein weiteres Hindernis in der Reindarstellung bedeuten. *Tabelle 2* (S. 107) gibt eine Übersicht über die bisher isolierten natürlich vorkommenden Verbindungen.

Literaturverzeichnis: SS. 138—148.

Tabelle 2. N-freie Pregnanderivate des Asclepiadaceen.

Aglykone	Zusammensetzung	Vorkommen	Säuren	Ermittelte Zucker
Δ⁵-Pregnen-3β-ol-20-on (202)..	$C_{21}H_{32}O_2$	Xysmalobium undulatum	—	D-Glucose
5α-Pregnan-3β-ol-20-on (202).	$C_{21}H_{34}O_2$	Xysmalobium undulatum	—	D-Glucose
Benzoylramanon (114, 115) ..	$C_{28}H_{36}O_4$	Metaplexis japonica	Benzoesäure	D-Digitoxose und D-Cymarose
Pergularin (118, 114—116) ...	$C_{21}H_{32}O_5$	Cynanchum caudatum	—	D-Cymarose
Utendin (1)	$C_{21}H_{34}O_5$	Cynanchum caudatum	—	D-Cymarose
Tomentogenin (128)..........	$C_{21}H_{36}O_5$	Marsdenia tomentosa	—	D-Digitoxose und D-Glucose
Lineolon (1)	$C_{21}H_{32}O_5$	Pachycarpus lineolatus		D-Cymarose und D-Thevetose
Cynanchogenin (121-123,126)	$C_{28}H_{42}O_6$	Cynanchum caudatum	3.4 Dimethyl-2 pentensäure	D Cymarose
Sarcostin (30, 31, 127, 137) ...	$C_{21}H_{34}O_6$	Sarcostemma australe / Marsdenia tomentosa / Metaplexis japonica	Benzoesäure und Zimtsäure im Zuckerteil	D-Glucose
Penupogenin (124)	$C_{30}H_{40}O_7$	Cynanchum caudatum	Zimtsäure	D-Cymarose
Tayloron (57, 126).........	$C_{21}H_{32}O_6$	Gongromena taylorii	—	—
Metaplexigenin (114)	$C_{23}H_{34}O_7$	Metaplexis japonica / Cynanchum caudatum	Essigsäure	D-Digitoxose und D-Glucose
Drevogenin A (157, 211)	$C_{28}H_{42}O_7$	Dregea volubilis	Essigsäure, Isovaleriansäure	D-Cymarose
Drevogenin B.............	$C_{23}H_{34}O_6$	Dregea volubilis	Essigsäure	D-Cymarose
Drevogenin D	$C_{21}H_{34}O_5$	Dregea volubilis		D-Cymarose
Drevogenin P	$C_{21}H_{32}O_5$	Dregea volubilis		D Cymarose
Kondurangogenin A (204) ...	$C_{32}H_{42}O_8$	Marsdenia cundurango	Essigsäure und Zimtsäure	D-Glucose, D-Thevetose, D-Cymarose

Die Struktur der einzelnen Aglykone.

a) Δ^5-Pregnen-3β-ol-20-on (8): Dieses Aglykon wurde von Tschesche und Snatzke (*202*) aus den Wurzeln von *Xysmalobium undulatum* R. Br. nach Hydrolyse mit einer β-Glucosidase in sehr kleiner Menge neben der entsprechenden 5α H-Verbindung *Allopregnanolon* isoliert. (Wegen der in sehr großen Mengen vorhandenen Cardenolidglykoside war eine Isolierung dieses Glucosids selbst nicht aussichtsreich.) Damit wurde das erste Mal Pregnenolon im Pflanzenreich aufgefunden, dessen wichtige Rolle als Relaissubstanz bei der Biogenese der tierischen Steroidhormone gesichert ist.

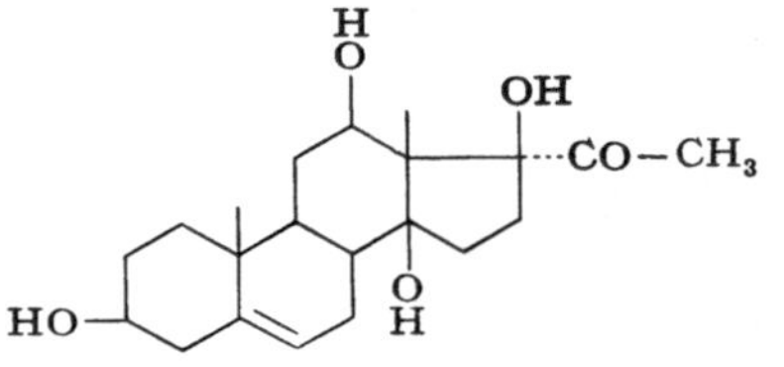

(10)
Pergularin.

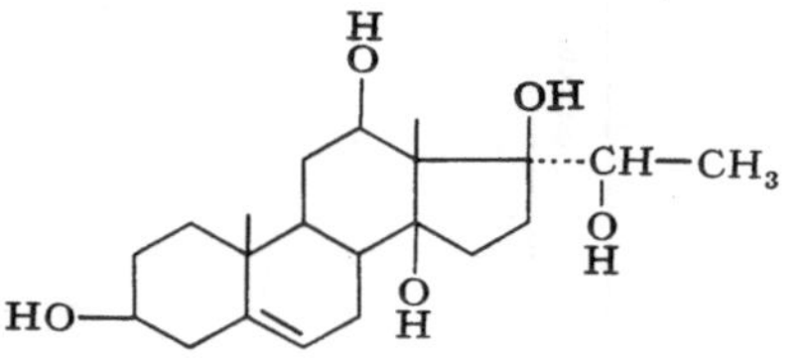

(11)
Utendin.

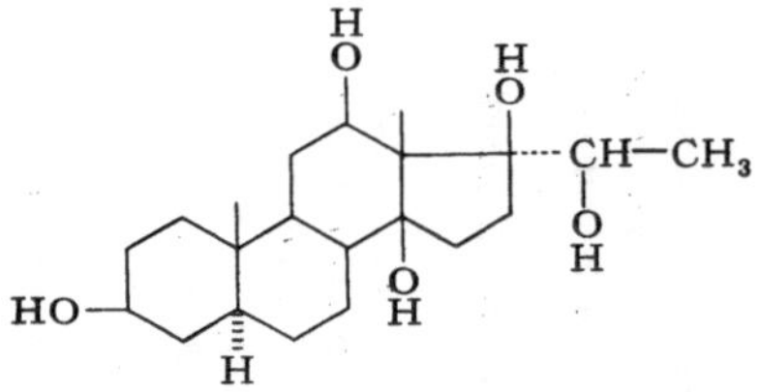

(12)
Tomentogenin.

b) Weitere Aglykone ohne OH an $C_{(8)}$. *Benzoylramanon* (9) wurde nach Säurehydrolyse von Mitsuhashi und Mitarb. (*114, 115*) aus den Wurzeln von *Metaplexis japonica* Makino isoliert. Durch Verseifung ging es in die Ramanone über, die mit dem Iso-, bzw. Digipurpurogenin II identisch sind (s. S. 104) (*115*). *Pergularin* (10) und *Utendin* (11) entstammen den Zweigen und Blättern der gleichen Pflanze und wurden ebenfalls nach Hydrolyse gewonnen (*114, 115, 118*). Durch Reduktion mit NaBH₄ lieferte Pergularin Utendin (*102, 116*), und dieses konnte durch katalytische Hydrierung der Δ^5-Doppelbindung in *Tomentogenin* (12) übergeführt werden. Dieses Aglykon entstammt nach Mitsuhashi und Mitarb. (*128*) *Marsdenia tomentosa* Decne. Tomentogenin (*120*) in Form des Diacetats ließ sich durch Serini-Reaktion (Erhitzen in Xylol mit aktiviertem Zink) in Diacetyl-dihydroramanon umwandeln (*119, 120*). Damit sind diese vier Aglykone untereinander in chemische Beziehung gebracht worden.

Literaturverzeichnis: SS. 138—148.

c) Die Aglykone mit einer Hydroxylgruppe an $C_{(8)}$ sind zwar untereinander chemisch, jedoch noch nicht mit einem bekannten Steroid verknüpft worden. Ihre Strukturaufklärung beruht vorwiegend auf physikalischen Messungen (*57, 126*). *Cynanchogenin* (13) ist das Aglykon eines
nicht isolierten Glykosides aus *Cynanchum caudatum* Max.; es enthält Δ^2-3,4-Dimethylpentensäure, die vermutlich an das Hydroxyl an $C_{(12)}$
esterartig gebunden ist. Cynanchogenin wurde ebenfalls vom Arbeitskreis MITSUHASHIS aufgefunden (*121, 123*). Die Verseifung führte u. a. zu
Lineolon (14), das daher ein Desacylcynanchogenin darstellt. Diese Ver-

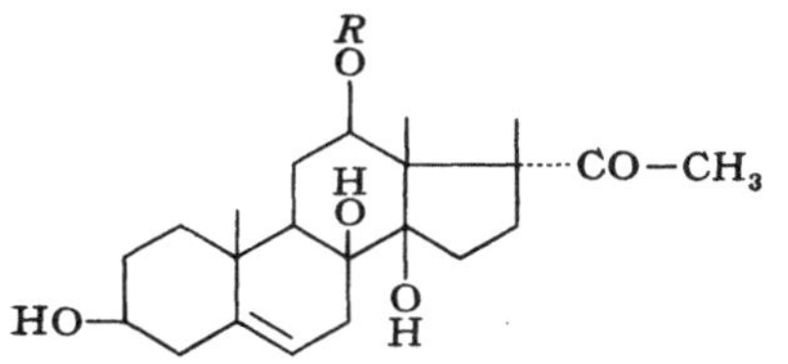

(13) Cynanchogenin:
$R = \Delta^2$-3,4-Dimethylpentenyl.
(14) Lineolon: $R =$ H.
(15) Substanz G: $R = C_6H_5$—CO.

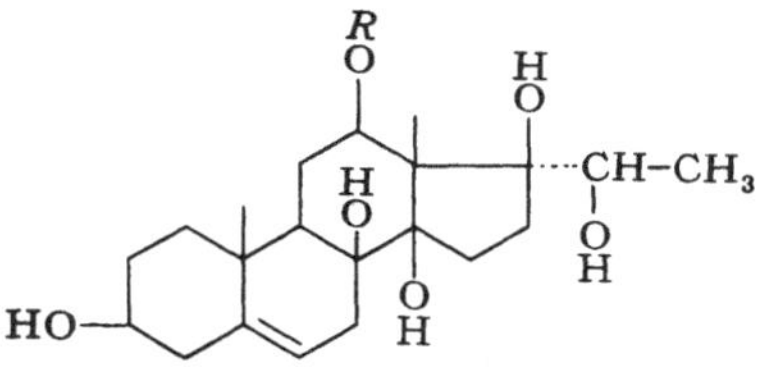

(16) Sarcostin: $R =$ H.
(17) Penupogenin: $R =$ Cinnamoyl.
(19) Dihydrosarcostin:
Doppelbindung Δ^5 hydriert (5αH).

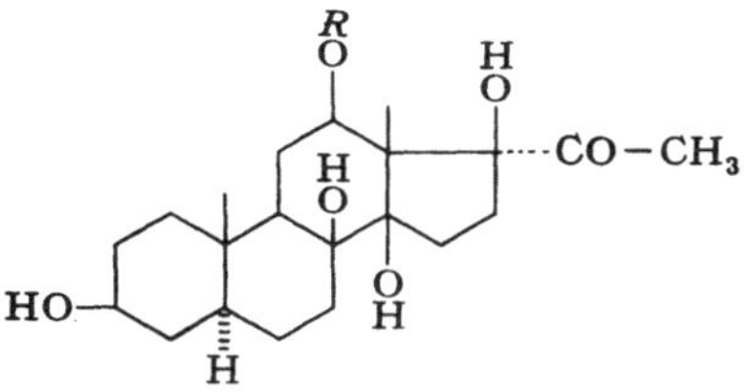

(18) Tayloron: $R =$ H.
(20) Metaplexigenin: $R = CH_3CO$ und Doppelbindung Δ^5.

bindung wurde von ABISCH, TAMM und REICHSTEIN (*1*) aus *Pachycarpus
lineolatus* (Decne) Bullock erhalten. Daneben fand sich eine als Substanz G (15) bezeichnete Verbindung (*1*), welche das 12-Benzoylderivat
des Lineolons ist.

Die am längsten bekannte Verbindung dieser Gruppe stellt *Sarcostin* (16) dar. Es wurde von CORNFORTH und EARL (*30*) aus *Sarcostemma
australe* R. Br. durch Hydrolyse gewonnen. Bei der Serinireaktion seines
Acetylderivates wurde nach Verseifung Lineolon erhalten, damit konnten
beide Verbindungen miteinander in Beziehung gesetzt werden (*122*).
Penupogenin (17) ist Cinnamoyl-sarcostin und entstammt ebenfalls
Cynanchum caudatum Max. (*124*). *Tayloron* (18) wurde von JAEGGI im Arbeitskreis von REICHSTEIN aus *Gongronema taylorii* (Schltr. et Rendle)
Bullock nach Hydrolyse erhalten; daneben fand sich *5αH-Dihydro-*

sarcostin (**19**) (*57*). In dieses ging Tayloron bei der Reduktion mit NaBH$_4$ über. *Metaplexigenin* (**20**) sowie sein Desacylderivat wurden von dem Arbeitskreis um Mitsuhashi (*114*) aus *Metaplexis japonica* Makino nach Hydrolyse erhalten (*117*); letzteres konnte auch aus *Cynanchum caudatum* Max. (*121, 123*) gewonnen werden. Es stellt 5-Dehydro-tayloron dar und konnte mit NaBH$_4$ in Sarcostin übergeführt werden. Nach Nascimento, Tamm, Jäger und Reichstein (*137*) sind die Wurzeln von *Asclepias glaucophylla* Schlechter ein besonders gutes Ausgangsmaterial für Sarcostin (1% der trockenen Wurzeln), das sich dort in geringer Menge auch in freier Form findet. Auch Lineolon neben einer unbekannten Verbindung U (verwandt mit diesem) konnte isoliert werden.

Für Cynanchogenin und Sarcostin waren zunächst Strukturvorschläge erwogen worden, bei denen der Ring *C* der Steroide zum Fünfring verengt und der Ring *D* zum Sechsring erweitert ist, wodurch ein System

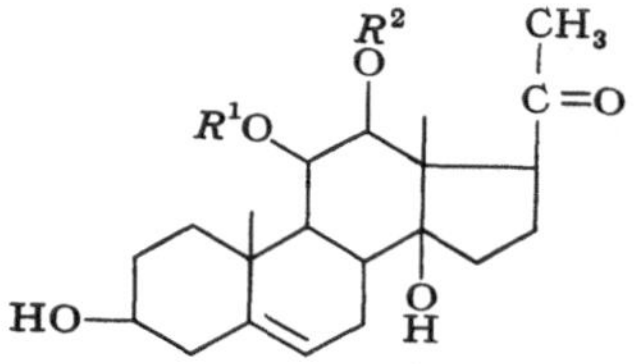

(**21**) Drevogenin P: $R^1 = R^2 = $ H.
(**22**) Drevogenin A: $R^1 = $ Isovaleryl, $R^2 = $ Acetyl.
(**23**) Drevogenin B: $R^1 = $ H, $R^2 = $ Acetyl.
(**24**) Drevogenin D = Dihydro-drevogenin P: OH an C$_{(20)}$.

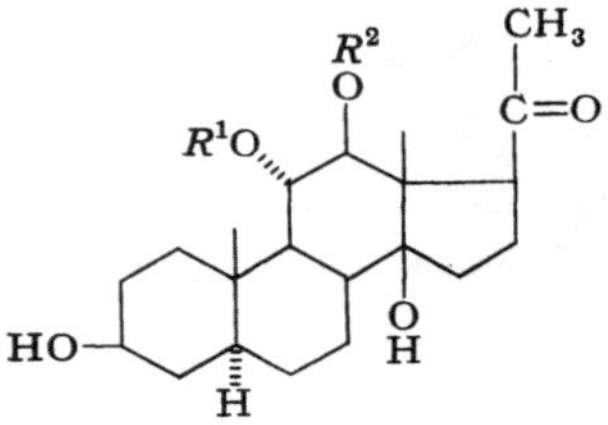

(**25**)
Kondurangogenin A: $R^1 = $ Cinnamoyl, $R^2 = $ Acetyl.

der vier Ringe analog dem entstand, wie es von den Alkaloiden Jervin und Veratramin angenommen wird. Bestimmend hierfür war die Bildung des Jacobsschen Kohlenwasserstoffes bei der Selendehydrierung (*29, 57, 125, 126, 136*). Hierbei tritt jedoch eine Umlagerung auf, wenn im Ring *C* an C$_{(12)}$ eine β-ständige OH-Gruppe vorhanden ist. Eine entsprechende Umlagerung ist unter anderen Bedingungen schon von Hirschmann, Snoddy, Hiskey und Wendler (*51*) an 12β-Mesyloxy-derivaten entdeckt worden.

d) Aglykone mit einer Glykolgruppierung an $C_{(11)}$ *und* $C_{(12)}$. Die *Drevogenine A, B, D und P* wurden von Reichstein und Mitarb. (*211*) aus *Dregea volubilis* (L.) Benth. nach Säurehydrolyse erhalten. Ihre Konstitution wurde vornehmlich mit modernen physikalischen Methoden ermittelt. Von diesen kann Drevogenin P (**21**) als der Grundkörper aufgefaßt werden. Alle enthalten u. a. zwei Hydroxylgruppen an C$_{(11)}$ (β) und C$_{(12)}$ (β). Ist im Drevogenin P das OH an C$_{(11)}$ mit Isovaleriansäure

Literaturverzeichnis: SS. 138—148.

und das OH an $C_{(12)}$ mit Essigsäure verestert, so liegt Drevogenin A vor (**22**). B (**23**) ist das Desisovaleryl-monoacetylderivat von P, während D (**24**) statt der Ketogruppe an $C_{(20)}$ eine Hydroxylfunktion trägt und keine esterartig gebundenen Säurereste enthält. In allen Drevogeninen ist die Seitenkette an $C_{(17)}$ β-orientiert, eine Behandlung mit Alkali führt bei den 20-Ketonen daher zu einem Gemisch der $C_{(17)}$-Isomeren (*157*).

Kondurangogenin A (**25**) wurde von TSCHESCHE, WELZEL und SNATZKE (*204*) nach vorsichtiger Säurehydrolyse aus den Rindenglykosiden von *Marsdenia cundurango* Rchb. fil. isoliert. Die Strukturbestimmung stützt sich auch hier vorwiegend auf physikalische Messungen. Kondurangogenin A ist mit Drevogenin P nahe verwandt und stellt bis auf die unterschiedliche Stellung der OH-Gruppe an $C_{(11)}$ dessen Dihydroderivat (5 α H) dar. An Stelle der Isovaleriansäure an $C_{(11)}$ ist Kondurangogenin A mit Zimtsäure verestert; mit dieser Anordnung der Säurereste sind die massenspektrometrischen Messungen am besten zu vereinen (*203*).

Vermutlich werden die nächsten Jahre weitere Aglykone vom Pregnantyp aus dieser Pflanzenfamilie erbringen.

IV. N-freie Pregnanderivate aus Apocynaceen.

Diese Gruppe umfaßt zur Zeit nur wenige Vertreter; die Ursache dürfte darin zu suchen sein, daß man auf sie als Vorstufen der zugehörigen Alkaloide erst kürzlich aufmerksam geworden ist und die in den Pflanzen vorkommenden Mengen gering sind. Sie liegen bisher alle in nicht glykosidierter Form vor.

a) Als besonders interessante Entdeckung kann die Auffindung von *Progesteron* (**26**) in *Holarrhena floribunda* gelten, das LEBOEUF, CAVÉ und GOUTAREL (*103*) damit zum ersten Mal auch im Pflanzenreich nachwiesen. Es kann als direkte Vorstufe zahlreicher Alkaloide des Pregnantyps angesehen werden.

b) Der erste Vertreter der Digitanole, der in der Familie der Apocynaceen aufgefunden wurde, war *Holadyson* (**27**), isoliert aus der Rinde von *Holarrhena antidysenterica* [TSCHESCHE, MÖRNER und SNATZKE (*197*)]. Nachdem die Struktur mit Hilfe moderner physikalischer Methoden ermittelt worden war, gelang seine Partialsynthese auf folgendem Weg: 11 α-Acetoxyprogesteron wurde in der Stellung 20 zur Hydroxyverbindung reduziert und anschließend die Ketogruppe an $C_{(3)}$ in das Äthylenketal verwandelt. Diese Verbindung wurde der „Hypojodit-Reaktion" nach WETTSTEIN und Mitarb. (*113*) unterworfen, und es wurde ohne Trennung der Reaktionsprodukte mit Chromsäure weiteroxydiert. Das in Stellung 18 eingeführte Jod konnte mit Silberacetat/Methanol gegen OH ausgetauscht werden. Die Schutzgruppe an $C_{(3)}$ ließ sich mit wäßriger Essigsäure wieder entfernen. Um die Doppelbindung Δ^1 einzuführen, erwies sich ein Schutz der OH-Gruppen an $C_{(11)}$ und $C_{(18)}$ durch Acetylierung notwendig. Nunmehr gelang es, die gewünschte Dehydrierung in Δ^1 mit 2,3-Dichlor-5,6-dicyan-benzochinon-1,4 durchzuführen. Nach Entacetylierung mit Alkali wurde Holadyson erhalten *(192)*.

CH₃

(26)
Progesteron.

(27)
Holadyson.

Als Beispiel für die Strukturermittlung mit modernen physikalischen Methoden seien die *Abb. 1—4* angeführt, die sich auf Holadyson beziehen. Für eine Beschreibung des NMR-Spektrums s. *(197)*.

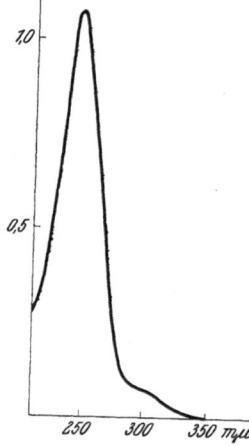

Abb. 1. Ultraviolettspektrum des Holadysons in Methanol.

c) Aus *Paravallaris microphylla* Pitard haben unter Benutzung von Girard-Reagenz Potier, Kan und Le Men *(145)* drei N-freie Pregnenderivate (Ketone) isolieren können, die in engster Beziehung zu den Alkaloiden des Paravallarin-Typs (mit Lactonring *E*) stehen. Es handelt sich um Δ^4-Pregnen-20 S-ol-3-on-18-carbonsäurelacton-($\rightarrow$ 20) (28), Δ^4-Pregnen-16α,20 S-diol-3-on-18-carbonsäurelacton-($\rightarrow$ 20) (29) und $\Delta^{4,6}$-Pregnadien-20 S-ol-3-on-18-carbonsäurelacton-($\rightarrow$ 20) (30). Die Verbindungen (28) und (29) wurden aus den Alkaloiden Paravallarin und Paravallaridin (s. S. 116) durch Ersatz der N-Methylaminogruppe an $C_{(3)}$ durch die Ketogruppe nach der Methode von Ruschig hergestellt. Hierbei wird ein primäres oder sekundäres Amin mit HOCl oder Chlorsuccinimid in das Chloramin übergeführt, das mit Natriumalkoholat das Ketimin ergibt. Dessen Hydrolyse führt zum entsprechenden Keton *(36, 152)*. Das Lacton (30) entsteht aus Paravallarin nach Ersatz der HNCH₃-Gruppe an $C_{(3)}$ durch OH beim Kochen mit Chloranil in siedendem Dioxan.

Die Pflanze soll noch weitere N-freie Steroidderivate dieses Typs enthalten.

Literaturverzeichnis: SS. 138—148.

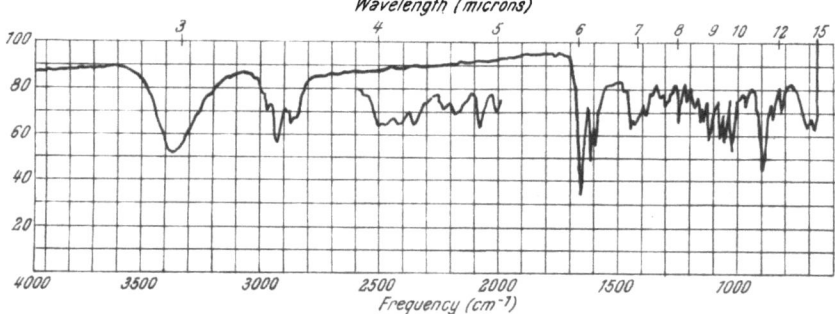

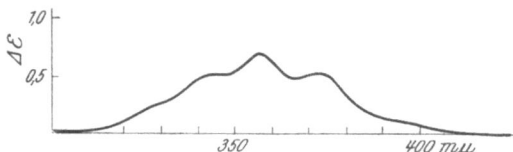

Abb. 2. Infrarotspektrum des Holadysons in KBr.

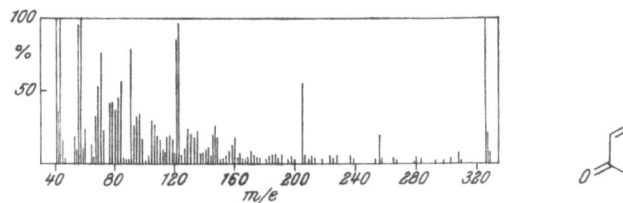

Abb. 3. Circulardichroismuskurve des Holadysons in Dioxan.

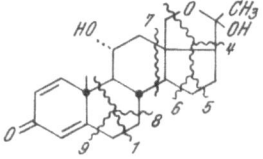

Abb. 4. Massenspektrum des Holadysons.

V. N-haltige Pregnanderivate aus Apocynaceen.

Die mehr als 80 bisher bekannten Alkaloide dieser Gruppe lassen sich am einfachsten in ihrer Bildung verstehen, wenn man Pregnen-3β-ol-20-on bzw. Progesteron als Muttersubstanz ansieht. Durch Aminierung in der 3-, bzw. 20-Stellung oder in 3 und in 20 entsteht eine Vielzahl von Alkaloiden, wobei noch die sterischen Verhältnisse an diesen C-Atomen variiert sein können. Durch Absättigung der Doppelbindung zu 5αH-Derivaten werden die Möglichkeiten weiter erhöht, zumal auch der Methylierungsgrad am N alle möglichen Formen bis zur quartären Ammoniumverbindung annehmen kann (Malouetin). Eine neue Variation wird durch Oxydation der CH_3-Gruppe an $C_{(18)}$ gewonnen. Dabei treten alle Oxydationsstufen auf. Durch Ringschluß der Carboxylgruppe $C_{(18)}$ mit dem Hydroxyl an $C_{(20)}$ entstehen die Alkaloide des Paravallarintyps. Bleibt die Oxydation auf der Aldehydstufe stehen, so kann diese bei Dimethylierung der Aminogruppe an $C_{(20)}$ erhalten bleiben (Malouphyllin) oder aber es kommt zum Ringschluß mit der NH_2-Gruppe unter Ausbildung einer Doppelbindung $\Delta^{18,N}$ (Conkurchin). Deren Hydrierung führt weiter zu den Alkaloiden vom Conarrhimintyp, dessen bekanntester Vertreter

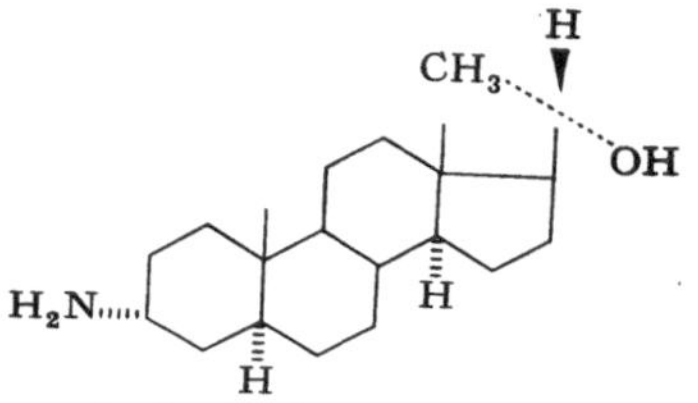

(31) Funtumidin.
(33) Funtumin = 20-Ketofuntumidin.

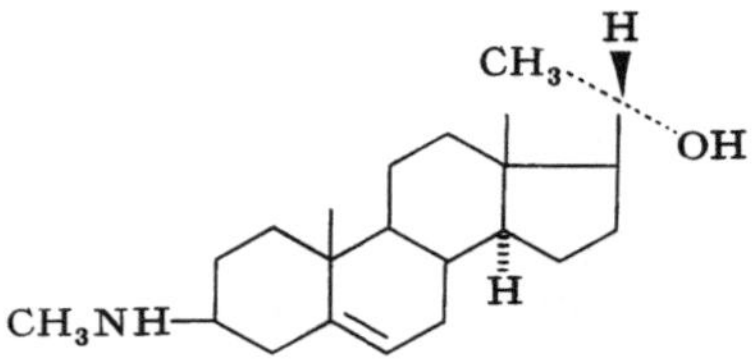

(32)
Holaphyllidin.

Conessin ist. Weitere Möglichkeiten der Variation sind durch Einführung von Hydroxylen gegeben, solche wurden bisher an $C_{(2)}$, $C_{(4)}$, $C_{(5)}$, $C_{(7)}$, $C_{(11)}$, $C_{(12)}$ und an $C_{(16)}$ beobachtet, wobei auch eine Oxydation sekundärer OH-Gruppen zum Keton eintreten kann. Wiederum sind von den Hydroxylderivaten verschiedene sterische Anordnungen möglich. Schließlich können Amino- oder Hydroxylgruppen durch unterschiedliche Säurereste substituiert werden. Damit kommen eine Fülle von Strukturen in Betracht, von denen in der Natur bisher nur ein Teil aufgefunden wurde.

a) An $C_{(3)}$ aminierte Derivate, OH an $C_{(20)}$. Funtumidin (31) wurde neben seinem 20-Ketoderivat Funtumin (33) von Janot, Khuong-Huu und Goutarel (62, 63) aus *Funtumia latifolia* isoliert. Seine Struktur ergibt sich aus folgenden Umsetzungen: Die Oxydation von Funtumidin mit Chromsäure ergibt Funtumin. Dieses wird mit Formaldehyd/Ameisensäure (Eschweiler-Clarke-Reaktion) ins Dimethylderivat übergeführt und anschließend nach Wolff-Kishner reduziert. Dabei entsteht 3α-Dimethylamino-5α-pregnan. Wird Funtumin nach dem Verfahren von Ruschig desaminiert, so kann 5α-Pregnan-3,20-dion erhalten werden. Wird Funtumin mit $NaBH_4$ oder mit katalytisch erregtem Wasserstoff reduziert, so entsteht Isofuntumidin, dagegen bildet sich mit Natrium in siedendem Äthanol vorwiegend Funtumidin. Daraus ergibt sich nach den

Literaturverzeichnis: SS. 138—148.

Erfahrungen bei derartigen Reduktionen für Funtumidin die 20 S-ol-Konfiguration, dieses ist also ein 3α-Amino-5α-pregnan-20 S-ol. Die entsprechende 3β-Methylamino-$\varDelta^5$-pregnen-20 S-ol-Verbindung ist *Holaphyllidin* (32), die von LEBOEUF, CAVÉ und GOUTAREL (*103*) aus *Holarrhena floribunda* erhalten werden konnte. Schließlich ist zu dieser Gruppe *Holadysamin* zu zählen, das aus den Blättern von *H. antidysenterica* stammt (*40, 72*). Die Struktur wurde vorwiegend mit physikalischen Methoden ermittelt; nach JANOT, LONGEVIALLE und GOUTAREL (*72*) soll es ein 3-Methylamino-pregnadien-20-ol sein, eine der Doppelbindungen dürfte $\varDelta^5$ sein. Die sterische Anordnung an $C_{(3)}$ und an $C_{(20)}$ ist nicht bekannt.

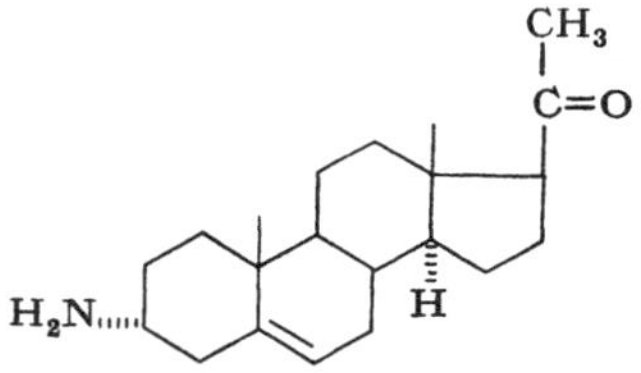

(34) Holamin.
(35) Holaphyllamin: NH_2 an $C_{(3)}$ (β).
(36) Holaphyllin: $HNCH_3$ an $C_{(3)}$ (β).

b) Von dem Typ mit *Ketofunktion an $C_{(20)}$ (an $C_{(3)}$ aminiert)* sind bisher vier Vertreter gefunden worden. Die Konstitution des *Funtumins* (33) aus *Funtimia latifolia* ergibt sich aus seiner Bildung bei der Oxydation von Funtumidin mit Chromsäure (S. 114). *Holamin* (34), *Holaphyllamin* (35) und *Holaphyllin* (36) erhielten JANOT, CAVÉ und GOUTAREL (*58, 59*) aus den Blättern von *Holarrhena floribunda*. Holamin geht bei der Hydrierung mit Pt/H_2 in Funtumin über, das ein 3α-Amino-5α-Derivat ist. Die nicht übliche Absättigung eines 3α-Aminoderivates mit H_2 zu einer 5α-Verbindung, wird von GOUTAREL (*42*) auf die freie NH_2-Gruppe zurückgeführt. Holamin wurde aus 3β-Tosyloxy-$\varDelta^5$-pregnen-20-on in folgender Weise synthetisiert: Der Austausch der Tosyloxygruppe durch den Azidrest in Dimethylsulfoxid führte erwartungsgemäß in die 3α-Reihe. Reduktion mit $LiAlH_4$ in Äther ergab nach anschließender Oxydation an $C_{(20)}$ Holamin. Der Austausch einer β-Tosyloxygruppe durch den Azidrest tritt bei gesättigten Steroiden unter sterischer Umkehr ein, bei $\varDelta^5$-Derivaten in hydroxylhaltigen Lösungsmitteln jedoch bleibt die Konfiguration erhalten (*42*). Bei der Desaminierung liefert Holamin Progesteron. Es sollte daher die Struktur eines 3α-Amino-$\varDelta^5$-pregnen-20-on haben. Die Reduktion seines Dimethylderivates nach Wolff-Kishner ergibt denn auch 3α-Dimethylamino-$\varDelta^5$-pregnen.

Holaphyllin (36) liefert bei der Reduktion nach Wolff-Kishner 3β-Methylamino-Δ^5-pregnen. Da ferner die Reduktion mit KBH_4 zum Holaphyllinol führt, das nach der Hydrierung und Methylierung 3β-Dimethylamino-5α-pregnan-20 S-ol ergibt, muß Holaphyllin 3β-Methylamino-Δ^5-pregnen-20-on sein. Holaphyllamin (35) ist das 3β-Isomere zum Funtumin (33), da es bei der Methylierung dasselbe Dimethylderivat ergibt, das auch aus Holaphyllin unter den gleichen Bedingungen entsteht.

c) An $C_{(3)}$ aminierte Derivate, mit weiteren O-Funktionen. Die drei Alkaloide *Kurchilin* (37), *Kurchiphyllin* (38) und *Kurchiphyllamin* (39) wurden von Janot, Longevialle und Goutarel (72) aus den Blättern von *Holarrhena antidysenterica* indischer Herkunft isoliert. Ihre Konsti-

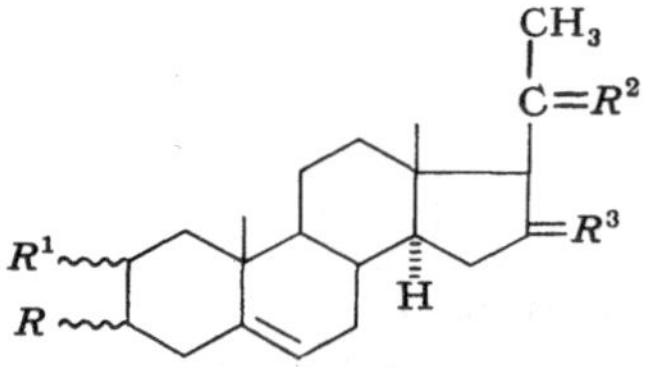

(37) Kurchilin: $R = N(CH_3)_2$ (α), $R^1 = OH$ (α), $R^2 = {>}O$, $R^3 = H_2$.
(38) Kurchiphyllin: $R = N(CH_3)_2$ (ξ), $R^1 = OH$ (ξ), $R^2 = H_2$, $R^3 = {>}O$.
(39) Kurchiphyllamin: $R = HN(CH_3)$ (ξ), $R^1 = OH$ (ξ) $R^2 = H_2$, $R^3 = {>}O$.
(40) Holadysin: $R = HN(CH_3)$ (ξ), $R^1 = H$, $R^2 = H_2$, $R^3 = {>}O$.

tution wurde bisher nur mit physikalischen Methoden bestimmt. *Holadysin* (40) entstammt den Blättern der gleichen Pflanze, gewachsen in Vietnam (*40, 42, 72*).

d) *Paravallarin-Typ* (Lactonring E); Δ^5. *Paravallarin* (41) und *Paravallaridin* (42) wurden von Le Men (8, 105, 106) in den Blättern, Stengeln und Wurzeln von *Paravallaris microphylla* aufgefunden. Paravallarin liefert bei der Desaminierung nach Ruschig ein α,β-ungesättigtes Keton. Dieselbe Reaktion am Dihydroderivat ausgeführt, erhalten durch katalytische Hydrierung in Eisessig, ergab ein gesättigtes Keton. Dieses war identisch mit 20 S-Hydroxy-3-keto-5α-pregnan-18-säure-lacton-($\rightarrow$20), das von Cainelli und Mitarb. (20) synthetisch gewonnen worden ist. Ebenso waren die entsprechenden 3β-Acetoxyderivate gleich. Da die katalytische Hydrierung der Δ^5-Doppelbindung im Paravallarin zur 5αH-Verbindung geführt hat, wird auf eine 3β-Orientierung der $HNCH_3$-Gruppe in der Stellung 3 geschlossen. Ram und Bhattacharyya (*148*) konnten Strukturbeziehungen zwischen Paravallarin und Holarrhimin (S. 122) herstellen (*4*, vgl. auch *90*). Paravallaridin enthält eine zusätzliche OH-Gruppe, die an $C_{(16)}$ α-ständig gebunden ist. Ihre Position ergibt

Literaturverzeichnis: SS. 138—148.

sich daraus, daß N-Methyl- bzw. N-Acetylparavallaridin zu einem Keton oxydierbar ist. Schützt man weiter diese $>$ CO-Gruppe durch Bildung eines Dioxolans, so kann nunmehr der Lactonring mit $LiAlH_4$ reduzierend geöffnet werden. Damit wird die bisher am Lactonring beteiligte OH-Gruppe an $C_{(20)}$ frei und kann nunmehr, nach Abspaltung der Schutzgruppe, mit Thionylchlorid zu einem α,β-ungesättigten Keton umgesetzt werden. Die zusätzliche OH-Gruppe des Paravallaridins muß daher die 16-Stellung einnehmen. Ihre sterische Orientierung ergibt sich u. a. daraus, daß die Reduktion der Ketogruppe $C_{(16)}$ mit $LiAlH_4$ in tert. Butylalkohol zum β-Isomeren führt. Auch die physikalischen Messungen sind mit der α-Stellung der OH-Gruppe an $C_{(16)}$ in Einklang.

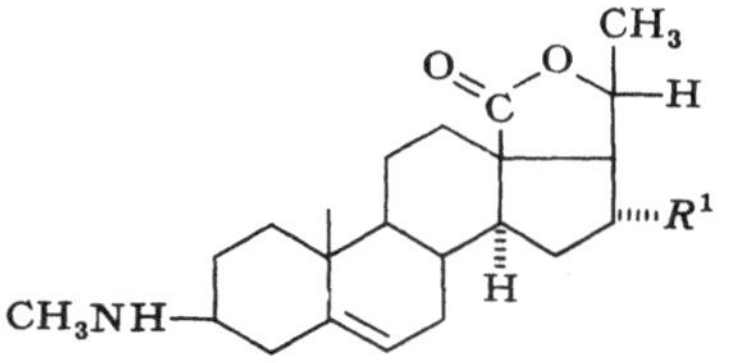

(41) Paravallarin: $R^1 = H.$
(42) Paravallaridin: $R^1 = OH\ (\alpha).$
(43) Kibatalin: $N(CH_3)_2$ an $C_{(3)}$ (α), $R^1 = H.$

Kibatalin (43) entstammt den Blättern und der Rinde von *Kibatalia gitingensis* Woods und wurde von CAVÉ, POTIER, CAVÉ und LE MEN (21) isoliert. Es ist das 3α-Isomere des Paravallarins, nur enthält es statt einer $HNCH_3$- eine $N(CH_3)_2$-Gruppe. Dihydroparavallarin konnte über das 3-Keton, sein Oxim und dessen Hydrierung in das 3α-Aminodihydro-paravallarin umgewandelt werden, dessen Methylierung ergab Dihydrokibatalin ($5\alpha H$). Auch über das 3β-Tosyloxyderivat, Austausch des Tosyloxyrestes durch die Azidgruppe (Inversion an $C_{(3)}$) und nachfolgende Hydrierung konnte Dihydrokibatalin erhalten werden. Die Lage der Doppelbindung Δ^5 ergibt sich aus dem Massenspektrum.

e) Zu den *allein in der Stellung 20 aminierten Derivaten* gehören solche mit 5α-Struktur wie mit Δ^5-Doppelbindung, vom erstgenannten Typ sind auch zwei 3-Ketoderivate als Pflanzeninhaltsstoffe bekannt: Die *Funtuphyllamine A, B und C* (44, 45 und 46) aus den Blättern von *Funtumia africana* (64, 67, 70) stellen das 20 S-Amino-, -Methylamino- und -Dimethylamino-derivat des 5α-Pregnan-3β-ols dar. Sie können aus 5α-Pregnan-3β-ol-20-on hergestellt werden, dadurch, daß man zunächst die Ketogruppe oximiert und dann mit Pt/H_2 reduziert. Dabei entsteht Funtuphyllamin A. Wird dieses formyliert und die Formylgruppe mit $LiAlH_4$ zu Methyl reduziert, so bildet sich Funtuphyllamin B. Die Methylierung mit Formaldehyd/Ameisensäure führt zu

Funtuphyllamin C. Die Art der Herstellung bestätigt zugleich die 20 S-Konfiguration. Oxydiert man Funtuphyllamin B oder C mit Chromsäure, so entsteht *Funtumafrin B* (47) bzw. *C* (48), die aus *F. africana (64)* stammen.

Den Funtuphyllaminen A, B und C entsprechen bei Vorliegen einer Δ^5-Doppelbindung die Alkaloide *Holafebrin* (49) (NH_2- an $C_{(20)}$), *Irehamin* (51) ($HNCH_3$- an $C_{(20)}$) und *Irehin* (52) [$N(CH_3)_2$- an $C_{(20)}$]. Holafebrin wurde aus *Holarrhena febrifuga (71)*, Irehamin und Irehin aus *F. latifolia* (75, 182) bzw. Irehamin auch aus *F. elastica* (65, 183) isoliert. Holafebrin geht bei der katalytischen Hydrierung in Funtuphyllamin A über, es muß daher 20 S-Amino-Δ^5-pregnen-3β-ol sein. Holafebrin findet sich als β-D-Glucosid *Conopharyngin* (50) in *Conopharyngia pachysiphon*

(44) Funtuphyllamin A: $R = NH_2$.
(45) Funtuphyllamin B: $R = HNCH_3$.
(46) Funtuphyllamin C: $R = N(CH_3)_2$
(47) Funtumafrin B: $R = HNCH_3$, an $C_{(3)} > O$.
(48) Funtumafrin C: $R = N(CH_3)_2$, an $C_{(3)} > O$.
(49) Holafebrin: $R = NH_2$, Δ^5.
(50) Conopharyngin = β-D-Glucosid von Holafebrin.
(51) Irehamin: $R = HNCH_3$, Δ^5.
(52) Irehin: $R = N(CH_3)_2$, Δ^5.
(53) Terminalin: $R = N(CH_3)_2$, OH an $C_{(3)}$ (β), OH an $C_{(4)}$ (α), 5 αH.

und wurde von Lucas und Mitarb. (33, 107) erhalten. Durch Umsetzung von N-Trifluoracetyl-holafebrin mit Acetobromglucose kann es auch synthetisch bereitet werden. Es ist bei parenteraler Anwendung durch blutdrucksenkende Eigenschaften ausgezeichnet. Dieses Glykosid stellt einen interessanten Übergang zu den N-freien Glykosiden der Digitanolgruppe dar. *Irehamin* geht durch Hydrierung in Funtuphyllamin B über, ferner konnte es auch aus N-Carbäthoxyholafebrin durch Reduktion mit LiAlH$_4$ erhalten werden. *Irehin* entsteht aus Holafebrin durch Methylierung nach der Methode von Eschweiler, es geht durch katalytische Hydrierung in Funtuphyllamin C über. Es muß daher ein N-Dimethyl-holafebrin sein. Irehin wurde auch in *Buxus sempervirens* aufgefunden und zunächst als Buxomegin bezeichnet (207). Irehin und Irehamin konnten auch aus Holafebrin durch Methylierung gewonnen werden (183).

Terminalin (53) wurde von Kikuchi und Mitarb. (85, 86) aus *Pachysandra terminalis* Sieb. et Zucc. isoliert; es ist ein 3,4-Diol-derivat. Da die Glykolgruppierung kein Acetonid lieferte, dürfte ein *trans*-Diol vorliegen. Durch Reduktion des entsprechenden Diosphenolderivates (erhalten durch Abbau aus Ergosterin) mit Natrium in siedendem Amylalkohol, konnte Terminalin partialsynthetisch hergestellt werden.

Literaturverzeichnis: SS. 138—148.

f) Unter den in den *Stellungen 3 und 20 aminierten Verbindungen* ohne weitere O-Funktionen in dem Molekül ist als das am längsten bekannte Alkaloid dieses Typs *Chonemorphin* (54) zu erwähnen, das DAS und PILLAY *(32)* aus *Chonemorpha macrophylla* G. Don. isoliert haben. Es findet sich auch in *Ch. fragrans* (Moon) und in *Ch. penangensis* Ridl. *(25, 26)*. Chonemorphin geht mit salpetriger Säure in Funtuphyllamin C (S. 117) über. Seine primäre NH_2-Gruppe muß daher an $C_{(3)}$ sitzen. Andererseits läßt sich Funtumafrin C (S. 118) über das Oxim und dessen Reduktion mit Natrium in Äthanol in Chonemorphin überführen. Diese Partialsynthese sichert gleichzeitig die 3β-Stellung der NH_2-Gruppe.

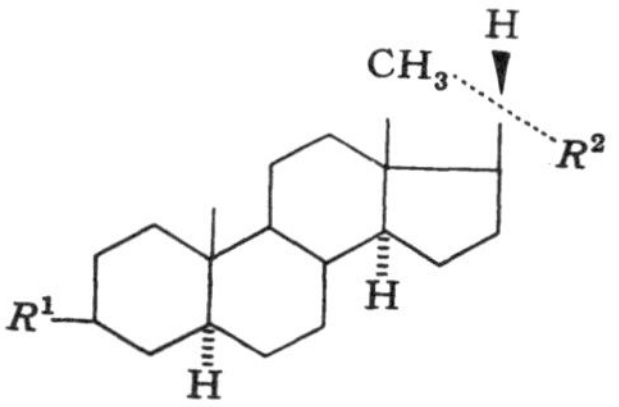

(54) Chonemorphin: $R^1 = NH_2$ (β), $R^2 = N(CH_3)_2$ (S).
(55) Dictyophlebin: $R^1 = HNCH_3$ (ξ), $R^2 = N(CH_3)_2$ (S).
(56) Malouetin: $R^1 = N^{\oplus}(CH_3)_3$ (β), $R^2 = N^{\oplus}(CH_3)_3$ (S).
(57) Epipachysamin C: $R^1 = HNCH_3$ (β), $R^2 = HNCH_3$ (S).
(58) Epipachysamin A: $R^1 = N(CH_3)_2$ (β), $R^2 = CH_3N—COCH_3$ (S).
(59) Epipachysamin B: $R^1 = HN$-nicotinyl, $R^2 = N(CH_3)_2$ (S).
(60) Pachysamin A: $R^1 = HN(CH_3)$ (α), $R^2 = N(CH_3)_2$ (S).
(61) N-Methylpachysamin A: $R^1 = N(CH_3)_2$, $R^2 = N(CH_3)_2$ (S).
(62) Pachysamin B: CH_3N-dimethylacryl (β), $R^2 = N(CH_3)_2$ (S).
(62 a) Epipachysamin D = 3-Benzoyl-chonemorphin.
(62 b) Epipachysamin F: $R^1 = N(CH_3)_2$ (β), $R^2 = NH_2$ (S).
(62 c) Epipachysamin E: β,β-Dimethylacryl-chonemorphin.

Dictyophlebin (55) wurde von MONSEUR und OP DE BEECK (vgl. *41*) aus *Dictyophleba lucida* in sehr kleinen Mengen gewonnen, seine Konstitution wurde mit physikalischen Methoden bestimmt, die Konfiguration an $C_{(3)}$ ist ungeklärt.

Malouetin (56) ist das bisquartäre Ammoniumderivat des Chonemorphins und wurde von JANOT, LAINÉ und GOUTAREL *(67)* aus den Blättern von *Malouetia bequaertiana* (E. Woodson) isoliert; daneben finden sich Funtuphyllamin B (S. 117) und Funtumafrin C (S. 118). Es entsteht aus Chonemorphin über das Bis-dimethylderivat und weitere Umsetzung mit Jodmethyl. Malouetin hat eine ausgesprochene Curare-Wirksamkeit. Die entsprechende Bis-monomethylverbindung wurde in Form des 20-Acetylderivates als natürlich vorkommendes Alkaloid in *Pachysandra terminalis* Sieb. et Zucc. von KIKUCHI und NISHINAGA *(85, 86,* vgl. *88)* aufgefunden und *Epipachysamin C* (**57**) genannt. Es wurde als Dictyodiamin auch aus *Dictyophleba lucida* (K. Sch.) Pierre erhalten. Dieses konnte aus dem in der gleichen Pflanze vorkommenden *Epipachysamin A* (**58**) durch Umsetzung mit BrCN nach v. Braun und Hydrolyse mit Alkali hergestellt werden. Die Acetylierung ergab danach dasselbe Diacetat, das auch aus dem Epipachysamin C erhalten werden konnte. Epipachysamin A dürfte daher die 3-Dimethylamino-verbindung sein. Es konnte ferner aus *Saracococa pruniformis* von Chatterjee und Mitarb. (*27*) isoliert werden, wobei zunächst die Bezeichnung Saracodin gewählt wurde. *Epipachysamin B* (**59**) ist das Nikotinsäureamidderivat des Chonemorphins und kann aus diesem mit Nikotinsäure nach üblichen Verfahren erhalten werden (*86*). Weiter sind kürzlich bekannt geworden (*87*): *Epipachysamin D*, das ein 3-Benzoyl-derivat des Chonemorphins ist (**62a**), *Epipachysamin F*, ein 20 S-Amino-3 β-dimethylamino-5 α-pregnan (**62b**) und *Epipachysamin E*, das β,β-Dimethylacrylsäure-derivat des Chonemorphins (**62c**).

In dem *Pachysamin A* (**60**), *N-Methyl-pachysamin A* (**61**) und *Pachysamin B* (**62**) liegt im Gegensatz zu den Epipachysaminen die 3 α-Form vor. Pachysamin A ist identisch befunden worden mit 3 α,20 S-Bisdimethylamino-5 α-pregnan; N-Methylpachysamin A wurde erst nach Hydrolyse aus Mutterlaugen gewonnen, es kann daher als Acylderivat in der Pflanze vorgelegen haben. Ein solches ist Pachysamin B; die in ihm gebundene Säure ist Dimethylacrylsäure (*181*).

g) An $C_{(3)}$ *und* $C_{(20)}$ *aminierte Derivate, 5 α und weitere O-Funktionen.* *Malouphyllin* (**63**) entstammt ebenfalls den Blättern von *Malouetia bequaertiana* (*41, 68*) und enthält u. a. eine Aldehyd- und eine Acetylgruppe. Bei der Reduktion nach Wolff-Kishner wird Chonemorphin (S. 119) erhalten. Damit ist die Struktur bis auf die Stellung der Oxogruppe und des Acetylrestes geklärt. Bei der Reduktion mit LiAlH$_4$ wird sowohl die Aldehydgruppe zu einem primären Hydroxyl reduziert als auch der Acetylrest in Äthyl umgewandelt. Wird diese Base mit Jodmethyl in die diquartäre Bis-ammoniumverbindung übergeführt und diese anschließend pyrolysiert, so entsteht Tetramethyldihydroholarrhimin (S. 123). Dieses hat eine primäre OH-Gruppe an $C_{(18)}$. Weiter bildet sich als Nebenprodukt eine Verbindung mit einem cyclischen Äther zwischen $C_{(18)}$ und $C_{(20)}$, wobei gleichzeitig eine Inversion am $C_{(20)}$ erfolgt sein muß. Die Isolierung von Chonemorphin zeigt ferner, daß der Acetylrest nur am N an $C_{(3)}$ gesessen haben kann und klärt damit die sterischen Verhältnisse an $C_{(3)}$ und $C_{(20)}$.

Die weiteren Basen dieses Typs mit zusätzlichen O-Funktionen entstammen alle *Pachysandra terminalis* (*180*). *Pachysandrin A* (**64**) enthält drei N-Methylgruppen und liefert bei der Behandlung mit starkem

Literaturverzeichnis: SS. 138—148.

Alkali ein Desacetyl-desbenzoyl-derivat. Einer der Säurereste muß daher als O-Ester vorgelegen haben. Die Chromsäureoxydation ergibt denn auch ein Ketoderivat. Seine Reduktion nach Wolff-Kishner entfernt nicht nur diese Ketogruppe, sondern auch gleichzeitig eine Methylaminogruppe, und es verbleibt ein 20 S-Dimethylamino-5 α-pregnan, das aus Bis-norallocholansäure bereitet werden konnte. Wird das Keton mit Alkali in Äthanol behandelt, so entsteht ein Diosphenol, das auch

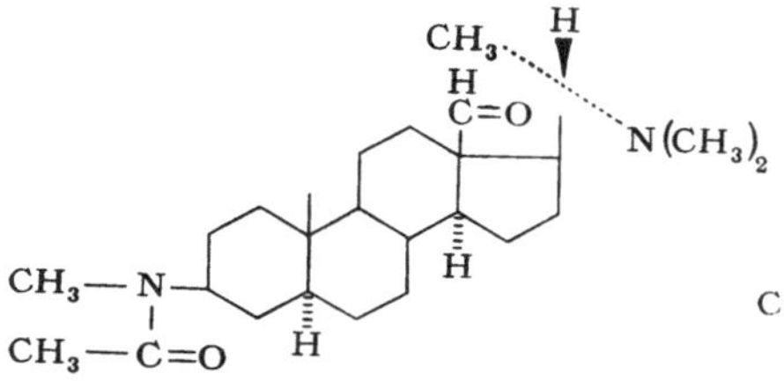

(63)
Malouphyllin.

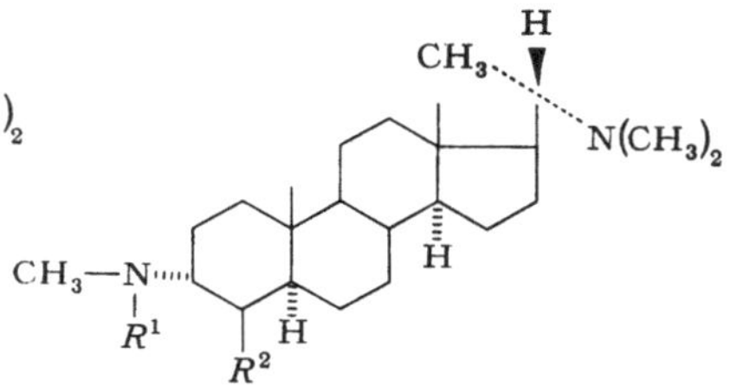

(64) Pachysandrin A: $R^1 = OCOC_6H_5$, $R^2 = OCOCH_3(\beta)$.
(65) Pachysandrin B: $R^1 = OCOCH=C(CH_3)_2$, $R^2 = OCOCH_3(\beta)$.
(66) Pachysandrin C: $R^1 = HNCH_3$ an $C_{(3)}(\alpha)$, $R^2 = OH(\alpha)$.
(67) Base XIII: $R^1 = CH_3$, $R^2 = OH(\beta)$.
(68) O-Desacetyl-pachysandrin B:
$R^1 = OCOCH=C(CH_3)_2$, $R^2 = OH(\beta)$.

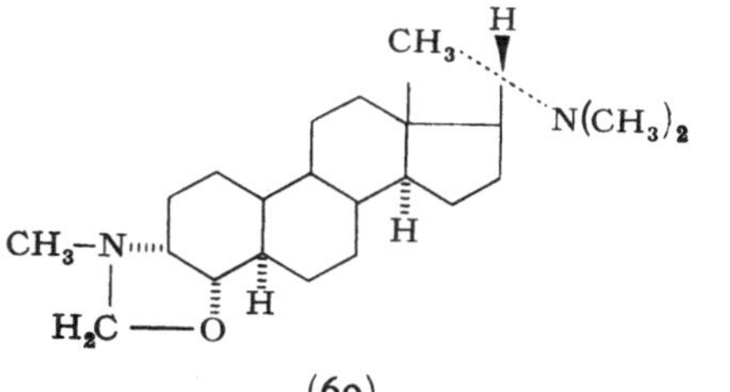

(69)
Base XI.

durch Luftoxydation von 20 S-Dimethylamino-5β-pregnan-3-on erhalten werden kann. Da der Acetylrest im Pachysandrin A schon mit verdünnterem Alkali abgespalten wird, sollte er an die an $C_{(4)}$ angenommene OH-Gruppe gebunden sein. Pachysandrin A dürfte daher 20 S-Dimethylamino-3-N,N-methyl-benzoylamino-4-acetoxy-5 α-pregnan sein. Die 3 α,4β-Konfiguration wurde vorwiegend mit modernen physikalischen Methoden gesichert.

Im *Pachysandrin B* (65) ist der Benzoylrest durch Dimethylacrylsäure ersetzt. Bei der alkalischen Hydrolyse wird nur der Essigsäurerest entfernt, erst mit konz. Salzsäure ließ sich auch die Dimethylacrylsäure abspalten. Hierbei tritt jedoch Inversion an $C_{(4)}$ zum 4-Epi-alkohol ein. *Pachysandrin C* (66), ein Desacylpachysandrin A, jedoch mit 3 α-Stellung der HNCH$_3$-Gruppe, wurde erst nach alkalischer Hydrolyse aus dem Pflanzenextrakt erhalten, so daß nicht sicher ist, ob es genuin vorkommt. *Pachysandrin D* ist der O-Dimethylacrylsäureester von Pachysandrin C, wie sich aus seiner leichten Hydrolysierbarkeit ergibt. *Base XIII* (67) wurde als O,N-Desacyl-N-methyl-pachysandrin A erkannt; es stellt vielleicht ein Hydrolyseprodukt dar, ebenso wie *O-Desacyl-pachysandrin B* (68). *Base XI* (69) wurde ebenfalls nach alkalischer Hydrolyse des Pflanzenextraktes gewonnen; sie entsteht mit Formaldehyd/Ameisensäure aus der entacylierten 3 α,4 α-Verbindung, während das 3 β,4 α-Derivat in normaler Reaktion die 3-Dimethylaminoverbindung liefert (*86*).

h) Diese Gruppe umfaßt *Δ⁵-Pregnenderivate, aminiert an C$_{(3)}$ und C$_{(20)}$,* ohne weitere O-Funktionen in dem Molekül: *Irehdiamin A* (70) wurde zusammen mit *Irehdiamin B* (71) aus den Blättern von *Funtumia elastica* von Truong-Ho, Khuong-Huu und Goutarel (*182*) isoliert. A enthält zwei primäre Aminogruppen und geht bei der Desaminierung nach Ruschig in Progesteron über. Bei der Methylierung nach Eschweiler entsteht das bekannte 3 β,20 S-Bis-dimethylamino-Δ⁵-pregnen, und dessen katalytische Hydrierung führt zu 3 β,20 S-Bis-dimethylamino-5 α-pregnan. Irehdiamin A muß daher 3 β,20 S-Diamino-Δ⁵-pregnen sein.

Irehdiamin B enthält eine Methylaminogruppe in der 3-Stellung; seine Struktur ergibt sich aus der Partialsynthese seines Dihydroderivates aus Holaphyllin (S. 115) über das Oxim und dessen katalytische Hydrierung. Irehdiamin B wurde auch aus der Rinde von *Holarrhena antidysenterica* isoliert (*205*) und erhielt zunächst den Namen Kurchamin, der nunmehr entfallen kann.

Kurchessin (72) wurde von Tschesche und Otto (*200*) aus der Rinde von *H. antidysenterica* erstmals isoliert und von Chatterjee und Mitarb. (*27*) auch in *Saracococa pruniformis* aufgefunden. Lábler und Šorm (*100*) zeigten, daß es die Struktur des 3 β,20 S-Bis-dimethylamino-Δ⁵-pregnens hat. *Saracoccin* (73) geht in Saracodin (S. 120) durch katalytische Hydrierung über; die Massenspektroskopie zeigte, daß es die zugehörige Δ⁵-ungesättigte Verbindung ist. *3 α,20 S-Bis-dimethylamino-Δ⁵-pregnen* (74) wurde nach Methylierung ebenfalls aus Kurchirinde erhalten; es selbst oder eine seiner methylärmeren Vorstufen kommt daher ebenfalls in der Natur vor (*200*). Seine Struktur wurde von Lábler und Mitarb. (*100*) geklärt.

i) In dieser Gruppe werden die *an C$_{(18)}$ hydroxylierten Derivate des Bis-diamino-Δ⁵-pregnens* behandelt: *Holarrhimin* (75) entstammt der Rinde von *Holarrhena antidysenterica* und wurde erstmals von Siddiqui und Pillay (*171, 172*) isoliert. Seine Struktur ergibt sich aus folgenden

Untersuchungen: Holarrhimin läßt sich katalytisch mit Pt/H_2 zu einem Dihydroderivat hydrieren, das nach Eschweiler zu einer Bis-dimethyl-amino-verbindung umgesetzt werden kann. Durch Oxydation wird nunmehr die primäre OH-Gruppe über den Aldehyd zu einer Säure oxydiert. Die Reduktion des Aldehyds nach Wolff-Kishner gibt $3\beta,20$ S-Bis-dimethylamino-5α-pregnan. Damit sind Stellung und Konfiguration der basischen Gruppen geklärt. Die gleiche Verbindung haben Šorm und Mitarb. (23) auch von der 3β-Acetoxy-bisnorallocholan-säure her bereitet. Mit der Methode von Curtius wird deren Carboxyl-

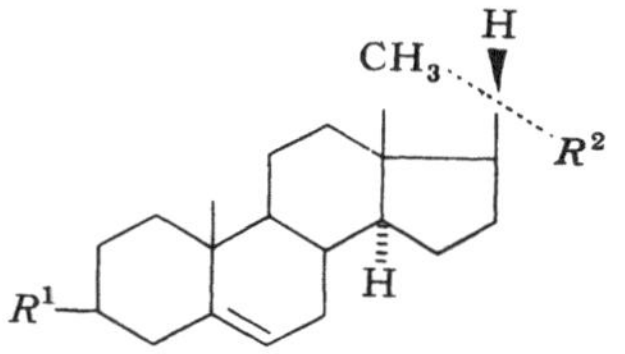

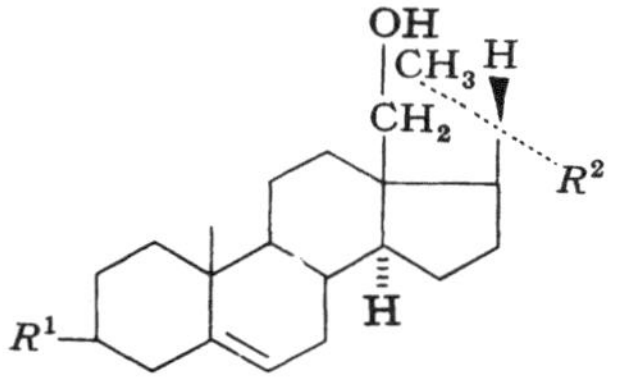

(70) Irehdiamin A: $R^1 = NH_2$, $R^2 = NH_2$.

(71) Irehdiamin B: $R^1 = HNCH_3$, $R^2 = NH_2$.

(72) Kurchessin: $R^1 = N(CH_3)_2$, $R^2 = N(CH_3)_2$.

(73) Saracoccin: $R^1 = N(CH_3)_2$, $R^2 = CH_3N—COCH_3$.

(74) $3\alpha,20$ S-Bis-dimethylamino-Δ^5-pregnen.

(75) Holarrhimin: $R^1 = NH_2$, $R^2 = NH_2$.

(76) Methylholarrhimin I: $R^1 = NH_2$, $R^2 = HNCH_3$.

(77) Methylholarrhimin II: $R^1 = HNCH_3$, $R^2 = NH_2$.

(78) Tetramethyl-holarrhimin: $R^1 = N(CH_3)_2$, $R^2 = N(CH_3)_2$.

(79) Holarrhidin: $R^1 = NH_2(\alpha)$, $R^2 = NH_2$.

gruppe ins primäre Amin verwandelt und dieses nach Eschweiler di-methyliert. Nach Verseifung der Acetoxygruppe in der 3-Stellung konnte zum Keton oxydiert werden, das über das Oxim zum primären 3β-Amin reduzierbar ist. Die weitere Methylierung lieferte das gleiche $3\beta,20$ S-Bis-dimethylamino-5α-pregnan, das auch aus Holarrhimin erhalten werden konnte.

Die Stellung der Doppelbindung in Δ^5 stützt sich auf die Bildung eines $\Delta^{3,5}$-Diens beim Hofmannschen Abbau des Tetramethylholarrhimins. Daneben entsteht noch eine Verbindung, bei der auch die zweite Di-methylaminogruppe entfernt worden ist und es zur Ausbildung eines Sauerstoffringes zwischen $C_{(18)}$ und $C_{(20)}$ gekommen ist. Das ließ er-warten, daß vom Tetramethyl-dihydro-holarrhimin ein Übergang in die Reihe des Conessins möglich sein müßte. Diesen Ringschluß haben Lábler, Černý und Šorm (24, 92, 93) folgendermaßen erreicht: Dihydro-tetramethyl-holarrhimin wurde mit Tosylchlorid in Monometho-dihydro-conessin-p-toluolsulfonat verwandelt, das mit NaJ und CH_3J Dihydro-conessin-dijodmethylat ergab. Bei der Eschweiler-Methylierung ent-stehen außer dem Tetramethylholarrhimin stets auch kleine Mengen Conessin (205).

Holarrhimin und ebenso Paravallarin stellen ein geeignetes Ausgangs-material für die Herstellung von in Stellung 18 hydroxylierten N-freien Pregnanderivaten dar. Derartige Arbeiten wurden vor allem von Šorm und Mitarb. (90, 95—99). ausgeführt.

Tschesche und Wiensz (205) isolierten aus der Rinde von *Holarrhena antidysenterica* zwei Monomethylholarrhimine, I und II. Die Arbeits-gruppe um Lábler (80) konnte später zeigen, daß *Monomethyl-holarrhimin I* (76) die 20-Methylamino-verbindung darstellt, während *Monomethylholarrhimin II* (77) das entsprechende Methylamino-derivat in Stellung 3 ist. *Tetramethyl-holarrhimin* (78) ist ebenfalls als natürlich vorkommendes Alkaloid bekannt (205) und aus Holarrhimin durch Methylierung nach Eschweiler erhältlich.

Holarrhidin (79) ist von Lábler und Černý (91) aus *H. antidysenterica* isoliert worden, es ist das 3α-Isomere des Holarr-himins. Da Holarrhidin und Holarrhimin beim Hofmann-Abbau das gleiche $\Delta^{3,5}$-Dien liefern, kann der Unterschied beider Alkaloide nur in der Konfiguration an $C_{(3)}$ beruhen. Durch den Vergleich der mole-kularen Drehungsdifferenzen ließ sich für Holarrhidin die 3α-Struktur nachweisen.

Schließlich gelang es Lábler, Hora und Černý (94) auch, den chemi-schen Beweis für die 3α-Stellung der NH_2-Gruppe zu führen. Zunächst wurde Tetramethyl-holarrhimin mit Tosylchlorid unter Ringschluß zum Ring E in das quartäre Tosylat verwandelt und durch Erhitzen in Äthanolamin in 3α-Dimethylamino-Δ^5-conenin übergeführt. Die gleiche Verbindung konnte auch aus Conessin erhalten werden. Dieses wurde in 3β-Tosyloxy-N-desmethyl-N-cyan-Δ^5-conenin umgewandelt und durch Erhitzen mit Natriumazid (in Dimethylsulfoxid) umgesetzt. Die rohe Mischung der Azide ergab nach Reduktion mit $LiAlH_4$, Behandlung mit Alkali und Methylierung nach Eschweiler das gleiche 3α-Dimethyl-amino-Δ^5-conenin (neben etwas Conessin und 3β-Hydroxy-Δ^5-conenin), das schon aus Holarrhimin erhalten worden war. Diese Reaktion ist in Übereinstimmung mit der Feststellung, daß äquatoriale Tosyloxy-ester mit NaN_3 in polaren protonenfreien Lösungsmitteln vorwiegend axiale Azide ergeben.

j) Diese Gruppe von Alkaloiden hat einen *angegliederten N-haltigen Ring E*, der durch Ringschluß zwischen $C_{(18)}$ und dem N an $C_{(20)}$ gebildet worden ist. Eine Aminofunktion an $C_{(3)}$ fehlt.

Latifolin (80) ist in kleiner Menge in der Rinde von *Funtumia latifolia* enthalten und erstmals von Janot, Khuong-Huu und Goutarel (65) isoliert worden. Beim Hofmann-Abbau verbleibt die basische Gruppe im Molekül, und es entsteht eine Doppelbindung in der Seitenkette. Dieses Verhalten entspricht demjenigen, das bei Conanin-Derivaten üblich ist.

Literaturverzeichnis: SS. 138—148.

Die Struktur des Latifolins wurde durch Synthese bewiesen. Δ^4-Conenin--3-on, von JOHNSON (*78*) bei der Partialsynthese von Conessin erhalten, ließ sich über das Enol-acetat durch Reduktion mit $NaBH_4$ in Latifolin überführen. Weitere Synthesen sind von PAPPO (*139, 140*) und von NAGATA und Mitarb. (*130*) entwickelt worden.

Norlatifolin (81) entstammt ebenfalls *F. latifolia*; bei der Methylierung nach Eschweiler geht es in Latifolin über, es fehlt ihm die CH_3-Gruppe am N. *Latifolinin* (82), aus der gleichen Pflanze (*83*), ist das Δ^4-3-Ketoderivat

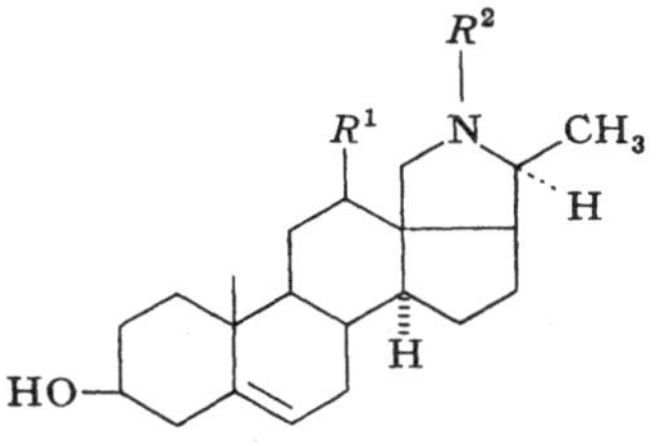

(80) Latifolin: $R^1 = H, R^2 = CH_3.$
(81) Norlatifolin: R^1 und $R^2 = H.$
(83) Funtulin: $R^1 = OH, R^2 = CH_3.$

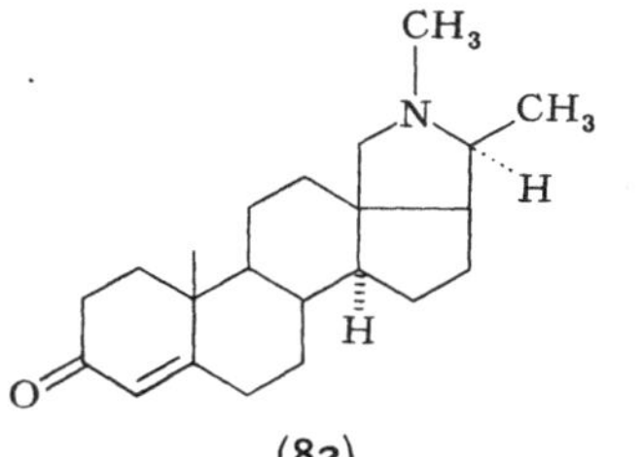

(82)
Latifolinin.

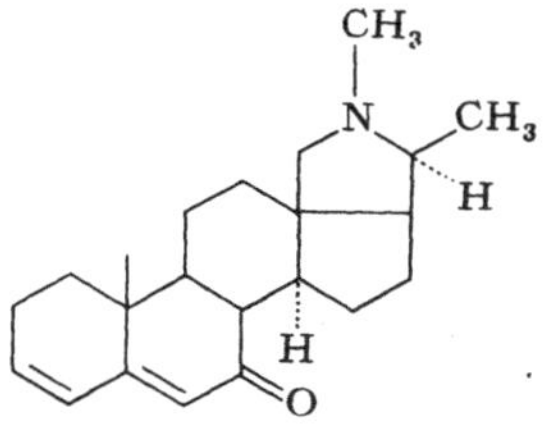

(84)
Funtudienin.

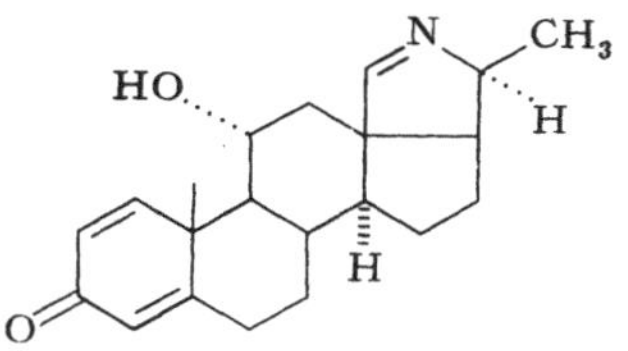

(85)
Holonamin.

zum Latifolin und als Zwischenstufe von JOHNSON bei der Synthese des Conessins erhalten worden. *Funtulin* (83), ein weiterer natürlicher Begleiter dieser Alkaloide, ist das 12β-Hydroxy-latifolin. Die Struktur wurde vornehmlich mit physikalischen Methoden bestimmt, es konnte aber auch vom Holarrhenin (S. 130) partialsynthetisch durch Ersatz von dessen Dimethylaminogruppe am C-Atom 3 durch OH erhalten werden. Hierbei wurde zunächst die Hydroxylgruppe an $C_{(12)}$ acetyliert und dann mit BrCN nach v. Braun eine der CH_3-Gruppen der Dimethylaminogruppe des Holarrhenins entfernt. Die Desaminierung nach Ruschig lieferte weiter die entsprechende Ketoverbindung, die über das Enolacetat mit $NaBH_4$ in das Diacetat des Funtulins übergeführt werden konnte. Anschließende Verseifung lieferte Funtulin (66).

Funtudienin (84), ebenfalls aus *F. latifolia*, ist $\Delta^{3,5}$-Conadienin-7-on. Es ließ sich durch Partialsynthese aus Latifolin gewinnen. Ausgehend von dessen Acetat, wurde mit Chromsäure in Stellung 7 eine Ketogruppe eingeführt, die Abspaltung der Acetoxygruppe an $C_{(3)}$ ließ sich mit *p*-Toluolsulfonsäure in Benzol durchführen. Das erhaltene Dienon erwies sich mit Funtudienin als identisch (*82*).

Holonamin (85) wurde von Tschesche und Ockenfels (*198, 199*) aus *Holarrhena antidysenterica*-Rinde isoliert. Seine Struktur wurde vornehmlich mit physikalischen Methoden bestimmt. Es enthält zwei Doppelbindungen zusätzlich zu der in Δ^4. Eine davon gehört einer Dienonstruktur an, während die zweite im Fünfring *E* wie im Irehlin (S. 132) angeordnet ist. Holonamin gibt daher auch alle Reaktionen, die von diesem Alkaloid beschrieben worden sind. Ferner ist ein Hydroxyl an $C_{(11)}$ (α) vorhanden. Holonamin steht im Aufbau in naher Beziehung zu dem N-freien Holadyson (S. 111), das vielleicht als eine Vorstufe anzusehen ist.

k) Diese Gruppe umfaßt die *5 α-Conanin-derivate mit einer Aminofunktion in der 3-Stellung: Malouphyllamin* (86) wurde ebenfalls von der französischen Arbeitsgruppe um Goutarel (*69*) isoliert, und zwar aus *Malouetia bequaertiana*. Seine Struktur ergibt sich daraus, daß es mit $LiAlH_4$ zu einem sekundären Amin reduzierbar ist ($—CO—CH_3$ in C_2H_5), welches bei der Desaminierung nach Ruschig ein Keton ergibt, das auch durch Desaminierung von Dihydro-isoconessimin (S. 120) erhalten werden kann. Hierbei wurde nach der Methode v. Brauns mit BrCN zunächst eine Methylgruppe entfernt und anschließend nach Ruschig desaminiert. Dieses Keton muß daher 5 α-Conanin-3-on sein. Es kann über das Oxim und dessen Hydrierung zum primären Amin durch anschließende Acetylierung in Malouphyllamin zurückverwandelt werden. Malouphyllamin konnte auch mit dem in der gleichen Pflanze vorkommenden Malouphyllin (S. 120) verknüpft werden. Letzteres wurde mit KBH_4 zu Malouphyllinol reduziert und die entstandene primäre OH-Gruppe in den *p*-Toluolsulfosäureester verwandelt. Auf einem Weg analog der Überführung von Dihydro-holarrhimin in Dihydroconessin entstand mit NaJ das Jodmethylat des Malouphyllamins. Diese Reaktionen sichern zugleich die 3β-Position der acylierten NH_2-Gruppe an $C_{(3)}$.

Dihydro-conessimin (87) und *Dihydro-conessin* (88) wurden aus der Rinde von *Holarrhena antidysenterica* isoliert (*22, 200*); sie sind auch durch katalytische Hydrierung von Conessimin und Conessin zugänglich. Desgleichen wurde auch *Dihydro-conkuressin* (89) aufgefunden, das ebenfalls durch Anlagerung von H_2 aus Conkuressin erhältlich ist (*100*). Es ist das 3α-Epimere des Dihydroconessins.

Funtessin (90) wurde von der Arbeitsgruppe um Goutarel (*84*) aus der Rinde von *Funtumia latifolia* erhalten. Seine Konstitution wurde

Literaturverzeichnis: SS. 138—148.

sowohl durch Anwendung von magnetischer Kernresonanz und Massenspektroskopie als auch auf chemischem Weg geklärt. Es gelang, Holarrhenin (S. 131) mit v. Brauns Methode zu entmethylieren und dann nach Ruschig zu desaminieren. Das erhaltene Keton lieferte über das Oxim und dessen Reduktion mit Natrium in Äthanol Funtessin.

l) Dieser Abschnitt behandelt die *Δ^5-Conenin-derivate mit einer zweiten basischen Gruppe.* Von dem 3β-aminierten Typ sind alle nach dem Methylierungsgrad möglichen Vertreter als natürlich vorkommende Alkaloide bekannt. Die am längsten bekannte und wichtigste Base ist Conessin. Weiter gehören hierzu einige 3α-aminierte Derivate sowie solche, die zusätzliche Hydroxylgruppen, zum Teil verestert, enthalten.

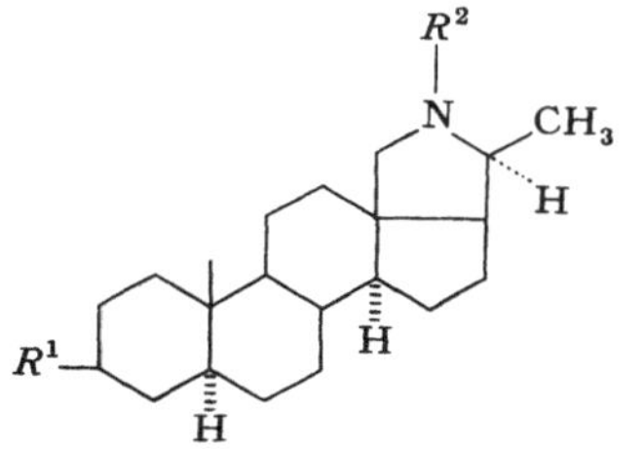

(86) Malouphyllamin: $R^1 = HN-COCH_3$, $R^2 = CH_3$.
(87) Dihydro-conessimin: $R^1 = N(CH_3)_2$, $R^2 = H$.
(88) Dihydro-conessin: $R^1 = N(CH_3)_2$, $R^2 = CH_3$.
(89) Dihydro-conkuressin: $R^1 = N(CH_3)_2(\alpha)$, $R^2 = CH_3$.
(90) Funtessin: $3\beta,12\beta$-Dihydroxy-5 α-conanin.

Conarrhimin (91, S. 131) wurde zuerst von SIDDIQUI (*169*) in nicht ganz reinem Zustand aus der Rinde von *H. antidysenterica* erhalten. BERTHO (*6*) hat später Conkurchin (S. 132) hydriert und so dessen Dihydroderivat hergestellt, das mit Conarrhimin identisch sein muß. Schließlich wurde es aus N-Acetyl-conamin (S. 131) durch Umsetzung mit BrCN bereitet. Dabei entsteht am Ringstickstoff das Cyanamidderivat, dessen Hydrolyse mit alkoholischem KOH unter Druck bei 180° Conarrhimin ergibt.

Conamin (92) wurde ebenfalls von SIDDIQUI (*169*) in der gleichen Rinde entdeckt und später von TSCHESCHE und ROY (*201*) wieder aufgefunden; seine Methylierung nach Eschweiler liefert Conessin. Da es ein primäres Amin darstellt, muß die vorhandene CH_3-Gruppe an den Stickstoff des Heteroringes gebunden sein. Conamin wurde von JARREAU, KHUNG-HUU und GOUTAREL (*76*) partialsynthetisch aus Conessin nach der Methode von PAPPO (*139, 140*) hergestellt. Hydroborierung und anschließende Chromsäureoxydation lieferten 6-Keto-dihydroconessin. Dessen Monojodmethylat wird weiter nach Hofmann behandelt, wodurch 3,5-Cyclo-conanin-6-on entsteht, das mit KBH_4 in den Alkohol verwandelt wird, der mit N_3H in Benzol in Gegenwart von BF_3 3β-Azido-Δ^5-conenin ergibt (*77*). Dessen Reduktion mit $LiAlH_4$ führte zu Conamin, das in den Konstanten jedoch von dem Naturprodukt abwich. Es ist zu vermuten, das letzteres durch Verwendung von Aceton ganz oder teilweise als Isopropyliden-Verbindung vorgelegen hat (*76*).

Conessimin (93) (*171*, *172*) geht ebenso wie die vorgenannten beiden Alkaloide aus *H. antidysenterica* bei der Methylierung nach Eschweiler in Conessin über. Da es eine tertiäre und eine sekundäre Aminogruppe enthält, bildet es ein Nitrosoderivat, und da die Struktur von Isoconessimin schon vergeben ist, bleibt für Conessimin nur die Konstitution mit einer $(CH_3)_2$N-Gruppe am $C_{(3)}$ übrig.

Eine Partialsynthese des Conessimins stammt von Bhattacharyya und Mitarb. (*10*). Hierbei wird das N-Oxid des Conessins mit BrCN behandelt, es entsteht so das Cyanamidderivat am N des Heterocyclus. Nunmehr wird die N-Oxidgruppierung mit $NaBH_4$ wieder reduziert und das Cyanamid mit alkoholischem KOH zu Conessimin gespalten.

Conimin (94), ebenfalls von Siddiqui (*168*) isoliert, ist eine disekundäre Base, muß also eine CH_3-Gruppe am N des $C_{(3)}$ enthalten. Behandelt man Conessin mit BrCN nach v. Braun unter energischeren Bedingungen, so bildet sich nach Siddiqui (*173*) das Bis-dicyanamidderivat, dessen Hydrolyse Conimin ergibt.

Isoconessimin (95) enthält je eine CH_3-Gruppe an beiden N-Atomen des Moleküls. Wird nach Siddiqui (*168*) Conessin mit BrCN behandelt, so reagiert eine der tertiären Aminogruppen sofort. Es ist diejenige am $C_{(3)}$, die ins Cyanamidderivat umgewandet wird, dessen Hydrolyse dann Isoconessimin ergibt (*173*). Durch Methylierung nach Eschweiler wird es in Conessin zurückverwandelt. Conimin und Isoconessimin sind ebenfalls Alkaloide aus *H. antidysenterica*.

Conessin (96), das Hauptalkaloid der Kurchirinde *(Holarrhena antidysenterica)* (*44*, *178*, *208*), wurde auch aus *H. africana* (*144*), *H. congolensis* (*144*, *147*), *H. febrifuga* (*170*), *H. mitis* (*9*), *H. floribunda* (*141*) und *H. wulfsbergi* (*50*) erhalten. Die umfangreiche Literatur über dieses Alkaloid soll hier vornehmlich unter dem Gesichtspunkt der Strukturbestimmung behandelt werden. Im übrigen sei auf die Zusammenfassung von Goutarel (*41*) hingewiesen.

Isoconessimin, in Form des Acetylderivates aus Conessin zugänglich, kann nach Haworth und Mitarb. (*45*) durch Hofmann-Abbau unter Aufsprengung des Ringes *E* in ein Methin übergeführt werden, das eine Vinylgruppe am $C_{(17)}$ enthält. Erneute Methylierung mit CH_3J und weiterer Abbau nach Emde mit Na-amalgam liefert N-Acetyl-N-methyl-$\Delta^{5,20}$-pregnadien, in dem mit Pt/H_2 die Doppelbindung Δ^{20} leichter als di' in Δ^5 abgesättigt werden kann. Nach Verseifung der Acetylgruppe entsteht 3-Methylamino-Δ^5-pregnen (*46*).

Geht man vom Conessin selbst aus und bereitet aus ihm das Methin, so wird nunmehr beim Hofmann-Abbau die Dimethylaminogruppe an $C_{(3)}$ entfernt, und der anschließende Emde-Abbau ergibt $\Delta^{3,5,20}$-Pregnatrien (*176*). Erhitzt man Conessinmethin mit alkalischer Glykollösung, so entsteht unter Rückbildung von Ring *E* Heteroconessin (*47*), das mit

Literaturverzeichnis: SS. 138—148.

Conessin an $C_{(20)}$ isomer ist. Beide geben mit Quecksilberacetat nach FAVRE und MARINIER (*38*) das gleiche Dehydroderivat. Andererseits läßt sich Dihydro-conessinmethin durch Kochen in Essigsäure in das Metho-acetat des Dihydro-conessins zurückverwandeln. Da für Conessin die 20 α- oder 20 S-Konfiguration angenommen wird, muß Hetero-conessin die 20 β- oder 20 R-Struktur haben. Diese Zuordnung der beiden Isomeren an $C_{(20)}$ wird durch den Vergleich der molekularen Drehungsdifferenzen (*143*) gestützt und ergibt sich auch aus den Überlegungen von HAWORTH (*45*) über die Beständigkeit der beiden Isomeren. Conessin muß daher die Struktur (**97**) und Heteroconessin (**98**) haben (*41, 48*).

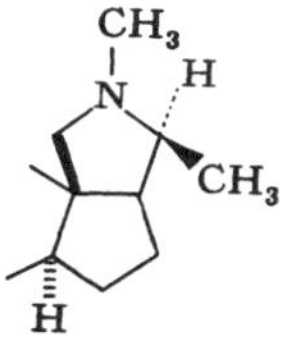

(**97**)
Conessin.

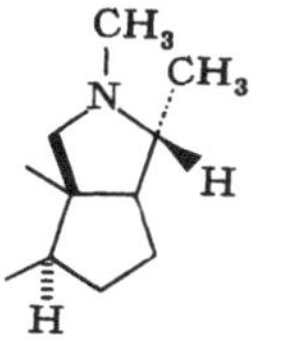

(**98**)
Heteroconessin.

(**99**)
Conanin.

Konfiguration des Ringes *E*.

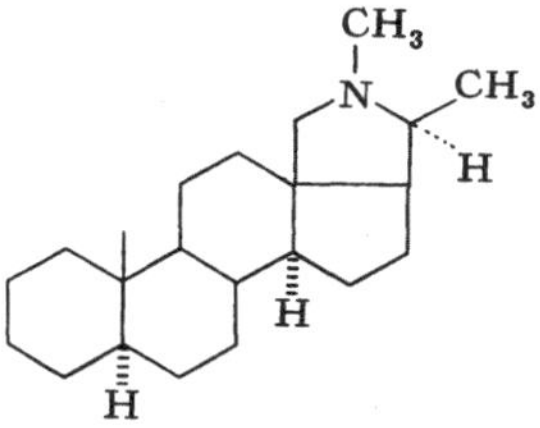

Die Konstitution des Conessins ist durch mehrere Partialsynthesen gesichert. JEGER und Mitarb. (*17*) führten den Aufbau des Ringes *E* aus 3-Desketo-funtumafrin B (S. 118) durch, indem sie es mit N-Chlor-succinimid umsetzten und das entstandene Chloramin mit einer Mischung von konz. Schwefelsäure und Eisessig behandelten. Sie erhielten so unter Anwendung der sogenannten Loeffler-Freytag-Reaktion Conanin (**99**). Entsprechend konnten COREY und HERTLER (*28*, s. *78*) in 90%iger Schwefelsäure unter Bestrahlen mit UV-Licht aus 3-Dimethylamino-20-N-methylchloramino-5 α-pregnan Dihydroconessin erhalten.

Ebenfalls eine photochemische Reaktion benutzen BARTON und MORGAN (*2*, vgl. dagegen *3*). Sie zersetzen 3β,20α-Bisazido-Δ^5-pregnen durch Photolyse und reduzieren das Reaktionsprodukt mit LiAlH$_4$; die anschließende Eschweiler-Methylierung ergibt in einer Ausbeute von 4,5% Conessin.

Die Stellung der Doppelbindung Δ^5 im Conessin folgt weiter aus der Partialsynthese von JOHNSON, BAUER und FRANCK (*78*), die von Δ^4-Co-nenin-3-on ausgeht. Dieses ist sowohl aus Isoconessimin zugänglich, kann aber auch aus Conessin mikrobiologisch erhalten werden. Als Organismen können dabei *Gloecosponum cyclaminis* oder *Hypomyces haematococcus* dienen (*39*). Die Kondensation dieses Ketons durch Erhitzen mit Dimethylamin in Gegenwart von $MgSO_4$ und Spuren von p-Toluolsulfosäure im geschlossenen Gefäß ergibt ein Enamin, das mit $NaBH_4$ Conessin liefert, wobei gleichzeitig Wanderung der Doppelbindung nach $\varDelta^5$ eintritt.

Totalsynthesen des d,l-Conessins wurden mehrere ausgearbeitet: Marshall und Johnson (*108*) gehen von 5-Methoxy-2-tetralon aus, Stork und Mitarb. (*179*) von 5-Methyl-6-methoxy-1-tetralon, und Nagata und Terasawa (*129*) verwenden 6-Methoxy-1-tetralon als Ausgangsmaterial.

Über Verfahren, Conessin zu N-freien Pregnen-derivaten abzubauen, vgl. Buzzetti et al. (*19*), Pappo (*139, 140*) sowie McNiven (*112*).

3 α-Amino-$\varDelta^5$-conenin (**100**) wurde von Černý, Dolejš und Šorm (*22*) aus *H. antidysenterica* isoliert und seine Struktur mittels physikalischer Methoden und durch Vergleich seines Acetylderivates mit partialsynthetisch gewonnenem 3 α-Acetylamino-$\varDelta^5$-conenin gesichert. Möglicherweise liegen auch noch methylierte Derivate des 3 α-Amino-$\varDelta^5$-conenins in der Pflanze vor.

Die bisher bekannten hydroxylierten 3-Amino-$\varDelta^5$-conenin-derivate leiten sich vom Conessimin und Conessin ab. *Holafrin* (**101**) wurde von Rostock und Seebeck (*151*) aus *H. floribunda* isoliert; es ist der Pyroterebinsäure-ester des 12 β-Hydroxy-conessimins. Durch alkalische Verseifung entsteht der freie Alkohol, der bei der Methylierung Holarrhenin (S. 130) ergibt. Offen bleibt nur noch die Stellung der beiden CH_3-Gruppen am Stickstoff. Hydroxy-conessimin liefert ein Nitrosoderivat, es muß daher ein sekundäres Amin sein. Der Eintritt der Nitrosogruppe erfolgt am N des Heterocyclus, da der Abbau von Holarrhenin nach Hofmann Trimethylamin ergibt, das nur aus der Dimethylaminogruppe an $C_{(3)}$ stammen kann.

Holarrhenin (**102**) ist ein 12 β-Hydroxy-conessin und wurde von Pyman (*147*) in der Rinde von *H. congolensis* entdeckt. Die Strukturbestimmung erfolgte durch Uffer (*206*). Holarrhenin läßt sich mit Chromsäure zu einem Keton oxydieren, dessen Äthylendithioketal mit Raneynickel nach Mozingo Conessin ergibt. Damit bleibt nur noch die Stellung der Hydroxylgruppe zu ermitteln. Der Hofmannsche Abbau des Holarrhenins ergibt ein Apoderivat, das zwei Doppelbindungen in Konjugation in den Ringen *A* und *B* und weiter eine Vinylgruppe an $C_{(17)}$ enthält. Die sekundäre OH-Gruppe bleibt bei der Reaktion erhalten. Wird das Jodmethylat des Apoholarrhenins mit Li in flüssigem Ammoniak behandelt und anschließend katalytisch hydriert, so entsteht ein 12-Hydroxy-5 α-pregnan, das auch vom Hecogenin aus zugänglich ist (*52*). Damit steht die Stellung der OH-Gruppe an $C_{(12)}$ fest; ihre sterische Anordnung als β ergibt sich aus der Beobachtung, daß die Reduktion des 12-Ketoderivates mit $LiAlH_4$ wieder zu Holarrhenin zurückführt. Weiter zeigt

Literaturverzeichnis: SS. 138—148.

das Tosylderivat des Holarrhenins mit Alkali eine Umlagerung (Verengung des Ringes *C* zum Fünfring und Erweiterung von *D* zum Sechsring), die schon vom Mesylderivat des Rockogenins bekannt ist (*51*).

Holarrhetin (103) ist der Pyroterebinsäureester des Holarrhenins und wurde von ROSTOCK und SEEBECK (*151*) aus *H. floribunda* isoliert. Die Verseifung führt zu Holarrhenin und Pyroterebinsäure.

7 α-Hydroxy-conessin (104) wurde von TSCHESCHE und OCKENFELS (*198*) aus der Rinde von *H. antidysenterica* erhalten. Es gibt bei der Chromsäureoxydation ein α,β-ungesättigtes Keton, woraus die Allyl-

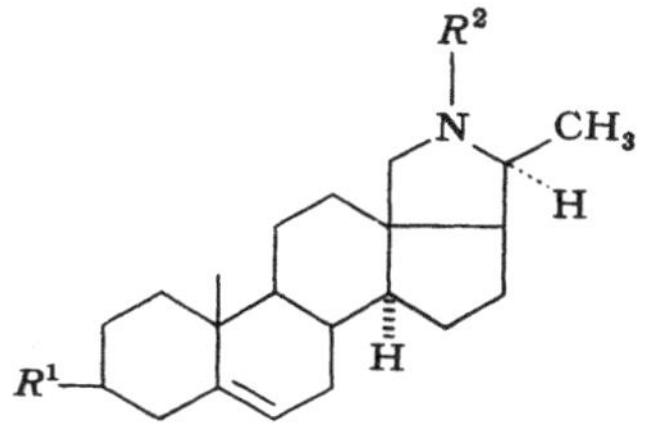

(91) Conarrhimin: $R^1 = NH_2$, $R^2 = H$.
(92) Conamin: $R^1 = NH_2$, $R^2 = CH_3$.
(93) Conessimin: $R^1 = N(CH_3)_2$, $R^2 = H$.
(94) Conimin: $R^1 = HNCH_3$, $R^2 = H$.
(95) Isoconessimin: $R^1 = HNCH_3$, $R^2 = CH_3$.
(96) Conessin: $R^1 = N(CH_3)_2$, $R^2 = CH_3$.
(100) 3 α-Amino-Δ^5-conenin.

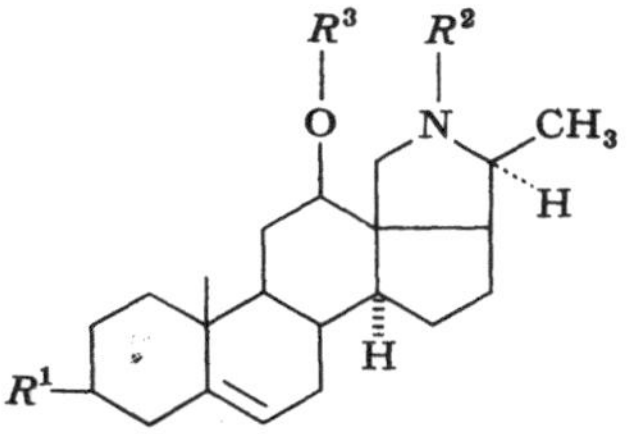

(101) Holafrin: $R^1 = N(CH_3)_2$, $R^2 = H$, $R^3 = CO—CH_2—CH=C(CH_3)_2$.
(102) Holarrhenin: $R^1 = N(CH_3)_2$, $R^2 = CH_3$, $R^3 = H$.
(103) Holarrhetin: $R^1 = N(CH_3)_2$, $R^2 = CH_3$, $R^3 = CO—CH_2—CH=C(CH_3)_2$.

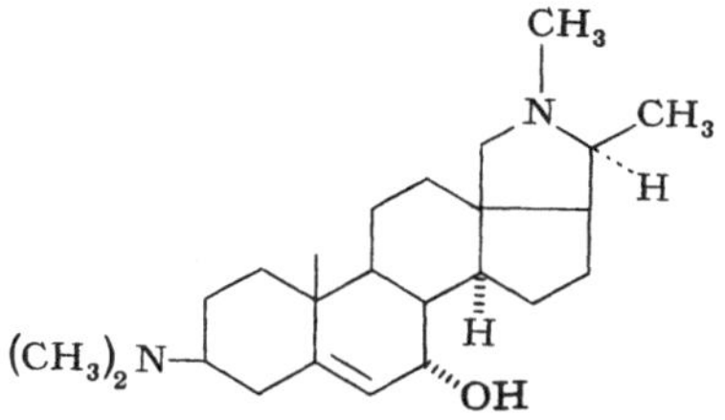

(104)
7 α-Hydroxyconessin.

stellung der Doppelbindung zum Hydroxyl erkennbar wird. Das Keton konnte aus Conessin durch Oxydation mit t-Butylchromat in CCl_4 in Gegenwart von Essigsäure/Essigsäureanhydrid erhalten werden. Schließlich wurde die Acetoxylierung von Conessin in der 7-Stellung direkt versucht. Mit Perbenzoesäure-t-butylester in Eisessig mit Cuprooxid als Katalysator entstand ein Isomerengemisch, das nach Verseifung chromatographisch getrennt werden konnte. Es entstand $^2/_3$ 7 α- und $^1/_3$ 7 β-Isomeres.

7 α- und *7 β-Hydroxy-conessin* sind auch durch mikrobiologische Oxydation mit *Aspergillus ochraceus* (*89a*) oder mit **Cunninghamella echinulata** (*142*) zugänglich.

m) Dieser Abschnitt behandelt die *doppelt ungesättigten Conanin-derivate*, die zu der Doppelbindung Δ^5 noch eine weitere im hetero-cyclischen Ring *E* enthalten. Die Lage dieser Doppelbindung ist lange Zeit zweifelhaft gewesen und konnte erst kürzlich einwandfrei geklärt werden.

Die beiden bekannten Vertreter *Conkurchin* (105) und *Conessidin* (106) sind schon von Bertho und Mitarb. (5, 7) aus *H. antidysenterica* isoliert worden; die Konstanten von Conkurchin waren jedoch nicht völlig korrekt, da man vermutlich nicht beachtet hatte, daß durch Aceton Conkurchin auf Grund der in ihm vorhandenen primären NH_2-Gruppe leicht ein Isopropylidenderivat bildet. So konnten Janot, Truong-Ho, Khuong-Huu und Goutarel (75) aus den Blättern von *Funtumia elastica* ein Alkaloid Irehlin gewinnen, dessen Identität mit Conkurchin zu be-weisen zunächst nicht ohne Schwierigkeiten möglich war (60, 61). Es sei darauf hingewiesen, daß die gleiche Doppelbindung sich auch im Holonamin (S. 126) wiederfindet. Conkurchin ist nach Conessin das zweite Hauptalkaloid von *H. antidysenterica*.

Conkurchin gibt leicht eine Salicylidenverbindung, die gut kristal-lisiert und zur Abscheidung der Base benutzt werden kann; sie sollte aus einer primären NH_2-Gruppe entstanden sein. Von den beiden Doppel-bindungen ist die im Heterocyclus leicht hydrierbar, die andere wesent-lich schwerer, so daß die Absättigung in Stufen erfolgen kann. Dihydro-conkurchin ist mit Conarrhimin identisch. Seine Methylierung nach Eschweiler führt zu Conessin, während die Methylierung von Tetra-hydro-conkurchin Dihydroconessin ergibt. Damit verengt sich die Lösung des Strukturproblems auf die Ermittlung der Stellung der zusätzlichen, leicht hydrierbaren Doppelbindung.

Schon Bertho hatte gezeigt, daß Conkurchin ein Δ^1-Pyrrolin sein muß, es liefert nach der Methode von Schöpf (158) mit *o*-Aminobenz-aldehyd ein Dihydrochinazoliniumderivat. Damit stimmt überein, daß die magnetische Kernresonanz eine Gruppierung R—CH=N—R' mit einem olefinischen Proton fordert, das bei $\tau = 2,42$ gefunden wird (75). Die Benzoylierung führt zur Öffnung des heterocyclischen Ringes auf Grund der Imin-struktur; es entsteht ein Dibenzoylderivat, das eine Aldehydgruppe enthält. Dieses ist durch den Nachweis eines aldehydi-schen Protons bei $\tau = 0,70$ gesichert. Reduziert man diese Ver-bindung zum Alkohol und spaltet die Benzoylreste mit alkoholischer KOH bei 200° unter Druck ab, so wird Holarrhimin gebildet. Conkurchin ist daher die Schiffsche Base des Aldehyds $C_{(18)}$ mit der NH_2-Gruppe an $C_{(20)}$. In diesem Zusammenhang muß an Malouphyllin (S. 120) erinnert werden, das ebenfalls eine Aldehydgruppe an $C_{(18)}$ enthält, das aber wegen der Dimethylierung der Aminogruppe an $C_{(20)}$ keine Schiffsche Base zu bilden vermag. Die Reduktion der Imin-doppelbindung mit KBH_4

Literaturverzeichnis: SS. 138—148.

führt zu einem Pyrrolidinderivat, das als Monoacetylderivat nach Esch-weiler methyliert N-Acetyl-conamin ergibt. Weiter führt die direkte Methylierung von Conkurchin zu Conessin; das früher beschriebene Trimethylconkurchin ist mit Conessin identisch. Ferner lassen sich Holarrhimin und Methylholarrhimin mit Chromsäure zu Conkurchin und Conessidin oxydieren (81).

Conessidin (106) stellt ein Conkurchin-derivat dar, das an der Amino-gruppe in 3-Stellung monomethyliert ist. Die Methylierung nach Esch-weiler führt denn auch zu Conessin. Die Hydrierung der Doppelbindung im Heterocyclus ergibt Conimin (S. 128); damit stimmt überein, daß die Umsetzung mit o-Aminobenzaldehyd nach Schöpf ein Dihydro-china-zolinium-Derivat liefert.

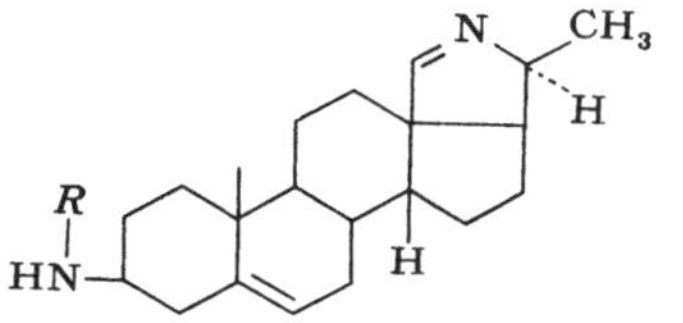

(105) Conkurchin: $R = H.$
(106) Conessidin: $R = CH_3.$

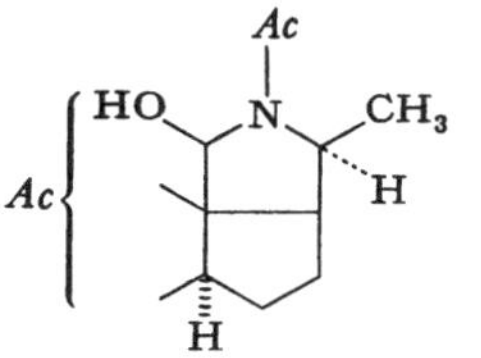

(107)
Diacetat.

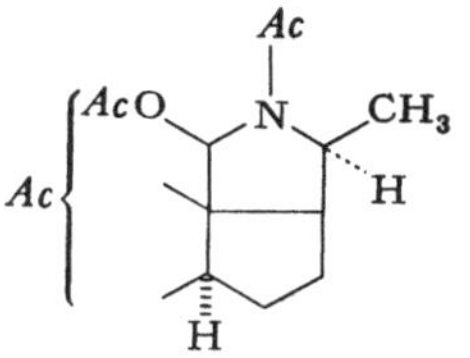

(108)
Triacetat.

Bemerkenswert ist noch das Verhalten der Imin-doppelbindung bei der Acetylierung mit Pyridin/Essigsäureanhydrid. Während die Benzoylierung nach GOUTAREL (75) zu einem Dibenzoylderivat unter Ringöffnung führt, entsteht bei dem entsprechend gebauten Holonamin (S. 126) ein Diacetat, bei der Aufarbeitung unter Ausschluß von Wasser jedoch ein Triacetat; dieses Verhalten wird durch die Formeln (107) und (108) erklärt.

Über die Darstellung von am $C_{(18)}$ substituierten Pregnanderivaten durch Oxydation von Conkurchin und Derivaten über Oxazirane vgl. JANOT, LUSINCHI und GOUTAREL (74). Es können so $C_{(18)}$-Oximo- und $C_{(13)}$-Nitriloderivate sowie solche des Paravallarins erhalten werden.

n) *Kurcholessin* (109) wurde von TSCHESCHE und Mitarb. (196) aus *H. antidysenterica* gewonnen und vor allem mittels physikalischer Metho-den (magnetische Kernresonanz, Massenspektrometrie usw.) in der Struktur geklärt. Es enthält eine zusätzliche CH₃-Gruppe in der 4-Stellung (α) und zwei weitere OH-Gruppen an $C_{(5)}(\alpha)$ und $C_{(7)}(\beta)$. Im übrigen liegt die Struktur des Conessins vor. Die zusätzliche Methylgruppe am $C_{(4)}$ wird als Relikt der Biogenese aufgefaßt; solche an $C_{(4)}$ erhalten gebliebene C-Atome sind schon von den Steroidderivaten Lophenol (*34, 35*), Citrostadienol (*109—111, 209*) und 24-Methylen-lophenol (*138, 159, 210*) bekannt.

Andererseits bedeutet Kurcholessin chemisch einen Übergang zu den zahlreichen Buxusalkaloiden (*11—16, 89, 177*), von denen hier *Cyclobuxin* (110) angeführt sei. Sie entstammen *Buxus sempervirens* L. und

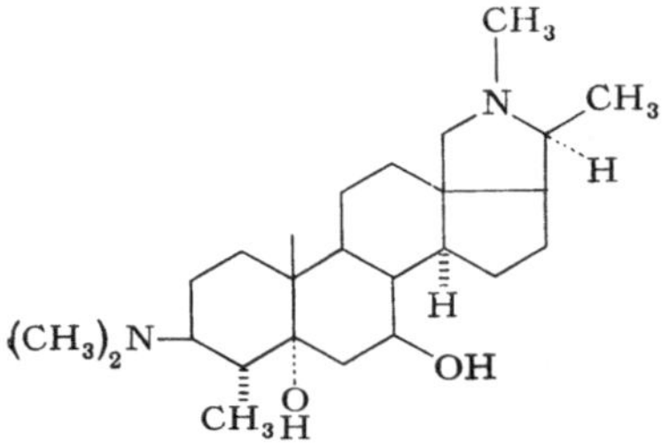

(109)
Kurcholessin.

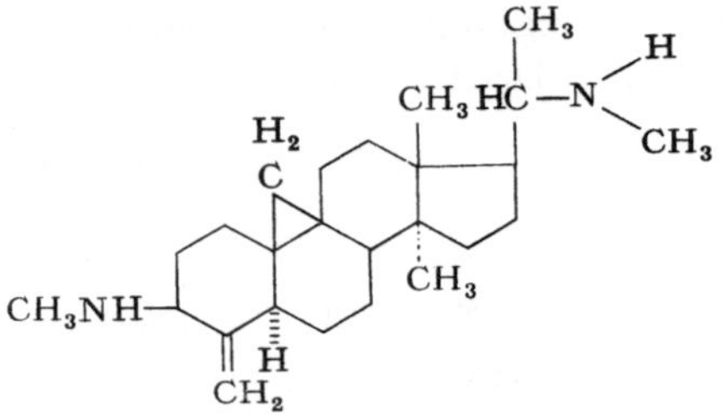

(110)
Cyclobuxin.

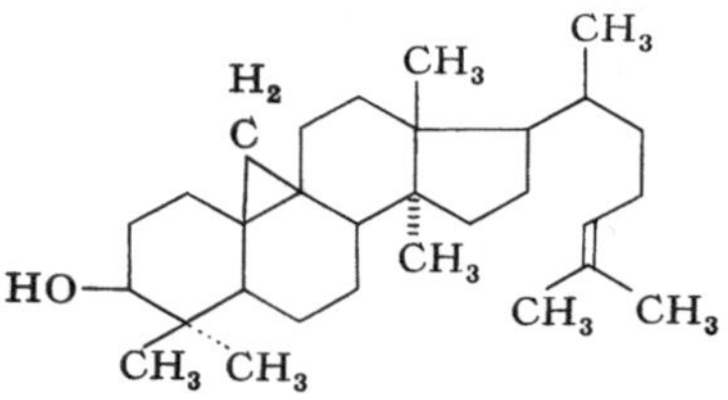

(111)
Cycloartenol.

Pachysandra microphylla Sieb. et Zucc. (*131—135*). Ihre biologische Muttersubstanz dürfte jedoch kein echtes Steroid, sondern Cycloartenol (111) oder eine nahe verwandte Substanz sein. Diese Alkaloide sollen daher hier nicht näher behandelt werden.

o) Diese Gruppe umfaßt zwei *Alkaloide mit D-Homo-androstanstruktur*, die aus *Dictyophleba lucida* von Janot und Mitarb. (*73*) isoliert wurden; fünf weitere Basen der gleichen Pflanze haben das normale Pregnangerüst.

Dictyolucidin (112) und *Dictyolucidamin* (113) ergeben bei der Methylierung nach Eschweiler dasselbe Dimethylaminoderivat, in dem außerdem zwei benachbarte Hydroxyle einen Methylendioxy-ring gebildet haben. Da Dictyolucidin nur eine CH_3-Gruppe am N enthält, während

Dictyolucidamin deren zwei hat, muß das erstere die Monomethyl-verbindung zu letzterem sein. Die Konstitution beider Verbindungen ließ sich durch Partialsynthese sichern: 17α-Hydroxy-progesteron wird mit äthanolischer Kalilauge zu 17aβ-Hydroxy-3,17-diketo-17aα-methyl-Δ^4-D-homo-androsten umgelagert. Dieses gibt nach der Methode von JOHNSON (78) mit Dimethylamin an $C_{(3)}$ ein Enamin, das mit KBH_4 in Methanol ein Gemisch der beiden Diole 17α und 17β ergibt, von denen 3β-Dimethylamino-17β,17aβ-dihydroxy-17α-methyl-Δ^5-D-homo-androsten überwiegt. Die weitere Behandlung dieser Mischung nach Esch-weiler mit Formaldehyd/Ameisensäure führt zu einem Methylenderivat zwischen den beiden OH-Gruppen an $C_{(17)}\beta$ und $C_{(17a)}\beta$, beide Hydroxyle nehmen die cis-Stellung ein. Die anschließende Hydrierung mit Palla-dium/Kohle als Katalysator liefert die gleiche Methylenverbindung, die schon aus Dictyolucidin und Dictyolucidamin erhalten worden ist.

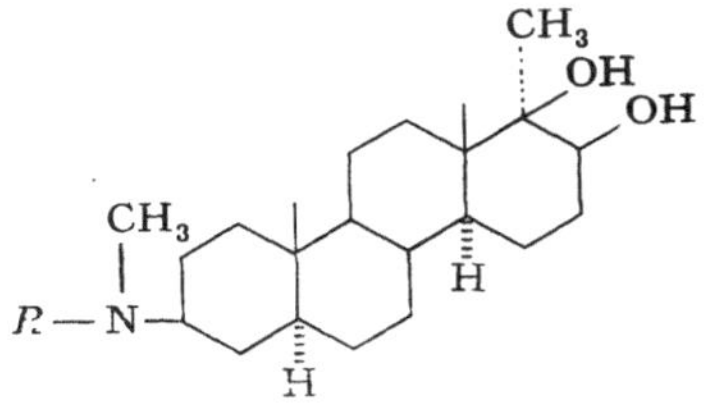

(112) Dictyolucidin: R = H.
(113) Dictyolucidamin: R = CH_3.

Das gleichzeitige Vorkommen von normalen Pregnanderivaten und D-Homo-Verbindungen in derselben Pflanze deutet daraufhin, daß diese aus ersteren im Stoffwechsel der Pflanze gebildet werden.

VI. Pregnenolon als biogenetische Vorstufe für Digitanole, Cardenolide, Bufadienolide und Aminopregnan-Derivate.

Die Auffindung von Δ^5-Pregnen-3β-ol-20-on und seines 5αH-Derivates (Allopregnanolon) (S. 108) im Pflanzenreich läßt vermuten, daß diese relativ einfach gebauten Pregnan-Derivate die Vorstufen aller der bisher behandelten Digitanol-derivate bzw. ihrer N-haltigen Analoga sind. Ein Beweis durch mit Isotopen markierte Verbindungen ist jedoch bisher hierfür nicht erbracht worden.

Anders liegen die Verhältnisse für die um zwei bzw. drei C-Atome reicheren pflanzlichen Cardenolide bzw. Bufadienolide, die ebenfalls Steranderivate sind. Das gemeinsame Vorkommen von Pregnenolon- und Allopregnanolon-glucosiden in kleiner Menge neben den zur Haupt-sache sich vorfindenden Uzarigeninglucosiden (Cardenolidderivat) (114) in *Xysmalobium undulatum* R. Br. läßt es möglich erscheinen, daß erstere die Vorläufer für die Biogenese der letzteren sind. Schon GREGORY und LEETE (43) haben gezeigt, daß Digitalis-Pflanzen mit ^{14}C im Carboxyl markierte Essigsäure beim Aufbau des Digitoxigenins (115) verwenden, wobei die C-Atome 21 und 22 frei von Radioaktivität bleiben. Sie folgerten daraus, daß entweder ein Ätiansäurederivat mit 20 C-Atomen mit 2 Mol Essigsäure unter Verlust von CO_2 den Butenolidring bildet oder aber ein Pregnanderivat mit 21 C-Atomen mit 1 Mol Essigsäure reagiert und auf diesem Weg Cardenolide mit 23 C-Atomen entstehen. von Euw und Reichstein (*37*, siehe auch *104* und *149*) haben diese Ergebnisse mit an $C_{(3)}$ mit ^{14}C markierter Mevalonsäure nachgeprüft und finden ebenfalls die C-Atome 21, 22 und 23 des gebildeten Digitoxigenins frei von ^{14}C; sie schließen sich damit der Pregnanhypothese von Gregory und Leete an.

Den Beweis, daß Pregnenolon Vorstufe der Cardenolide sein kann, führten Tschesche und Lilienweiss (*193*) dadurch, daß sie an $C_{(21)}$ mit ^{14}C markiertes Pregnenolon-β-D-glucosid Pflanzen oder überlebenden Blättern von *Digitalis lanata* anboten. 20—30 Tage später konnten nach Hydrolyse radioaktives Digitoxigenin sowie auch Gitoxigenin (116) und Digoxigenin (117) isoliert werden, neben dem in *D. lanata* bisher nicht aufgefundenen Xysmalogenin (118). Damit ist ein Pregnanderivat als Vorstufe der Cardenolide nachgewiesen. Digitalispflanzen können Pregnenolonglucosid für diese Biosynthese verwerten. Die relativ hohe Ausbeute von mehr als 6% unterstreicht diese Annahme. Da bei länger laufenden Versuchen der Anteil an Gitoxigenin- und Digoxigenin-glykosiden ansteigt, kann vielleicht auf eine spätere Hydroxylierung von primär gebildeten Digitoxigenin geschlossen werden. Vermutlich besitzt *D. lanata* 12β- und 16β-Hydroxylasen analog ähnlichen Fermenten im Tierreich; dort sind solche für die Einführung von Hydroxyl an 11α, 17α und 21 bekannt. Ferner ist zu bemerken, daß Digitalispflanzen eine Hydrierung der Δ^5-Doppelbindung zum 5β-Derivat durchführen können. Auf welcher Stufe die 14β und die an $C_{(21)}$ notwendige Hydroxylierung bei der Biosynthese der Cardenolide eintritt, ist ungeklärt.

Mit einer entsprechenden Versuchsanordnung haben Tschesche und Brassat (*184*) auch die Biogenese der Bufadienolide geprüft. Mit dem gleichen an $C_{(21)}$ mit ^{14}C markiertem Pregnenolon-glucosid ließ sich mit dem Versuchsobjekt *Helleborus atrorubens* die Bildung von radioaktivem Hellebrigenin (119) nachweisen. Auch hier fand die Biosynthese wie die der Cardenolide in den Blättern statt. Nach 16 Tagen konnte eine Ausbeute von mehr als 1% erzielt werden. Das gebildete Hellebrin wurde nach Acetylierung der Ozonspaltung unterworfen und nachgewiesen, daß die Ringe *A* bis *D* des Moleküls frei von Radioaktivität waren, ein Abbau des Pregnenolons zu radioaktiver Essigsäure und daraus ein erneuter Aufbau fand also nicht statt. Bemerkenswert sind außerdem die bei der Bildung des Hellebrigenins auftretenden Hydroxylierungen und die Umwandlung der CH_3-Gruppe an $C_{(10)}$ in eine Aldehydfunktion.

Literaturverzeichnis: SS. 138—148.

Während für den Aufbau des Butenolidringes der Cardenolide vermutlich aktivierte Malonsäure eine Rolle spielt, bleibt für den Lactonring der Bufadienolide die Verwendung von aktiver Oxalessigsäure eine zu prüfende Möglichkeit. In beiden Fällen würde ein C-Atom beim Einbau als CO_2 abgespalten werden.

Diese Beobachtungen hinsichtlich der Biogenese der Cardenolide und Bufadienolide, zusammen mit dem Nachweis zahlreicher Pregnanderivate in den Familien der Scrophulariaceen, Asclepiadaceen und Apocynaceen läßt die Frage auftauchen, ob dem Pregnenolon oder einer nahe verwandten Verbindung im Pflanzenreich dieselbe große Bedeutung als Relaissubstanz zukommt wie im Tierreich. Cardenolid- und Bufadienolid-

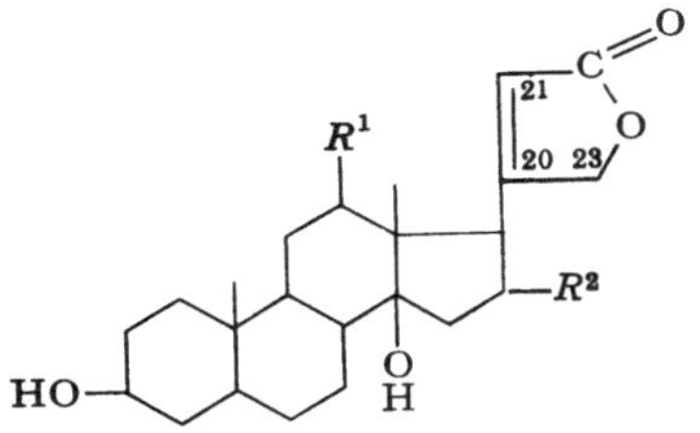

(**114**) Uzarigenin, 5α-Derivat: $R^1 = R^2 = H$.
(**115**) Digitoxigenin, 5β-Derivat: $R^1 = R^2 = H$.
(**116**) Gitoxigenin, dgl.: $R^1 = H$, $R^2 = OH$.
(**117**) Digoxigenin, dgl.: $R^1 = OH$, $R^2 = H$.
(**118**) Xysmalogenin, Δ^5-Derivat: $R^1 = R^2 = H$.

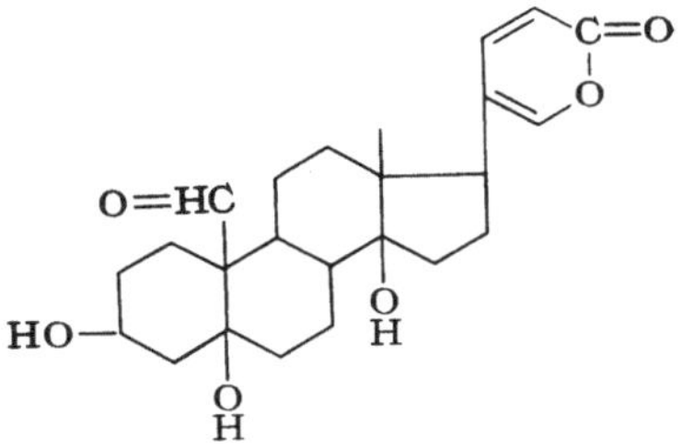

(**119**)
Hellebrigenin.

glykoside sind auch noch aus anderen Pflanzenfamilien bekannt, so den Liliaceen, Cruciferen und gelegentlich den Moraceen, Papilionaceen und Tiliaceen. Im Tierreich entstehen die androgenen und oestrogenen Hormone, wie wir heute wissen, ebenfalls über Pregnenolon. Nun ist aber Oestradiol von BUTENANDT (*18*) im Palmkernöl aufgefunden worden, und SKARZYNSKI (*175*) wies Östratriol in weiblichen Weidenkätzchen nach. Diese Befunde sind in Zweifel gezogen (*56*), andererseits auch wieder bestätigt worden (*49*). Es ist in diesem Zusammenhang jedoch bemerkenswert, daß verschiedene Stämme von *Nocardia restrictus* 19-Hydroxysteroide in Oestron umwandeln können (*174*). Die Aromatisierung des Ringes *A* in Steroiden ist daher auch im Pflanzenreich nachgewiesen. Dazu kommt die kürzlich bekannt gewordene Isolierung des Androstanderivates 3β,16α,17α-Trihydroxy-5α-androstan in *Aplopappus heterophyllus* Blake durch ZALKOW und Mitarb. (*212*). Es sprechen daher noch weitere Befunde dafür, daß dem Δ^5-Pregnen-3β-ol-20-on auch im Pflanzenreich eine erhebliche Bedeutung zukommt.

Literaturverzeichnis.

1. Abisch, E., Ch. Tamm und T. Reichstein: Die Glykoside der Wurzeln von *Pachycarpus lineolatus* (Decne.) Bullock (oder *P. Schweinfurthii* [N. E. Br.] Bullock). Helv. Chim. Acta **42**, 1014 (1959).

2. Barton, D. H. R. and L. R. Morgan, Jr.: The Synthesis of Conessine. Proc. Chem. Soc. (London) **1961**, 206.

3. Barton, D. H. R. and A. N. Starrat: Photochemical Transformations. Part XVIII. Some Experiments Related to the Synthesis of Conessine. J. Chem. Soc. (London) **1965**, 2444.

4. Bellon, R.: Steroids Derived from Paravallaridine. Belg. Pat. 627351 (1963).

5. Bertho, A.: Kurchi-Alkaloide, II. Mitt. Über die Gewinnung von Conessin und dessen Nebenbasen. Arch. Pharmaz. **277**, 237 (1939).

6. — Kurchi-Alkaloide, III. Über das wirksame Hauptalkaloid Conkurchin. Liebigs Ann. Chem. **555**, 214 (1944).

7. Bertho, A., G. v. Schuckmann und W. Schönberger: Kurchi-Alkaloide, I. Mitt. Über einige neue Basen aus *Holarrhena antidysenterica*. Ber. dtsch. chem. Ges. **66**, 786 (1933).

8. Beugelmans, R., C. Kan et J. Le Men: Préparation de dérivés désaminés et structure de la paravallaridine: alcaloïde stéroïdique du *Paravallaris microphylla* Pitard (Apocynacées). Bull. soc. chim. France **1963**, 1306.

9. Bhavanandan, V. P. and G. P. Wannigama: Isolation of Conessine from *Holarrhena mitis* R. Br. J. Chem. Soc. (London) **1960**, 2368.

10. Bhattacharyya, P. K., B. D. Kulkarni, S. Kanthamani and C. R. Narayanan: Synthesis of Conessimine from Conessine. Chem. and Ind. **1962**, 1377.

11. Brown, K. S., Jr. and S. M. Kupchan: The Structure of Cyclobuxine. J. Amer. Chem. Soc. **84**, 4590 (1962).

12. — — The Configuration of Cyclobuxine and its Interrelation with Cycloeucalenol. J. Amer. Chem. Soc. **84**, 4592 (1962).

13. — — Buxus Alkaloids. III. The Structure of Cyclobuxine. J. Amer. Chem. Soc. **86**, 4414 (1964).

14. — — Buxus Alkaloids. IV. The Configuration of Cyclobuxine and its Interrelation with Cycloeucalenol. J. Amer. Chem. Soc. **86**, 4424 (1964).

15. — — Buxus Alkaloids. V. The Constitution of Cyclobuxamine, a 4β-Monomethyl Cyclosteroid Alkaloid. J. Amer. Chem. Soc. **86**, 4430 (1964).

16. — — Buxus Alkaloids. VI. The Constitution of Cyclovirobuxine-D. Tetrahedron Letters **1964**, 2895.

17. Buchschacher, P., J. Kalvoda, D. Arigoni and O. Jeger: Direct Introduction of a Nitrogen Function at C-18 in a Steroid. J. Amer. Chem. Soc. **80**, 2905 (1958).

18. Butenandt, A. und H. Jacobi: Über die Darstellung eines kristallisierten pflanzlichen Tokokinins (Thelykinins) und seine Identifizierung mit dem α-Follikelhormon. Untersuchungen über das weibliche Sexualhormon, 10. Mitt. Z. physiol. Chem. **218**, 104 (1933).

19. Buzzetti, F., W. Wicki, J. Kalvoda und O. Jeger: Über Steroide und Sexualhormone, 209. Mitt. Einführung einer Sauerstofffunktion an die quaternäre Methylgruppe C-18 im intakten Steroidgerüst. Helv. Chim. Acta **42**, 388 (1959).

20. Cainelli, G., M. Lj. Mihailović, D. Arigoni und O. Jeger: Über Steroide und Sexualhormone, 211. Mitt. Direkte Einführung einer Sauerstofffunktion in die Methylgruppe C-18 im intakten Steroidgerüst. Helv. Chim. Acta **42**, 1124 (1959).

21. CAVÉ, ADRIEN, P. POTIER, ANDRÉ CAVÉ et J. LE MEN: Structure de la kibata-line: alcaloïde du *Kibatalia gitingensis* (Elm.) Woods (Apocynacées). Alcaloïdes stéroïdiques des Apocynacées. 6. Bull. soc. chim. France 1964, 2415.

22. ČERNÝ, V., L. DOLEJŠ and F. ŠORM: On Steroids. LXXXVII. Dihydroisoco-nessimine and 3α-Aminoconan-5-ene, New Alkaloids from *Holarrhena anti dysenterica* Wall. Collect. Czech. Chem. Comm. 29, 1591 (1964).

23. ČERNÝ, V., L. LÁBLER and F. Šorm: On Steroids. XXIV. The Structure of Holarrhimine. Collect. Czech. Chem. Comm. 22, 76 (1957).

24. ČERNÝ, V. and F. ŠORM: The Nature of the Hydroxyl Group in Holarrhimine. Chem. Listy 49, 909 (1955).

25. CHATTERJEE, A. and B. DAS: Chemistry of Chonemorphine: A Steroidal Alkaloid of *Chonemorphia macrophylla* G. Don. and *Chonemorphia penangensis*. Chem. and Ind. 1959, 1445.

26. — — Chemistry of Chonemorphine, the Steroidal Alkaloid of *Chonemorphia macrophylla* G. Don and *Chonemorphia penangensis*. Chem. and Ind. 1960, 290.

27. CHATTERJEE, A., B. DAS, C. P. DUTTA and K. S. MUKHERJEE: Steroid Alkaloids of *Saracocca pruniformis* Lindl. Tetrahedron Letters 1965, 67.

28. COREY, E. J. and W. R. HERTLER: The Synthesis of Dihydroconessine. A Method for Functionalizing Steroids at C-18. J. Amer. Chem. Soc. 80, 2903 (1958).

29. CORNFORTH, J. W.: On the Structure of Sarcostin. Chem. and Ind. 1959, 602.

30. CORNFORTH, J. W. and J. C. EARL: The Saponin of *Sarcostemma australe* R. Br. J. Chem. Soc. (London) 1939, 737.

31. — — Sarcostin. Part. I. A Preliminary Study of its Behaviour with Reagents. J. Chem. Soc. (London) 1940, 1443.

32. DAS, K. G. and P. P. PILLAY: Chonemorphine, the Chief Alkaloid of *Chonemorphia macrophylla*. J. Sci. Indust. Res. (India) 13 B, 602 (1954).

33. DICKEL, D., R. LUCAS and H. B. MACPHILLAMY: A New Hypotensive Steroid Alkaloid from *Conopharyngia pachysiphon*. J. Amer. Chem. Soc. 81, 3154 (1959).

34. DJERASSI, C., G. W. KRAKOWER, A. J. LEMIN, L. H. LIU, J. S. MILLS and R. VILLOTTI: The Neutral Constituents of the Cactus *Lophocereus schottii*. The Structure of Lophenol – 4α-Methyl-Δ^7-cholesten-3β-ol – a Link in Sterol Bio-genesis. J. Amer. Chem. Soc. 80, 6284 (1958).

35. DJERASSI, C., J. S. MILLS and R. VILLOTTI: The Structure of the Cactus Sterol Lophenol. A Link in Sterol Biogenesis. J. Amer. Chem. Soc. 80, 1005 (1958).

36. EHRHART, G., H. RUSCHIG und W. AUMÜLLER: Neue Synthesen in der Sterin-reihe. Angew. Chem. 52, 363 (1939).

37. EUW, J. V. und T. REICHSTEIN: Die Biosynthese des Digitoxigenins, Herkunft des C-20. Glykoside und Aglykone, 253. Mitt. Helv. Chim. Acta 47, 711 (1964).

38. FAVRE, H. et B. MARINIER: Sur la transformation de la dihydrohétéroconessine en dihydroconessine. Canad. J. Chem. 36, 429 (1958).

39. FLINES, J. DE, A. F. MARX, W. F. VAN DER WAARD and D. VAN DER SIJDE: Microbiological Conversion of Conessine. Tetrahedron Letters 1962, 1257.

40. GOUTAREL, R.: Stéroïdes aminés naturels des Apocynacées. Tetrahedron 14, 126 (1961).

41. — Les alcaloïdes stéroïdiques des Apocynacées. Actualités scientifiques et industrielles, p. 1302. Paris: Hermann. 1964.

42. GOUTAREL, R., A. CAVÉ, L. TAN et M. LEBOEUF: Alcaloïdes stéroïdes, XII. Synthèse de l'holaphyllamine et de l'holamine. Bull. soc. chim. France 1962, 646.

43. GREGORY, H. and E. LEETE: Biogenesis of Digitoxigenin. Chem. and Ind. 1960, 1242.

44. Haines, R.: Conessine (Wrightine or Nereine) an Alkaloid in the Bark of *Wrightia antidysenterica*. Pharmac. J. [2] **6**, 432 (1865).

45. Haworth, R. D. and J. McKenna: Structure of Conessine. Chem. and Ind. **1957**, 1510.

46. Haworth, R. D., J. McKenna and R. G. Powell: The Constitution of Conessine. Part V. Synthesis of Some Basic Steroids. J. Chem. Soc. (London) **1953**, 1110.

47. Haworth, R. D., J. McKenna and G. H. Whitfield: The Constitution of Conessine. Part. II. J. Chem. Soc. (London) **1949**, 3127.

48. — — — The Constitution of Conessine. Part IV. The Position of the Double Bond and Dimethylamino Group. J. Chem. Soc. (London) **1953**, 1102.

49. Heftmann, E., S.-T. Ko and R. D. Bennett: Identification of Estrone in Date Seeds by Thin-Layer Chromatography. Naturwiss. **52**, 431 (1965).

50. Henry, T. A. and H. C. Brown: Reputed Dysentery Remedies. Trans. Roy. Soc. Tropical Med. Hyg. **17**, 378 (1924) [Chem. Abstr. **18**, 2043 (1924)].

51. Hirschmann, R., C. S. Snoddy, Jr., C. F. Hiskey and N. L. Wendler: The Rearrangement of the Steroid C/D Rings. J. Amer. Chem. Soc. **76**, 4013 (1954).

52. Hove, L. van: Contribution à l'étude des alcaloïdes de l'*Holarrhena congolensis* Stapf: Structure de l'holarrhénine. Tetrahedron **7**, 104 (1959).

53. Ishii, H.: Structure of Purpnigenin. Chem. Pharmac. Bull. (Tokyo) **9**, 411 (1961).

54. — Studies on Digitalis Glycosides. XIV. The Structure of Purpnigenin. 1. Chem. Pharmac. Bull. (Tokyo) **10**, 351 (1962).

55. — Studies on Digitalis Glycosides. XV. The Structure of Purpnigenin. 2. Chem. Pharmac. Bull. (Tokyo) **10**, 354 (1962).

56. Jacobsohn, G. M., M. J. Frey and R. B. Hochberg: The Absence of Steroid Estrogens in Plants. Steroids **6**, 93 (1965).

57. Jaeggi, K. A., Ek. Weiss und T. Reichstein: Sarcostin, vermutliche Struktur, vorl. Mitt. Glykoside und Aglykone, 248. Mitt. Helv. Chim. Acta **46**, 694 (1963).

58. Janot, M.-M., A. Cavé et R. Goutarel: Togholamine, holaphyllamine et holaphylline, trois nouveaux alcaloïdes retirés des feuilles de l'*Holarrhena floribunda* (G. Don.) Dur. et Schinz. Alcaloïdes stéroliques, 3. Bull. soc. chim. France **1959**, 896.

59. — — — Alcaloïdes stéroïdes. Holaphyllamine et holamine, alcaloïdes de l'*Holarrhena floribunda* (G. Don.) Dur. et Schinz. C. R. hebd. séances Acad. Sci. **251**, 559 (1960).

60. Janot, M.-M., F.-X. Jarreau, M. Truong-Ho, Q. Khuong-Huu et R. Goutarel: Alcaloïdes stéroïdiques, XXVI. Conarrhimine et alcaloïdes du groupe de la conkurchine. Structures de la conkurchine et de la conessidine. Identité de la conkurchine et de l'irehline. Bull. soc. chim. France **1964**, 1555.

61. — — — Alcaloïdes steroïdiques. Identité de la conkurchine et de l'irehline. C. R. hebd. séances Acad. Sci. **258**, 2089 (1964).

62. Janot, M.-M., Q. Khuong-Huu et R. Goutarel: Deux nouveaux alcaloïdes stéroliques, la funtumine et la funtumidine. C. R. hebd. séances Acad. Sci. **246**, 3076 (1958).

63. — — — Structure de la funtumidine. C. R. hebd. séances Acad. Sci. **248**, 982 (1959).

64. — — — Alcaloïdes stéroïdes. Funtuphyllamines A, B et C, funtumafrines B et C, alcaloïdes du *Funtumia africana* (Benth.) Stapf. C. R. hebd. séances Acad. Sci. **250**, 2445 (1960).

65. — — — Alcaloïdes stéroïdes. Structure de la latifoline, nouvel alcaloïde retiré des écorces de *Funtumia latifolia* Stapf. C. R. hebd. séances Acad. Sci. **254**, 1326 (1962).

66. JANOT, M.-M., Q. KHUONG-HUU, J. YASSI et R. GOUTAREL: Alcaloïdes stéroïdiques, XXIV. Structure de la funtuline, nouvel alcaloïde des écorces du *Funtumia latifolia* Stapf. Bull. soc. chim. France 1964, 787.

67. JANOT, M.-M., F. LAINÉ et R. GOUTAREL: Alcaloïdes stéroïdes, V. Alcaloïdes du *Malouetia bequaertiana* E. Woodson (Apocynacées). La funtuphyllamine B et la malouétine. Ann. pharmac. franç. 18, 673 (1960).

68. — — — Alcaloïdes stéroïdes, XIII. Structure de la malouphylline, alcaloïde des feuilles du *Malouetia bequaertiana* E. Woodson. Bull. soc. chim. France 1962, 648.

69. JANOT, M.-M., F. LAINÉ, R. GOUTAREL et M.-J. MAGDELEINE: Alcaloïdes stéroïdes, XVI. La malouphyllamine, alcaloïde du *Malouetia bequaertiana* E. Woodson (Apocynacées). Bull. soc. chim. France 1963, 641.

70. JANOT, M.-M., F. LAINÉ, Q. KHUONG-HUU et R. GOUTAREL: Alcaloïdes stéroïdes, IX. Dérivés amino-20 et diamino-3,20 du pregnane-5α, alcaloïdes du *Funtumia africana* (Benth.) Stapf et du *Malouetia bequaertiana* Woods. Bull. soc. chim. France 1962, 111.

71. JANOT, M.-M., X. MONSEUR, C. CONREUR et R. GOUTAREL: Alcaloïdes stéroïdes, X. L'holafébrine, amino-20α-pregnène-5-ol-3β, stéroïde aminé naturel, retiré de l'*Holarrhena febrifuga* Klotzsch et du *Kibatalia arborea* G. Don. Bull. soc. chim. France 1962, 285.

72. JANOT, M.-M., P. LONGEVIALLE, R. GOUTAREL et C. CONREUR: Alcaloïdes stéroïdiques, XXXI. Alcaloïdes des feuilles de l'*Holarrhena antidysenterica* (Roxb.) Wall. Structure de la kurchiline; synthèse des quatre hydroxy-2-diméthylamino-3-pregnane-5α diastéréoïsomères. Bull. soc. chim. France 1964, 2158.

73. JANOT, M.-M., C. MONNERET, X. MONSEUR, Q. KHUONG-HUU et R. GOUTAREL: Alcaloïdes stéroïdiques. Structure de la dictyolucidine et de la dictyolucidamine, nouveaux alcaloïdes du *Dictyophleba lucida* (K. Sch.) Pierre, Apocynacées. C. R. hebd. séances Acad. Sci. 260, 6118 (1965).

74. JANOT, M.-M., X. LUSINCHI et R. GOUTAREL: Alcaloïdes stéroïdes. Stéroïdes substitués en position 18, à partir des alcaloïdes du groupe de la conanine. C. R. hebd. séances Acad. Sci. 256, 2627 (1963).

75. JANOT, M.-M., M. TRUONG-HO, Q. KHUONG-HUU et R. GOUTAREL: Alcaloïdes stéroïdiques, XX. L'irehline, nouvel alcaloïde retiré des feuilles du *Funtumia elastica* (Preuss) Stapf. Bull. soc. chim. France 1963, 1977.

76. JARREAU, F.-X., Q. KHUONG-HUU et R. GOUTAREL: Alcaloïdes stéroïdiques. XIX. Nouvelle méthode de synthèse d'amino-3β,Δ5-stéroïdes. Bull. soc. chim. France 1963, 1861.

77. JARREAU, F.-X., C. MONNERET, Q. KHUONG-HUU et R. GOUTAREL: Alcaloïdes stéroïdiques, XXX. Mécanisme de formation des azido-3β,Δ5 et azido-6β cyclo-3α,5α stéroïdes. Bull. soc. chim. France 1964, 2155.

78. JOHNSON, W. S., V. J. BAUER and R. W. FRANCK: The Synthesis of Conessine from the Corresponding 3-Keto-Δ4-unsaturated System. Tetrahedron Letters 1961, 72.

79. KARRER, W.: Untersuchungen über herzwirksame Glucoside. Festschrift für E. C. Barell, S. 238. Basel. 1936 [Chem. Zbl. 107, II, 2727 (1936)].

80. KASAL, A., A. POLÁKOVÁ, A. V. KAMERNITZKY, L. LÁBLER and V. ČERNÝ: On Steroids. LXXVI. N³-Methylholarrhimine and N²⁰-Methylholarrhimine. Collect. Czech. Chem. Comm. 28, 1189 (1963).

81. KHUONG-HUU, Q., L. LÁBLER, M. TRUONG-HO et R. GOUTAREL: Alcaloïdes stéroïdiques, XXVII. Obtention de la conkurchine et de la conessidine, à partir de l'holarrhimine. Bull. soc. chim. France 1964, 1564.

82. Khuong-Huu, Q., C. Monneret, J. Yassi et R. Goutarel: Alcaloïdes stéroïdiques, XXXIII. Structure de la funtudiénine, nouvel alcaloïde des écorces du *Funtumia latifolia* Stapf. Bull. soc. chim. France **1964**, 2169.

83. Khuong-Huu, Q., J. Yassi et R. Goutarel: Alcaloïdes stéroïdiques, XXIII. Structure de la latifolinine et de la nor-latifoline, deux nouveaux alcaloïdes retirés des écorces du *Funtumia latifolia*. Bull. soc. chim. France **1963**, 2486.

84. Khuong-Huu, Q., J. Yassi, C. Monneret et R. Goutarel: Alcaloïdes stéroïdiques, XXXVIII. Structure de la funtessine, nouvel alcaloïde des écorces du *Funtumia latifolia* Stapf, Apocynacées. Bull. soc. chim. France **1965**, 1831.

85. Kikuchi, T., S. Uyeo, Jr., M. Ando and A. Yamamoto: Studies on the Alkaloids of *Pachysandra terminalis* Sieb. et Zucc. (3). Systematic Separation and Characterization of Alkaloids. Tetrahedron Letters **1964**, 1817.

86. Kikuchi, T., S. Uyeo and T. Nishinaga: Studies on the Alkaloids of *Pachysandra terminalis* Sieb. et Zucc. (4). Structure of Epipachysamine-B, -C and Terminaline. Tetrahedron Letters **1965**, 1993.

87. — — Studies on the Alkaloids of *Pachysandra terminalis* Sieb. et Zucc. (5). Structure of Epipachysamine-D, -E and -F. Tetrahedron Letters **1965**, 3169.

88. Knaack, W. F., Jr. and T. A. Geissman: Steroid Alkaloids of *Pachysandra terminalis* (Buxaceae). Tetrahedron Letters **1964**, 1381.

89. Kupchan, S. M. and W. L. Asbun: Buxus Alkaloids. VII. The Structure of Buxenine-G, a New and Novel Steroidal Alkaloid from *Buxus sempervirens* L. Tetrahedron Letters **1964**, 3145.

89a. Kupchan, S. M., C. J. Sih, S. Kubota and A. M. Rahim: Microbiological Transformations. IV. Hydroxylation of Conessine. Tetrahedron Letters **1963**, 1767.

90. Lábler, L.: On Steroids. LVI. An Alternative Route for Preparation of (20 R)-20-Hydroxy-3-oxo-pregn-4-en-18-oic Acid (18 → 20) Lactone. Collect. Czech. Chem. Comm. **26**, 724 (1961).

91. Lábler, L. and V. Černý: Steroids. XXXIII. Holarrhidine, a New Alkaloid from the Bark of *Holarrhena antidysenterica*. Chem. Listy **51**, 2344 (1957).

92. Lábler, L., V. Černý and F. Šorm: The Structure of Holarrhimine. Chem. and Ind. **1955**, 1119.

93. — — — On Steroids. XIX. Structural Correlation of Holarrhimine with Conessine. Collect. Czech. Chem. Comm. **20**, 1484 (1955).

94. Lábler, L., J. Hora and V. Černý: On Steroids. LXXIX. Synthesis of 3α-Dimethylaminoconan-5-ene. Corroboration of the Structure of Holarrhidine. Collect. Czech. Chem. Comm. **28**, 2015 (1963).

95. Lábler, L. and F. Šorm: 3β : 18-Dihydroxy*allo*pregnan-20-one (18 → 20 *cyclo*Hemiketal) from Holarrhimine. Chem. and Ind. **1958**, 1661.

96. — — On Steroids. XLIV. 3β,18-Dihydroxy-5α-pregnan-20-one (18 → 20 *cyclo*Hemiketal) from Holarrhimine. Collect. Czech. Chem. Comm. **24**, 2975 (1959).

97. — — On Steroids. XLVIII. Partial Synthesis of 18-Hydroxyprogesterone from Holarrhimine. Collect. Czech. Chem. Comm. **25**, 265 (1960).

98. — — On Steroids. LIV. Some Derivatives of (20 R)-3β,20-Dihydroxypregn-5-en-18-oic Acid (18 → 20) Lactone. Collect. Czech. Chem. Comm. **25**, 2855 (1960).

99. — — (20 R)-20-Hydroxy-3-oxo-pregn-4-en-18-oic Acid (18 → 20) Lactone from Holarrhimine. Chem. and Ind. **1960**, 935.

100. — — On Steroids. LXXXI. The Structure of Concuressine and of some Less Polar Alkaloids from *Holarrhena antidysenterica* Wall. Collect. Czech. Chem. Comm. **28**, 2345 (1963).

101. LAOS, I. and R. PAPPO: 18-Dimethylamino Steroids. U. S. Pat. 2 933 511 (1960).

102. LARDON, A., W. KLYNE, E. ISELI and T. REICHSTEIN: IUPAC Symposium, Kyoto, 1964.

103. LEBOEUF, M., A. CAVÉ et R. GOUTAREL: Alcaloïdes stéroïdiques. Présence de la progestérone dans les feuilles de l'*Holarrhena floribunda* (G. Don.) Dür. et Schinz. C. R. hebd. séances Acad. Sci. **259**, 3401 (1964).

104. LEETE, E., H. GREGORY and E. G. GROS: Biosynthesis of Plant Steroids. I. The Origin of the Butenolide Ring of Digitoxigenin. J. Amer. Chem. Soc. **87**, 3475 (1965).

105. LE MEN, J.: Paravallarine, alcaloïde stéroïdique du *Paravallaris microphylla* Pitard (**A**pocynacées). Bull. soc. chim. France **1960**, 860.

106. LE MEN, J., C. KAN et R. BEUGELMANS: Structure de la paravallaridine, alcaloïde stéroïdique du *Paravallaris microphylla* Pitard (Apocynacées). Bull. soc. chim. France **1963**, 597.

107. LUCAS, R. A., D. F. DICKEL, R. L. DZIEMIAN, M. J. CEGLOWSKI, B. L. HENSLE and H. B. MACPHILLAMY: Some Hypotensive Amino Steroid Glycosides. J. Amer. Chem. Soc. **82**, 5688 (1960).

108. MARSHALL, J. A. and W. S. JOHNSON: Total Synthesis of Racemic Conessine. J. Amer. Chem. Soc. **84**, 1485 (1962).

109. MAZUR, Y. and F. SONDHEIMER: Synthesis of 4α-Methyl-Δ⁷-steroids. The Interrelationship of Cholesterol, Citrostadienol and Lophenol. J. Amer. Chem. Soc. **80**, 6296 (1958).

110. MAZUR, Y., A. WEIZMANN and F. SONDHEIMER: The Structure of Citrostadienol, a Natural 4α-Methylsterol. J. Amer. Chem. Soc. **80**, 1007 (1958).

111. — — — Steroids and Triterpenoids of Citrus Fruit. III. The Structure of Citrostadienol, a Natural 4α-Methylsterol. J. Amer. Chem. Soc. **80**, 6293 (1958).

112. McNIVEN, N. L.: Some Simple 18-Oxygenated Steroids. The Constitution of Conessine. Chem. and Ind. **1957**, 1296.

113. MEYSTRE, CH., K. HEUSLER, J. KALVODA, P. WIELAND, G. ANNER und A. WETT-STEIN: Reaktionen von Steroid-Hypojoditen, II. Über die Herstellung 18-oxygenierter Pregnanverbindungen. Helv. Chim. Acta **45**, 1317 (1962).

114. MITSUHASHI, H. and T. NOMURA: On the Structure of Metaplexigenin and Benzoylramanone. Chem. Pharmac. Bull. (Tokyo) **11**, 1333 (1963).

115. — — On the Structure of Ramanone. Steroids **3**, 271 (1964).

116. — — On the Structure of Pergularin. Chem. Pharmac. Bull. (Tokyo) **12**, 1523 (1964).

117. — — Studies on the Constituents of Asclepiadaceae Plants. XV. On the Components of *Metaplexis japonica* Makino. II. Chem. Pharmac. Bull. (Tokyo) **13**, 274 (1965).

118. MITSUHASHI, H., T. NOMURA, Y. SHIMIZU, I. TAKEMORI and E. YAMADA: Studies on the Constituents of Asclepiadaceae Plants. VIII. On the Components of *Metaplexis japonica* Makino. Chem. Pharmac. Bull. (Tokyo) **10**, 811 (1962).

119. MITSUHASHI, H., T. SATO, T. NOMURA and I. TAKEMORI: On the Structure of Tomentogenin. Chem. Pharmac. Bull. (Tokyo) **12**, 981 (1964).

120. — — — — On the Structure of Tomentogenin. Chem. Pharmac. Bull. (Tokyo) **13**, 267 (1965).

121. MITSUHASHI, H. and Y. SHIMIZU: On the Structure of Cynanchogenin. Chem. Pharmac. Bull. (Tokyo) **7**, 949 (1959).

122. — — Structural Relationship between Desacylcynanchogenin and Sarcostin. Chem. Pharmac. Bull. (Tokyo) **10**, 433 (1962).

123. MITSUHASHI, H. and Y. SHIMIZU: Studies on the Constituents of Asclepiadaceae Plants. IV. The Structure of Cynanchogenin. 3. Chem. Pharmac. Bull. (Tokyo) **10**, 719 (1962).

124. — — Studies on the Constituents of Asclepiadaceae Plants. V. The Isolation and Structure of Penupogenin. Chem. Pharmac. Bull. (Tokyo) **10**, 725 (1962).

125. — — On the Selenium Dehydrogenation of $3\beta,12\beta,20\beta$-Trihydroxy-5α-pregnane and a C-Nor-*D*-homopregnane Derivative. Tetrahedron Letters **1962**, 909.

126. — — Structure of Cynanchogenin and Sarcostin. Steroids **2**, 373 (1963).

127. MITSUHASHI, H., Y. SHIMIZU, T. NOMURA, T. YAMADA and E. YAMADA: Studies on the Constituents of Asclepiadaceae Plants. XI. Separation of New Aglycones from *Cynanchum caudatum* Max. Chem. Pharmac. Bull. (Tokyo) **11**, 1198 (1963).

128. MITSUHASHI, H., I. TAKEMORI, Y. SHIMIZU, T. NOMURA and E. YAMADA: Studies on the Constituents of Asclepiadaceae Plants. VI. Components of *Marsdenia tomentosa* Decne. Chem. Pharmac. Bull. (Tokyo) **10**, 804 (1962).

129. NAGATA, W., T. TERASAWA and T. AOKI: Stereospecific Total Synthesis of Racemic 5α-Pregnan-3β-ol-20-one. Tetrahedron Letters **1963**, 865.

130. — — — Stereospecific Total Synthesis of Racemic Latifoline and Conessine. Tetrahedron Letters **1963**, 869.

131. NAKANO, T. and M. HASEGAWA: Buxus Alkaloids. Part III. The Structure of "Alkaloid L" and its Correlation with Cyclomicrophylline-A. Tetrahedron Letters **1964**, 3679.

132. NAKANO, T. and S. TERAO: Buxus Alkaloids. Part I. The Structure of three new Alkaloids, Cyclomicrophylline-A, B, and C, from *B. microphylla* Sieb. et Zucc. Tetrahedron Letters **1964**, 1035.

133. — — Buxus Alkaloids. Part II. The Structures and the Stereochemistry of Cyclomicrophylline-A, B, and C. Tetrahedron Letters **1964**, 1045.

134. — — Buxus Alkaloids. Part IV. Isolation and Structure Elucidation of Eight New Alkaloids, Cyclomicrophylline-A, -B, and -C, Dihydrocyclomicrophylline-A and -F, Cyclomicrophyllidine-A, Dihydrocyclomicrophyllidine-A, and Cyclomicrobuxine, from *B. microphylla* Sieb. et Zucc., *var. suffruticosa* Makino. J. Chem. Soc. (London) **1965**, 4512.

135. — — Buxus Alkaloids. Part V. The Constitution of Cyclobuxoxazine, a New Skeletal Alkaloid Containing a Tetrahydro-oxazine Ring. J. Chem. Soc. (London) **1965**, 4537.

136. NASCIMENTO, J. M. do, Jr., H. JÄGER, CH. TAMM und T. REICHSTEIN: Die Dehydrierung von Sarcostin mit Selen. Glykoside und Aglykone, 197. Mitt. Helv. Chim. Acta **42**, 661 (1959).

137. NASCIMENTO, J. M. do, Jr., CH. TAMM, H. JÄGER und T. REICHSTEIN: Die Glykoside der Wurzeln von *Asclepias glaucophylia* Schlechter. Glykoside und Aglykone, 257. Mitt. Helv. Chim. Acta **47**, 1775 (1964).

138. OSSKE, G. und K. SCHREIBER: Sterine und Triterpenoide. VI. 24-Methylenlophenol, ein neues 4α-Methyl-sterin aus *Saccharum officinarum* L. und *Solanum tuberosum* L. Tetrahedron **21**, 1559 (1965).

139. PAPPO, R.: Synthesis of 18-Oxygenated Progesterones. J. Amer. Chem. Soc. **81**, 1010 (1959).

140. — 18-Dimethylamino Steroids and Intermediates from Conessine. U. S. Pat. 2913455 (1959).

141. PARIS, R.: *Holarrhena africana* A. DG. Bull. sci. pharm. **45**, 543 (1938).

142. PATTERSON, E. L., W. W. ANDRES and R. E. HARTMAN: The Microbiological Oxidation of Conessine. Experientia **20**, 256 (1964).

143. PHILLIPS, D. D. and A. W. JOHNSON: The Structure of Conessine. Chem. and Ind. **1957**, 1211.

144. POHLSDORFF, K. und P. SCHIRMER: Über Conessin. Ber. dtsch. chem. Ges. **19**, 78 (1886).

145. POTIER, P., C. KAN et J. LE MEN: Présence de stéroïdes cétoniques dans les feuilles du *Paravallaris microphylla* Pitard (Apocynacées). Tetrahedron Letters **1964**, 1671.

146. PRESS, J. und T. REICHSTEIN: Teilsynthesen des 14-*iso*-17-*iso*-Allopregnans und seine Identifizierung mit Diginan. Glykoside und Aglykone, 27. Mitt. Helv. Chim. Acta **30**, 2127 (1947).

147. PYMAN, F. L.: The Alkaloids of *Holarrhena congolensis* Stapf. J. Chem. Soc. (London) **115**, 163 (1919).

148. RAM, M. and P. K. BHATTACHARYYA: Transformations of *Kurchi* Alkaloids. IV. Structural Correlation of the *Kurchi* Alkaloid, Holarrhimine and the Apocynaceae Alkaloids, Paravallarines. Indian J. Chem. **2**, 41 (1964).

149. RAMSTAD, E. and J. L. BEAL: Mevalonic Acid, Precursor of Digitoxigenin. Chem. and Ind. **1960**, 177.

150. REICHSTEIN, T. und O. SCHINDLER: Privatmitteilung.

151. ROSTOCK, H. und F. SEEBECK: Holafrin und Holarrhetin, zwei unbekannte Esteralkaloide aus *Holarrhena africana* A. DC. Helv. Chim. Acta **41**, 11 (1958).

152. RUSCHIG, H., W. FRITSCH, J. SCHMIDT-THOMÉ und W. HAEDE: Über die Herstellung von 17α-Oxy-20-keto-Steroiden aus 17(20)-En-20-acetamino-Steroiden. Chem. Ber. **88**, 883 (1955).

153. SATOH, D.: Studies on Digitalis Glycosides. Structure of Digipronin and Digiprogenin. Chem. Pharmac. Bull. (Tokyo) **8**, 270 (1960).

154. SATOH, D. and M. HORIE: Studies on Digitalis Glycosides. The Configuration of Digiprogenin. Chem. Pharmac. Bull. (Tokyo) **12**, 979 (1964).

155. SATOH, D., H. ISHII and Y. OYAMA: Studies on Digitalis Glycosides. Structure of Purpnigenin and Purprogenin. Chem. Pharmac. Bull. (Tokyo) **8**, 657 (1960).

156. SATOH, D., H. ISHII, Y. OYAMA, T. WADA and T. OKUMURA: Studies on Digitalis Glycosides. VI. Isolation of Odoroside H, Digiproside and Digitalonine. Chem. Pharmac. Bull. (Tokyo) **4**, 284 (1956).

157. SAUER, H. H., EK. WEISS und T. REICHSTEIN: Struktur der Drevogenine. Glykoside und Aglykone, 267. Mitt. Helv. Chim. Acta **48**, 857 (1965).

158. SCHÖPF, C. und F. OECHLER: Zur Frage der Biogenese des Vasicins (Peganins). Die Synthese des Desoxyvasicins unter physiologischen Bedingungen. Liebigs Ann. Chem. **523**, 1 (1936), s. spez. S. 25.

159. SCHREIBER, K. und G. OSSKE: Isolierung von 4α-Methyl-5α-stigmasta-7,24(28)-dien-3β-ol aus *Solanum tuberosum*, sowie über die Identität dieser Verbindung mit α_1-Sitosterin. Experientia **19**, 69 (1963).

160. SHOPPEE, C. W.: Diginin, 3. Mitt. Abbau des Diginigenins zu einem Kohlenwasserstoff Diginan $C_{21}H_{36}$. Helv. Chim. Acta **27**, 246 (1944).

161. — Diginin und Diginigenin, 4. Mitt. Helv. Chim. Acta **27**, 426 (1944).

162. SHOPPEE, C. W. and R. E. LACK: Steroids. Part XXI. The Structure of Digacetigenin. J. Chem. Soc. (London) **1964**, 3611.

163. — — Steroids. Part XXII. The Configuration of 14α-Digiprogenin. J. Chem. Soc. (London) **1964**, 3619.

164. SHOPPEE, C. W., R. E. LACK and A. V. ROBERTSON: Steroids. Part XVII. The Structure of Diginin and Diginigenin. J. Chem. Soc. (London) **1962**, 3610.

165. — — — The Structure of Diginin and Diginigenin. Proc. Chem. Soc. (London) **1962**, 65.

166. Shoppee, C. W., R. E. Lack and S. Sternhell: Steroids. Part XIX. The Structure of Digifolein and Digifologenin. J. Chem. Soc. (London) **1963**, 3281.

167. Shoppee, C. W. und T. Reichstein· Diginin, 1. Mitt. Helv. Chim. Acta **23**, 975 (1940).

168. Siddiqui, S.: The Alkaloids of *Holarrhena antidysenterica*. II. Two Further New Alkaloids from the Bark and the Seeds of Indian *Holarrhena* and their Constitutional Relationship to Conessine. J. Indian Chem. Soc. **11**, 283 (1934).

169. — The Alkaloids of *Holarrhena antidysenterica*. IV. The Occurence of Two Further New Bases in the Bark of Indian *Holarrhena* and their Relationship to Conessine and Holarrhimine. Proc. Indian Acad. Sci., Sect. A **3**, 249 (1936) [Chem. Zbl. **107**, II, 1734 (1936)].

170. Siddiqui, S., S. C. Misra and V. N. Sharma: Chemical Examination of *Holarrhena febrifuga* Klotzsch. I. J. Sci. Indust. Res. (India) **3**, 555 (1945) [Chem. Abstr. **39**, 5399 (1945)].

171. Siddiqui, S. and P. P. Pillay: The Alkaloids of *Holarrhena antidysenterica*. I. Three New Alkaloids from the Bark of Indian *Holarrhena* and New Methods of Isolation and Further Purification of Conessine. J. Indian Chem. Soc. **9**, 553 (1932).

172. — — Preliminary Chemical Examination of the Bark of *Holarrhena antidysenterica*. J. Indian Chem. Soc. **10**, 673 (1933).

173. Siddiqui, S. and R. H. Siddiqui: The Alkaloids of *Holarrhena antidysenterica*. III. Studies in the Action of BrCN on Conessine and its N-Demethylation to *iso*Conessimine and Conimine. J. Indian Chem. Soc. **11**, 787 (1934).

174. Sih, C. J. and K. C. Wang: A New Route to Estrone from Sterols. J. Amer. Chem. Soc. **87**, 1387 (1965).

175. Skarzynski, B.: An Estrogenic Substance from Plant Material. Nature **131**, 766 (1933).

176. Späth, E. und O. Hromatka: Zur Konstitution des Conessins. Ber. dtsch. chem. Ges. **63**, 126 (1930).

177. Stauffacher, D.: Über Buxamin, Nor-buxamin und Buxaminol, neue Alkaloide aus *Buxus sempervirens* L. Helv. Chim. Acta **47**, 968 (1964).

178. Stenhouse, J.: Pharmac. J. [2] **5**, 493 (1864); zitiert nach Nr. *41*.

179. Stork, G., S. D. Darling, I. T. Harrison and P. S. Wharton: A Stereospecific Synthesis of 18-Substituted Steroids. Application to the Synthesis of *dl*-Conessine. J. Amer. Chem. Soc. **84**, 2018 (1962).

180. Tomita, M., S. Uyeo, Jr. and T. Kikuchi: Studies on the Alkaloids of *Pachysandra terminalis* Sieb. et Zucc. Structure of Pachysandrine A and B. Tetrahedron Letters **1964**, 1053.

181. — — — Studies on the Alkaloids of *Pachysandra terminalis* Sieb. et Zucc. (2). Structure of Pachysamine A and B. Tetrahedron Letters **1964,** 1641.

182. Truong-Ho, M., Q. Khuong-Huu et R. Goutarel: Alcaloïdes stéroïdiques, XV. Les irehdiamines A et B, alcaloïdes du *Funtumia elastica* (Preuss) Stapf. Bull. soc. chim. France **1963**, 594.

183. Truong-Ho, M., X. Monseur, Q. Khuong-Huu et R. Goutarel: Alcaloïdes stéroïdiques, XXI. Structures de l'iréhine et de l'iréhamine. Bull. soc. chim. France **1963**, 2332.

184. Tschesche, R. und B. Brassat: Zur Biogenese von Steroidderivaten im Pflanzenreich, 2. Mitt. Bufadienolid-Biogenese aus Δ^5-Pregnenol-3β-on-20-glucosid. Z. Naturforsch. **20 b**, 707 (1965).

185. Tschesche, R. und G. Brügmann: Digitanolglykoside, X. Zur Konstitution des Diginigenins und Digifologenins. Tetrahedron **20**, 1469 (1964).

186. Tschesche, R., G. Brügmann, H.-W. Marquardt und H. Machleidt: Über Digitanolglykoside, VII. Die Konstitution des Digipurpurogenins. Liebigs Ann. Chem. **648**, 185 (1961).

187. Tschesche, R., G. Brügmann und G. Snatzke: Über Digitanolglykoside, IX. Die Konstitution des Digipurpurogenins. Tetrahedron Letters **1964**, 473.

188. Tschesche, R. und G. Buschauer: Über Digitanolglykoside, I. Zur Konstitution von Diginin, Digifolein und Lanafolein. Liebigs Ann. Chem. **603**, 59 (1957).

189. Tschesche, R. und G. Grimmer: Über pflanzliche Herzgifte, XXX. Mitt. Neue Glykoside aus den Blättern von *Digitalis purpurea* und *Digitalis lanata*. Chem. Ber. **88**, 1569 (1955).

190. Tschesche, R., W. Hammerschmidt und G. Grimmer: Über Digitanolglykoside, III. Digacetinin, ein neues Digitanol-Glykosid. Liebigs Ann. Chem. **614**, 136 (1958).

191. Tschesche, R., W. Hammerschmidt und G. Snatzke: Über Digitanolglykoside, VI. Zur Konstitution des Digacetigenins. Liebigs Ann. Chem. **642**, 199 (1961).

192. Tschesche, R., V. Knittel und G. Snatzke: Über Digitanolglykoside, XI. Partialsynthese des Holadysons. Chem. Ber. **98**, 1974 (1965).

193. Tschesche, R. und G. Lilienweiss: Cardenolid-Biosynthese aus Pregnenolonglucosid. Z. Naturforsch. **19 b**, 265 (1964).

194. Tschesche, R. und G. Lipp: Über Digitanolglykoside, IV. Weitere Untersuchungen am Digifolein, Lanafolein und Digipronin. Liebigs Ann. Chem. **615**, 210 (1958).

195. Tschesche, R., G. Lipp und G. Grimmer: Über Digitanolglykoside, II. Zur Kenntnis des Digipronins und Purpnins. Liebigs Ann. Chem. **606**, 160 (1957).

196. Tschesche, R., W. Meise und G. Snatzke: Über Kurchi-Alkaloide, VII. Die Konstitution des Kurcholessins. Tetrahedron Letters **1964**, 1659.

197. Tschesche, R., I. Mörner und G. Snatzke: Über Digitanol-Glykoside, VIII. Holadyson, ein neues Digitanol aus *Holarrhena antidysenterica* Wall. Liebigs Ann. Chem. **670**, 103 (1963).

198. Tschesche, R. und H. Ockenfels: Über Kurchi-Alkaloide, V. 7α-Hydroxyconessin und Holonamin, zwei neue Basen aus Kurchirinde. Chem. Ber. **97**, 2316 (1964).

199. — — Über Kurchi-Alkaloide, VI. Die Strukturaufklärung des Holonamins. Chem. Ber. **97**, 2326 (1964).

200. Tschesche, R. und P. Otto: Über Kurchi-Alkaloide, IV. Weitere Basen aus Kurchirinde Chem. Ber. **95**, 1144 (1962).

201. Tschesche, R. und A. C. Roy: Über Kurchi-Alkaloide, II. Die Konstitution des Conessidins. Chem. Ber. **89**, 1288 (1956).

202. Tschesche, R. und G. Snatzke: Über Digitanol-glykoside, V. Δ^5-Pregnenol-(3β)-on-(20)- und 5α-Pregnanol-(3β)-on-(20)-Glucoside aus der Uzara-wurzel. Liebigs Ann. Chem. **636**, 105 (1960).

203. Tschesche, R., P. Welzel und H.-W. Fehlhaber: Digitanolglykoside, XIII. Massenspektrometrische Untersuchungen am Kondurangogenin A. Tetrahedron **21**, 1797 (1965).

204. Tschesche, R., P. Welzel und G. Snatzke: Digitanolglykoside, XII. Die Konstitution von Kondurangogenin A, dem Aglykon eines Esterglykosides aus der Kondurangorinde. Tetrahedron **21**, 1777 (1965).

205. Tschesche, R. und K. Wiensz: Über Kurchi-Alkaloide, III. Neue Basen aus Kurchirinde. Chem. Ber. **91**, 1504 (1958).

206. Uffer, A.: Holarrhenin aus *Holarrhena congolensis* Stapf. Helv. Chim. Acta **39**, 1834 (1956).

207. Votický, Z. and J. Tomko: Alkaloids from *Buxus sempervirens* L. II. Structure of Buxomegine and its Identity with Irehine. Collect. Czech. Chem. Comm. **30**, 348 (1965).

208. Warnecke, H.: Über Wrightin. Ber. dtsch. chem. Ges. **19**, 60 (1886).

209. Weizmann, A. and Y. Mazur: Steroids and Triterpenoids of Citrus Fruit. II. Isolation of Citrostadienol. J. Organ. Chem. (USA) **23**, 832 (1958).

210. Williams, B. L., L. J. Goad and E. I. Mercer: Triterpenoid Components of Grapefruit Peel. Biochem. J. **96**, 31 P (1965).

211. Winkler, R. E. und T. Reichstein: Die Glykoside der Samen von *Dregea volubilis* (L.) Benth. ex Hook. Glykoside und Aglykone, 131. Mitt. Helv. Chim. Acta **37**, 721 (1954).

212. Zalkow, L. H., N. I. Burke and G. Keen: Occurrence of 5α-Androstane-3β,16α,17α-triol in "Rayless Goldenrod" (*Aplopappus heterophyllus* Blake). Tetrahedron Letters **1964**, 217.

(Eingelaufen am 22. Oktober 1965.)

Cyclite: Biosynthese, Stoffwechsel und Vorkommen.

Von H. KINDL und O. HOFFMANN-OSTENHOF, Wien.

Inhaltsübersicht.

Die im vorliegenden Übersichtsreferat beschriebenen Untersuchungen aus dem Laboratorium der Verfasser wurden durch Förderungsbeiträge der Ludwig-Boltzmann-Gesellschaft, Wien, sowie der Sektion IV (Verstaatlichte Betriebe) des Bundeskanzleramts der Republik Österreich in großzügiger Weise unterstützt, wofür wir auch an dieser Stelle unseren Dank aussprechen.

I. Einleitung.

Unter dem Sammelbegriff *Cyclite* versteht man nach Micheel (*171*), welcher diesen Namen geprägt hat, isozyklische Polyalkohole, deren Hydroxylgruppen an die Ringkohlenstoffe gebunden sind.

Substanzen dieser Klasse sind schon seit sehr langer Zeit als Naturstoffe bekannt. Es scheint, daß bereits 1849 Braconnot (s. *217*) L-Quercit aus Eicheln isolierte; er hielt das Produkt allerdings für Lactose. Ein Jahr später konnte aber Scherer (*236*) eine Substanz aus dem Fleischsaft isolieren, die er als Muskelzucker bzw. Inosit (von ις, ινος griechisch Fleisch) bezeichnete; diese Verbindung trägt heute die Namen myo-Inosit oder meso-Inosit. Dessaignes (*65*), der 1851 den von Braconnot aus den Eicheln isolierten Stoff wieder untersuchte, konnte ihn als eine selbständige Verbindung erkennen und wenige Jahre später berichteten Staedeler und Frerichs (*252*) über eine dem Inosit von Scherer anscheinend nahe verwandte Substanz aus den Organen von Knorpelfischen, welcher sie den Namen Scyllit gaben.

Aus den Ausschwitzungen von *Pinus lambertiana* isolierte Berthelot (*19*) den D-Pinit, einen Monomethyläther des D-chiro-Inosits, eines der beiden möglichen optisch aktiven Hexahydroxycyclohexane. Durch

Literaturverzeichnis: SS. 192—205.

Entmethylierung mit Hilfe von Jodwasserstoff erhielt GIRARD (*91*) aus D-Pinit den Grundkörper D-chiro-Inosit, eine Substanz, die sich ebenfalls im freien Zustand in manchen Pflanzen findet. GIRARD (*90*) konnte auch aus dem Kautschuksaft L-Bornesit, einen Methyläther des myo-Inosits, isolieren. Ein weiterer Methyläther eines Cyclits, der L-Quebrachit, wurde 1889 von TANRET (*256*) aus der Quebrachorinde gewonnen, wobei es dem Autor gelang, die Beziehung der Substanz zum zweiten optisch-aktiven Hexahydroxycyclohexan, dem L-chiro-Inosit, nachzuweisen.

So waren bereits im neunzehnten Jahrhundert die sechs am häufigsten vorkommenden natürlichen Cyclite bekannt; später wurde noch eine Anzahl weiterer Substanzen dieser Klasse, die meisten durch die mühevolle Arbeit von PLOUVIER (*212*), in den Organismen nachgewiesen. Eine vollständige Liste aller bis jetzt in biologischen Materialien aufgefundenen und ausreichend charakterisierten Cyclite findet sich in *Formelübersicht 1*.

Die Aufklärung der Struktur der Cyclite war allerdings mit großen Schwierigkeiten verbunden. Wohl konnte bereits SCHERER (*236*) die richtige Bruttoformel des myo-Inosits aufstellen und nach Vorarbeiten von VOHL (*265*) legte schließlich MAQUENNE (*169*) die Natur dieser Substanz als ein Cyclohexanhexol fest. Obwohl aber schon 1894 BOUVEAULT (*25*) erkannt hatte, daß es insgesamt neun stereoisomere Cyclohexanhexole geben müßte (vgl. *Formelübersicht 2*), waren zur Bestimmung der Konfiguration der einzelnen Isomeren langwierige Arbeiten erforderlich, und erst 1942 konnte dem myo-Inosit eindeutig seine korrekte Konfiguration zugeschrieben werden (*63, 216*).

Für den Biochemiker gewannen die Cyclite zum ersten Mal Interesse, als EASTCOTT (*69*) nachweisen konnte, daß myo-Inosit ein essentieller Bestandteil des seinerzeit von WILDIERS (*273*) entdeckten Hefewuchsstoffs „Bios" ist. In der Folge wurden auch Befunde erhoben, nach welchen myo-Inosit für bestimmte Säugetiere (*275*) und auch für manche Stämme von Zellkulturen menschlichen Ursprungs (*68*) ein Vitamin bzw. einen Wuchsstoff darstellt. Nach manchen, allerdings teilweise widersprüchlichen Berichten [vgl. dazu POSTERNAK (*217*)] wird dem myo-Inosit auch eine lipotrope Wirkung zugeschrieben.

Für den Biochemiker bedeutsam ist aber auch, daß myo-Inosit ein wesentlicher Bestandteil einer Klasse von Phospholipiden, der sogenannten Phosphoinositide, ist. Schon 1930 konnten ANDERSON und ROBERTS (*10*) myo-Inosit als Bestandteil von Phosphatiden aus dem Tuberkelbazillus identifizieren; 1939 berichteten KLENK und SAKAI (*134*) als erste über myo-Inosit-haltige Verbindungen in den Lipidfraktionen aus pflanzlichen Materialien, und FOLCH und WOOLLEY (*85*) fanden auch bald darauf Phosphoinositide als Teil der Phosphatidfraktion in tierischen Geweben. Heute besteht wohl kein Zweifel darüber, daß in den Phospholipidfraktionen aus allen biologischen Materialien

(1)
myo-Inosit.

(2)
L-Bornesit.

(3)
D-Bornesit.

(4)
Sequoyit.

(5)
D-Ononit.

(6)
Dambonit.

(7)
Liriodendrit.

(8)
scyllo-Inosit.

(9)
D-chiro-Inosit.

(10)
D-Pinit.

(11)
L-chiro-Inosit.

(12)
L-Quebrachit.

(13)
L-Pinit.

(14)
1-O-Methyl-muco-inosit.

(15)
myo-Inosit-hexaphosphat.

(16)
scyllo-Inosit-hexaphosphat.

(17)
neo-Inosit-hexaphosphat.

Literaturverzeichnis: SS. 192—205.

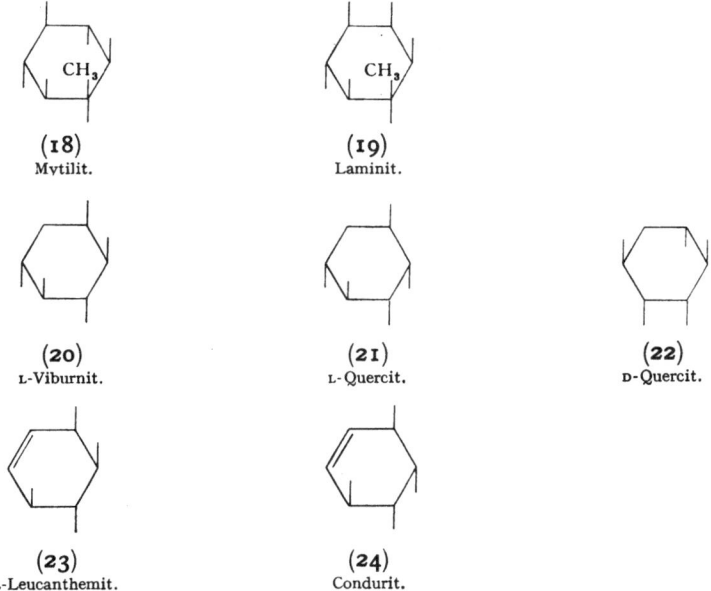

Formelübersicht 1. Natürlich vorkommende Cyclite.

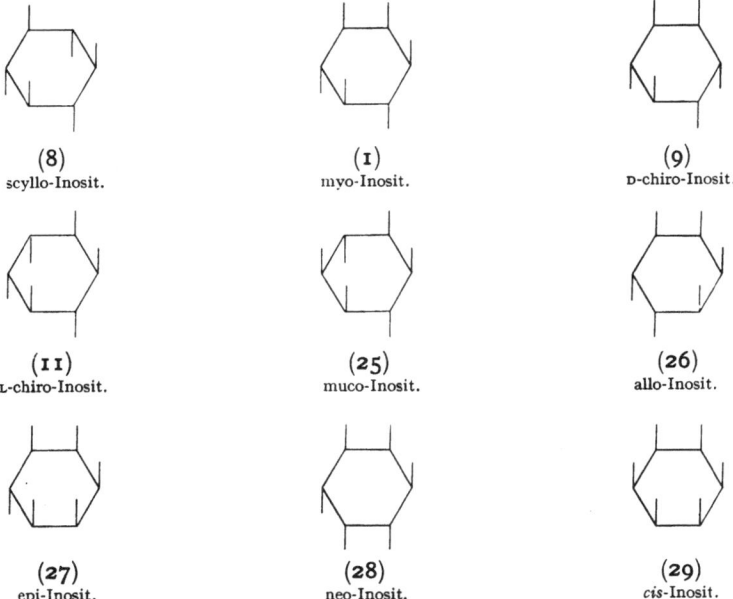

Formelübersicht 2. Die isomeren Inosite.

Phosphoinositide vorkommen. Allerdings sind trotz intensiver Bearbeitung des Problems die Funktionen der Phosphoinositide bis heute noch nicht klar umreißbar, obwohl bereits Befunde über eine spezifische Beteiligung bei der ATP-abhängigen Kontraktion der Mitochondrien (*262, 263*) sowie über eine Funktion bei Permeationsvorgängen (vgl. HOKIN und HOKIN, *108a*) vorliegen; jedenfalls handelt es sich hier ohne Zweifel um Substanzen, denen biologisch eine große Bedeutung zukommt.

Es ist weder sinnvoll noch möglich, im Rahmen eines Übersichtsreferats der hier vorliegenden Art alle Aspekte der Biochemie der Cyclite ausführlich zu behandeln. Wir haben uns deshalb auf die Besprechung der drei im Titel genannten Teilgebiete beschränkt. Diese Auswahl wurde vor allem deshalb getroffen, weil gerade hier in letzter Zeit neue Erkenntnisse erarbeitet wurden, die bisher noch nirgends zusammenhängend behandelt wurden, aber auch, weil es sich um diejenigen Probleme handelt, über die im Laboratorium der Verfasser seit Jahren gearbeitet wird.

Die an sich so interessanten neueren Ergebnisse über Phosphoinositide wurden kürzlich in einer Monographie über Phospholipide [ANSELL und HAWTHORNE (*13*)] behandelt. In der vorliegenden Arbeit sollen nur Befunde berichtet werden, welche das Vorkommen und den unmittelbaren Stoffwechsel der Cyclite betreffen; der Stoffwechsel und die Funktionen der Phosphoinositide werden demnach nicht berührt. Für die allgemeine Chemie der Cyclite sei auf die ausgezeichnete Monographie von POSTERNAK (*217*) verwiesen, in der übrigens auch eine kritische Darstellung der bisher vorliegenden Befunde über myo-Inosit als Vitamin, Wuchsstoff und lipotropes Agens zu finden ist.

II. Nomenklatur der Cyclite.

Die Nomenklatur der Cyclite bietet zahlreiche schwierige Probleme, von denen einige bis heute noch keiner einvernehmlichen Lösung zugeführt wurden.

Das ursprüngliche auf MAQUENNE zurückgehende System ist auch jetzt noch die Grundlage für die Bezeichnung der Cyclite; allerdings machte die Entwicklung der Cyclitchemie einige Modifikationen erforderlich (*11, 12, 83, 217*). In den letzten Jahren hat sich eine Unterkommission der IUPAC-IUB-Kommission für biochemische Nomenklatur damit beschäftigt, eine derartige Modifikation des Systems von MAQUENNE auszuarbeiten. In den meisten Punkten wurde Übereinstimmung erzielt. Da darüber kein Zweifel besteht, daß die vorliegenden Ergebnisse dieser aus hervorragenden Fachleuten auf dem Gebiet zusammengesetzten Gruppe die Grundlage jeder zukünftigen Cyclitnomenklatur darstellen werden, sollen die so entwickelten Nomenklaturvorschläge hier kurz referiert werden; auch im Rahmen der vorliegenden Arbeit werden die einzelnen Cyclite und ihre Derivate nach den im folgenden geschilderten Prinzipien benannt.

Literaturverzeichnis: SS. 192—205.

A. Schreibweise der Formeln.

Für die Cyclite wird allgemein vorgeschlagen, daß der Ring als Ebene senkrecht zur Papierebene betrachtet wird, wobei die Substituenten und Wasserstoffatome sich oberhalb oder unterhalb des Ringes befinden. Beim Schreiben der Formeln werden die Wasserstoffatome dann völlig ausgelassen, während die Hydroxylgruppen durch senkrechte Striche nach oben oder nach unten bezeichnet werden. Bei modifizierten Hydroxylgruppen (—OCH_3, CH_3COO— usw.) und anderen Substituenten (—NH_2, —SH, usw.) wird die entsprechende Gruppe ausgeschrieben; in diesem Falle repräsentieren die senkrechten Striche Bindungen.

B. Bezifferung der Ringkohlenstoffatome.

1. Wenn nur Hydroxylgruppen oder modifizierte Hydroxylgruppen als Substituenten vorliegen, gelten folgende Regeln:
a) Die größte Anzahl von Substituenten und die größte Anzahl aufeinander folgender zueinander cis-ständiger Substituenten soll oberhalb des Ringes stehen; b) Ziffer 1 kann nur ein C-Atom erhalten, das einen Substituenten trägt; das Kohlenstoffatom einer —CH_2-Gruppe erhält somit die höchste mögliche Zahl; c) Die Ziffer 1 erhält dasjenige Kohlenstoffatom, dem die größtmögliche Anzahl von Kohlenstoffatomen mit cis-ständigen Substituenten folgt. d) Bei einer Unterbrechung der Reihenfolge von Kohlenstoffatomen mit cis-ständigen Substituenten durch C-Atome mit trans-ständigen Substituenten oder solchen ohne Substituenten soll das nach der Unterbrechung folgende nächste Kohlenstoffatom mit zum ersten Kohlenstoffatom cis-ständigem Substituenten die niedrigste mögliche Zahl erhalten.

2. Wenn außer Hydroxylgruppen noch andere Substituenten vorkommen, welche jedoch nach den Regeln der organischen Nomenklatur bei der Anführung der Suffixe gegenüber der Hydroxylgruppe keine Priorität besitzen, erhalten diese Substituenten (—SH, —NH_2) unter Berücksichtigung der Punkte 1 a, 1 b und 1 c die niedrigsten möglichen Nummern.

3. Bei Vorliegen von Substituenten im Ring, welche nach den Regeln der organischen Nomenklatur gegenüber der Hydroxylgruppe Priorität bei der Anführung der Suffixe innehaben, erhalten die Kohlenstoffatome des Rings, die solche Substituenten tragen, die Nummer 1.

C. Benennung der Verbindungen.

1. Cyclohexanhexole werden als Inosite bezeichnet. Für alle anderen Polyhydroxycycloalkane folgt die Nomenklatur den Beispielen: Cyclohexanpentol, Cycloheptantriol, Cyclohexentetrol.

2. Die Stellung der Hydroxylgruppen wird mit Hilfe eines Bruchausdrucks dargestellt, wobei die Nummern derjenigen Kohlenstoffatome, deren Hydroxylgruppen oberhalb des Ringes liegen, in den Zähler und die Nummern derjenigen Kohlenstoffatome, deren Hydroxylgruppen unterhalb des Ringes liegen, in den Nenner geschrieben werden. Beispiele: 1,2,3,5/4,6-Inosit (myo-Inosit) (1, S. 152); 1,2,4/3-Cyclohexentetrol (L-Leucanthemit) (23, S. 153).

3. Modifizierte Hydroxylgruppen werden unter Angabe der Stellung vor den Bruchausdruck geschrieben. Beispiel: 5-O-Methyl-1,2,3,5/4,6-inosit (Sequoyit) (4).

4. Amino- und Sulfhydrylgruppen werden an den Bruchausdruck anschließend dem Namen vorangesetzt; innerhalb des Bruchausdrucks wird die der Substitution entsprechende Zahl durch das Symbol N oder S gekennzeichnet. Beispiele: 1,2 N,3/4,5-Aminocyclopentantetrol,   2 S,3 N/1,5-Aminomercaptocyclohexandiol, 1,2 N,3,5/4,6-Aminodesoxyinosit.

5. Bei Vorliegen von Substituenten, welche nach den Regeln der organischen Nomenklatur gegenüber der Hydroxylgruppe Priorität bei der Anführung der Suffixe besitzen, wird das der entsprechenden Gruppe zukommende Suffix dem Namen nachgesetzt, wobei im Gegensatz zur sonstigen Übung die Hydroxylgruppen dem Namen vorangehen. Beispiel: 2,4,6/3,5-Pentahydroxycyclohexanon (44, S. 169).

D. Bezifferung der optisch aktiven Cyclite; Richtungssinn der Bezifferung.

In diesem Bereich bestehen noch Divergenzen.

Nach dem Vorschlag der einen Gruppe, welcher sich stärker auf das ursprüngliche System von MAQUENNE beruft, erfolgt die Numerierung grundsätzlich nur im Sinne des Uhrzeigers. In Verbindung mit den systematischen Namen definiert diese Bezifferung eindeutig die Stellung der Substituenten im Ring. Auch die absolute Stereochemie der Substituenten wird festgelegt, ohne daß Symbole wie D, L, R oder S oder Präfixe wie cis- oder trans- verwendet werden. Das System selbst gibt allerdings keine Angabe über eine etwaige optische Aktivität der Verbindungen; Enantiomere lassen sich nicht ohne weiteres erkennen, da die vor den Namen gestellten Bruchausdrücke verschieden sind. Die Enantiomeren werden durch den vor den Bruchausdrücken angegebenen Drehsin (+) oder (—) gekennzeichnet. Beispiel: Die beiden optisch aktiven Inosite werden als (+)-1,2,5/3,4,6-Inosit (9) und (—)-1,2,4/3,5,6-Inosit (11) bezeichnet.

Nach den Vorschlägen der anderen Gruppe kann die Bezifferung der Formeln sowohl im Uhrzeigersinn als auch im entgegengesetzten Sinn erfolgen. Dadurch ergibt sich gleiche Bezifferung für enantiomere Verbindungen. Zur Bezeichnung der absoluten Konfiguration wird eine vertikale Fischer-Projektion mit dem Kohlenstoffatom 1 an der Spitze betrachtet. Dabei ist die Richtung des Substituenten am aktiven Zentrum mit der niedrigsten Nummer konfigurationsbestimmend. Die entsprechenden Symbole (D und L) werden vor den Bruchausdruck gesetzt. Die beiden optisch aktiven Inosite werden nach diesem System z. B. als D-1,2,4/3,5,6-Inosit (9) und L-1,2,4/3,5,6-Inosit (11) bezeichnet. An Stelle der Konfigurationsangabe 1,2,4/3,5 kann der Ausdruck „chiro" verwendet werden. Bei Anwendung dieses Systems muß aber Punkt B-1 erweitert werden: e) Modifizierte Hydroxylgruppen (Ester, Äther, Acetale usw.) sollen unter Berücksichtigung von B-1a bis B-1d die niedrigsten möglichen Nummern erhalten.

In der vorliegenden Arbeit bedienen wir uns des zuletzt beschriebenen Systems, das uns für die Beschreibung der als Naturstoffe vorkommenden Cyclite als besonders geeignet erscheint.

Der hier kurz zusammengefaßten Nomenklatur liegt die Schreibweise des Cyclohexanringes als gleichseitiges Sechseck zugrunde. Diese Bevorzugung der planaren Darstellung gegenüber der Schreibweise in Sesselform, in der die meisten Cyclohexanverbindungen vorliegen, bei nomenklatorischen Fragen, führte neben anderen Gründen dazu, daß wir bei der Darstellung der Formeln im folgenden Text die hexagonale, flächige Darstellung verwenden. Es wäre denkbar, daß bei einer Bevorzugung der Darstellung in Sesselform nicht die Stellungen oberhalb oder unterhalb des Ringes zur Bezifferung herangezogen werden, sondern die Lage der axialen bzw. äquatorialen Substituenten angegeben wird.

III. Biosynthese.

Schon MAQUENNE (169), der als erster die Struktur des myo-Inosits als Cyclohexanhexol erkannt hatte, vermutete eine biogenetische Beziehung zwischen den einfachen Monosacchariden und myo-Inosit; der

Literaturverzeichnis: SS. 192—205.

Cyclit sollte seiner Vorstellung nach etwa durch eine Aldolkondensation zwischen den Kohlenstoffatomen 1 und 6 einer Hexose entstehen. Nachdem dann 1942 die letzten Zweifel über die Konfiguration des myo-Inosits beseitigt worden waren und es damit offensichtlich wurde, daß die Konfiguration von vier benachbarten asymmetrischen Zentren in myo-Inosit (1) und D-Glucose (30) die gleiche ist, postulierte H. O. L. FISCHER (*81*), daß D-Glucose der Vorläufer des myo-Inosits in den Organismen ist und durch eine Ringschlußreaktion in diesen übergeht (*Reaktionsschema 1*). Zur Stützung dieser Hypothese trug auch bei, daß es demselben Autor gelungen war, durch eine alkalische Aldolkondensation von 6-Desoxy-6-nitro-D-glucose eine Mischung von Nitroverbindungen von myo-Inosit und anderen Inositen zu erhalten (*97*).

$$
\begin{array}{c}
\text{OH} \\
|\\
\text{CH}_2\!-\!\!\!\!\!\\
\text{O}\!=\!\text{CH}
\end{array}
\qquad \longrightarrow
$$

(30)
D-Glucose.

(1)
myo-Inosit.

Reaktionsschema 1. Bildung von myo-Inosit aus D-Glucose nach FISCHER (*81*).

Während somit seit längerer Zeit ziemlich wohlbegründete Vorstellungen über die Biosynthese von myo-Inosit bestanden, erweckte die Bildung der anderen Cyclite in den Organismen erst in allerletzter Zeit das Interesse der Bearbeiter des Gebiets.

A. Biosynthese des myo-Inosits.

1. Vorversuche.

Etwa zur gleichen Zeit, als FISCHER (*81*) seine Hypothese über die Biosynthese des myo-Inosits aus D-Glucose formulierte, berichteten FERNÁNDEZ und Mitarb. (*78*, *79*) über die Umwandlung von freier D-Glucose und D-Glucosephosphaten in myo-Inosit in den Blättern von *Prunus armeniaca*. Dieselben Autoren beschrieben auch ein aus den Blättern der gleichen Pflanze isoliertes Enzym, welches sie „Cyclase" nannten und das die Zyklisierung von D-Glucose und ihrer Phosphatester zu myo-Inosit katalysieren sollte. Diese Befunde wurden mit einfachen kolorimetrischen Methoden, d. h. ohne Verwendung radioaktiv markierter Vorstufen oder der Papierchromatographie erhoben. Da wir heute wissen, daß die in einem derartigen Versuch zu erwartenden Umsätze sehr klein sind und kaum mit so einfachen analytischen Hilfsmitteln verfolgt werden können, werden die Ergebnisse der spanischen Autoren hier mit allem Vorbehalt referiert. 1949 beobachtete KURSANOV (*143a*) eine Zunahme des Gehalts an myo-Inosit nach Gabe von Zuckern an höhere Pflanzen; dies konnte später nicht reproduziert werden (*244a*).

Es war von vornherein sicher, daß alle höheren Pflanzen zur Synthese von myo-Inosit imstande sind, was auch durch Versuche über die Bildung von markiertem myo-Inosit aus $^{14}CO_2$ in photosynthetisierenden Pflanzen erhärtet werden konnte (5, 125, 131, 173, 227, 239). In Anbetracht der für manche Tiere und auch für verschiedene Arten von Mikroorganismen postulierten Vitamin- bzw. Wuchsstoff-Funktion des myo-Inosits mußte vorerst die Frage entschieden werden, ob höhere Tiere überhaupt zu einer Biosynthese von myo-Inosit befähigt sind oder ob sie ihren Bedarf an diesem Cyclit ausschließlich aus der Nahrung oder über die in ihnen symbiotisch lebende Darmflora decken. DAUGHADAY und Mitarb. (64) injizierten D-Glucose-U-^{14}C* in das Chorion von vorinkubierten Hühnereiern und erhielten einen Einbau in myo-Inosit mit einer radiochemischen Ausbeute von 0,01%. HALLIDAY und ANDERSON (100) beobachteten dieselbe Reaktion in der Ratte; die radiochemische Ausbeute war hier 0,07%. Der Nachweis, daß es das tierische Gewebe selbst und nicht die Darmflora ist, welches in der Ratte für die Bildung des myo-Inosits verantwortlich ist, gelang FREINKEL und DAWSON (87) bei Versuchen mit keimfreien Ratten, welche ebenfalls eindeutig einen Einbau der Aktivität von D-Glucose-U-^{14}C in myo-Inosit zeigten. Auch über den Ort der Synthese im höheren Tier wurde gearbeitet. Nachdem MANN (167a) bereits 1954 gezeigt hatte, daß die Samenflüssigkeit des Ebers einen besonders hohen Gehalt an myo-Inosit (bis 2%) aufweist, fanden EISENBERG und BOLDEN (72) eine ähnlich hohe Konzentration an myo-Inosit in der Samenflüssigkeit männlicher Ratten und konnten zeigen, daß dort die Biosynthese in den Tubuli der Hoden erfolgt. Damit ist aber die Frage noch nicht völlig geklärt; es ist weder sicher, daß in den männlichen Tieren der Hoden das einzige Gewebe ist, welches myo-Inosit produziert, noch wissen wir etwas über den Ort der Bildung des Cyclits in weiblichen Tieren.

In diesem Zusammenhang müssen auch die Untersuchungen der Schule von EAGLE (67, 68) erwähnt werden. Bestimmte Stämme von menschlichen Zellkulturen, von denen bekannt ist, daß sie myo-Inosit als Wuchsstoff in ihrem Nährmedium benötigen, sind ebenso wie Mäusefibroblasten, die keinen myo-Inosit brauchen, dazu imstande, den Cyclit aus D-Glucose aufzubauen. Diese Befunde scheinen gemeinsam mit den vorher berichteten Ergebnissen dahin zu deuten, daß Tiere und tierische Zellen, welche von außen zugeführten myo-Inosit als Nährstoff benötigen, an sich die zur Synthese dieser Substanz notwendigen Enzyme besitzen. Eine Vitaminfunktion des myo-Inosits ließe sich hier nur so erklären, daß die endogene Synthese den Bedarf an myo-Inosit nicht zu decken vermag. Inwieweit analoge Überlegungen auch für diejenigen Mikroorganismen, welche myo-Inosit als Wuchsstoff benötigen, Gültigkeit besitzen, ist noch unbekannt.

------

* U = gleichmäßig markiert.
Literaturverzeichnis: SS. 192—205.

2. Versuche mit selektiv markierten Vorstufen.

Es wurde bald erkannt, daß Einbauversuche mit spezifisch markierten Vorstufen am besten für die Klärung des Biosynthese-mechanismus bei myo-Inosit geeignet sind. Dem Vorhergesagten entsprechend, lag es nahe, an verschiedenen Kohlenstoffatomen selektiv mit ^{14}C markierte D-Glucose einzusetzen. Dabei war die primäre Fragestellung: Wird die intakte Kohlenstoffkette von D-Glucose nach den Vorstellungen von MAQUENNE (*169*) und FISCHER (*81*) in den myo-Inosit eingebaut oder erfolgt vorher ein Abbau bzw. eine Fragmentierung des Zuckermoleküls. Zur Entscheidung dieser Frage wurden zwei Methoden herangezogen:

1. Es wurde geprüft, ob alle Kohlenstoffatome des Monosaccharids gleich gut in myo-Inosit eingebaut werden. Ein gleichmäßiger Einbau wurde als Hinweis für die Überführung des intakten Kohlenstoffskeletts, ein verschieden starker Einbau hingegen als Argument für eine vorhergehende Fragmentierung aufgefaßt. 2. Der biosynthetisch aus selektiv markierten Molekülen von D-Glucose erhaltene myo-Inosit wurde in solcher Weise abgebaut, daß man die Aktivität der einzelnen Kohlenstoffatome des Cyclitmoleküls bestimmen konnte. Rückblickend kann man feststellen, daß es fast ausschließlich die zweite der genannten Methoden war, welche schließlich zur Klärung des Problems führte.

a) Frühe Versuche über die Entstehung von myo-Inosit in Mikroorganismen. Es ist unzweifelhaft das Verdienst von CHARALAMPOUS (*39, 40, 43*) als erster beide geschilderten Methoden zur Aufklärung der Biosynthese von myo-Inosit angewandt zu haben. Bei der Untersuchung der Verhältnisse in *Candida utilis* setzte dieser Autor sowohl markierte Ameisensäure und Essigsäure als auch D-Glucose-1-^{14}C, -2-^{14}C und -6-^{14}C ein. Er beobachtete einen verschieden starken Einbau der verschiedenen selektiv markierten D-Glucose-Präparate in den myo-Inosit. Ein von ihm ausgearbeiteter selektiver Abbau zur Bestimmung der Aktivität der einzelnen Kohlenstoffatome im myo-Inosit verläuft nach Öffnung des Cyclohexanrings mit Bleitetraacetat über DL-Idarsäure (*31*, S. 160), die nach Trennung der optischen Antipoden zu Glyoxylsäure und Ameisensäure oxydiert wird *(Reaktionsschema 2)*.

Die mit diesem Abbauweg erhaltenen Ergebnisse und das bereits erwähnte Verhalten der verschiedenen selektiv markierten D-Glucose-Präparate wurden vom Autor in dem Sinne gedeutet, daß bei der Biosynthese von myo-Inosit D-Glucose in C_2- und C_4-Bruchstücke aufgespalten wird, die ihrerseits in einem Gleichgewicht miteinander stehen. Aus dem „pool" dieser C_2- und C_4-Körper erfolge dann der Aufbau des myo-Inosits. Diese Vorstellungen sind auf Grund aller heute vorliegenden Ergebnisse als falsch anzusehen und wurden auch vom Autor selbst zurückgezogen (*51*).

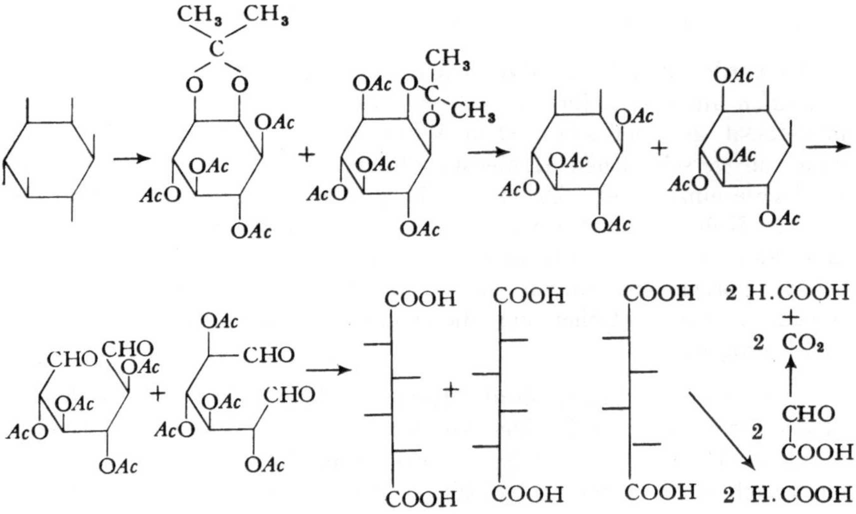

$$CH_3 \quad CH_3$$

(31)
DL-Idarsäure.

Reaktionsschema 2. Selektiver Abbau von myo-Inosit nach CHARALAMPOUS (40).

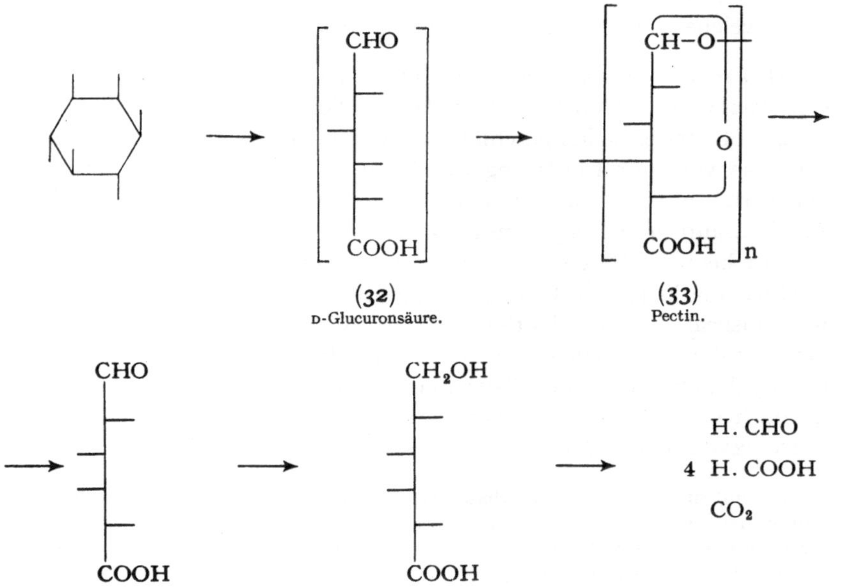

(32)
D-Glucuronsäure.

(33)
Pectin.

(34)
L-Galaktonsäure.

H. CHO

4 H. COOH

CO_2

Reaktionsschema 3. Selektiver Abbau von myo-Inosit nach LOEWUS and KELLY (158).
Literaturverzeichnis: SS. 192—205.

b) Untersuchungen über die Entstehung von myo-Inosit aus D-*Glucose in höheren Pflanzen.* Die Biosynthese von myo-Inosit wurde unter Einsatz selektiv markierter D-Glucose in zwei verschiedenen Pflanzen mit

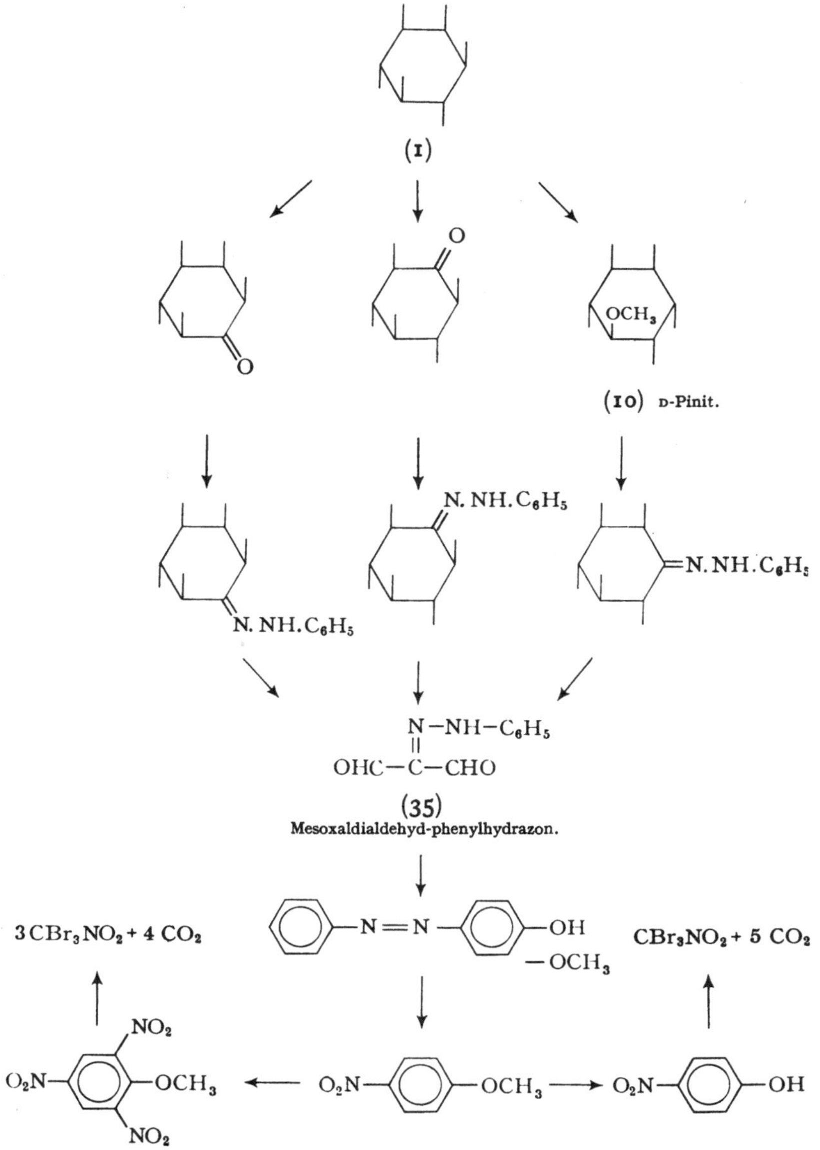

(I)

(10) D-Pinit.

N. NH. C_6H_5

N. NH. C_6H_5

N. NH. C_6H_5

$$\begin{array}{c} N-NH-C_6H_5 \\ \parallel \\ OHC-C-CHO \end{array}$$

(35)
Mesoxaldialdehyd-phenylhydrazon.

$3\,CBr_3NO_2 + 4\,CO_2$ ⟶ $N = N$ —OH $CBr_3NO_2 + 5\,CO_2$
—OCH_3

O_2N —OCH_3 ⟵ O_2N —OCH_3 ⟶ O_2N —OH
NO_2 / NO_2

Reaktionsschema 4. Selektiver Abbau von myo-Inosit nach KINDL und HOFFMANN-OSTENHOF (*128, 129*).

sehr unterschiedlichen Abbaumethoden untersucht. LOEWUS und KELLY (*158*) verwendeten Blätter von *Petroselium* für ihre Einbauversuche und bauten den aus D-Glucose-1-^{14}C entstandenen myo-Inosit unter teilweiser Benützung biologischer Reaktionen selektiv ab *(Reaktionsschema 3)*. Zu diesem Zwecke wurde der biosynthetisch erhaltene myo-Inosit in unreife Erdbeeren infundiert, wobei daraus — wahrscheinlich auf dem Wege über D-Glucuronsäure (**32**) — die D-Galakturonsäure-Einheiten des Pectins (**33**) gebildet werden. Die nach Hydrolyse des Pectins und Reduktion entstandene L-Galaktonsäure (**34**) wird mit Perjodat oxydativ abgebaut. Die Ergebnisse zeigten, daß der aus D-Glucose-1-^{14}C entstandene myo-Inosit fast ausschließlich im Kohlenstoffatom 6 markiert war.

In unserem Laboratorium (*127, 128*) wurde die Bildung des myo-Inosits aus D-Glucose-1-^{14}C und -2-^{14}C in der Senfpflanze *(Sinapis alba)* untersucht. Zur Lokalisierung der Aktivität innerhalb des myo-Inosits wurde ein ausschließlich auf chemischen Reaktionen beruhender Abbau *(Reaktionsschema 4*, S. 161) entwickelt. Dabei ist der erste Schritt die Oxydation des myo-Inosits zu einem Pentahydroxycyclohexanon („Inosose"). Wie aus Reaktionsschema 4 ersichtlich ist, stehen dabei Methoden zur Herstellung mehrerer Pentahydroxycyclohexanone aus myo-Inosit zur Verfügung. Aus dem entstandenen Pentahydroxycyclohexanon wird das entsprechende Phenylhydrazon hergestellt, das dann mit Perjodat zu einem C$_3$-Bruchstück (**35**) oxydiert wird. Dieses wird mit Aceton kondensiert und auf dem im Schema beschriebenen Weg weiter abgebaut. Mit Hilfe dieser Methode ist es möglich, die einzelnen C-Atome des durch Biosynthese entstandenen myo-Inosits voneinander zu trennen und ihre Aktivität isoliert zu bestimmen. Ein weiterer Vorteil des Abbauwegs besteht darin, daß er nicht nur für myo-Inosit, sondern ganz generell für alle Cyclite anwendbar ist, da diese Substanzen durchwegs in meist sehr einfachen Reaktionen in Ketone überführt werden können (*7, 107*), die in analoger Weise weiterbehandelt werden.

Im Falle von *Sinapis alba* konnte mit Hilfe der geschilderten Methode gezeigt werden, daß D-Glucose-1-^{14}C mit einer Einbaurate von 0,2% in myo-Inosit umgewandelt wird, in welchem sich die Aktivität fast ausschließlich im Kohlenstoffatom 6 findet. Unter gleichen Bedingungen entsteht in derselben Pflanze aus D-Glucose-2-^{14}C myo-Inosit-5-^{14}C (*128*).

Aus den Befunden von LOEWUS und KELLY (*158, 159*) einerseits und denjenigen von KINDL und HOFFMANN-OSTENHOF (*127, 128*) andererseits kann man schließen, daß ganz allgemein die Biosynthese des myo-Inosits in höheren Pflanzen durch direkten Ringschluß aus D-Glucose erfolgt und somit den Theorien von MAQUENNE und FISCHER entspricht.

c) Untersuchungen über die Biosynthese von myo-Inosit aus D-Glucose in Tieren. Versuche über den Einbau von verschieden markierten

D-Glucose-Präparaten ergaben vorerst ziemlich wechselnde Resultate. Während POSTERNAK und Mitarb. (*220*) sowie IMAI (*115*) einen fast gleich starken Einbau von D-Glucose-1-^{14}C, -2-^{14}C und -6-^{14}C in myo-Inosit in Ratten feststellten, berichteten HAUSER und FINELLI (*103*), daß D-Glucose--1-^{14}C und D-Glucose-2-^{14}C in Gewebeschnitten aus Rattennieren fast doppelt so gut in myo-Inosit überführt werden als D-Glucose-6-^{14}C. Die zuletzt berichteten Befunde können allerdings kaum als Hinweis auf eine Fragmentierung des Monosaccharids vor Einbau in den myo-Inosit gewertet werden, da im verwendeten Gewebe sowohl Ausgangsstoff als auch Endprodukt einem starken Stoffwechsel unterliegen.

EISENBERG und BOLDEN (*71*) gelang es, Homogenate aus Rattentestes herzustellen, welche zur Überführung von D-Glucose in myo-Inosit imstande waren. Unter Anwendung der oben berichteten Methoden zur Erforschung der Biosynthese des myo-Inosits in *Petroselium* konnte gezeigt werden (*74*), daß aus D-Glucose-1-^{14}C in diesem System vorwiegend myo-Inosit-6-^{14}C entsteht, während aus D-Glucose-6-^{14}C ein myo-Inosit gebildet wird, der in Stellung 1 markiert ist. D-Glucose-2-^{14}C ergibt einen myo-Inosit, dessen Hauptaktivität in den Kohlenstoffatomen 3, 4 oder 5 lokalisiert ist, wobei keine weitere Differenzierung vorgenommen wurde.

Diese Ergebnisse lassen sich nur in dem Sinn deuten, daß auch beim höheren Tier eine den Theorien von MAQUENNE und FISCHER entsprechende Ringschlußreaktion der D-Glucose zum myo-Inosit führt.

d) Neuere Untersuchungen über die Umwandlung von D-Glucose in myo-Inosit in Mikroorganismen. Die zuletzt berichteten an höheren Pflanzen und Tieren erhobenen Befunde, nach welchen der Mechanismus der Biosynthese in den untersuchten Organismen grundsätzlich von demjenigen verschieden war, den CHARALAMPOUS (*40*) auf Grund seiner Versuche mit dem Sproßpilz *Candida utilis* postuliert hatte, waren der Anlaß zu einer Wiederaufnahme der Untersuchungen an diesem Organismus. Dabei erwies es sich von besonderem Vorteil, daß es in der Zwischenzeit CHEN und CHARALAMPOUS (*50, 52*) gelungen war, ein zellfreies Präparat des Enzymsystems aus *Candida utilis* herzustellen, welches in Anwesenheit von NAD$^+$ und Mg^{++} D-Glucose-6-phosphat in myo-Inosit überführt, während für die Umwandlung von freier D-Glucose in myo-Inosit die Anwesenheit von ATP als weiterer Cofactor erforderlich ist*.

Vorerst konnten NEUBACHER, KINDL und HOFFMANN-OSTENHOF (*179*) zeigen, daß das nach CHEN und CHARALAMPOUS (*50*) hergestellte Enzympräparat D-Glucose-2-^{14}C ohne Fragmentierung in myo-Inosit-5-^{14}C überführt. Bei eingehenderen Untersuchungen von KINDL, BIEDL-NEUBACHER

---

* Für die hier verwendete Nomenklatur der Enzyme und Coenzyme vgl. (*74b*).

und HOFFMANN-OSTENHOF (*126*) fand die bereits bei der Besprechung der Biosynthese von myo-Inosit in höheren Pflanzen berichtete Methode des selektiven Abbaus über verschiedene Pentahydroxycyclohexanone Anwendung. Es konnte eindeutig bewiesen werden, daß auch bei diesem Mikroorganismus ein direkter Ringschluß ohne vorhergehende Fragmentierung der Bildung von myo-Inosit zugrunde liegt.

Ähnliche Ergebnisse führten auch CHEN und CHARALAMPOUS (*51*) zur Ansicht, daß bei der Biosynthese des myo-Inosits in *Candida utilis* keine Fragmentierung des Kohlenstoffskeletts der D-Glucose erfolgt, und damit zur Zurückziehung der seinerzeit von CHARALAMPOUS aufgestellten Hypothese der C_2-C_4-Kondensation. CHEN und CHARALAMPOUS (*51*) oxydierten den mit Hilfe ihrer Enzym-Präparate aus *Candida utilis* aus selektiv markierter D-Glucose gewonnenen myo-Inosit mit der myo-Inosit-Oxygenase aus Rattenniere (vgl. S. 176) zu D-Glucuronsäure (*32*), die dann entsprechend *Reaktionsschema 5* weiter abgebaut wurde.

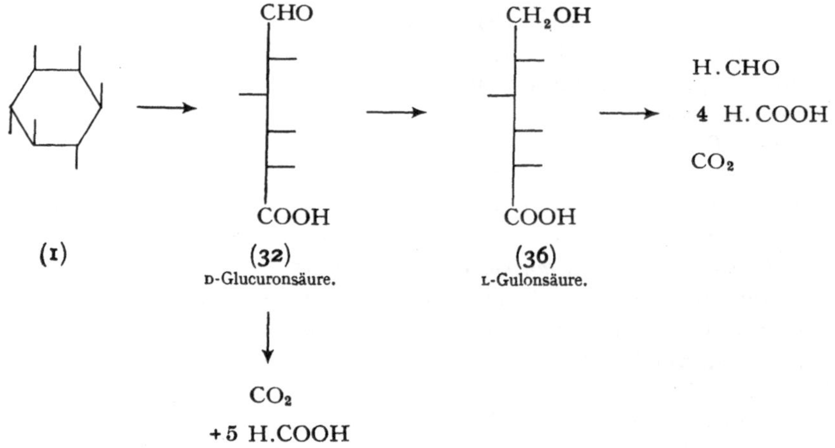

$$
\begin{array}{ccccc}
& & \text{CHO} & & \text{CH}_2\text{OH} \\
& & | & & | \\
& & | & & | \\
& \longrightarrow & | & \longrightarrow & | & \longrightarrow & \begin{array}{l} \text{H.CHO} \\ 4\ \text{H.COOH} \\ \text{CO}_2 \end{array} \\
& & | & & | \\
& & \text{COOH} & & \text{COOH} \\
(\text{1}) & & (\text{32}) & & (\text{36}) \\
& & \text{D-Glucuronsäure.} & & \text{L-Gulonsäure.}
\end{array}
$$

$$\downarrow$$

$$\text{CO}_2$$
$$+5\ \text{H.COOH}$$

Reaktionsschema 5. Selektiver Abbau von myo-Inosit nach CHEN und CHARALAMPOUS (*51*).

e) Untersuchungen über den Mechanismus der Umwandlung von D-*Glucose in myo-Inosit und über Zwischenprodukte dieser Ringschlußreaktion.* Die in den letzten Abschnitten berichteten Ergebnisse scheinen eindeutig zu zeigen, daß es einen einheitlichen Weg der Biosynthese des myo-Inosits gibt, wobei D-Glucose durch eine Ringschlußreaktion im Sinne der Theorien von MAQUENNE und FISCHER in den Cyclit überführt wird.

Die Bildung von Ringsystemen durch Entstehung einer Kohlenstoff-Kohlenstoffbindung ist eine in der Biochemie nur selten aufzufindende Reaktion. Derjenige Vorgang, der der myo-Inosit-Entstehung aus D-Glu-

Literaturverzeichnis: SS. 192—205.

cose am ähnlichsten ist, scheint die Überführung von 3-Desoxy-D-*arabino*-heptulosonsäure-7-phosphat (37) in Dehydrochinasäure (38) (*251*) zu sein, welche eine wichtige Zwischenreaktion bei der Bildung der aromatischen Körper auf dem sogenannten Shikimisäureweg darstellt *(Reaktionsschema 6)*. Die Analogie geht allerdings, wie man gleich sehen wird, nicht sehr weit.

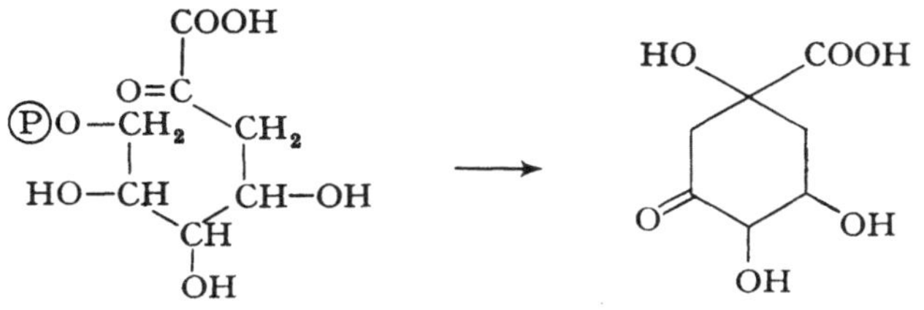

(37)
3-Desoxy-D-*arabino*-heptulosonsäure-7-phosphat.

(38)
Dehydrochinasäure.

Reaktionsschema 6. Ringschluß bei der Bildung von Dehydrochinasäure (*251*).

Wie bereits oben erwähnt, konnten CHEN und CHARALAMPOUS (*51, 52*) als erste zeigen, daß D-Glucose-6-phosphat (39, S. 166) ein Zwischenprodukt bei der Bildung des myo-Inosits ist. Auch KINDL und Mitarb. (*126*) beobachteten beim Enzymsystem aus *Candida utilis*, daß selektiv markiertes D-Glucose-6-phosphat wesentlich besser und mit geringerer Verschmierung („randomization") der Aktivität in myo-Inosit eingebaut wird als selektiv markierte freie D-Glucose. Beim tierischen Enzymsystem aus Rattentestes beobachtete EISENBERG (*70*), daß durch Hitzedenaturierung nur das D-Glucose phosphorylierende Enzym zerstört wird, weshalb das Präparat nach einer solchen Behandlung nur mehr D-Glucose-6-phosphat, aber nicht mehr freie D-Glucose in myo-Inosit überführte. Auch das aus dem Homogenat erhaltene lösliche Enzymsystem derselben Autoren zeigte die gleiche Substratspezifität und den gleichen Cofactor-Bedarf wie das Enzym aus *Candida utilis*.

Das Ausgangsprodukt der Synthese des myo-Inosits ist somit ebenso wie dasjenige der Synthese der Dehydrochinasäure phosphoryliert. Während aber bei der Cyclisierung zur Dehydrochinasäure (38) eine Phosphateliminierung stattfindet, ist im System der myo-Inosit-Bildung die Dephosphorylierung eine erst nach der Cyclisierung erfolgende gesonderte Reaktion. CHEN und CHARALAMPOUS (*53*) konnten mit ihrem Präparat aus *Candida utilis* durch Weglassen von Mg++ aus den sonst kompletten Ansätzen die Akkumulation einer phosphathaltigen Verbindung beobachten, die sie als L-1-*O*-Phospho-myo-inosit (*40*) identifizierten. Auch EISENBERG und BOLDEN (*73*) konnten mit dem tierischen Enzymsystem die Bildung des gleichen Phosphats nachweisen.

Das Enzymsystem, welches D-Glucose zu myo-Inosit umsetzt, besteht somit aus der Hexokinase, einem oder mehreren Enzymen, welche für die eigentliche Cyclisierung verantwortlich sind und einer Phospho-esterase, welche Mg^{++} benötigt. Über das Cyclisierungssystem besitzen wir zur Zeit nur Hinweise, von denen wohl der wichtigste der Bedarf der Reaktion für NAD$^+$ ist.* Dies deutet auf eine Oxydoreduktion, von der wir allerdings noch nicht wissen, auf welche Weise sie erfolgt und ob sie von einem oder von zwei Enzymen katalysiert wird.

In diesem Zusammenhang sind Versuche von Kindl (124) zu erwähnen. Bei Untersuchungen über die Biosynthese von myo-Inosit in Blättern von *Chrysan-themum leucanthemum* (Compositae) wurde D-Glucose-6-^{14}C-6-^{3}H eingesetzt. Das Ver-hältnis ^{3}H zu ^{14}C verringerte sich im entstandenen myo-Inosit auf 80% des in der ein-gesetzten D-Glucose vorgelegenen Verhältnisses, wobei fast 50% der Tritiumaktivität in der 2-Stellung lokalisierbar war. Dies könnte in dem Sinne gedeutet werden, daß Tritium von D-Glucose teilweise auf NAD$^+$ übertragen wird, das seinerseits dann eine Ketogruppe, die ursprünglich in Stellung 5 der D-Glucose und nach dem Ringschluß in Stellung 2 des myo-Inosits liegen sollte, reduziert und damit tritiiert. Im Zu-sammenhang mit den vorher erwähnten Ergebnissen sollte dies für eine Zwischen-stufenfunktion von D-2-Phospho-2,4,6/3,5-pentahydroxycyclohexanon (41) sprechen (*Reaktionsschema 7*).

(30)   (39)   (41)   (40)   (1)
D-Glucose.   D-Glucose-6-phosphat.   D-2-O-Phosphoryl-   L-1-O-Phosphoryl-   myo-Inosit.
                                        2,4,6/3,5-pentahydroxy-   myo-inosit.
                                        cyclohexanon

Reaktionsschema 7. Wahrscheinlicher Biosyntheseweg des myo-Inosits.

B. Biosynthese der Methyläther des myo-Inosits.

Die Biosynthese dieser Verbindungen ist nur in wenigen Fällen untersucht.

Die Methylierung von myo-Inosit zu Sequoyit (4, S. 152) konnte bisher bei den Leguminosen in *Trifolium incarnatum* (2,2% Einbau) (239), *Ononis spinosa* (0,7% Einbau) (130) und *Phaseolus vulgaris* (1,5% Einbau) (125), sowie in *Hoya carnosa* (Asclepiadaceae) (0,5% Einbau) (129), *Thuja occidentalis* (Cupressaceae) (2,5% Einbau) (131 d) und *Artemisia dracunculus* (Compositae) (1,6% Einbau) (240) eindeutig nachgewiesen werden. Der Einbau des radioaktiven myo-Inosits ist in allen Fällen bedeutend höher als derjenige von D-Glucose-U-^{14}C.

D-Ononit (5) wird in *Ononis spinosa* (Leguminosae) unmittelbar aus myo-Inosit gebildet. Durch Einsatz von myo-Inosit-2-^{3}H konnten vorstellbare Zwischenstufen aus der scyllo-Inosit-Reihe ausgeschlossen werden (130).

---

* NAD$^+$ = Nicotinamid-adenin-dinucleotid; NADH = dessen reduzierte Form, NADP$^+$ = Nicotinamid-adenin-dinucleotid-phosphat.
Literaturverzeichnis: SS. 192—205.

D-Bornesit (3, S. 152) wird in *Borago officinalis*, *Lathyrus odoratus* und *Myosotis caucasica* aus myo-Inosit aufgebaut (*131b*). Untersuchungen in *Nerium oleander* zeigten, daß D-Bornesit (3), der dort auch aus myo-Inosit entsteht, in Dambonit (6) überführt wird (*131c*). Die Biosynthese des zweiten in der Natur vorkommenden Dimethyläthers des myo-Inosits, des Liriodendrits (7) wurde bisher noch nicht untersucht.

Die Tatsache, daß von den sechs Hydroxylgruppen des myo-Inosits fünf biologisch methyliert werden können, was durch das Vorkommen der vorgenannten Methyläther angenommen werden muß, läßt auf die Existenz von mehreren spezifischen Methyltransferasen für myo-Inosit als Akzeptor schließen, deren Spezifitätsverhältnisse eine Untersuchung rechtfertigen sollten.

C. Biosynthese von Glykosiden des myo-Inosits.

Aus *Pisum sativum* konnte ein zellfreies Enzym-Präparat gewonnen werden, welches die Bildung von Galaktinol (42) aus Uridindiphosphat-D-Galaktose und myo-Inosit katalysiert (*Formelübersicht 3*) (88). Ähnliche Enzym-Präparate anderer Autoren (96) aus *Sporobolomyces singularis* bewirken die Bildung von β-D-Glucosyl-myo-inosit (43) aus Disacchariden und myo-Inosit, wobei die 1- und 5-Position des myo-Inosit glykosidiert wird.

D. Die Herkunft des myo-Inosit-Restes in Phytinsäure und in den Phosphoinositiden.

Nach einer Untersuchung von AHUJA (5) an *Pisum sativum* wird markierter myo-Inosit, der in die Pflanze infundiert wird, in Phytinsäure (15, S. 152) eingebaut; als Einbaurate wird 1—10% angegeben. Versuche aus unserem Laboratorium, die an verschiedenen höheren Pflanzen durchgeführt wurden (*129*), ergaben allerdings Einbauraten, die etwa um eine Größenordnung niedriger sind. Es mußte zunächst festgestellt werden, ob myo-Inosit ohne Änderung der Ringstruktur in Phytinsäure überführt wird oder ob der Einbau der Aktivität von Bruchstücken stammt, die durch den Abbau von myo-Inosit entstanden sind und nun ihrerseits zur Bildung der Phytinsäure verwendet wurden. Versuche hierzu ergaben, daß selektiv markierter myo-Inosit (myo-Inosit-2-^{3}H) zur Bildung von Phytinsäure führte, die im Cyclitring eine gegenüber dem vorgegebenen myo-Inosit geänderte Aktivitätsverteilung aufwies; in der Stellung 2 der gebildeten Phytinsäure waren je nach der verwendeten Pflanze sehr verschiedene Mengen der Gesamtaktivität lokalisiert. In der Folge konnte gezeigt werden, daß das Cyclitskelett der Phytinsäure in der gleichen Weise aus D-Glucose entsteht wie myo-Inosit; D-Glucose-6-^{14}C führte zur Bildung von myo-Inosit-1-^{14}C, und D-Glucose-2-^{14}C ergab myo-Inosit-5-^{14}C. Beide Ergebnisse zusammen

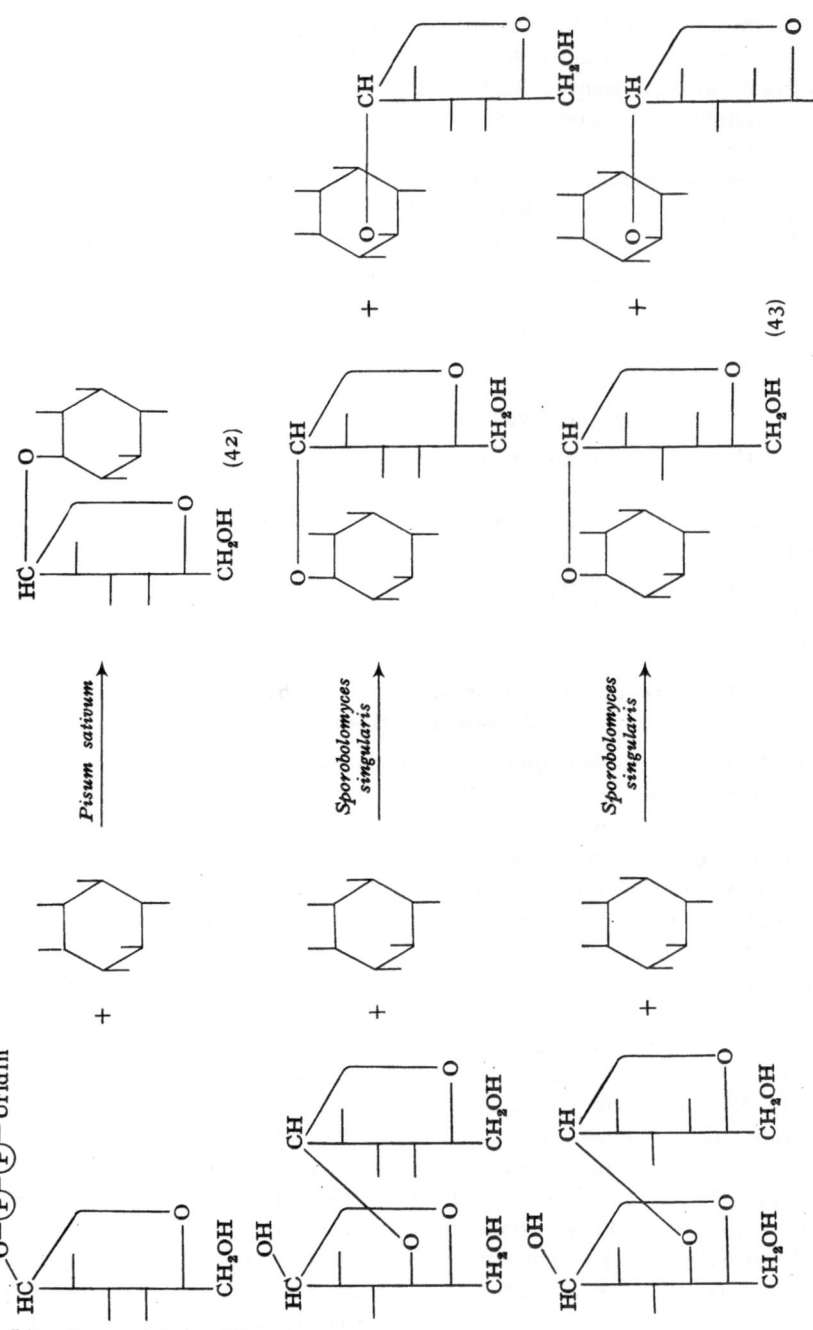

Formelübersicht 3. Bildung von Glykosiden des myo-Inosits.

sprechen dafür, daß Phytinsäure ähnlich wie myo-Inosit gebildet wird, aber nicht nur der freie myo-Inosit, sondern vermutlich auch eine bereits cyclische Vorstufe dieses Cyclits zur Phytinsäure phosphoryliert wird. Hier wäre nach dem Vorhergesagten (vgl. S. 165 f.) vor allem an L-1-O-Phospho-myo-inosit (40) zu denken (74a).

Im Gegensatz zu den Verhältnissen bei der Phytinsäure ist es im Fall der Phosphoinositide absolut sichergestellt, daß freier myo-Inosit intakt in diese Phospholipide eingebaut wird (29, 183). Bei der Bildung der Phosphoinositide entsteht primär aus Phosphatidsäure und Cytidin-triphosphat (CTP) ein Cytidindiphosphat-Diglycerid, das dann mit freiem myo-Inosit unter Abspaltung von Cytidinmonophosphat (CMP) reagiert. Die für diese Reaktionsfolge verantwortlichen Enzyme sind bekannt (183).

E. Bildung von scyllo-Inosit.

Bereits 1957 beobachtete HELLEU (104) eine deutliche Scylliturie bei Menschen, welche mit einer stark myo-Inosit-haltigen Kost ernährt worden waren, und postulierte, daß im Menschen ein Übergang von myo-Inosit (1) in scyllo-Inosit (8) stattfinde. In Versuchen mit intakten Ratten konnten POSTERNAK und Mitarb. (220) eine solche Überführung auch tatsächlich nachweisen. Durch Verwendung von markiertem myo-Inosit und markiertem scyllo-Inosit wurde gezeigt, daß diese beiden Verbindungen ineinander umgewandelt werden, wobei 2,4,6/3,5-Penta-hydroxycyclohexanon als Zwischenstufe fungiert.

Ähnliche Untersuchungen wurden auch in unserem Laboratorium an höheren Pflanzen durchgeführt (242). In Calycanthus occidentalis (Calycanthaceae) stellt myo-Inosit einen wesentlich besseren Vorläufer des scyllo-Inosits dar als D-Glucose. Bei Applikation von markiertem 2,4,6/3,5-Pentahydroxycyclohexanon entstehen etwa 30% scyllo-Inosit neben 5% myo-Inosit.

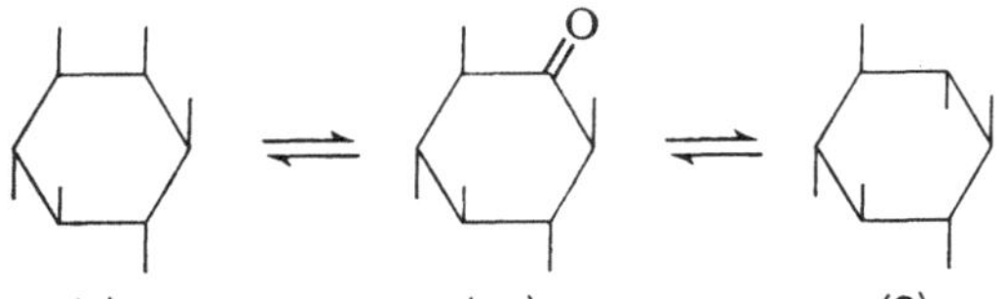

(1)   (44)   (8)
myo-Inosit.   2,4,6/3,5-Pentahydroxycyclohexanon.   scyllo-Inosit.

Reaktionsschema 8. Biogenetische Beziehungen zwischen myo-Inosit und scyllo-Inosit.

Die berichteten Ergebnisse lassen darauf schließen, daß es sowohl in der Ratte als auch in der höheren Pflanze ein System von zwei stereospezifischen Oxydoreductasen gibt, welche das in Reaktionsschema 8 gezeigte Gleichgewicht katalysieren.

F. Bildung von D-chiro-Inosit und D-Pinit.

In bezug auf seine Konfiguration verhält sich D-chiro-Inosit (9) zu D-Galaktose gleichartig wie myo-Inosit zu D-Glucose. Dementsprechend wäre vorstellbar, daß D-chiro-Inosit aus D-Galaktose durch einen ähnlichen Cyclisierungsmechanismus entsteht, wie er für den Übergang von D-Glucose in myo-Inosit nachgewiesen wurde. Untersuchungen über den Einbau von radioaktiv markierten möglichen Vorstufen an Blättern von *Trifolium incarnatum*, welche von Scholda, Billek und Hoffmann-Ostenhof (*239, 241, 244*) durchgeführt wurden, ergaben allerdings, daß D-Galaktose und auch D-Glucose um etwa zwei Größenordnungen weniger gut in D-Pinit (10) und D-chiro-Inosit eingebaut werden als myo-Inosit. Es konnte weiter durch Einbauversuche mit markiertem Sequoyit, D-chiro-Inosit und D-Pinit festgestellt werden, daß myo-Inosit nicht direkt zu D-chiro-Inosit epimerisiert wird, sondern primär eine Methylierung zu Sequoyit erfolgt, welcher dann über die Zwischenstufe D-5-*O*-Methyl-2,3,5/4,6-pentahydroxycyclohexanon (45) zu D-Pinit epimerisiert wird. D-chiro-Inosit (9) entsteht ausschließlich durch Entmethylierung von D-Pinit (10). In *Reaktionsschema 9* findet sich eine Darstellung des Bildungsweges von D-Pinit und D-chiro-Inosit in *Trifolium incarnatum*; eine gleichartige Entstehung der beiden Cyclite konnte auch in anderen Pflanzen (*Ononis spinosa, Hoya carnosa*) (*130, 131*) verifiziert werden.

| (1) | (4) | (45) | (10) | (9) |
| myo-Inosit. | Sequoyit. | D-5-*O*-Methyl-2,3,5/4,6-pentahydroxycyclohexanon. | D-Pinit. | D-chiro-Inosit. |

Reaktionsschema 9. Bildung von Sequoyit, D-Pinit und D-chiro-Inosit aus myo-Inosit.

Innerhalb der beschriebenen Reaktionskette stellt der erste Schritt, eine Methylierung einer Hydroxylgruppe, eine in der Natur sehr häufige Reaktion dar, während der letzte Vorgang, eine Entmethylierung, eher selten beobachtet wird. Von besonderem Interesse ist aber die Epimerisierung von Sequoyit (4) zu D-Pinit (10), ein Reaktionstypus, für welchen es in der Biochemie nur wenige gut untersuchte Beispiele gibt.

Wenn wir die natürlichen Epimerisierungsreaktionen auf dem Gebiet der Kohlenhydrate näher betrachten, so lassen sich drei Gruppen unterscheiden: Der erste Reaktionstypus, der bei der Biosynthese von L-Rhamnose oder bei der von der D-Ribulose-5-phosphat-3-epimerase (5.1.3.1) katalysierten Umlagerung vorliegt, betrifft eine Epimerisierung an einem Kohlenstoffatom, das in α-Stellung zu einer Carbonylgruppe steht; da an der Reaktion Wasser teilnimmt, kann man ein Enol

Literaturverzeichnis: SS. 192—205.

als Zwischenstufe annehmen. Als zweiter Reaktionstypus lassen sich Umlagerungen wie diejenigen, die von L-Ribulose-5-phosphat-4-epimerase (5.1.3.4) und von UDP-D-Glucose-4-epimerase* (5.1.3.2) bewirkt werden, zusammenfassen. Hier ist Wasser nicht an der Reaktion beteiligt, exogenes NAD^+ ist nicht erforderlich, hingegen dürften beide Enzyme festgebundenes NAD^+ enthalten, welches als Wasserstoffakzeptor bzw. in seiner reduzierten Form als Donor fungiert. Die Epimerisierung von Sequoyit (4) zu D-Pinit (10) gehört einem dritten Typus an, der auch bei der Umlagerung von D-Milchsäure in L-Milchsäure vorliegt. Hier läßt sich eine Ketoverbindung als Zwischenstufe nachweisen: Bei der Milchsäure-Epimerisierung ist Brenztraubensäure der Intermediärkörper, während bei der Umwandlung von Sequoyit in D-Pinit das D-5-O-Methyl-2,3,5/4,6-pentahydroxycyclohexanon (45) als Zwischenstufe gefaßt werden konnte.

Diese Epimerisierung erhält dadurch besondere Bedeutung, weil es gelungen ist, die beiden dafür verantwortlichen Enzyme nachzuweisen und auf das etwa 100fache anzureichern (139).

Dabei wurde von einem Acetontrockenpulver aus *Trifolium incarnatum* ausgegangen; der daraus gewonnene Extrakt wurde mit Hilfe von partieller Säuredenaturierung, Ammoniumsulfatfällung, Adsorption an DEAE-Cellulose und mit Sephadex G 100 gereinigt.

Die Enzyme, welche beide D-5-O-Methyl-2,3,5/4,6-pentahydroxycyclohexanon (45) reduzieren, unterscheiden sich durch ihre Reaktionsprodukte, durch die erforderlichen Wasserstoffdonatoren, durch ihre Stabilität, durch ihre pH-Optima und auch durch die Tatsache, daß das eine von ihnen, welches den Ketonkörper mit Hilfe von NADH zu Sequoyit reduziert, nur in Anwesenheit von Cystein oder anderen Sulfhydrylverbindungen aktiv ist, während das andere, das das Keton mit NADPH zu D-Pinit reduziert, ohne Sulfhydrylgruppen wirksam ist.

G. Bildung von L-chiro-Inosit und seiner Methyläther.

Untersuchungen über die Biosynthese von L-chiro-Inosit (11) und seiner Methyläther, L-Quebrachit (12) und L-Pinit (13), in *Artemisia vulgaris* und *A. dracunculus* wurden vor kurzem von SCHOLDA, BILLEK und HOFFMANN-OSTENHOF (240, 243) durchgeführt. Bei Einbauversuchen mit markierten möglichen Vorläufern ergab sich, daß myo-Inosit wesentlich besser in diese Cyclite eingebaut wird als D-Glucose. Der Weg von myo-Inosit zu L-chiro-Inosit und seinen Methyläthern unterscheidet sich allerdings wesentlich von demjenigen zu D-chiro-Inosit (9, S. 152) und D-Pinit (10), wie er oben beschrieben wurde. Während dort eine Methylierung des myo-Inosits vor der Epimerisierung erforderlich ist, scheint bei der Bildung der Cyclite der L-Reihe die Epimerisierung direkt am myo-Inosit oder zumindest vor einer Methylierung zu erfolgen. Der Reaktionsweg, dessen enzymologische Aufklärung nicht ohne Interesse sein sollte, ist in *Reaktionsschema 10* dargestellt.

---

* UDP = Uridindiphosphat.

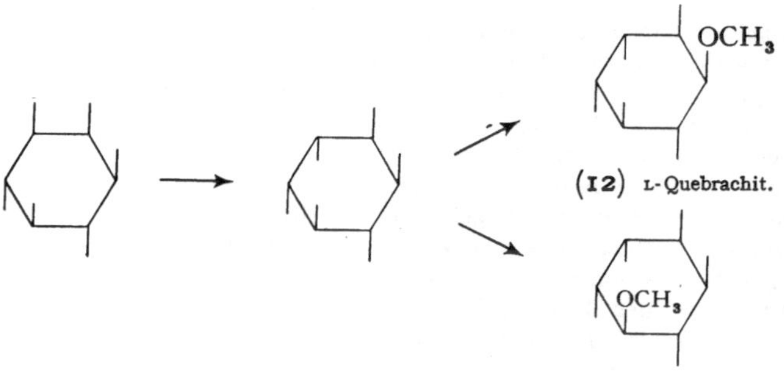

(12) L-Quebrachit.

(13)
L-Pinit.

(1)
myo-Inosit.

(11)
L-chiro-Inosit.

Reaktionsschema 10. Bildung von L-chiro-Inosit, L-Quebrachit und L-Pinit aus myo-Inosit.

H. Bildung von L-Quercit, L-Viburnit und L-Leucanthemit.

Als PLOUVIER (*210*) aus *Chrysanthemum*-Arten erstmalig L-Leucanthemit (**23**, S. 153) isolierte und auf das gemeinsame Vorkommen dieses Cyclohexentetrols mit L-chiro-Inosit (**11**) und L-Viburnit (**20**) aufmerksam wurde, diskutierte er die Möglichkeit biogenetischer Beziehungen zwischen diesen drei Cycliten. Versuche in unserem Laboratorium (*131, 131a, 131e*) zeigten aber, daß myo-Inosit, welcher ohne Zweifel die Vorstufe von L-chiro-Inosit darstellt, nicht in L-Viburnit, L-Quercit oder L-Leucanthemit eingebaut wird. Die drei letztgenannten Substanzen werden aber, wie entsprechende Versuche ergaben, in den Pflanzen aus CO_2 bei Belichtung oder aus D-Glucose etwa gleich schnell aufgebaut wie myo-Inosit. Es sollten somit Biosynthesewege für die Cyclohexanpentole und auch für die Cyclohexentetrole existieren, welche zumindest in ihren letzten Schritten von demjenigen für den myo-Inosit unabhängig sind.

Um den Übergang von Kohlenhydraten — insbesondere von D-Glucose — in die Cyclohexanpentole studieren zu können, wurde der in *Reaktionsschema 11* angeführte Abbauweg ausgearbeitet *(131a)*. L-Quercit bzw. L-Viburnit werden dabei in Isopropylidenverbindungen überführt, die mit Perjodat oxydativ gespalten werden. Die über L-*erythro*-Tetrahydroxypentan (**46**) bzw. D-*arabino*-Pentahydroxyhexan (**47**) erhaltene β-Hydroxypropionsäure wird nach UNRAU und CANVIN (*261a*) weiter abgebaut.

Bei diesen Versuchen wurde in *Quercus*-Arten gefunden, daß D-Glucose-6-^{3}H in L-Quercit-2-[3]H übergeht; bei analogen Experimenten in *Chrysanthemum leucanthemum* konnte eine Überführung von D-Glucose-1-^{14}C in L-Viburnit-3-^{14}C, von D-Glucose-2-^{14}C in L-Viburnit-2-^{14}C und von D-Glucose-6-^{3}H in L-Viburnit-4-^{3}H festgestellt werden. Es scheint somit, daß bei der Bildung von Cyclohexanpentolen D-Glucose ohne Aufspaltung des Kohlenstoffskeletts in diese Cyclite eingebaut wird. Die Art des Ringschlusses ist aber von derjenigen, die bei der Biosynthese der Cyclohexanhexole beobachtet wurde (vgl. S. 166), grundsätzlich verschieden, wobei vorhergehende Epimerisierungs-Reaktionen postuliert werden müssen.

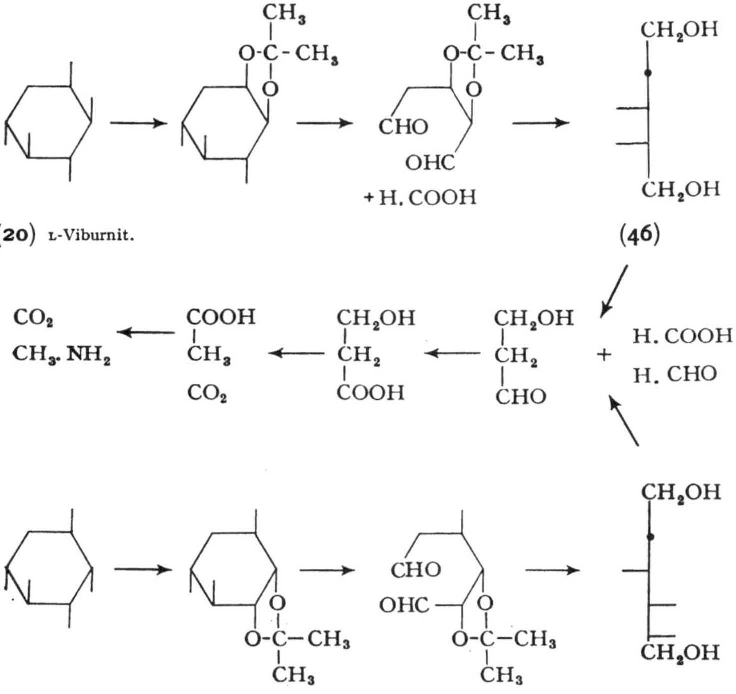

Reaktionsschema 11. Selektiver Abbau von L-Viburnit und L-Quercit.

I. Bildung von Amino-desoxycycliten.

Es konnte vor kurzem gezeigt werden, daß der Streptamin-Anteil des Streptomycins (**52**, S. 182) in *Streptomyces griseus* aus myo-Inosit gebildet wird (*110*). WALKER und WALKER (*269a*) gelang es, die Entstehung von 1-Desoxy-1-guanidino-2-O-phosphoryl-scyllo-inosit und Streptidinphosphat in Extrakten von *Streptomyces bluensis* und anderen *Streptomyces*-Arten zu zeigen. Mit Hilfe eines gereinigten Amidinotransferase-Präparates konnten die Autoren die Übertragung der Guanidinogruppe von Guanidino-L-arginin auf 1-Amino-1-desoxy-2-O-phosphoryl-scyllo-inosit, Streptaminphosphat und 2-Desoxy-streptamin-phosphat nachweisen. Man kann somit annehmen, daß 1-Amino-1-desoxy-

2-O-phosphoryl-scyllo-inosit, der wahrscheinlich über scyllo-Inosit aus myo-Inosit entsteht, die Vorstufe für das Bluensidinphosphat und das Streptidinphosphat darstellt.

IV. Katabolischer Stoffwechsel.

Schon bald nach Entdeckung des myo-Inosits (1) durch Scherer (236) wurden Untersuchungen über den Stoffwechsel dieser Verbindung unternommen. Bei Gärversuchen mit verschiedenen Mikroorganismen erhielt man aliphatische Carbonsäuren, wie Essigsäure, Weinsäure und Milchsäure [vgl. dazu Posternak (217)], ohne aber über die für die Bildung dieser Substanzen verantwortlichen Reaktionen etwas aussagen zu können. Erst in viel späterer Zeit wurden exakte Untersuchungen über den Abbau von myo-Inosit mit verschiedenen Organismen durchgeführt, die uns in manchen Fällen bereits ein ziemlich klares Bild dieser recht unterschiedlichen Reaktionswege vermittelten.

A. Oxydationsreaktionen unter Erhaltung der Cyclitstruktur.

Bei Untersuchungen über das Vermögen des Mikroorganismus *Pseudomonas beijerincki*, myo-Inosit zu oxydieren, konnten Kluyver, Hof und Boezaardt (135) Tetrahydroxychinon (48) und Rhodizonsäure (49) als Produkt identifizieren *(Reaktionsschema 12)*. Es handelt sich hier um eine weitgehende Oxydation unter Erhaltung des Cyclohexanringes.

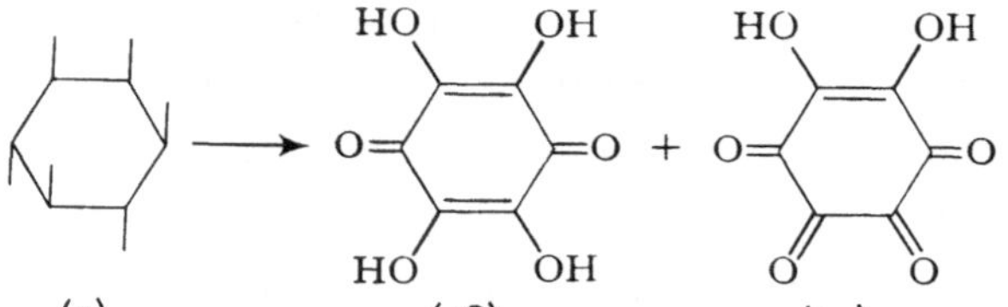

(1)	(48)	(49)
myo-Inosit.	Tetrahydroxybenzochinon.	Rhodzionsäure.

Reaktionsschema 12. Oxydation von myo-Inosit in *Pseudomonas beijerincki (135)*.

Mit dem Stamm ATCC 621 von *Acetobacter suboxydans* konnten Magasanik, Chargaff und Mitarb. (164, 165) verschiedene Cyclite in kontrollierten Reaktionen oxydieren, wobei sie ganz bestimmte Regelmäßigkeiten feststellten, die eine gewisse Analogie zur Bertrandschen Regel aufweisen. Die Gesetzmäßigkeiten lassen sich in folgenden drei Sätzen zusammenfassen: 1. Es werden nur axiale Hydroxylgruppen der Cyclite oxydiert. 2. An dem zur oxydierbaren Gruppe in 3-Stellung (*meta*-Stellung) stehenden C-Atom muß sich eine äquatoriale Hydroxyl-

Literaturverzeichnis: SS. 192—205.

gruppe befinden. 3. Dasselbe gilt auch für das in 4-Stellung (*para*-Stellung) befindliche C-Atom. Diese Regeln wurden bis auf wenige Ausnahmen von ANDERSON und Mitarb. (*8*, *9*) bestätigt.

Demgegenüber fanden POSTERNAK und Mitarb. (*218*, *219*), welche den Stamm von KLUYVER und DE LEEUW verwendeten, einige Abweichungen von den obigen Regeln 2 und 3. Bei den ausgedehnten Untersuchungen dieser Autoren wurden neben den Inositen auch Pentahydroxycyclohexanone, Cyclohexanpentole, Cyclohexantetrole und Cyclohexantriole einbezogen. Aus *Acetobacter suboxydans* konnten auch bereits zellfreie Enzympräparate erhalten werden, welche myo-Inosit dehydrieren (*43*, *86*); allerdings wurde offenbar noch keine weitergehende Reinigung der für diese Reaktion verantwortlichen Dehydrogenase unternommen.

Durch myo-Inosit im Kulturmedium kann in *Aerobacter aerogenes* die Bildung eines Enzymsystems induziert werden (*178*), das außer myo-Inosit (1) auch D-chiro-Inosit (9), 2,4,6/3,5-Pentahydroxycyclohexanon (44) (myo-Inosose-2), und D-2,3,5/4,6-Pentahydroxycyclohexanon (myo-Inosose-3) unter O_2-Aufnahme oxidiert (*162*, *163*). Aus dem Organismus konnten gereinigte Enzympräparate gewonnen werden, welche myo-Inosit unter Mitwirkung von NAD+ zum 2,4,6/3,5-Pentahydroxycyclohexanon (44) und weiter zum 3,5/4,6-Tetrahydroxycyclohexandion-(1,2) (50) dehydrieren; ein weiteres Enzym dehydratisiert 2,4,6/3,5-Pentahydroxycyclohexanon zur Enolform des 3,5/4-Trihydroxycyclohexandions-(1,2) (51) (*Reaktionsschema 13*) (*18a*, *93*). Es konnte schließlich im selben Organismus ein Enzym nachgewiesen werden, das 3,5/4-Trihydroxycyclohexandion-(1,2) (51) zu einer 4-Desoxy-5-hexulosonsäure aufspaltet (*18b*).

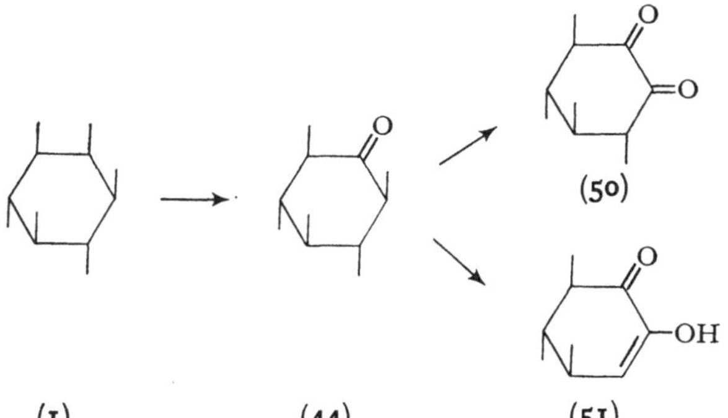

(1)
myo-Inosit.   (44)   (51)
2,4,6/3,5-Pentahydroxycyclohexanon.

Reaktionsschema 13. Oxydation von myo-Inosit in *Aerobacter aerogenes* (*44*, *50*, *93*).

LARNER und Mitarb. (*145*) berichteten über die teilweise Reinigung der induzierbaren myo-Inosit-Dehydrogenase (systematischer Name: myo-Inosit : NAD-Oxydoreductase; 1.1.1.18) aus *Aerobacter aerogenes*, wobei sie zu einer etwa 20-fachen Anreicherung gegenüber dem Rohextrakt kamen. Das Enzym ist für NAD$^+$ als Wasserstoffakzeptor spezifisch; NADP$^+$ ist wirkungslos. Die Autoren konnten das Produkt der Dehydrierung nicht identifizieren; mit NADH als Wasserstoffdonor reduziert das Enzym sowohl 2,4,6/3,5-Pentahydroxycyclohexanon als auch DL-2,3,4,6/5-Pentahydroxycyclohexanon ($\pm$-epi-myo-Inosose). Aus einem anderen Stamm von *Aerobacter aerogenes* isolierte Weissbach (*271*) die myo-Inosit-Dehydrogenase und konnte mit Hilfe dieses Enzyms eine spektrophotometrische Bestimmungsmethode für myo-Inosit ausarbeiten. Nach seinen Angaben ist die Dehydrogenase weitgehend spezifisch für myo-Inosit und NAD$^+$; schwache Aktivitäten gegenüber anderen Cycliten wie scyllo-Inosit, D-chiro-Inosit, neo-Inosit, D-Pinit und Sequoyit könnten durch Verunreinigungen der Präparationen dieser Stoffe mit myo-Inosit verursacht gewesen sein.

B. Oxydative Aufspaltung des Cyclohexanringes.

1. Abbau im tierischen Organismus oder durch Enzympräparate tierischer Herkunft.

Wenn man Ratten markierten myo-Inosit parenteral oder intraperitoneal appliziert, so findet sich die Aktivität nach kurzer Zeit vorwiegend in CO_2, im Glykogen und in den Phospholipiden (*4, 106, 173*). Bereits 1946 konnten Stetten und Stetten (*254*) auch die Ausscheidung von markierter D-Glucose im Harn nach Gabe von vermutlich 2-deuteriertem myo-Inosit zeigen. Genauere Untersuchungen des Abbaues des myo-Inosits mit darauffolgender Gluconeogenese im tierischen Organismus wurden von der Schule von Posternak (*47, 221, 222*) vorgenommen. Bei Gabe von myo-Inosit-2-^{3}H wurde D-Glucose-6-^{3}H, bei solcher von myo-Inosit-6-^{3}H aber D-Glucose-1,3-^{3}H gefunden. Bei einer ähnlichen Versuchsanordnung beobachteten Anderson und Coots (*6*) die Umwandlung von myo-Inosit-2-^{14}C in Glykogen, dessen D-Glucose-Einheiten in den Stellungen 1 und 6 markiert waren. Howard (*111*) kam zu weitgehend analogen Ergebnissen bei der Untersuchung des Abbaus von myo-Inosit-2-^{14}C in Rattennierenhomogenaten. Alle vorliegenden Ergebnisse scheinen jedenfalls dahinzudeuten, daß myo-Inosit primär zu D-Glucuronsäure oxydiert wird und diese dann über den sogenannten Glucuronsäureweg (*31, 259*) und den Pentosephosphatcyclus D-Glucose ergibt.

Es ist ohne Zweifel das Verdienst von Charalampous und Mitarb. (*41—45*), als erste das Enzym, welches die oxydative Ringspaltung des myo-Inosits zur D-Glucuronsäure katalysiert, aus Rattennieren isoliert zu haben. Die myo-Inosit-Oxygenase (systematischer Name: myo-Inosit: Sauerstoff-Oxydoreductase; 1.13.1.11) ist ein gegen Schwermetallionen

Literaturverzeichnis: SS. 192—205.

besonders empfindliches Enzym, das zweiwertiges Eisen enthält. Die Stelle der Ringöffnung konnte dadurch festgelegt werden, daß man die Überführung von myo-Inosit-2-^{14}C in D-Glucuronsäure-5-^{14}C nachwies, was eine Sprengung des Ringes zwischen den C-Atomen 1 und 6 beweist. *(Reaktionsschema 14.)*

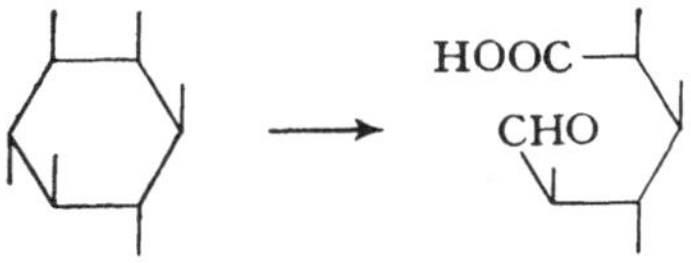

(1)   (32)
myo-Inosit.   D-Glucuronsäure.

Reaktionsschema 14. Oxydative Aufspaltung von myo-Inosit durch myo-Inosit-Oxygenase.

Die Angabe von CHARALAMPOUS (*45*), daß neben diesem Enzym in den Rattennieren auch ein zweites myo-Inosit oxydierendes Enzym vorhanden sein soll, welches den Cyclit zu L-Glucuronsäure oxydiere, und welches erst bei der Reinigung der myo-Inosit-Oxygenase entfernt würde, konnte allerdings von anderen Bearbeitern des Gebietes nicht bestätigt werden (*32, 258*). '

Die D-Glucuronsäure produzierende myo-Inosit-Oxygenase aus Rattenniere wurde von CHARALAMPOUS (*41*) auf das 450-fache angereichert; sein Präparat verhält sich sowohl in der Elektrophorese als auch in der Ultrazentrifuge homogen und hat, berechnet aus seinem Eisengehalt, ein Mindestmolekulargewicht von 68000. Das pH-Optimum der Wirkung liegt beim Neutralpunkt, die Michaelis-Konstante K_m für myo-Inosit beträgt 22 mMol. Es handelt sich um ein SH-Enzym. Die Spezifität ist nicht absolut; außer myo-Inosit werden auch D-chiro-Inosit und myo-Inosit-monophosphat, nicht aber L-chiro-Inosit, scyllo-Inosit, D-Pinit, 2,4,6/3,5-Pentahydroxycyclohexanon oder D- und L-2,3,5/4,6-Pentahydroxycyclohexanon oxydiert. Der Nachweis für die Bildung von D-Glucuronsäure aus myo-Inosit durch ein Enzym aus Rattennieren wurde auch von RICHARDSON und AXELROD (*227*) erbracht.

2. Oxydative Aufspaltung durch Mikroorganismen und durch Enzyme mikrobiellen Ursprungs.

Sowohl *Aerobacter*-Arten (*163, 266*) als auch *Propionibacterium pentosaceum* (*267*) sind imstande, myo-Inosit durch oxydative Ringöffnung abzubauen, wobei die Produkte sich von denjenigen des D-Glucose-Stoffwechsels unterscheiden. Die dafür verantwortlichen Zwischenreaktionen sind bisher aber noch unbekannt. Auch *Mycobacterium tuberculosis avium* baut myo-Inosit unter Sauerstoffaufnahme sehr schnell ab (*15*), während in *Mycobacterium BCG* myo-Inosit und auch andere Cyclite die Bildung eines myo-Inosit-oxydierenden Enzyms induzieren (*181*).

Ebenso findet man in manchen anderen Mikroorganismen einen schnellen Abbau von myo-Inosit, wobei häufig aus den entstehenden Bruchstücken bevorzugt gebildete Inhaltsstoffe aufgebaut werden. So überführen *Fusarium*-Arten myo-Inosit in Lycopersen (*138*), während *Penicillium chrysogenum* die Bruchstücke von myo-Inosit in Benzylpenicillin einbaut (*116*). *Xanthomonas*-Arten liefern beim Abbau von myo-Inosit Tartrat, Glyoxalat, Glykolat, Oxalat, Formiat und CO_2 (*182*). Wesentlich besser sind die Verhältnisse beim Abbau von myo-Inosit durch die Hefeart *Schwanniomyces occidentalis* bekannt. Nachdem JANKE und Mitarb. (*118*) gezeigt hatten, daß der genannte Sproßpilz ebenso wie einige andere Hefearten auf myo-Inosit als der einzigen Kohlenstoffquelle wachsen kann und später die Bildung eines durch myo-Inosit induzierbaren Enzyms für den oxydativen Abbau nachgewiesen wurde, konnten SIVAK und HOFFMANN-OSTENHOF (*247*) in diesem Organismus nach Züchtung auf einem myo-Inosit-haltigen Medium sämtliche Enzyme nachweisen, welche einen Abbau des myo-Inosits in der gleichen Weise wie im Säugetiergewebe, d. h. über D-Glucuronsäure, den Glucuronsäureweg und den Pentosephosphatzyklus, katalysieren können. Daß dies auch tatsächlich der Abbauweg des myo-Inosits in *S. occidentalis* ist, welcher sich in dieser Hinsicht ganz analog zum Säugetier und demnach stark abweichend von den vorher besprochenen Mikroorganismen verhält, konnte durch Einsatz von myo-Inosit-U-^{14}C in Kurzzeitversuchen nachgewiesen werden; es wurden L-Gulonsäure, Xylit und Xylulose als Zwischenprodukte sowie D-Glucose und D-Gluconsäure isoliert (*66*). Die so weitgehende Übereinstimmung zwischen einem Mikroorganismus und dem Säugetier ist wohl überraschend; es darf aber daran erinnert werden, daß schon seit langem die Analogien zwischen der Glykolyse im Muskel und dem Abbau der Glucose durch die Hefegärung bekannt sind.

Auch das Enzym des Primärangriffs auf myo-Inosit, die myo-Inosit-Oxygenase, aus *S. occidentalis* wurde näher untersucht (*258*); sie zeigt ein sehr ähnliches Verhalten wie das aus Rattenniere isolierte, oben besprochene Enzym. Das Enzym weist aber eine etwas andere Spezifität auf, da auch L-chiro-Inosit und L-2,3,5/4,6-Pentahydroxycyclohexanon angegriffen werden.

3. Oxydative Aufspaltung in höheren Pflanzen.

Bei der Untersuchung des Stoffwechsels von myo-Inosit-2-^{3}H und myo-Inosit-2-^{14}C durch Blätter von *Petroselium* (Umbelliferae) und unreife Früchte von *Fragaria* (Rosaceae) wurden Ergebnisse erhalten, welche es sehr wahrscheinlich machen, daß in den genannten Pflanzen myo-Inosit primär ebenso wie im Säugetiergewebe und in *Schwanniomyces occidentalis* durch oxydative Aufspaltung des Ringes zwischen den Kohlenstoffatomen 1 und 6 abgebaut wird (*155, 156, 158—160*). In der unreifen Erdbeere wird die Aktivität vorwiegend im Pectin wiedergefunden, dessen D-Galakturonsäure-Reste im $C_{(5)}$ markiert sind. Außerdem läßt sich

Aktivität in den L-Arabinose- und D-Xylose-Resten der Hemicellulosen sowie in D-Xylose, L-Gulonsäure und D-Glucuronsäure nachweisen. Der Abbau des myo-Inosits erfolgt relativ schnell; bereits nach 3 Tagen sind mehr als 40% in andere Produkte umgewandelt. Die Umwandlung von myo-Inosit in Pectin konnte auch in weiteren Pflanzen gefunden werden (*157*).

Diese Resultate von LOEWUS und Mitarb. zeigen deutlich, daß die höheren Pflanzen myo-Inosit-Oxygenasen — ähnlich denjenigen der Rattenniere und von *Schwanniomyces occidentalis* — besitzen, die den Cyclit primär in D-Glucuronsäure umwandeln. Über das Enzym ist allerdings noch nichts bekannt; ebensowenig wissen wir über die Enzyme, welche für den Umbau der D-Glucuronsäure in die verschiedenen Bestandteile der Pectine und Hemicellulosen verantwortlich sind. Auch nach älteren Untersuchungen von KURSANOV (*143a*) und SCHRAUDOLF (*244a*) unterliegt myo-Inosit in höheren Pflanzen einem Abbau, dessen Produkte allerdings nicht identifiziert wurden.

Chrysanthemum leucanthemum (Compositae) führt myo-Inosit-U-^{14}C in D-Glucose, Fructose, Saccharose, Glycerin und D-Xylose über (*125*). Nach 4-tägigem Stoffwechsel treten bereits 12% der eingesetzten Aktivität in der D-Glucose auf. Auch O-Methyl-inosite, wie D- und L-Bornesit, sowie Cyclohexanpentole, wie L-Viburnit (*20*, S. 153) und L-Quercit (*21*), sind einem starken Stoffwechsel unterworfen (*125, 131b, 131c*).

V. Vorkommen der Cyclite.

Cyclite und komplexe Cyclitverbindungen sind in der Natur weit verbreitet. Das Vorkommen und die Häufigkeit dieser Substanzen scheinen mit deren Stabilität parallel zu gehen. So finden wir bevorzugt Cyclite mit wenigen axialen Hydroxylgruppen; ebenso sind bisher nur solche Methyläther der Cyclite bekannt, bei denen sich die Methoxylgruppe in äquatorialer Stellung befindet.

Obwohl in den letzten Jahren — vor allem aus taxonomischem Interesse — verstärkte Anstrengungen gemacht wurden, die Verteilung der Cyclite auf die Organismen kennenzulernen, ist auf diesem Gebiet noch sehr viel zu tun. So gibt es z. B. zahlreiche Pflanzenfamilien, die bisher überhaupt noch nicht auf Cyclite untersucht wurden. Dazu kommt, daß die bisherigen Untersuchungen meist nur einen Cyclit als Inhaltsstoff ergeben haben; bei Verwendung moderner Verfahren kann man häufig weitere Cyclite in geringeren Mengen nachweisen.

Im Vergleich zu älteren Zusammenstellungen über das Vorkommen der Cyclite (*16, 62, 82, 84*) und auch zu einer neueren Übersicht, bei der zum erstenmal taxonomische Gesichtspunkte in Betracht gezogen wurden (*212*), sollen im vorliegenden Abschnitt zahlreiche Arbeiten aus

jüngster Zeit berücksichtigt werden; weiters sollen die vor kurzem erhaltenen, bereits in einem früheren Abschnitt behandelten Erkenntnisse
über die biogenetischen Beziehungen verschiedener Cyclite untereinander
in die Betrachtung einbezogen werden. Wir hoffen, daß die vorgelegten
Tabellen (S. 184—191) eine gute Übersicht vermitteln und einen weiter-
gehenden taxonomischen Aussagewert besitzen.

A. myo-Inosit.

Das Vorkommen von myo-Inosit (**1**, S. 152) scheint absolut ubiquitär
zu sein (*11, 49, 82*); aus diesem Grunde und auch deshalb, weil myo-Inosit
als Vorstufe aller anderer Cyclohexanhexole erkannt wurde, wird hier
darauf verzichtet, die in der Literatur zu findenden Angaben über seine
Verbreitung einzeln zu referieren. Verschiedene Phosphorsäureester des
myo-Inosits, darunter vor allem die Phytinsäure (**15**), das myo-Inosit-
hexaphosphat, sind im Pflanzenreich und auch im Tierreich sehr weit
verbreitet (*10, 104, 60, 235*). myo-Inosit ist ferner ein Bauelement der
Phosphoinositide (*13*) und der Phosphatidopeptide (*113*). An dieser Stelle
soll auch das Vorkommen eines Glykosids des myo-Inosits, des Galaktinols
(**42**) (Galaktosyl-myo-inosit) erwähnt werden, das bisher in *Beta vulgaris*
(*30, 121*) und in *Pisum sativum* (*88*) nachgewiesen wurde. LABARCA und
Mitarb. (*144*) berichteten eine teilweise Charakterisierung von Indol-
acetyl-myo-inosit aus *Zea mays* (Gramineae).

B. Methyläther des myo-Inosits.

Die bereits oben berichteten biogenetischen Beziehungen zwischen
Sequoyit (**4**), D-Pinit (**10**) und D-chiro-Inosit (**9**), die in *Trifolium incar-
natum* (*244*) und *Ononis spinosa* (*130*) vorgefunden wurden, erteilen dem
Sequoyit eine gewisse Sonderstellung unter den Methyläthern des myo-
Inosits. Dies rechtfertigt es, in *Tabelle 1* (S. 184) das Vorkommen von
Sequoyit, D-Pinit und D-chiro-Inosit zusammenzufassen. Aus Tabelle 1
ersieht man, daß Sequoyit und D-Pinit sehr häufig — besonders bei
Gymnospermen — gemeinsam vorkommen; es darf hier angemerkt werden,
daß sie oft einen beträchtlichen Anteil der sekundären Inhaltsstoffe der
Pflanzen repräsentieren. Die Vergesellschaftung der genannten Cyclite
in so vielen Pflanzen mag auch als Hinweis dahin dienen, daß ihr Biosyntheseweg in den verschiedenen Pflanzen gleich ist.

Das Vorkommen der anderen Monomethyläther des myo-Inosits,
D-Bornesit (**3**), L-Bornesit (**2**) und D-Ononit (**5**) ist in *Tabelle 2* (S. 186),
dasjenige der Dimethyläther des myo-Inosits, Dambonit und Lirio-
dendrit, ist in *Tabelle 3* (S. 187) zusammengefaßt.

Eine Substanz, die die Autoren für einen Monomethyläther eines Inosits hielten,
wurde aus dem Seestern *Asterias rutens* isoliert (*1*). POSTERNAK (*217*) konnte aber
zeigen, daß es sich dabei um α-Methyl-D-glucosid handelt.

Literaturverzeichnis: SS. 192—205.

C. scyllo-Inosit.

scyllo-Inosit (8) stellt ebenfalls einen Inhaltsstoff vieler Pflanzen dar
(*Tabelle 4*, S. 187); dieser Inosit wurde schon sehr früh in einzelnen
Organen von Fischen gefunden (*252*). Manche höheren Tiere und auch der
Mensch scheiden die Substanz im Harn aus (*166*). COSGROVE (*58*) konnte
das Hexaphosphat dieses Cyclits (16) im Boden finden.

D. D-chiro-Inosit und seine Methyläther.

Wie bereits erwähnt, bestehen zwischen dem 5-Monomethyläther des
myo-Inosits, Sequoyit (4, S. 152), D-Pinit (10) sowie D-chiro-Inosit (9)
biogenetische Beziehungen, weshalb das Vorkommen dieser drei Cyclite
in Tabelle 1 zusammengefaßt wurde.

Ein Methyläther des D-Inosits mit noch unbekannter Stellung der Methylgruppe soll in *Caesalpinia bonducella* vorkommen (*224, 225*).

E. L-chiro-Inosit und seine Methyläther.

L-chiro-Inosit (11) und L-Quebrachit (12) sind in einigen Familien
der Angiospermen (Aceraceae, Apocynaceae, Compositae) sehr stark
verbreitet (*Tabelle 5*, S. 188); hingegen wurde L-Pinit (13) bisher nur in
einer Species, *Artemisia dracunculus*, gefunden (*203*).

F. DL-chiro-Inosit.

In durchwegs älteren Veröffentlichungen wird über das Vorkommen
von DL-chiro-Inosit in *Rauwolfia caffra* (*137*), *Viscum alba* (*256*) und
Triclisia gilletii (*38*) berichtet. Es wäre wünschenswert, diese Angaben
nachzuprüfen. Wenn das Racemat tatsächlich natürlich vorkommt,
wäre dies auch vom Standpunkt der Biogenese von einigem Interesse.
Ein DL-chiro-Inosithexaphosphat wurde im Boden gefunden (*33, 58*).

G. Methyläther des muco-Inosits.

Aus *Phyllocladus trichomannoides* (Podocarpaceae) wurde vor kurzem
das erste in der Natur vorkommende Derivat des muco-Inosits, der
L-1-O-Methyl-muco-inosit (14), isoliert (*3*).

H. Derivate des neo-Inosits.

COSGROVE und TATE (*59*) gelang es, in einer Phytatfraktion aus dem
Boden neo-Inosithexaphosphat nachzuweisen.

I. C-Methylinosite.

Das Vorkommen der beiden bekannten C-Methylinosite Laminit (19)
und Mytilit (18) scheint auf Braunalgen und Rotalgen beschränkt zu
sein (*Tabelle 6*, S. 189).

J. Cyclohexanpentole und Cyclohexentetrole.

L-Viburnit (**20**, S. 153) und L-Quercit (**21**) wurden bisher in mehreren Familien der Angiospermen — in einigen Fällen auch nebeneinander — nachgewiesen (*Tabelle 7*, S. 190). D-Quercit (**22**) wurde als Inhaltsstoff von *Eucalyptus populnea* (Myrtaceae) aufgefunden (*209*). Condurit (**24**) wurde bisher nur in Asclepiadaceae nachgewiesen, während L-Leucanthemit (**23**) häufiger angetroffen wird (*Tabelle 8*, S. 191).

Nach einer älteren, seither niemals reproduzierten Angabe (*154*) soll Betit, ein Cyclohexantetrol unbekannter Struktur, in *Beta vulgaris* vorkommen.

K. Amino-desoxycyclite.

Cyclite, bei denen eine oder zwei Hydroxylgruppen durch Aminogruppen ersetzt sind, finden sich als Strukturelemente verschiedener als Antibiotica bekannter mikrobieller Stoffwechselprodukte verschiedener *Streptomyces*-Arten *(Formelübersicht 4)*. So ist Streptamin (**52**), 1,3-Diamino-1,3-didesoxy-scyllo-inosit, ein Baustein des Streptomycins, woraus es auch durch Hydrolyse erhalten werden kann (*147*, *217*).

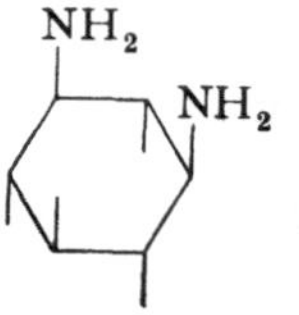

(**52**)
Streptamin.

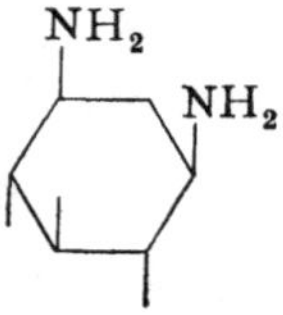

(**53**)
Desoxystreptamin.

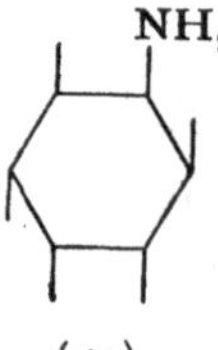

(**54**)
2-Amino-2-desoxy-neo-inosit.

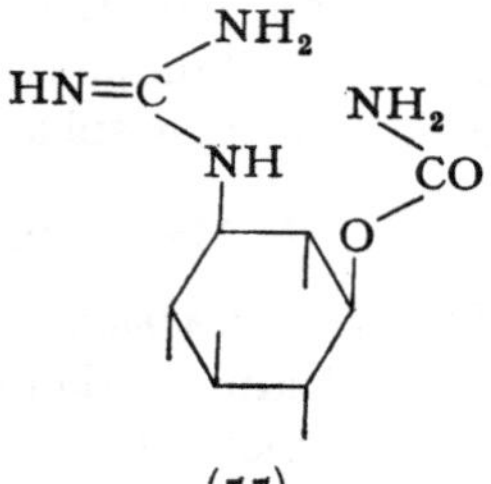

(**55**)
Glebidin, Bluensidin.

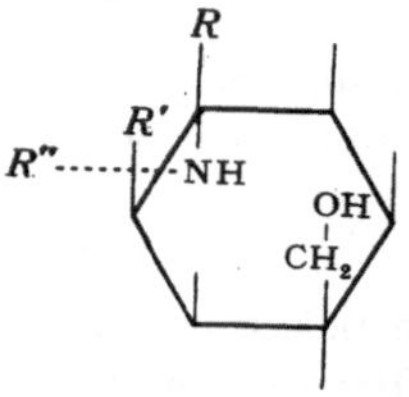

(**56**) Cyclohexanring des Tetrodotoxins.

Formelübersicht 4. Inosamine als Strukturelemente von Antibiotica.
Literaturverzeichnis: SS. 192—205.

Bei der Hydrolyse zahlreicher Antibiotica wird ein Desoxystreptamin erhalten, das als 1,3-Diamino-1,2,3-tridesoxy-scyllo-inosit (53) erkannt wurde. Folgende Antibiotica enthalten diese Verbindung als Strukturelement: Neomycin (36, 108, 143, 231, 232), Kanamycin A (61, 108, 161, 261), Kanamycin B (237), Kanamycin C (176), Paromomycin (102, 108), Zygomycin A$_1$ und A$_2$ (109, 257).

2-Amino-2-desoxy-neo-inosit (54) ist ein Baustein des aus Hygromycin entstehenden Hyosamins (167, 176, 274).

Bei der Hydrolyse von Glebomycin bzw. Bluensomycin wird 3-O-Carbamoyl-1-desoxy-1-guanidino-scyllo-inosit (Glebidin, Bluensidin, 55) erhalten (17a, 17b, 177).

Actinamin, ein Bestandteil des Actinospectatins, wurde als 1,3-bis-[Dimethylamino]-1,3-didesoxy-myo-inosit erkannt (118a).

Schließlich darf hier erwähnt werden, daß die Gonaden des giftigen Bowlfisches und anderer Tetraodontidae eine überaus giftige Verbindung enthalten, welche kristallisiert werden konnte und als Tetrodotoxin bezeichnet wird. Mit Tetrodotoxin identisch ist der Giftstoff des California-Salamanders (Taricha torosa) und anderer Salamandridae (173a). Diese toxische Substanz, welche die Permeabilität verschiedener Membranen stark beeinflußt und das Atemzentrum lähmt, enthält in ihrer Struktur einen an Amino-desoxycyclite erinnernden Cyclohexanring (56, 236a, 259a, 260).

VI. Tabellen

Tabelle 1. Vorkommen von Sequoyit (s), D-chiro-Inosit (Di) und D-Pinit (Dp).

Divisio	Classis	Familia (Tribus)	Genus/Species
Gymnospermae	Cycadopsida	Cycadaceae	*Ceratozamia mexicana* [Dp (215)], [s (215)]; *Cycas:* 3 Species [Dp (195, 215)], 2 Species [s (215)]; *Encephalartos:* 5 Species [Dp (215)], 5 Species [s (215)]; *Macrozamia:* 2 Species [s (215, 228)]
	Coniferopsida	Ginkgoaceae	*Ginkgo biloba* [s (205)], [Dp (195)]
		Araucariaceae	*Agathis:* 2 Species [s (215)]; *Araucaria:* 5 Species [s (205)]
		Cephalotaxaceae	*Cephalotaxus:* 3 Species [s (205)], 4 Species [Dp (194, 195, 205)]
		Cupressaceae	*Chamaecyparis:* 2 Species [Dp (205)]; *Cupressus:* 3 Species [s (205)], 2 Species [Dp (194, 195, 205, 207)]; *Juniperus:* 4 Species [s (205, 207)]; *Libocedrus decurrens* [Dp (207)]; *Thuja gigantea* [Dp (194, 195, 207)]; *T. orientalis* [s (205)]; *Thujopsis:* 2 Species [Dp (194, 195)]
		Pinaceae	*Abies:* 2 Species [s (211)], 3 Species [Dp (194, 195, 205)]; *Cedrus:* 2 Species [Dp (194, 195, 207)]; *Larix europeae:* [s (207)], [Dp (207)]; *Picea:* 6 Species [s (208)], 6 Species [Dp (205, 207)]; *Pinus:* 9 Species [s (17, 180, 205, 207, 208, 211)], 10 Species [Dp (19, 75, 76, 152, 153, 194, 195, 205, 207)] [Di (211)]; *Pseudolarix kaempferi* [s (205)], [Dp (194, 195, 205, 207)]; *Pseudotsuga douglasii* [s (207)], [Dp (194, 195)]; *Tsuga:* 3 Species [s (92, 207, 208, 211)], 3 Species [Dp (92, 194, 195, 205, 207)]
		Podocarpaceae	*Dacrydium cupressinum* [s (35)]; *Phyllocladus trichomannoides* [s (3)], [Di (3)]; *Podocarpus:* 6 Species [s (28, 195)]
		Taxodiaceae	*Cryptomeria:* 2 Species [Dp (194, 195)]; *Cunninghamia sinensis* [Dp (205)]; *Glyptostrobus pensilis* [s (255)]; *Metasequoia glyptostroboides* [s (122)]; *Sciadopitys verticillata* [s (207)]; *Sequoia:* 2 Species [s (205, 207, 246)], 2 Species [Dp (194, 195, 205, 245)]; *Taxodium:* 2 Species [s (205)]
	Taxopsida	Taxaceae	*Taxus:* 2 Species [s (133, 205)], *Torreya:* 3 Species [s (205)]
Angiospermae	Dicotyledonae	Aizoaceae	*Drosanthemum floribundum* [Dp (213)]; *Tetragonia expansa*

Literaturverzeichnis: SS. 192—205.

Apocynaceae	*Landolphia madagascariensis* [Dp (*91*)]
Aristolochiaceae	*Aristolochia*: 3 Species [Dp (*202*)]
Caryophyllaceae (Alsineae)	*Arenaria*: 3 Species [Dp (*197*)]; *Cerastium*: 7 Species [Dp (*197, 204*)]; *Minuartia*: 2 Species [Dp (*197*)]; *Moehringia muscosa* [Dp (*197*)]; *Sagina subulata* [Dp (*197*)]; *Scleranthus annus* [Dp (*204*)]; *Stellaria*: 3 Species [Dp (*204*)]
(Diantheae)	*Dianthus*: 20 Species [Dp (*197, 204, 211*)]; *Gypsophila*: 3 Species [Dp (*197, 211*)]; *Saponaria*: 2 Species [Dp (*197, 211, 213*)]; *Tunica*: 2 Species [Dp (*197, 204*)]; *Vaccaria segetalis* [Dp (*197, 211*)]; *Velenzia rigida* [Dp (*204*)]
(Lychnideae)	*Lychnis*: 9 Species [Dp (*197, 204*)]; *Silene*: 13 Species [Dp (*197, 204*)]
(Paronychieae)	*Herniaria*: 3 Species [Dp (*197, 211*)]; *Paronychia*: 2 Species [Dp (*197, 204*)]
(Polycarpeae) (Sperguleae)	*Polycarpon tetraphyllum* [Dp (*204*)]; *Spergula*: 2 Species [Dp (*197, 204*)]; *Telephium imperati* [Dp (*197*)]
Cistaceae	*Halimium umbellatum* [Dp (*207*)]; *Helianthemum*: 3 Species [Dp (*207*)]
Compositae	*Artemisia dracunculus* [s (*240*)]
Euphorbiaceae	*Euphorbia sp.* [Dp (*276*)]
Guttiferae	*Frankenia laevis* [Dp (*213*)]; *Myricaria germanica* [Dp (*213*)]; *Tamarix*: 4 Species [Dp (*213*)]
Leguminosae	in zirka 200 Species (über 70 Genera, 12 Triben): [Dp (*46, 56, 80, 123, 130, 136, 146, 189, 192, 201, 211, 215, 184, 226, 234, 250, 269*)]; *Ononis spinosa* [s (*130*)]; *Trifolium incarnatum* [Di (*239*)], [s (*239*)]
Loranthaceae	*Viscum album* [Dp (*196*)]
Magnoliaceae	*Magnolia*: 10 Species [Dp (*202, 204*)]
Meliaceae	*Khaya senegalensis* [Dp (*215*)]
Nyctaginaceae	*Bougainvillea glabra* [Dp (*197*)]; *Mirabilis*: 2 Species [Dp (*197*)]; *Oxybaphus viscous* [Dp (*204*)]
Oleaceae	*Ximenia americana* [Dp (*80*)]
Phytoaccaceae	*Phytolacca americana* [Dp (*197, 213*)]
Zygophyllaceae	*Zygophyllum fabago* [Dp (*197*)]

Tabelle 2. Vorkommen von D-Bornesit (Db), L-Bornesit (Lb) und D-Ononit (o).

Divisio	Classis	Familia (Tribus)	Genus/Species
Angiospermae	Dicotyledonae	Apocynaceae	
		(Alstonieae)	*Amsonia:* 2 Species [Lb (*209*)]; *Dyera:* 2 Species [Lb (*12*)]; *Vinca:* 3 Species [Lb (*209, 215*)]
		(Ecdysanthereae)	*Apocynum:* 2 Species [Db (*209*)]; *Urceola:* 2 Species [Lb (*90*)]
		(Thevetieae)	*Tanghinia venenifera* [Lb (*215*)]; *Thevetia nereifolia* [Lb (*215*)]
		Asclepiadaceae	*Cynanchum vincetoxicum* [Db (*131*)]; *Hoya carnosa* [Db (*131*)]; *Marsdenia condurango* [Db (*131*)]
		Boraginaceae	in 15 Genera: [Db (*21, 206, 211*)]
		Flacourtiaceae	*Kiggelaria africana* [o (*207*)]
		Leguminosae	
		(Phaseoleae)	*Dolichos lablab* [o (*200*)]
		(Trifolieae)	*Medicago sativa* [o (*170*)]; *Ononis:* 3 Species [o (*130, 200*)]
		(Vincieae)	*Latyrus:* 10 Species [Db (*200, 206*)]; *Leucaena glauca* [o (*211*)]; *Vigna catjang* [o (*200*)]
		Protaceae	*Banksia integrifolia* [Db (*207, 211*)]; *Leucadendron argenteum* [Db (*214*)]; *Leucospermum reflexum* [Db (*214*)]; *Macadamia ternifolia* [Db (*207, 211*)]; *Stenocarpus sinnatus* [Db (*207, 211*)]
		Rhamnaceae	*Berchemia racemoas* [Db (*206, 211*)]; *Rhamnus:* 23 Species [Db (*206, 211*)]
	Monocotyledonae	Rubiaceae	*Sarcocephalus diderrichii* [Lb (*132*)]
		Palmae	*Caryota urens* [Db (*215*)]

Literaturverzeichnis: *SS. 192—205.*

Tabelle 3. Vorkommen von Dambonit (d) und Liriodendrit (l).

Divisio	Classis	Familia (Tribus)	Genus/Species
Angiospermae	Dicotyledonae	Apocynaceae (Alstonieae) (Carisseae) (Nerieae) (Parsonsieae)	*Dyera:* 2 Species [d (57, 120)]; *Vinca:* 2 Species [d (208, 209)] *Acokanthera spectabilis* [d (215)] *Funtumia elastica* [d (89)]; *Nerium oleander* [d (208, 209)] *Trachlospermum jasminoides* [d (208, 209)]
		Magnoliaceae	*Liriodendron:* 2 Species [l (199)]
		Moraceae	*Castilloa elastica* [d (270)]

Tabelle 4. Pflanzen, in welchen scyllo-Inosit als Inhaltsstoff nachgewiesen wurde.

Divisio	Classis	Familia	Genus/Species
Rhodophyta	Rhodophyceae	—	*Porphyra umbilicalis* [(148)]
	Rhodomelaceae	—	*Polysiphonia fastigiata* [(272)]
Angiospermae	Monocotyledoneae	Palmae	*Cocos nucifera* [(175)]; *C. plumosa* [(175)]
	Dicotyledoneae	Calycanthaceae	*Calycanthus floridus* [(168)]; *C. glaucus* [(168)]; *C. occidentalis* [(202)]; *Chimonanthus fragrans* [(202)]
		Compositae	*Veronia altissima* [(233)]
		Cornaceae	*Cornus florida* [(101)]
		Fagaceae	*Quercus:* 2 Species [(131e, 264)]
		Thamnaceae	*Helinus ovatus* [(94)]
		Tiliaceae	*Tilia tomentosa* [(172)]

Tabelle 5. Vorkommen von L-chiro-Inosit (Li), L-Quebrachit (q) und L-Pinit (Lp).

Divisio	Classis	Familia (Tribus)	Genus/Species
Angiospermae	Dicotyledonae	Aceraceae	in 2 der 3 Genera (*Acer, Negundo*) allgemein verbreitet: [q (*186, 188, 195*)]
		Apocynaceae (Alstonieae)	*Aspidosperma quebracho* [q (*256*)]; *Haplophyton cimicidum* [q (*55*)]; *Holarrhena antidysenterica* [Li (*213*)], [q (*213*)]
		(Carisseae)	*Carissa edulis* [q (*215*)]
		(Rauvolfieae)	*Rauwolfia caffra* [Li (*137*)]
		(Tabernaemontaneae)	*Conopharyngia durissima* [q (*209*)]
		Compositae (Anthemideae)	*Achillea ptarmica* [Li (*213*)]; *Artemis nobilis* [Li (*210*)]; *Artemisia*: 10 Species [q (*95, 190, 208*)], 2 Species [Li (*240*)], *A. dracunculus* [Lp (*203, 240*)]: *Chrysanthemum*: 3 Species [Li (*125, 210*)]
		(Astereae)	*Erigeron ramosum* [Li (*211*)]
		(Cichorieae)	*Crepis dioscuridis* [Li (*213*)]; *Lactuca scariola* [Li (*213*)]; *Mulgedium bourgeaei* [Li (*213*)]; *Picris hieracioides* [Li (*213*)]; *Sonchus arvensis* [Li (*253*)]; *Taraxacum officinale* [Li (*213*)]
		(Cynareae)	*Centaurea*: 3 Species [Li (*211, 213*)]; *Serratula coronata* [Li (*211*)]
		(Eupatorieae)	*Eupatorium cannabinum* [Li (*211*)]
		(Helenieae)	*Helenium autumnale* [Li (*213*)]; *Tagetes patula* [Li (*213*)]
		(Heliantheae)	*Bidens pilosa* [Li (*213*)]; *Coreopsis grandiflora* [Li (*213*)]; *Cosmos bipinnatus* [Li (*213*)]; *Silphium integrifolium* [Li (*213*)]
		(Inuleae)	*Helichrysum arenarium* [Li (*268*)]; *Inula helenium* [Li (*211*)]; *Pulicaria dysenterica* [Li (*211*)]
		(Senecioneae)	*Senecio*: 2 Species [Li (*213*)]
		(Veronieae)	*Veronia altissima* [Li (*233*)]
		Elaeagnaceae	*Elaeagnus*: 6 Species [q (*193*)]; *Hippophae*: 2 Species [q (*193*)]; *Shepherdia argentea* [q (*193*)]

Divisio		Familia	Genus/Species
Angiospermae	Dicotyledonae	Euphorbiaceae	*Acalypha indica* [q (229, 230)]; *Euphorbia*: 6 Species [Li (99, 213)]; *Hevea brasiliensis* [Li (191, 248)]; *Hura crepitans* [Li (213)]
		Hippocastanaceae	*Aesculus*: 5 Species [q (191, 208)]
		Loranthaceae	*Viscum album* [q (196, 256)]
		Menispermaceae	*Trichisia gilletii* [Li (37, 38)]
		Moraceae	*Cannabis sativa* [q (2)]; *Humulus lupulus* [q (208)]
		Protaceae	*Grevillea*: 2 Species [q (23, 213)]; *Hakea laurina* [q (24)]
		Sapindaceae	in 17 Species: [q (185—188, 191, 211, 213)]
		Ulmaceae	*Celtis*: 10 Species [q (207)]; *Pterocellis tatarinovii* [q (207)]

Tabelle 6. Vorkommen von Laminit (l) und Mytilit (m).

Divisio	Familia	Genus/Species
Phaeophyta	Phaeophyceae	*Desmarestia aculeata* [1 (26)]; *Fucus spiralis* [1 (26)]; *Laminaria clousoni* [1 (150, 151)];
Rhodophyta	Rhodophyceae	*Gelidium cartilagineum* [m (149)], [1 (149)]; *Polysiphonia fastigiata* [m (272)], [1 (272)]; *Porphyra umbilicalis* [1 (148)]

Tabelle 7. Vorkommen von L-Viburnit (v) und L-Quercit (q).

Divisio	Classis	Familia	Genus/Species
Angiospermae	Monocotyledonae	Palmae	*Chamerops humilis* [q (174, 175)]
		Asclepiadaceae	*Gymnema sylvestre* [v (223)]; *Marsdenia condurango* [v (131)]
		Caprifoliaceae	*Viburnum:* 3 Species [v (105, 215)], 2 Species [q (125)]
		Compositae	*Achillea:* 2 Species [v (208)]; *Chrysanthemum:* 3 Species [v (131a, 208)]; *C. leucanthemum* [q (131a)]; *Tanacetum vulgare* [v (208)]
	Dicotyledonae	Fagaceae	*Quercus:* 35 Species [q (27, 65, 198, 209, 238)], 2 Species [v (131e)]
		Leguminosae	*Pterocarpus lucens* [q (201)]
		Loganiaceae	*Strychnos toxifera* [q (22)]
		Magnoliaceae	*Talauma mexicana* [q (249)]
		Menispermaceae	*Cissampelos pareira* [q (20)]; *Coculus:* 2 Species [q (198)]; *Cyclea burmanii* [q (48)]; *Legnephora moorii* [q (114)]; *Menispermum canadense* [v (202)]; *Stephania hernandifolia* [v (77)]; *Tiliacora acuminata* [q (117)]; *Triglisia gilletti* [q (37, 38)]
		Myrsinaceae	*Embelia ribes* [q (140, 141)]; *Myrsine:* 2 Species [q (140, 141)]
		Myrtaceae	*Eugenia jambolana* [q (217)]
		Sapotaceae	*Achras sapota* [q (98)]; *Butyrospermum parkii* [q (18)]; *Mimusops elengi* [q (98)]

Literaturverzeichnis: SS. 192—205.

Tabelle 8. Vorkommen von L-Leucanthemit (l) und Condurit (c).

Divisio	Classis	Familia (Tribus)	Genus/Species
Gymnospermae	Coniferopsida	Cupressaceae	*Thuja occidentalis* [l (*131 d*)]
Angiospermae	Dicotyledonae	Apocynaceae (Nerieae)	*Nerium oleander* [l (*131 c*)]
		Asclepiadaceae (Asclepiadeae) (Tylophorae)	*Cynanchum vincetoxicum* [c (*131*)], [l (*131*)] *Gymnema sylvestre* [c (*167 b*)] *Hoya carnosa* [c (*131*)], [l (*131*)] *Marsdenia condurango* [c (*131*)], [l (*131*)]
		Boraginaceae (Anchuseae) (Lithospermeae)	*Borago officinalis* [l (*131 b*)] *Myosotis*: 2 Species [l (*131 b*)]
		Compositae (Anthemideae)	*Chrysanthemum*: 3 Species [l (*131 a, 210*)]; *Tanacetum vulgare* [l (*125*)]
		Fagaceae	*Quercus*: 2 Species [l (*131 e*)]
		Leguminosae	*Ononis*: 2 Species [l (*130*)]

Literaturverzeichnis.

1. Ackermann, D. und G. Hoppe-Seyler: Asterit, ein neuer biologischer Cyclit (Cyclohexanolmonomethyläther). Z. physiol. Chem. **336**, 1 (1964).
2. Adams, R., D. C. Pease and J. H. Clark: Isolation of Cannabinol, Cannabidiol and Quebrachitol from Red Oil of Minnesota Wild Hemp. J. Amer. Chem. Soc. **62**, 2194 (1940).
3. Adhikari, S. K., R. A. Bell and W. E. Harvey: Cyclitols from the Heartwood of *Phyllocladus trichomannoides*. J. Chem. Soc. (London) **1962**, 2829.
4. Agranoff, B. W., R. M. Bradley and R. O. Brady: The Enzymatic Synthesis of Inositol Phosphatide. J. Biol. Chem. **233**, 1077 (1958).
5. Ahuja, J. N.: Thesis, Michigan State Univ., East Lansing, 1962.
6. Anderson, L. and R. H. Coots: Metabolism of 2-^{14}C-myo-Inositol in the Rat. Biochim. Biophys. Acta **28**, 666 (1958).
7. Anderson, L. and G. G. Post: Catalytic Air Oxidation of Cyclitols and some Observations on the Hydrogenolysis of Inososes. Abstr. 134th Meeting Amer. Chem. Soc., Chicago, p. 12 D (1958).
8. Anderson, L., R. Takeda, S. J. Angyal and D. J. McHugh: Cyclitol Oxidation by *Acetobacter suboxydans*. II. Additional Cyclitols and the "Third Specificity Rule". Arch. Biochem. Biophys. **78**, 518 (1958).
9. Anderson, L., K. Tomita, P. Kussi and S. Kirkwood: On the Cyclitol Oxidizing Enzyme System of *Acetobacter suboxydans*. J. Biol. Chem. **204**, 769 (1953).
10. Anderson, R. J. and E. G. Roberts: The Chemistry of the Lipoids of Tubercle Bacilli. XXI. The Polysaccharide Occurring in the Phosphatide from the Human Tubercle Bacilli. J. Amer. Chem. Soc. **52**, 5023 (1930).
11. Angyal, S. J. and L. Anderson: The Cyclitols. Adv. Carbohydrate Chem. **14**, 135 (1959).
12. Angyal, S. J., P. T. Gilham and C. G. MacDonald: Methyl Ethers of *myo*-Inositol. J. Chem. Soc. (London) **1957**, 1417.
13. Ansell, G. B. and J. N. Hawthorne: Phospholipids. Amsterdam: Elsevier. 1964.
14. Appel, H. H. and E. Lobos: (+)-Pinitol in *Adesmia* Species from Chile. Scientia **29**, 33 (1962) [Chem. Abstr. **60**, 5613 (1964)].
15. Ata, S., J. Ito and N. Iso: The Effect of Inositol on the Growth and Metabolism of the Tubercle Bacillus. Nagoya J. Med. Sci. **14**, 77 (1951).
16. Ballou, C. E.: Inositol and Related Compounds (Cyclitols). In: W. Ruhland et al., Encyclopedia of Plant Physiology, Vol. X, p. 442. Berlin-Heidelberg-Göttingen: Springer. 1958.
17. Ballou, C. E. and A. B. Anderson: On the Cyclitols Present in Sugar Pine (*Pinus lambertiana* Dougl.). J. Amer. Chem. Soc. **75**, 648 (1953).
17a. Bannister, B. and A. D. Argoudelis: The Chemistry of Bluensomycin. I. The Structure of Bluensidine. J. Amer. Chem. Soc. **85**, 119 (1963).
17b. — — The Chemistry of Bluensomycin. II. The Structure of Bluensomycin. J. Amer. Chem. Soc. **85**, 234 (1963).
18. Bauer, K. H. und H. Moll: Über die Inhaltsstoffe der Sheanüsse, der Samen von *Butyrospermum Parkii*. Arch. Pharmaz. **280**, 37 (1942).
18a. Berman, T. and B. Magasanik: The Pathway of *myo*-Inositol Degradation in *Aerobacter aerogenes*. Dehydrogenation and Dehydration. J. Biol. Chem. **241**, 800 (1966).
18b. — — The Pathway of *myo*-Inositol Degradation in *Aerobacter aerogenes*. Ring Scission. J. Biol. Chem. **241**, 807 (1966).
19. Berthelot, M.: Sur quelques matières sucrées. Ann. chim. phys. **46**, 76 (1856).

20. BHATTACHARJI, S., V. N. SHARMA and M. L. DHAR: Chemical Examination of the Roots of *Cissampelos pareira*. J. Sci. Industr. Res. (India) 11 B, 81 (1952).

21. BIEN, S. and D. GINSBURG: The Structure of Bornesitol. J. Chem. Soc. (London) 1958, 3189.

22. BOEHM, R.: Über Curare und Curarealkaloide. Arch. Pharmaz. 235, 660 (1897).

23. BOURQUELOT, E. et A. FICHTENHOLZ: Présence de la québrachite dans les feuilles de *Grevillea robusta* A. Cunn. C. R. hebd. séances Acad. Sci. 155, 615 (1912).

24. BOURQUELOT, E. et H. HÉRISSEY: Application de la méthode biochimique à l'étude des feuilles d'*Hakea laurina*. Extraction d'un glucoside (arbutine) et de québrachite. C. R. hebd. scéances Acad. Sci. 168, 414 (1919).

25. BOUVEAULT, L.: De l'isomérie optique dans les corps à chaînes fermées. Bull. soc. chim. France [3] 11, 144 (1894).

26. BOUVENG, H. and B. LINDBERG: Low-molecular Carbohydrates in Algae. VII. Investigation of *Fucus spiralis* and *Desmarestia aculeata*. Acta Chem. Scand. 9, 168 (1955).

27. BRÄUTIGAM, W.: Über das Vorkommen von Vanillin im Korke. Pharmaz. Zentralh. 39, 685, 722 (1898).

28. BRIGGS, L. H., R. C. CAMBIE and J. L. HOARE: Chemistry of the Podocarpaceae. III. A New Lignan, *seco*-Isolariciresinol and Further Constituents of the Heartwood of *Podocarpus spicatus*. Tetrahedron 7, 262 (1959).

29. BROCKERHOFF, H. and C. E. BALLOU: Phosphate Incorporation in Brain Phosphoinositides. J. Biol. Chem. 237, 49 (1962).

30. BROWN, R. J. and R. F. SERRO: Isolation and Identification of O-α-D-Galactopyranosyl-*myo*-inositol and of *myo*-Inositol from Juice of the Sugar Beet (*Beta vulgaris*). J. Amer. Chem. Soc. 75, 1040 (1953).

31. BURNS, J. J. and G. ASHWELL: L-Ascorbic Acid. In: P. D. Boyer, H. Lardy and K. Myrbäck, The Enzymes, Vol. 3, p. 387. New York: Academic Press. 1960.

32. BURNS, J. J., N. TROUSOF, C. EVANS, N. PAPADOPOULOS and B. W. AGRANOFF: Conversion of *myo*-Inositol to D-Glucuronic Acid and L-Gulonic Acid in the Rat. Biochim. Biophys. Acta 33, 215 (1959).

33. CALDWELL, A. G. and C. A. BLACK: Inositol Hexaphosphate. I. Quantitative Determination in Extracts of Soils and Manures. Soil Sci. Soc. Amer., Proc. 22, 290 (1958).

34. CALDWELL, K. A.: Optically Active Inosamine and a Substituted Pinitol Glucoside. Univ. Microfilms (Ann Arbor, Mich.), L. C. Card No. Mic60-2420 [Chem. Abstr. 55, 2496 (1961)].

35. CAMBIE, R. C. and B. F. CAIN: Bark Extractives of *Dacrydium cupressinum*. New Zealand J. Sci. 3, 121 (1960).

36. CARTER, H. E., J. R. DYER, P. D. SHAW, K. L. RINEHART, Jr. and M. HICHENS: The Structure of Neamine. J. Amer. Chem. Soc. 83, 3723 (1961).

37. CASTAGNE, E.: Beitrag zur chemischen Untersuchung der Liane „Efiri". 2. Mitt. Über die Gegenwart von Cyclohexanpentol in den Stengeln von „Efiri". Congo 1934, I, 341 [Chem. Zbl. 1934, II, 76].

38. — Beitrag zur chemischen Untersuchung der Liane „Efiri". 3. Mitt. Über die Gegenwart von inaktivem spaltbaren Inosit in den Stengeln von „Efiri". Congo 1934, II, 357 [Chem. Zbl. 1935, I, 583].

39. CHARALAMPOUS, F. C.: Biochemical Studies on Inositol. II. Chemical Degradation of Inositol to Glyoxylic and Formic Acids. J. Biol. Chem. 225, 585 (1957).

40. — Biochemical Studies on Inositol. III. Biosynthesis of Inositol by Yeast. J. Biol. Chem. 225, 595 (1957).

41. Charalampous, F. C.: Biochemical Studies on Inositol. V. Purification and Properties of the Enzyme that Cleaves Inositol to D-Glucuronic Acid. J. Biol. Chem. **234**, 220 (1959).

42. — Biochemical Studies on Inositol. VI. Mechanism of Cleavage of Inositol to D-Glucuronic Acid. J. Biol. Chem. **235**, 1286 (1960).

43. Charalampous, F. C. and P. Abrahams: Biochemical Studies on Inositol. I. Isolation of *myo*-Inositol from Yeast and its Quantitative Enzymatic Estimation. J. Biol. Chem. **225**, 575 (1957).

44. Charalampous, F. C., S. Bumiller and S. Graham: The Site of Cleavage of *myo*-Inositol by Purified Enzymes of Rat Kidney. J. Amer. Chem. Soc. **80**, 2022 (1958).

45. Charalampous, F. C. and C. Lyras: Biochemical Studies on Inositol. IV. Conversion of Inositol to Glucuronic Acid by Rat Kidney Extracts. J. Biol. Chem. **228**, 1 (1957).

46. Charaux, C.: Sur la manne du Caroubier et le sucre retiré de cette manne. Identité de ce sucre avec la pinite ou méthyl-*d*-inosite. Bull. soc. chim. biol. (Paris) **4**, 597 (1922).

47. Charollais. E. et Th. Posternak: Recherches sur la biochimie des cyclitols. IX. Contribution à l'étude du métabolisme du *ms*-inositol chez le Rat. Helv. Chim. Acta **48**, 280 (1965).

48. Chaudhry, G. R. and M. L. Dhar: Chemical Examination of the Roots of *Cyclea burmanii*. J. Sci. Industr. Res. (India) **17 B**, 163 (1958).

49. Cheldelin, V. H. and R. J. Williams: The B Vitamin Content of Foods. Univ. Texas Publ. No. 4237, p. 105 (1942).

50. Chen, I. W. and F. C. Charalampous: A Soluble Enzyme System from Yeast which Catalyzes the Biosynthesis of Inositol from Glucose. Biochem. Biophys. Res. Comm. **12**, 62 (1963).

51. — — Mode of Conversion of Glucose-6-P to Inositol and the Role of DPN and NH$_4^+$ Ions. Biochem. Biophys. Res. Comm. **17**, 521 (1964).

52. — — Biochemical Studies on Inositol. VII. Biosynthesis of Inositol by a Soluble Enzyme System. J. Biol. Chem. **239**, 1905 (1964).

53. — — Inositol 1-Phosphate as Intermediate in the Conversion of Glucose 6-Phosphate to Inositol. Biochem. Biophys. Res. Comm. **19**, 144 (1965).

54. — — Biochemical Studies on Inositol. VIII. Purification and Properties of the Enzyme System which Converts Glucose 6-Phosphate to Inositol. J. Biol. Chem. **240**, 3507 (1965).

55. Clark, E. P.: The Occurrence of Quebrachite in the Stems of *Haplophyton cimicidum*. J. Amer. Chem. Soc. **58**, 1009 (1936).

56. Clark-Lewis, J. W., G. F. Katekar and P. I. Mortimer: Flavan Derivatives. IV. Teracacidin, a New Leucoanthocyanidin from *Acacia intertexta*. J. Chem. Soc. (London) **1961**, 499.

57. Comollo, A. J. and A. K. Kiang: Dambonitol: its Isolation from *Dyera lowii* and *Dyera costulata* and its Constitution. J. Chem. Soc. (London) **1953**, 3319.

58. Cosgrove, D. J.: Forms of Inositol Hexaphosphate in Soils. Nature **194**, 1265 (1962).

59. Cosgrove, D. J. and M. E. Tate: Occurrence of *neo*-Inositol Hexaphosphate in Soil. Nature **200**, 568 (1963).

60. Courtois, J. E.: Les esters phosphoriques de l'inositol. Bull. soc. chim. biol. (Paris) **33**, 1075 (1951).

61. Cron, M. J., D. L. Johnson, F. M. Palermiti, Y. Perron, H. D. Taylor, D. F. Whitehead and I. R. Hooper: Kanamycin. I. Characterization and Acid Hydrolysis Studies. J. Amer. Chem. Soc. **80**, 752 (1958).

62. DANGSCHAT, G.: In: K. Paech und M. V. Tracey, Moderne Methoden der Pflanzenanalyse. Bd. II, S. 64. Berlin-Heidelberg-Göttingen: Springer. 1955.

63. — Acetonierung und Konfiguration des *meso*-Inosits. Naturwiss. **30**, 146 (1942).

64. DAUGHADAY, W. H., J. LARNER and C. HARTNETT: The Synthesis of Inositol in the Immature Rat and Chick Embryo. J. Biol. Chem. **212**, 869 (1955).

65. DESSAIGNES, V.: Recherches sur une matière sucrée particulière, trouvée par M. Braconnot dans le gland du chêne. C. R. hebd. séances Acad. Sci. **33**, 308 (1851).

66. DWORSKY, P. und O. HOFFMANN-OSTENHOF: Abbau von meso-Inosit durch intakte Zellen von *Schwanniomyces occidentalis*. Biochem. Z. **343**, 394 (1965).

67. EAGLE, H., B. W. AGRANOFF and E. E. SNELL: The Biosynthesis of *meso*-Inositol by Cultured Mammalian Cells, and the Parabiotic Growth of Inositol-dependent and Inositol-independent Strains. J. Biol. Chem. **235**, 1891 (1960).

68. EAGLE, H., V. I. OYAMA, M. LEVY and A. E. FREEMAN: *myo*-Inositol as an Essential Growth Factor for Normal and Malignant Human Cells in Tissue Culture. J. Biol. Chem. **226**, 191 (1957).

69. EASTCOTT, E. V.: Wildiers' Bios. The Isolation and Identification of „Bios I". J. Physic. Chem. **32**, 1094 (1928).

70. EISENBERG, F., Jr.: Biosynthesis of Inositol in the Mammal. In: H. Kindl, Cyclitols and Phosphoinositides. Oxford: Pergamon. 1966.

71. EISENBERG, F., Jr. and A. H. BOLDEN: Biosynthesis of Inositol in Rat Testis Homogenate. Biochem. Biophys. Res. Commun. **12**, 72 (1963).

72. — — Reproductive Tract as Site of Synthesis and Secretion of Inositol in the Male Rat. Nature **202**, 599 (1964).

73. — — D-*myo*-Inositol-1-phosphate, an Intermediate in the Biosynthesis of Inositol in the Mammal. Biochem. Biophys. Res. Commun. **21**, 100 (1965).

74. EISENBERG, F., Jr., A. H. BOLDEN and F. A. LOEWUS: Inositol Formation by Cyclization of Glucose Chain in Rat Testis. Biochem. Biophys. Res. Comm. **14**, 419 (1964).

74a. ENGLISH, P. D., M. DIETZ and P. ALBERSHEIM: Myoinositol Kinase: Partial Purification and Identification of Product. Science **151**, 198 (1966).

74b. Enzyme Nomenclature — Recommendations 1964 of the International Union of Biochemistry. Amsterdam: Elsevier. 1965.

75. ERDTMAN, H.: The Phenolic Constituents of Pine Heartwood. IV. Membrane-forming Substance in Pine Heartwood. Svensk Papperstidn. **46**, 226 (1943).

76. — The Phenolic Constuents of Pine Heartwood. VI. The Heartwood of *Pinus cembra* L. Svensk Kem. Tidskr. **56**, 26 (1944).

77. EWING, J., G. K. HUGHES and E. RITCHIE: A New Source of *l*-Quercitol (Viburnitol). Austral. J. Sci. Res. **3A**, 514 (1950).

78. FERNÁNDEZ, O., G. IZQUIERDO and E. MARTÍNEZ: Biological Origin of Rubber — Cyclases. Farm. nueva (Madrid) **9**, 563 (1944) [Chem. Abstr. **40**, 4115 (1946)].

79. FERNÁNDEZ, O., M. DE MINGO and E. MARTÍNEZ: Cyclization of Glucose. Farm. nueva (Madrid) **10**, 541 (1945) [Chem. Abstr. **43**, 4229 (1949)].

80. FINNEMORE, H., J. M. COOPER, M. B. STANLEY, J. H. COBCROFT and L. J. HARRIS: The Cyanogenetic Constituents of Australian and Other Plants. J. Soc. Chem. Ind., Trans. **57**, 162 (1938).

81. FISCHER, H. O. L.: Chemical and Biological Relationships between Hexoses and Inositols. Harvey Lect. **40**, 156 (1944—45).

82. FLETCHER, H. G., Jr.: The Chemistry and Configuration of the Cyclitols. Adv. Carbohydrate Chem. **3**, 45 (1948).

83. FLETCHER, H. G., Jr., L. ANDERSON and H. A. LARDY: The Nomenclature of the Cyclohexitols and their Derivatives. J. Organ. Chem. (USA) **16**, 1238 (1951).

84. Fleury, P. et B. Balatre: Chimie et biochimie des inositols. Paris: Masson. 1947.
85. Folch, J. and D. W. Woolley: Inositol, a Constituent of a Brain Phosphatide. J. Biol. Chem. **142**, 963 (1942).
86. Franzl, R. E. and E. Chargaff: Bacterial Enzyme Preparations Oxidizing Inositol and their Inhibition by Colchicine. Nature **168**, 955 (1915).
87. Freinkel, N. and R. M. C. Dawson: The Synthesis of *meso*-Inositol in Germfree Rats and Mice. Biochem. J. **81**, 250 (1961).
88. Frydman, R. B. and E. F. Neufeld: Synthesis of Galactosylinositol by Extracts from Peas. Biochem. Biophys. Res. Comm. **12**, 121 (1963).
89. Girard, A.: Note sur un nouveau principe volatil et sucré trouvé dans le caoutchouc du Gabon. C. R. hebd. séances Acad. Sci. **67**, 820 (1868).
90. — Sur un nouveau principe volatil et sucré, trouvé dans le caoutchouc de Bornéo. C. R. hebd. séances Acad. Sci. **73**, 426 (1871).
91. — Sur une nouvelle matière volatile, extraite du caoutchouc de Madagascar. C. R. hebd. séances Acad. Sci. **77**, 995 (1873).
92. Goldschmid, O. and H. L. Hergert: Examination of Western Hemlock for Lignin Precursors. Tappi **44**, 858 (1961).
93. Goldstone, J. M. and B. Magasanik: Inositol Dehydrogenase and Keto-inositol Dehydrase from Inositol-adapted *Aerobacter aerogenes*. Federat. Proc. (Amer. Soc. Exp. Biol.) **13**, 218 (1954).
94. Goodson, J. A.: Constituents of the Leaves of *Helinus ovatus*. J. Chem. Soc. (London) **117**, 140 (1920).
95. — The Constituents of the Flowering Tops of *Artemisia afra* Jacq. Biochem. J. **16**, 489 (1922).
96. Gorin, P. A. J., K. Horitsu and J. F. T. Spencer: Formation of O-β-D-Glucopyranosyl- and O-β-D-Galactopyranosyl-*myo*-inositols by Glycosyl Transfer. Canad. J. Chem. **43**, 2259 (1965).
97. Grosheintz, J. M. and H. O. L. Fischer: Cyclization of 6-Nitrodesoxyaldo-hexoses to Nitrodesoxyinositols. J. Amer. Chem. Soc. **70**, 1479 (1948).
98. Haar, A. W. van der: Über das Vorkommen von *d*-Quercit in den Samenkernen von *Achras sapota* L. Rec. trav. chim. Pays-Bas **41**, 784 (1922).
99. Hallett, F. P. and L. M. Parks: The Isolation of *l*-Inositol from *Euphorbia pilulifera*. J. Amer. Pharm. Assoc. **40**, 474 (1951).
100. Halliday, J. W. and L. Anderson: The Synthesis of *myo*-Inositol in the Rat. J. Biol. Chem. **217**, 797 (1955).
101. Hann, R. M. and C. E. Sando: Scyllitol from Flowering Dogwood *(Cornus florida)*. J. Biol. Chem. **68**, 399 (1926).
102. Haskell, T. H., J. C. French and Q. R. Bartz: Paromomycin. I. Paromamine, a Glycoside of D-Glucosamine. J. Amer. Chem. Soc. **81**, 3480 (1959).
103. Hauser, G. and V. N. Finelli: The Biosynthesis of Free and Phosphatide *myo*-Inositol from Glucose by Mammalian Tissue Slices. J. Biol. Chem. **238**, 3224 (1963).
104. Helleu, C.: Étude de l'élimination urinaire du méso-inositol en fonction du régime alimentaire. Bull. soc. chim. biol. (Paris) **39**, 633 (1957).
105. Hérissey, H. et G. Poirot: Extraction, des feuilles de *Viburnum tinus* L., d'un principe immédiat cristallisé, le viburnitol. C. R. hebd. séances Acad. Sci. **203**, 466 (1936).
106. Herken, H., D. Maibauer und F. Weygand: Über den Stoffwechsel von *meso*-Inosit. Z. Naturforsch. **12 b**, 598 (1957).
107. Heyns, K. und H. Paulsen: Oxydative Umwandlungen an Kohlenhydraten. VIII. Katalytische Oxydation von *meso*-Inosit zu *scyllo-meso*-Inosose. Chem. Ber. **86**, 833 (1953).

108. HICHENS, M. and K. L. RINEHART, Jr.: Chemistry of the Neomycins. XII. The Absolute Configuration of Deoxystreptamine in the Neomycins, Paromomycins and Kanamycins. J. Amer. Chem. Soc. **85**, 1547 (1963).

108a. HOKIN, L. E. and M. R. HOKIN: The Chemistry of Cell Membranes. Scient. American **213**, No. 4, 78 (1965).

109. HORII, S.: Chemistry of Zygomycin A. Structure of Zygomycin A_1, Zygomycin A_2 and Dextromycin. J. Antibiotics (Tokyo) **A15**, 187 (1962) [Chem. Abstr. **58**, 9217 (1963)].

110. HORNER, W. H.: Biosynthesis of Streptomycin. II. myo-Inositol, a Precursor of the Streptidine Moiety. J. Biol. Chem. **239**, 2256 (1964).

111. HOWARD, C. F.: The Catabolism of myo-Inositol-2-^{14}C by Rat Kidney Homogenates and Slices. Thesis, Univ. Wisconsin, Madison. 1963 [Chem. Abstr. **60**, 16303 (1964)].

112. HÜBSCHER, G. and J. N. HAWTHORNE: The Isolation of Inositol Monophosphate from Liver. Biochem. J. **67**, 523 (1957).

113. HUGGINS, C. G. and D. V. COHN: Studies Concerning the Composition, Distribution and Turnover of Phosphorus in a Phosphatido-Peptide Fraction from Mammalian Tissue. J. Biol. Chem. **234**, 257 (1959).

114. HUGHES, G. K., F. P. KAISER, N. MATHESON and E. RITCHIE: The Constituents of *Legnephora moorei*. Austral. J. Chem. 6, 90 (1953).

115. IMAI, Y.: Biosynthesis of myo-Inositol in the Rat. J. Biochemistry (Tokyo) **53**, 50 (1963).

116. IRANI, R. J. and K. GANAPATHI: myo-Inositol in the Biosynthesis of Benzylpenicillin by the Mycelial Suspensions of *Penicillium chrysogenum*. Experientia **15**, 22 (1959).

117. ITALLIE, L. VAN and A. J. STEENHAUER: Investigations of the Bark of *Tiliacora acuminata* Miers. Pharm. Weekbl. **59**, 1381 (1922).

118. JANKE, R. G., C. JUNGWIRTH, I. B. DAWID und O. HOFFMANN-OSTENHOF: Über den oxydativen Abbau von myo-Inosit durch einige Sproßpilzarten. Monatsh. Chem. **90**, 382 (1959).

118a. JOHNSON, A. L., R. H. GOURLAY, D. S. TARBELL and R. L. AUTREY: The Chemistry of Actinamine. J. Organ. Chem. (USA) **28**, 300 (1963).

119. JONG, A. W. K. DE: La présence de québrachite dans le latex de *Hevea brasiliensis*. Rec. trav. chim. **25**, 48 (1906).

120. — La présence de la diméthylinosite inactive dans le latex de Melaboeai de Sumatra. Rec. trav. chim. Pays-Bas **27**, 257 (1908).

121. KABAT, E. A., D. L. MacDONALD, C. E. BALLOU and H. O. L. FISCHER: On the Structure of Galactinol. J. Amer. Chem. Soc. **75**, 4507 (1953).

122. KARYONE, T., M. TAKAHASHI, K. ISOI and M. YOSHIKURA: Chemical Constituents of the Plant of Coniferae and Allied Orders. XX. Components of the Leaves of *Metasequoia glytostroboides* (1). J. Pharmac. Soc. Japan **78**, 801 (1958).

123. KEPPLER, H. H.: The Isolation and Constitution of Mollisacacidin, a New Leucoanthocyanidin from the Heartwood of *Acacia mollissima* Willd. J. Chem. Soc. (London) **1957**, 2721.

124. KINDL, H.: Biosynthesis of meso-Inositol in Microorganisms and Higher Plants. In: H. Kindl, Cyclitols and Phosphoinositides, p. 15. Oxford: Pergamon. 1966.

125. — Unveröffentlicht.

126. KINDL, H., J. BIEDL-NEUBACHER und O. HOFFMANN-OSTENHOF: Untersuchungen über die Biosynthese der Cyclite. IX. Überführung von D-Glucose und D-Glucose-6-phosphat in meso-Inosit durch einen zellfreien Extrakt aus *Candida utilis*. Biochem. Z. **341**, 157 (1965).

127. Kindl, H. und O. Hoffmann-Ostenhof: Untersuchungen über die Biosynthese der Cyclite. II. Bildung von meso-Inosit aus Glucose-1-^{14}C in *Sinapis alba* und selektiver Abbau des so entstandenen markierten Produktes. Biochem. Z. **339**, 374 (1964).

128. — — Untersuchungen über die Biosynthese der Cyclite. IV. Bildung von meso-Inosit-5-^{14}C aus Glucose-2-^{14}C. Monatsh. Chem. **95**, 548 (1964).

129. — — Untersuchungen über die Biosynthese der Cyclite. XI. Der Ursprung des meso-Inosit-Anteils der Phytinsäure in höheren Pflanzen. Biochem. Z. **345**, 454 (1966).

130. — — Untersuchungen über die Biosynthese der Cyclite. XII. Die Bildung von D-Ononit und anderen Cycliten in *Ononis spinosa*. Z. physiol. Chem. **345**, 257 (1966).

131. — — Untersuchungen über die Biosynthese der Cyclite. XIII. Vorkommen und Biosynthese von Cycliten in Asclepiadaceae. Phytochem. (im Druck).

131a. — — Untersuchungen über die Biosynthese der Cyclite. XIV. Die Bildung von L-Viburnit in *Chrysanthemum leucanthemum*. Phytochem. (im Druck).

131b. — — Untersuchungen über die Biosynthese der Cyclite. XVI. Bildung von D-Bornesit in Boraginaceae und Leguminosae. Monatsh. Chem. (im Druck).

131c. — — Untersuchungen über die Biosynthese der Cyclite. XVII. Bildung von D-Bornesit und Dambonit in *Nerium oleander*. Monatsh. Chem. (im Druck).

131d. Kindl, H., G. J. Kremlicka und O. Hoffmann-Ostenhof: L-Leucanthemit als Inhaltsstoff von Gymnospermae. Monatsh. Chem. (im Druck).

131e. Kindl, H., R. Scholda und O. Hoffmann-Ostenhof: Untersuchungen über die Biosynthese der Cyclite. XV. Zur Bildung von L-Quercit in *Quercus*-Arten. Phytochem. (im Druck).

132. King, F. E. and L. Jurd: The Chemistry of Extractives from Hardwoods. XII. The Cyclitols and Steroids from Opepe *(Sarcocephalus diderrichii)*. J. Chem. Soc. (London) **1953**, 1192.

133. King, F. E. and T. J. King: *iso*Taxiresinol (3′-Demethyl*iso*lariciresinol), a New Lignan Extracted from the Heartwood of the English Yew, *Taxus baccata*. J. Chem. Soc. (London) **1952**, 17.

134. Klenk, E. und R. Sakai: Inositmonophosphorsäure, ein Spaltprodukt der Sojabohnenphosphatide. Z. physiol. Chem. **258**, 33 (1939).

135. Kluyver, A. J., T. Hof and A. G. J. Boezaardt: On the Pigment of *Pseudomonas Beijerinckii*. Enzymologia **7**, 257 (1939).

136. Knowles, W. S. and R. C. Elderfield: Investigations on Loco Weeds. IV. A Preliminary Study of the Constituents of *Astragalus wootoni*. J. Organ. Chem. (USA) **7**, 389 (1942).

137. Koepfli, J. B.: Chemical Investigation of *Rauwolfia caffra*. I. Rauwolfine. J. Amer. Chem. Soc. **54**, 2412 (1932).

138. Kreitman, G., O. K. Sebek and F. F. Nord: On the Mechanism of Enzyme Action. XLIII. Chemistry and Interaction of Lycopersin in the Carbohydrate→ Fat Conversion by *Fusarium vasinfectum*. Arch. Biochem. Biophys. **28**, 77 (1950).

139. Kremlicka, G. J. und O. Hoffmann-Ostenhof: Untersuchungen über die Biosynthese der Cyclite. X. Reduktion von 5-O-Methyl-meso-inosose-3 durch Enzyme aus *Trifolium incarnatum*. Z. physiol. Chem. **344**, 261 (1966).

140. Krishna, S. and B. S. Varma: Active Principle of *Myrsine africana* Linn. J. Indian Chem. Soc. **13**, 115 (1936).

141. — — Active Principles of *Embelia robusta* Roxb., *Myrsine semiserrata* Wall, and *M. capitella* Wall. Forest Bull. (Dehra Dun) **1941**, 102.

142. Kubler, K.: Beiträge zur Chemie der Kondurangorinde. Arch. Pharmaz. **246**, 620 (1908).

143. KUEHL, F. A., Jr., M. N. BISHOP and K. FOLKERS: *Streptomyces* Antibiotics. XXIII. 1,3-Diamino-4,5,6-trihydroxycyclohexane from Neomycin A. J. Amer. Chem. Soc. **73**, 881 (1951).

143a. KURSANOV, A. L., M. VOROBEVA and E. VYSKREBENTSEVA: meso-Inositol in Tea Leaves and its Formation Paths. Doklady Akad. Nauk (USSR) **68**, 737 (1949) [Chem. Abstr. **44**, 1568 (1950)].

144. LABARCA, C., P. B. NICHOLLS and R. S. BANDURSKI: A Partial Characterization of Indoleacetylinositols from *Zea mays*. Biochem. Biophys. Res. Comm. **20**, 641 (1965).

145. LARNER, J., W. T. JACKSON, D. J. GRAVES and J. R. STAMER: Inositol Dehydrogenase from *Aerobacter aerogenes*. Arch. Biochem. Biophys. **60**, 352 (1956).

146. LEE, L. S. and N. J. MORRIS: Isolation and Identification of Pinitol from Peanut Flour. J. Agr. Food Chem. **11**, 321 (1963) [Chem. Abstr. **59**, 5701 (1963)].

147. LEMIEUX, R. U. and M. L. WOLFROM: The Chemistry of Streptomycin. Adv. Carbohydrate Chem. **3**, 337 (1948).

148. LINDBERG, B.: Low-molecular Carbohydrates in Algae. XI. Investigation of *Porphyra umbilicalis*. Acta Chem. Scand. **9**, 1097 (1955).

149. — Methylated Taurines and Choline Sulphate in Red Algae. Acta Chem. Scand. **9**, 1323 (1955).

150. LINDBERG, B. and J. McPHERSON: Low-molecular Carbohydrates in Algae. V. Investigation of *Laminaria cloustoni*. Acta Chem. Scand. **8**, 1547 (1954).

151. — — Low-molecular Carbohydrates in Algae. VI. Laminitol, a New C-Methyl Inositol from *Laminaria cloustonii*. Acta Chem. Scand. **8**, 1875 (1954).

152. LINDSTEDT, G.: Constituents of Pine Heartwood. XIV. The Heartwood of *Pinus monticola* Douglas. Acta Chem. Scand. **3**, 1147 (1949).

153. — Constituents of Pine Heartwood. XV. The Heartwood of *Pinus excelsa* Wall. Acta Chem. Scand. **3**, 1375 (1949).

154. LIPPMANN, E. O.: Über ein Vorkommen von Chinasäure. Ber. dtsch. chem. Ges. **34**, 1159 (1901).

155. LOEWUS, F. A.: Inositol Metabolism in Plants. II. The Absolute Configuration of D-Xylose-5-t_1 Derived Metabolically from *myo*-Inositol-2-t in the Ripening Strawberry. Arch. Biochem. Biophys. **105**, 590 (1964).

156. — Inositol Metabolism and Cell Wall Formation in Plants. Federat. Proc. (Amer. Soc. Exp. Biol.) **24**, 855 (1965).

157. — Analytical Studies Using Low Levels of Carbon-14 and Tritium: A Method of Determining the Labeling Pattern in *myo*-Inositol. Adv. Tracer Methodology **2**, 163 (1965).

158. LOEWUS, F. A. and S. KELLY: Conversion of Glucose to Inositol in Parsley Leaves. Biochem. Biophys. Res. Comm. **7**, 204 (1962).

159. — — Inositol Metabolism in Plants. I. Labeling Patterns in Cell Wall Polysaccharides from Detached Plants Given *myo*-Inositol-2-t or -2-[14]C. Arch. Biochem. Biophys. **102**, 96 (1963).

160. LOEWUS, F. A., S. KELLY and E. F. NEUFELD: Metabolism of *myo*-Inositol in Plants. Conversion to Pectin, Hemicellulose, D-Xylose and Sugar Acids. Proc. Nat. Acad. Sci. (USA) **48**, 421 (1962).

161. MAEDA, K., M. MURASE, H. MAWATARI and H. UMEZAWA: Degradation Studies on Kanamycin. J. Antibiotics (Tokyo) **A11**, 73 (1958). [Chem. Abstr. **53**, 20526 (1959)].

162. MAGASANIK, B.: Enzymatic Adaptation in the Metabolism of Cyclitols in *Aerobacter aerogenes*. J. Biol. Chem. **205**, 1007 (1953).

163. — The Pathway of Inositol Dissimilation in *Aerobacter aerogenes*. J. Biol. Chem. **205**, 1019 (1953).

164. Magasanik, B. and E. Chargaff: The Stereochemistry of an Enzymatic Reaction: The Oxidation of *l*-, *d*- and *epi*-Inositol by *Acetobacter suboxydans*. J. Biol. Chem. **174**, 173 (1948).

165. Magasanik, B., R. E. Franzl and E. Chargaff: The Stereochemical Specificity of the Oxidation of Cyclitols by *Acetobacter suboxydans*. J. Amer. Chem. Soc. **74**, 2618 (1952).

166. Malangeau, P.: Sur les cyclohexitols présents dans l'urine humaine. Bull. soc. chim. biol. (Paris) **38**, 729, (1956).

167. Mann, R. L. and D. O. Woolf: Hygromycin. III. Structure Studies. J. Amer. Chem. Soc. **79**, 120 (1957).

167a. Mann, T.: The Biochemistry of Semen, p. 189. London: Methuen. 1954.

167b. Manni, P. E. and J. E. Sinsheimer: Constituents from *Gymnema sylvestre* Leaves. J. Pharmac. Sci. **54**, 1541 (1965).

168. Manske, R. H. F.: A New Source of Cocositol. Canad. J. Res. **19B**, 34 (1941).

169. Maquenne, L.: Les sucres et principaux dérivés. Paris: Gauthier-Villars. 1900.

170. McComb, E. A. and V. V. Rendig: Ononitol (4-O-Methyl-*myo*-inositol) as a Constituent of *Medicago sativa*. Arch. Biochem. Biophys. **99**, 192 (1962).

171. Micheel, F.: Übergang von der Hexosereihe in die Cyclitreihe. Liebigs Ann. Chem. **496**, 77 (1932).

172. Mokranjac, M. S. and B. Medaković: Chemical Composition of Leaves of Lime Tree. Acta Pharmac. Jugosl. **4**, 143 (1954).

173. Moscatelli, E. A. and J. Larner: The Metabolism in the Rat of Photosynthetically Prepared *myo*-Inositol-14C. Arch. Biochem. Biophys. **80**, 26 (1959).

173a. Mosher, H. S., F. A. Fuhrman, H. D. Buchwald and H. G. Fischer: Tarichatoxin-Tetrodotoxin: A Potent Neurotoxin. Science **144**, 1100 (1964).

174. Müller, H.: The Occurrence of Quercitol in the Leaves of *Chamaerops humilis*. J. Chem. Soc. (London) **91**, 1766 (1907).

175. — Cocositol, a Constituent of the Leaves of *Cocos nucifera* and *Cocos plumosa*. J. Chem. Soc. (London) **91**, 1767 (1907).

176. Murase, M.: Structural Studies on Kanamycin C. J. Antibiotics (Tokyo) **A14**, 367 (1961) [Chem. Abstr. **57**, 9940 (1962)].

177. Naito, T.: Glebomycin, a New Member of the Streptomycin Class. IV. Structure of Glebomycin. Penishirin Sono Ta Koseibusshitsu **15**, 373 (1962) [Chem. Abstr. **60**, 4230 (1964)].

178. Neidhardt, F. C. and B. Magasanik: Effect of Mixtures of Substrates on the Biosynthesis of Inducible Enzymes in *Aerobacter aerogenes*. J. Bacteriol. **73**, 260 (1957).

179. Neubacher, J., H. Kindl and O. Hoffmann-Ostenhof: Transformation of [2-14C]Glucose into *meso*-[5-14C]Inositol in a Cell-free Extract from *Candida utilis*. Biochem. J. **92**, 56P (1964).

180. Nilsson, M.: Constituents of Pollen. I. Low-molecular Carbohydrates in Pollen from *Pinus montana* Mill. Acta Chem. Scand. **10**, 413 (1956).

181. Ottey, L. and F. Bernheim: A Comparison of the Factors which Affect the Formation of Adaptive Enzymes for Benzoic Acid and Inositol in a *Mycobacterium*. Enzymologia **17**, 279 (1956).

182. Paranjapye, V. N. and J. V. Bhat: Studies of Microbial Metabolism of *myo*-Inositol. I. Noncyclic Pathway of *myo*-Inositol Breakdown by *Xanthomonas panici*. J. Indian Inst. Sci. **43**, 104 (1961).

183. Paulus, H. and E. P. Kennedy: The Enzymatic Synthesis of Inositol Monophosphatide. J. Biol. Chem. **235**, 1303 (1960).

184. PEASE, D. C., M. J. REIDER and R. C. ELDERFIELD: Investigations on Loco Weeds. II. The Isolation of *d*-Pinite from *Astralgus earlei* and from *Oxytropis lambertii*. J. Organ. Chem. (USA) 5, 198 (1940).

185. PETRIE, J. M.: The Occurrence of Methyl-*l*-inositol in an Australian Poisonous Plant. Proc. Linnean Soc. N. S. Wales **43**, 850 (1918) [Chem. Abstr. **15**, 3508 (1921)].

186. PLOUVIER, V.: Sur la présence de québrachitol dans quelques Sapindacées et Acéracées. C. R. hebd. séances Acad. Sci. **224**, 1842 (1947).

187. — Sur la recherche des itols et du saccharose chez quelques Sapindales. C. R. hebd. séances Acad. Sci. **227**, 85 (1948).

188. — Sur la recherche du québrachitol et de l'allantoïne chez les Érables et le Platane. C. R. hebd. séances Acad. Sci. **227**, 225 (1948).

189. — Sur la présence de pinitol dans quelques Légumineuses. C. R. hebd. séances Acad. Sci. **228**, 859 (1949).

190. — Sur la présence de québrachitol dans quelques *Artemisia* (Composées). Ann. pharmac. franç. **7**, 192 (1949).

191. — Nouvelles recherches sur le québrachitol des Sapindacées et Hippo-castanacées, le dulcitol des Célastracées et le saccharose de quelques autres familles. C. R. hebd. séances Acad Sci. **228**, 1886 (1949).

192. — Nouvelles recherches sur le pinitol des Légumineuses. C. R. hebd. séances Acad. Sci. **230**, 125 (1950).

193. — Sur la présence de québrachitol chez les Elaeagnacées. Sa recherche dans quelques autres Myrtiflorae. C. R. hebd. séances Acad. Sci. **232**, 1239 (1951).

194. — Sur la recherche du pinitol chez quelques Conifères et plantes voisines. C. R. hebd. séances Acad. Sci. **234**, 362 (1952).

195. — Sur le pinitol des Conifères et le québrachitol des Acéracées; recherche de ces deux itols dans quelques autres familles. C. R. hebd. séances Acad. Sci. **236**, 317 (1953).

196. — Sur la recherche des itols et des hétérosides du Gui, *Viscum album* L. (Loranthacées). C. R. hebd. séances Acad. Sci. **237**, 1761 (1953).

197. — Sur la présence de pinitol chez les Caryophyllacées et quelques plantes de familles voisines. C. R. hebd. séances Acad. Sci. **239**, 1678 (1954).

198. — Sur la recherche du D-quercitol chez quelques Fagacées et autres plantes. C. R. hebd. séances Acad. Sci. **240**, 113 (1955).

199. — Sur un cyclitol nouveau, le liriodendritol, isolé des *Liriodendron* (Magno-liacées). C. R. hebd. séances Acad. Sci. **241**, 765 (1955).

200. — Sur deux cyclitols nouveaux, le L-bornésitol isolé de *Lathyrus vernus* Bernh., le D-ononitol isolé d'*Ononis Natrix* L. (Légumineuses). C. R. hebd. séances Acad. Sci. **241**, 983 (1955).

201. — Nouvelles recherches sur le pinitol des Légumineuses. Sur la présence de *d*-quercitol dans le *Pterocarpus lucens* Guill. et Perr. C. R. hebd. séances Acad. Sci. **241**, 1838 (1955).

202. — Sur la recherche de quelques cyclitols: Viburnitol, scyllitol, pinitol. C. R. hebd. séances Acad. Sci. **242**, 2389 (1956).

203. — Sur un cyclitol nouveau, le *l*-pinitol, isolé d'*Artemisia dracunculus* L. C. R· hebd. séances Acad. Sci. **243**, 1913 (1956).

204. — Sur la recherche du pinitol chez quelques Caryophyllacées, Magnoliacées et plantes de familles voisines. C. R. hebd. séances Acad. Sci. **244**, 382 (1957).

205. — Sur la recherche du séquoyitol et du pinitol chez quelques Gymnospermes. C. R. hebd. séances Acad. Sci. **245**, 2377 (1957).

206. — Sur la recherche du bornésitol chez les Rhamnacées, Borraginacées et quelques autres familles. C. R. hebd. séances Acad. Sci. **247**, 2190 (1958).

207. Plouvier, V.: Sur la recherche des éthers méthyliques des inositols dans quelques groupes botaniques. C. R. hebd. séances Acad. Sci. **247**, 2423 (1958).

208. — Nouvelles recherches de cyclitols dans quelques groupes botaniques; signification phylogénique du séquoyitol. C. R. hebd. scéances Acad. Sci. **251**, 131 (1960).

209. — Recherche de cyclitols chez quelques Apocynacées et Myrtacées; présence de l-quercitol dans l'*Eucalyptus populnea* F. Müll. C. R. hebd. scéances Acad. Sci. **253**, 3047 (1961).

210. — Sur un cyclitol nouveau, le ,,leucanthémitol", isolé de la Grande Marguerite, *Chrysanthemum leucanthemum* L. Sa recherche dans quelques autres Composées-Anthémidées. C. R. hebd. scéances Acad. Sci. **255**, 360 (1962).

211. — Nouvelles recherches de cyclitols dans quelques groupes botaniques; le l-inositol des Composées, le d-pinitol des Légumineuses. C. R. hebd. scéances Acad. Sci. **255**, 1770 (1962).

212. — Distribution of Aliphatic Polyols and Cyclitols. In: T. Swain, Chemical Plant Taxonomy, p. 313. London-New York: Academic Press. 1963.

213. — Recherche des l-inositol, l-québrachitol et d-pinitol dans quelques groupes botaniques. Présence de l'acide shikimique dans *Mammea americana* L. (Guttifères). C. R. hebd. scéances Acad. Sci. **258**, 2921 (1964).

214. — Sur la recherche des polyalcools et des hétérosides cyanogénétiques chez quelques Protéacées. C. R. hebd. scéances Acad. Sci. **259**, 665 (1964).

215. — Sur la recherche des polyalcools dans quelques groupes botaniques; le pinitol et le séquoyitol des Cycadacées. C. R. hebd. scéances Acad. Sci. **260**, 1003 (1965).

216. Posternak, Th.: Recherches dans la série des cyclites. VI. Sur la configuration de la méso-inosite, de la scyllite et d'un inosose obtenu par voie biochimique (scyllo-*ms*-inosose). Helv. Chim. Acta **25**, 746 (1942).

217. — The Cyclitols. Paris: Hermann. 1965.

218. Posternak, Th., A. Rapin et A.-L. Haenni: Recherches dans la série des cyclitols. XXIV. Sur les règles d'oxydation des cyclitols par *Acetobacter suboxydans*. Helv. Chim. Acta **40**, 1594 (1957).

219. Posternak, Th. et D. Reymond: Recherches dans la série des cyclitols. XVII. Sur l'oxydation de divers cyclitols par *Acetobacter suboxydans*. Helv. Chim. Acta **36**, 260 (1953).

220. Posternak, Th., W. H. Schopfer, B. Kaufmann-Boetsch et S. Edwards: Recherches sur la biochimie des cyclitols. VIII. Sur la biosynthèse du *méso*-inositol et du scyllitol chez le Rat. Helv. Chim. Acta **46**, 2676 (1963).

221. Posternak, Th., W. H. Schopfer et D. Reymond: Biochimie des cyclitols. I. Contribution à l'étude du métabolisme du *méso*-inositol chez le Rat. Helv. Chim. Acta **38**, 1283 (1955).

222. Posternak, Th., W. H. Schopfer, D. Reymond et C. Lark: Recherches sur la biochimie des cyclitols, V. Glucogenèse à partir de *méso*-inositols deutériés chez le Rat phloriziné. Helv. Chim. Acta **41**, 235 (1958).

223. Power, F. B. and F. Tutin: A Laevorotatory Modification of Quercitol. J. Chem. Soc. (London) **85**, 624 (1904).

224. Qudrat-i-Khuda, M̄., M. E. Ali and Q. A. Ahmed: *Caesalpinia bonducella*. II. Chemical Examination of the Leaves. Pakistan J. Sci. Ind. Res. **4**, 104 (1961) [Chem. Abstr. **58**, 12367 (1963)].

225. Qudrat-i-Khuda, M., M. E. Ali and A. Malek: *Caesalpinia bonducella*. IV. *d*-Inositol from Caesalpinitol. Sci. Res. (Dacca, Pakistan) **1**, 96 (1964) [Chem. Abstr. **61**, 8558 (1964)].

226. RENDIG, V. V. and E. A. McCOMB: Pinitol (5-O-Methyl-D-inositol) as a Constituent of *Medicago sativa*. Arch. Biochem. Biophys. **96**, 455 (1962).

227. RICHARDSON, K. E. and B. AXELROD: Changes in Inositol Content During Germination and Growth of Some Higher Plants. Plant Physiol. **32**, 334 (1957).

228. RIGGS, N. V.: The Nature of a Cyclitol Isolated from *Macrozamia Riedlei*. J. Chem. Soc. (London) **1949**, 3199.

229. RIMINGTON, C.: The Occurrence of Cyanogenetic Glucosides in South African Species of *Acacia*. II. Determination of the Chemical Constitution of Acacipetalin. Its Isolation from *Acacia stolonifera* Burch. Onderstepoort J. Vet. Sci. **5**, 445 (1935).

230. RIMINGTON, C. and G. C. S. ROETS: Chemical Investigation of the Plant *Acalypha indica*. Isolation of Triacetonamine, a Cyanogenetic Glucoside, and Quebrachitol. Onderstepoort J. Vet. Sci. **9**, 193 (1937).

231. RINEHART, K. L., Jr., W. S. CHILTON, M. HICHENS and W. v. PHILLIPSBORN: Chemistry of the Neomycins. XI. NMR Assignment of the Glycosidic Linkages. J. Amer. Chem. Soc. **84**, 3216 (1962).

232. RINEHART, K. L., Jr., M. HICHENS, A. D. ARGOUDELIS, W. S. CHILTON, H. E. CARTER, M. P. GEORGIADIS, C. P. SCHAFFNER and R. T. SCHILLINGS: Chemistry of the Neomycins. X. Neomycins B and C. J. Amer. Chem. Soc. **84**, 3218 (1962).

233. ROWE, E. J., A. A. HARWOOD and D. B. MYERS: The Isolation of Three Inositols from *Veronia altissima*. J. Amer. Pharmac. Assoc. **44**, 308 (1955).

234. SANNIÉ, C. et J. DUSSY: Sur la présence de pinitol dans les feuilles d'*Erythrophleum guineense* (G. Don). C. R. hebd. séances Acad. Sci. **224**, 1381 (1947).

235. SCHEFFER, F., R. KICKUTH und H. LORENZ: Die Bedeutung von Inositphosphat bei der Aufnahme von Kalium durch Maispflanzen. Naturwiss. **52**, 518 (1965).

236. SCHERER, J.: Über eine neue aus dem Muskelfleisch gewonnene Zuckerart. Liebigs Ann. Chem. **73**, 322 (1850).

236a. SCHEUER, P. J.: The Chemistry of Toxins Isolated from Some Marine Organisms. Fortschr. Chem. organ. Naturstoffe **22**, 265 (1964).

237. SCHMITZ, H., O. B. FARDIG, F. A. O'HERRON, M. A. ROUSCHE and I. R. HOOPER: Kanamycin. III. Kanamycin B. J. Amer. Chem. Soc. **80**, 2911 (1958).

238. SCHOLDA, R.: Dissert., Univ. Wien, 1965.

239. SCHOLDA, R., G. BILLEK and O. HOFFMANN-OSTENHOF: Untersuchungen über die Biosynthese der Cyclite. I. Bildung von D-Pinit, D-Inosit und Sequoyit aus meso-Inosit in Blättchen von *Trifolium incarnatum*. Z. physiol. Chem. **335**, 180 (1964).

240. — — — Untersuchungen über die Biosynthese der Cyclite. III. Bildung von Methyläthern des L-Inosits aus meso-Inosit in Blättchen von *Artemisia vulgaris* und *Artemisia dracunculus*. Monatsh. Chem. **95**, 541 (1964).

241. — — — Untersuchungen über die Biosynthese der Cyclite. V. Weitere Versuche über die Bildung der einzelnen Cyclite in Blättchen von *Trifolium incarnatum*. Z. physiol. Chem. **337**, 277 (1964).

242. — — — Untersuchungen über die Biosynthese der Cyclite. VI. Die Bildung von Scyllit in *Calycanthus occidentalis*. Z. physiol. Chem. **339**, 28 (1964).

243. — — — Untersuchungen über die Biosynthese der Cyclite. VII. Der Mechanismus der Umwandlung von meso-Inosit in Methyläther des L-Inosits in Blättchen von *Artemisia vulgaris* und *Artemisia dracunculus*. Monatsh. Chem. **95**, 1305 (1964).

244. — — — Untersuchungen über die Biosynthese der Cyclite. VIII. Der Mechnismus der Umwandlung von meso-Inosit in D-Pinit und D-Inosit in *Trifolium incarnatum*. Monatsh. Chem. **95**, 1311 (1964).

244a. Schraudolf, H.: Untersuchungen zum Stoffwechsel des m-Inosits. Dissert. Tübingen, 1956.

245. Sherrard, E. C. and E. F. Kurth: Occurrence of Pinitol in Redwood. Ind. Eng. Chem. **20**, 722 (1928).

246. — — Sequoyite, a Cyclose from Redwood *(Sequoia sempervirens)*. J. Amer. Chem. Soc. **51**, 3139 (1929).

247. Sivak, A. und O. Hoffmann-Ostenhof: Enzyme des meso-Inositabbaus in *Schwanniomyces occidentalis*. Biochem. Z. **336**, 229 (1962).

248. Smith, R. H.: The Phosphatides of the Latex of *Hevea brasiliensis*. III. Carbohydrate and Polyhydroxy Constituents. Biochem. J. **57**, 140 (1954).

249. Sodi Pallares, E. and H. M. Garza: Study of Yoloxochitl. Arch. inst. cardiol. Méx. **17**, 833 (1947).

250. Soine, T. O. and G. L. Jenkins: A Phytochemical Study of *Lupinus caudatus* Kellog. Pharmac. Arch. **12**, 65 (1941).

251. Srinivasan, P. R., J. Rothschild and D. B. Sprinson: The Enzymic Conversion of 3-Deoxy-D-*arabino*-heptulosonic Acid 7-Phosphate to 5-Dehydroquinate. J. Biol. Chem. **238**, 3176 (1963).

252. Staedeler, G. und T. Frerichs: Vorkommen von Harnstoff, Taurin und Scyllit in den Organen der Plagiostomen. J. prakt. Chem. **73**, 48 (1858).

253. Stern, F. und J. Zellner: Beiträge zur vergleichenden Pflanzenchemie. XI. Über *Sonchus arvensis* L. Monatsh. Chem. **46**, 459 (1925).

254. Stetten, M. R. and DeW. Stetten, Jr.: Biological Conversion of Inositol into Glucose. J. Biol. Chem. **164**, 85 (1946).

255. Takahashi, M., T. Ito and A. Mizutani: Chemical Constituents of the Plants of Coniferae and Allied Orders. XLIV. Structure of Distichin and the Components of Taxodiaceae Plants, *Metasequoia glyptostroboides* and Others. J. Pharmac. Soc. Japan **80**, 1557 (1960).

256. Tanret, C.: Sur deux sucres nouveaux retirés du québracho. C. R. hebd. séances Acad. Sci. **109**, 908 (1889).

257. Tatsuoka, S., S. Horii, T. Yamaguchi, H. Hitomi and A. Miyake: Chemistry of Zygomycin A: Structure of Zygomycins A_1 and A_2. Antimicrobial Agents Chemother. **1962**, 188 [Chem. Abstr. **59**, 11641 (1963)].

258. Thonet, E. und O. Hoffmann-Ostenhof: Über die meso-Inosit-Oxygenase aus dem Sproßpilz *Schwanniomyces occidentalis*. Monatsh. Chem. **97**, 107 (1966).

259. Touster, O.: Essential Pentosuria and the Glucuronate-Xylulose Pathway. Federat. Proc. (Amer. Soc. Exp. Biol.) **19**, 977 (1960).

259a. Tsuda, K.: Über Tetrodotoxin, Giftstoff der Bowlfische. Naturwiss. **53**, 171 (1966).

260. Tsuda, K., C. Tamura, R. Tachikawa, K. Sakai, O. Amakasu, M. Kawamura and S. Ikuma: Constitution and Configuration of Tetrodoic Acid. Chem. Pharmac. Bull. **11**, 1473 (1963).

261. Umezawa, S. and T. Tsuchiya: (6-Amino-6-deoxy-α-D-glucopyranosyl)-deoxystreptamine, an Antibacterial Degradation Product of Kanamycin. J. Antibiotics (Tokyo) **A15**, 51 (1962).

261a. Unrau, A. M. and D. T. Canvin: Biosynthesis of Plant Constituents I. The Complete Degradation of 2-Deoxy-D-ribose and Some 2-Deoxy-D-hexoses. Canad. J. Chem. **41**, 607 (1963).

262. Vignais, P. M., P. V. Vignais and A. L. Lehninger: Identification of Phosphatidylinositol as a Factor Required in Mitochondrial Contraction. J. Biol. Chem. **239**, 2011 (1964).

263. VIGNAIS, P. V., P. M. VIGNAIS and A. L. LEHNINGER: A Heat-stable Factor Required for Contraction of Pretreated Mitochondria. J. Biol. Chem. **239**, 2002 (1964).

264. VINCENT, C. et DELACHANAL: Sur un hydrate de carbone contenu dans le gland du chêne. C. R. hebd. séances Acad. Sci. **104**, 1855 (1887).

265. VOHL, H.: Über den Phaseomannit und seine Identität mit dem Inosit. Liebigs Ann. Chem. **101**, 50 (1857).

266. VOLK, W. A. and D. PENNINGTON: The Fermentation of Inositol. J. Bacteriol. **61**, 469 (1951).

267. — — The Fermentation of Inositol by *Propionibacterium pentosaceum.* J. Bacteriol. **64**, 347 (1952).

268. VRKOČ, J., V. HEROUT und F. ŠORM: Über Pflanzenstoffe. X. Isolierung der kristallinen Bestandteile der Sandstrohblume (*Helychrysum arenarium* MCH). Collect. Czech. Chem. Comm. **24**, 3938 (1959).

269. WAAL, H. L. DE: Investigations of *Lotononis laxa* E and Z. I. The Isolation of Pinitol, a Fatty Ester and Benzaldehyde. Onderstepoort J. Vet. Sci. **13**, 229 (1939).

269a. WALKER, M. S. and J. B. WALKER: Enzymic Studies on the Biosynthesis of Streptomycin. Transamidination of Inosamine and Streptamine Derivatives. J. Biol. Chem. **241**, 1262 (1966).

270. WEBER, C. O.: Zur Chemie des Kautschuks. III. Ber. dtsch. chem. Ges. **36**, 3108 (1903).

271. WEISSBACH, A.: The Enzymatic Determination of *myo*-Inositol. Biochim. Biophys. Acta **27**, 608 (1958).

272. WICKBERG, B.: Isolation of 2-L-Amino-3-hydroxy-1-propane Sulfonic Acid from *Polysyphonia fastigiata.* Acta Chem. Scand. **11**, 506 (1957).

273. WILDIERS, E.: Une nouvelle substance indispensable au developement de la levure. La Cellule **18**, 313 (1901).

274. WILEY, P. F., M. V. SIGAL, Jr. and O. WEAVER: Degradation Products of Hygromycin B. J. Organ. Chem. (USA) **27**, 2793 (1962).

275. WOOLLEY, D. W.: A New Dietary Factor Essential for the Mouse. J. Biol. Chem. **136**, 113 (1940).

276. YANAGITA, M.: Chemical Study of a Chinese Drug ,,Kan-zui". J. Pharmac. Soc. Japan **63**, 408 (1943).

(Eingelaufen am 16. Dezember 1965.)

The Chemistry of the Order Cupressales.

By **H. Erdtman** and **T. Norin**, Stockholm.

Contents.

Acknowledgement. We would like express our sincere gratitude to the late Professor R. FLORIN for the interest he always showed in our studies in the conifer field. Although not a chemist, he clearly saw the potentialities of the chemical approach to taxonomy. During innumerable discussions he gave us support and advice from his enormous knowledge about the conifers and their historical and geographical distribution.

I. Introduction.

Studies on lignans from spruce and pine as well as the investigation of phenolic constituents from the heartwood of Scot's pine and a large number of other pine species have indicated that a thorough study of the constituents of conifers might yield interesting results in relation to the classification and the phylogenesis of the conifers (*71–74*).

It was found that the strongly fungitoxic pinosylvin is a very characteristic constituent of the heartwood of the pines*. Apart from pinosylvin a series of flavonoids were found to occur in pine heartwoods. Moreover, it was observed in the course of this work that there are differences in the patterns of heartwood constituents by means of which one can distinguish chemically between the two pine subgenera *Haploxylon* and *Diploxylon*.

It was then of great interest to find out if in the large group of conifers of the cypress type, botanically classified as the family Cupressaceae or according to a newer classification the order Cupressales, similar examples of chemical characterisation of genera and subgenera etc. could be found. For this reason the heartwoods of a large number of conifers belonging to the order Cupressales have been investigated.

---

* The naturally occurring stilbenes have recently been reviewed by BILLEK in this Series (*22*).

The studies were initiated about 1947. Then the analytical methods were far less developed than they are to-day and many of the early investigations can only be said to be superficial. Therefore, a reexamination was needed in order to give anything like a complete picture of the various constituents. We were at first looking mainly for phenolic compounds but we came across so many unusual terpenoids that, unfortunately, the phenolic constituents became rather neglected.

II. Botanical Classification of the Order Cupressales.

The order Cupressales comprises about 150 species of the approximately 600 recent conifers. The order has been subdivided botanically in various ways. Generally two families are recognised, Taxodiaceae and Cupressaceae. The former family embraces a series of only distantly related genera which are either monotypic or contain only a few species. The recent species of the Taxodiaceae constitute the scattered remnants of old, larger genera some of which are known from the cretaceous period (100–125 million years ago). PILGER and MELCHIOR (212) recognise seven groups. The family Cupressaceae has three subfamilies.

I. Taxodiaceae:

1. **Sequoieae** with *Sequoia* (1 N)* and *Sequoiadendron* (1 N).
2. **Metasequoieae** with *Metasequoia* (1 N).
3. **Taxodieae** with *Taxodium* (3 N) and *Glyptostrobus* (1 N).
4. **Cryptomerieae** with *Cryptomeria* (1 N).
5. **Cunninghamieae** with *Cunninghamia* (2 N).
6. **Sciadoptityeae** with *Sciadopitys* (1 N).
7. **Athrotaxeae** with *Athrotaxis* (3 S) and *Taiwania* (1 N).

II. Cupressaceae:

1. **Cupressoideae** with *Cupressus* (20 N) and *Chamaecyparis* (7 N).
2. **Juniperioideae** with *Arceuthos* (1 N) and *Juniperus* (70 N[S]).
3. **Thujoideae** with: a) *Calocedrus* (3 N), *Fokienia* (3 N), *Microbiota* (1 N), *Platycladus (Biota)* (1 N), *Thuja* (5 N), *Thujopsis* (1 N).
   b) *Actinostrobus* (2 S), *Austrocedrus* (1 S), *Callitris* (16 S), *Diselma* (1 S), *Fitzroya* (1 S), *Libocedrus* (sensu stricto) (5 S), *Neocallitropsis (Callitropsis)* (1 S), *Papuacedrus* (3 S), *Pilgerodendron* (1 S), *Widdringtonia* (5 S) and *Tetraclinis* (1 N).

There are several other classifications indicating that botanists have not yet reached agreement concerning the mutual relationships between the genera. Of the two subfamilies Cupressoideae and Juniperoideae only one species of *Juniperus* (*J. procera*) is southern hemispheric (Kenya), and in the subfamily Thujoideae the a-group is entirely northern hemispheric and the b-group southern hemispheric, apart from *Tetraclinis* occurring in southern Spain and north-western Africa and generally supposed to be related to *Callitris*.

* The figures indicate number of species. N = northern, S = southern hemispheric.

References, pp. 275—287.

An extensive investigation of the chemical constituents of the various organs: wood, bark, leaves, fruits, seeds etc. of large genera such as *Cupressus* and *Juniperus* would indicate which compounds are really characteristic of the genera, and the degree of chemical variability within the genera. Genera of intermediate magnitude are less suitable for this purpose. The chemical investigation of small or monotypic genera can at the most give some additional evidence of interest to the taxonomist. This is particularly true if chemical constituents of unusual type are found which seem to indicate affinities with other genera. It is, however, dangerous to draw conclusions from the presence or absence of only a single particular compound in different genera. At the present stage, the utilisation of the chemical results must essentially be a task for the botanical taxonomist. The chemical evidence becomes more valuable when specific patterns of compounds are found to be characteristic of groups of plants, particularly if the compounds are formed along different biosynthetic pathways. Hence, it is important to investigate fatty acids, terpenoids, phenols etc. The scarcity of alkaloids in conifers is unfortunate since these compounds frequently appear to possess a relatively high taxonomic value (*114*).

III. Chemical Constituents of Cupressales.

1. Cyclitols and Simple Phenols.

The cyclitols sequoyitol and pinitol have been isolated from the green parts of several genera of Taxodiaceae and Cupressaceae (*216*), sequoyitol also from the wood of *Sequoia sempervirens* and from the pollen of *Metasequoia glyptostroboides* (*146*). Free pyrogallol, gallic acid, phloroglucinol, protocatechuic acid and possibly catechol have been reported to occur in a powdery solid accompanying the seeds of the cones of *Sequoia sempervirens* and *Sequoiadendron giganteum* (*147*). Eugenol methyl ether and isoeugenol methyl ether occur in the ether extract of the wood of *Sciadopitys* (ERDTMAN, NORIN and SUMIMOTO, unpublished).

There are several reports on the occurrence of small amounts of unidentified phenolic compounds in the extracts of steam-volatile fractions from Cupressales woods. However, the phenols (and quinones) (1–14) have been isolated from the Cupressaceae and definitely characterised (for most of them see *147*). They are frequently co-occurring and this as well as their structures indicate that they are biosynthetically interrelated and related to monoterpenes. Carvacrol (**8**), or its methyl ether, is almost invariably found in woods of species containing tropolones.

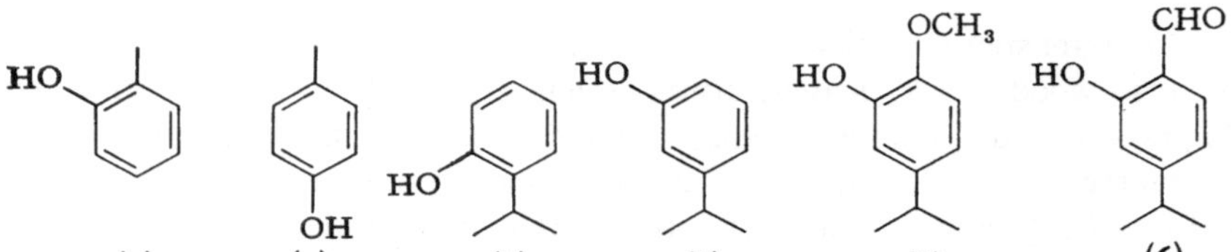

(1) (2) (3) (4) (5) (6)

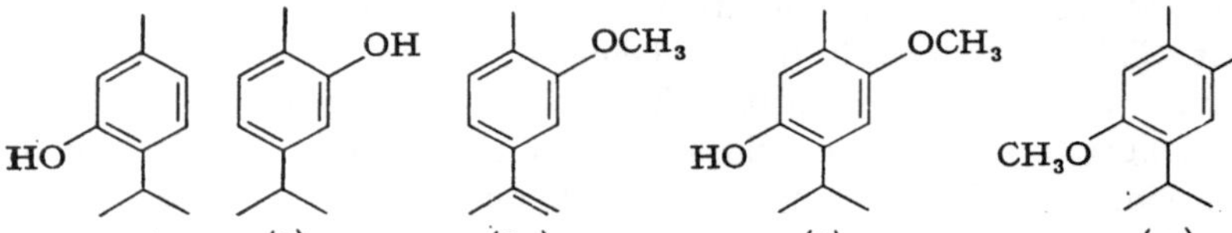

(7) (8) (8a) * (9) (10)

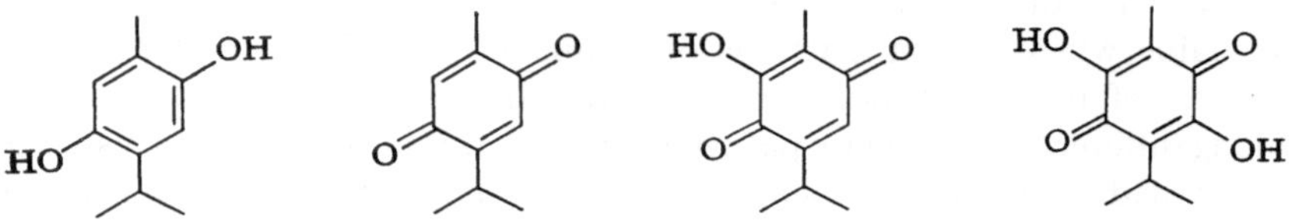

(11) (12) (13) (14)

* From needle oil (*254*).

The heartwood of *Calocedrus decurrens* (syn. *Libocedrus* or *Heyderia decurrens*) contains some interesting compounds derived from oxidative couplings of *p*-methoxythymol (9) and *p*-methoxycarvacrol (10), e. g. libocedrol (15), heyderiol (16) and 3-libocedroxythymoquinone (17). Libocedrol is obtained from the wood as an adduct (m. p. 91.5–92°) containing one mole of *p*-methoxycarvacrol which like *p*-methoxythymol is a constituent of the wood (*267–269*).

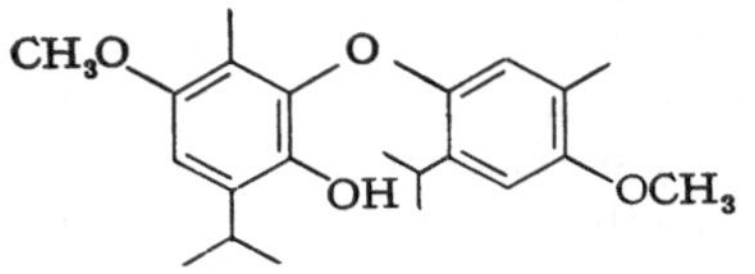

(15)
Libocedrol.

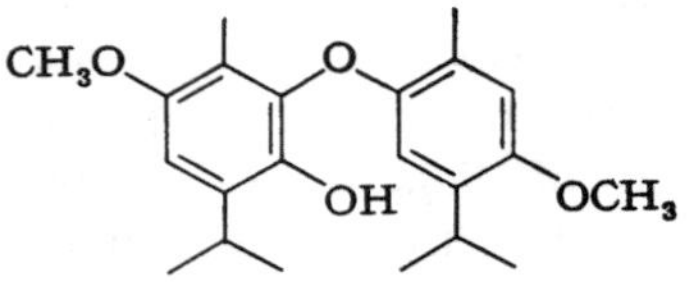

(16)
Heyderiol.

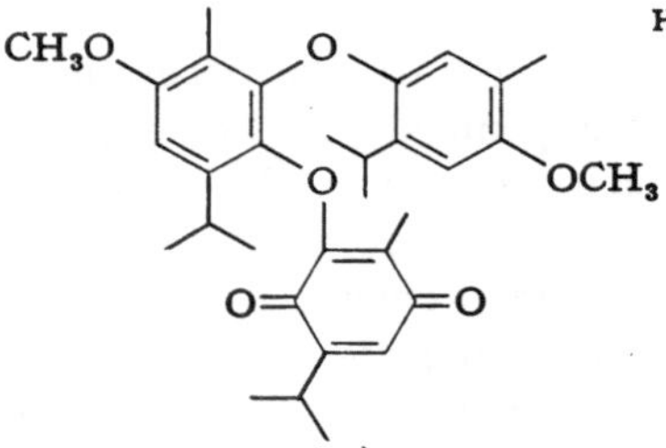

(17)
Libocedroxythymoquinone.

References, pp. 275—287.

Libocedrol is formed in vitro by oxidation of p-methoxythymol (265). Heyderiol together with libocedrol has been similarly obtained by oxidation of a mixture of p-methoxythymol and p-methoxycarvacrol. 3-Libocedroxythymoquinone is formed simply by aeration of sawdust impregnated with the libocedrol-p-methoxythymol adduct and thymoquinone (264). All these phenolic compounds contribute to the durability of the heartwood of *Calocedrus decurrens* but the coupled products are less potent fungicides than hydrothymoquinone and its monomethyl ethers (5).

From the wood of *Cryptomeria japonica* (Japanese cedar or "sugi") phenols containing seventeen carbon atoms were isolated, namely, sugiresinol (18) and the related hydroxysugiresinol (m. p. 212°, decomp.; $[\alpha]_D - 19.5°$), which contains four hydroxyl groups two of which belong to a catechol grouping (97). Hinokiresinol (19) is another related phenol which was isolated from the wood of the Japanese "hinoki", *Chamaecyparis obtusa* (124).

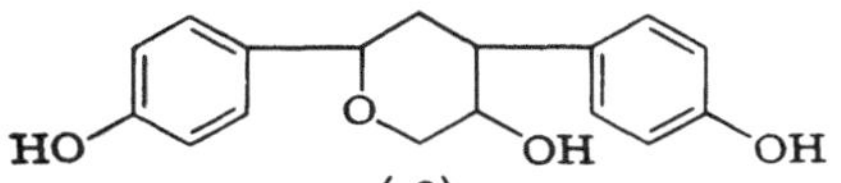

(18)
Sugiresinol.

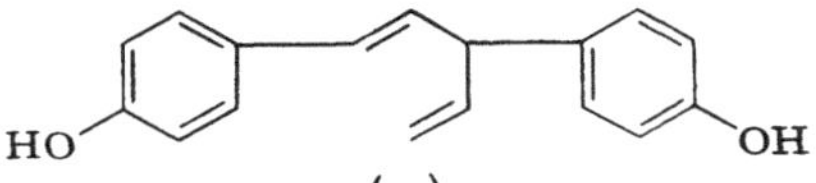

(19)
Hinokiresinol.

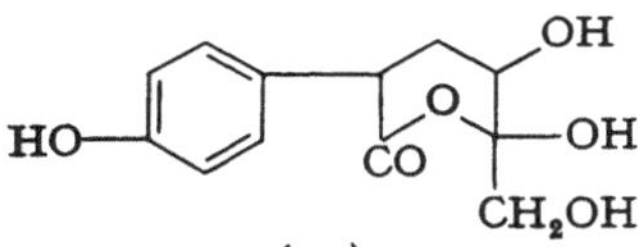

(20)
Sawarinin.

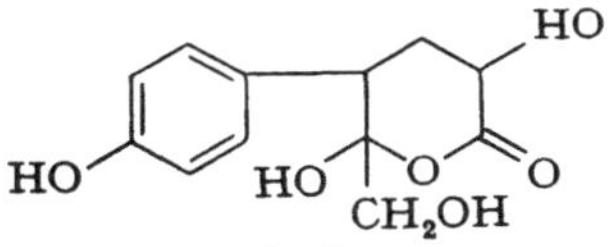

(21)

Several phenolic substances have been found in the heartwood of *Sequoia sempervirens*, e. g. sequirin-A ($C_{17}H_{18}O_4$), m. p. 245°; sequirin-B ($C_{17}H_{18}O_5$), m. p. 218°, and sequirin-C ($C_{17}H_{18}O_5$), m. p. 193° (16). They are all reported to be optically inactive. Sequirins-A und -B appear to be identical with sugiresinol and hydroxysugiresinol, respectively (ANDERSON, private communication). A wood constituent recently isolated from *Athrotaxis selaginoides* is identical with sugiresinol (ERDTMAN, NISHIMURA and NORIN, unpublished)*. The same species contains another probably related phenol, athrotaxin ($C_{17}H_{16}O_6$ + H_2O; m. p. 285°, decomp.; $[\alpha]_D - 275°$) (91). From the heartwood of *Chamaecyparis pisifera* a phenolic lactone sawarinin (20) has been isolated (129). The suggested structure appears to indicate that sawarinin is formed by oxidative fission of a diphenyl derivative. However, structure (21) accounts equally well for the experimental results and could indicate relations with the C_{17}-phenols.

---

* *A. selaginoides* contains also hinokiresinol (19) and agatharesinol (ERDTMAN, NISHIMURA and NORIN, unpublished). The latter compound has recently been found in *Agathis australis* (Araucariales) (70, 70a). (Added in Proof.)

2. Lignans.

Only a few lignans have been isolated from the order Cupressales (*Table 1*, p. 257). Juniper needles and berries, particularly from savin *(Juniperus sabina)*, have long been known to be toxic and extracts from

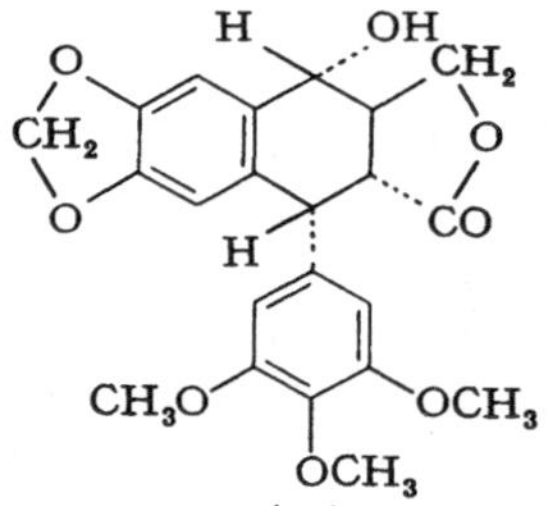

(22)
Podophyllotoxin.

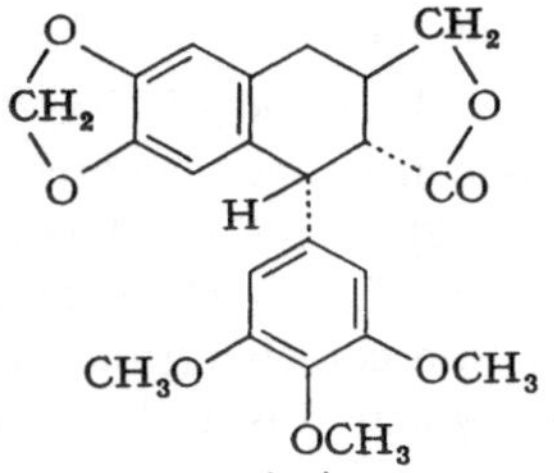

(23)
Deoxypodophyllotoxin.

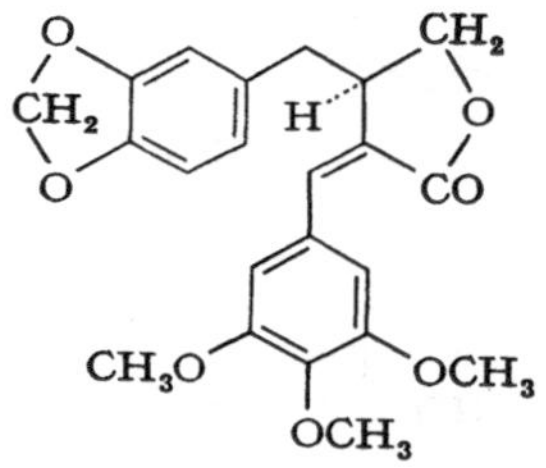

(24)
Savinin.

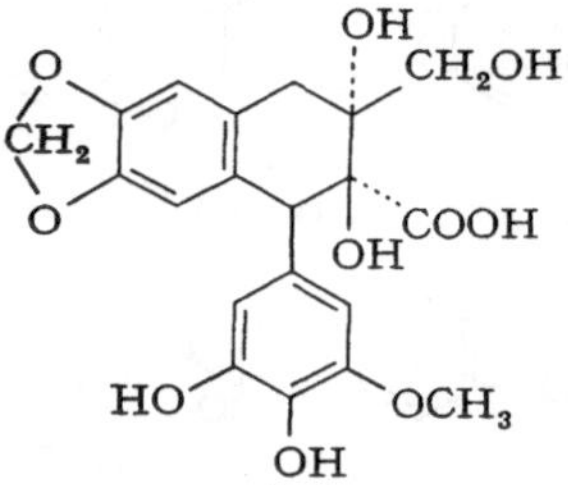

(25)
Plicatic acid.

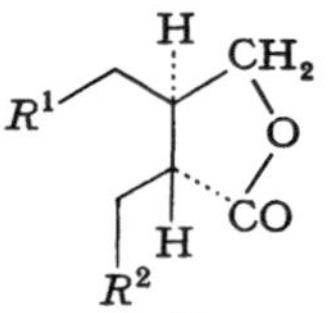

(26)
Thujaplicatin methyl ether.

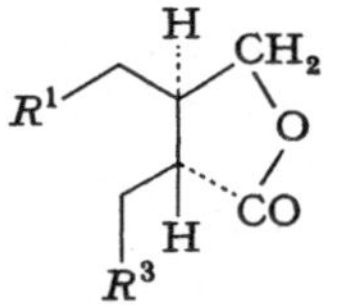

(27)
Thujaplicatin.

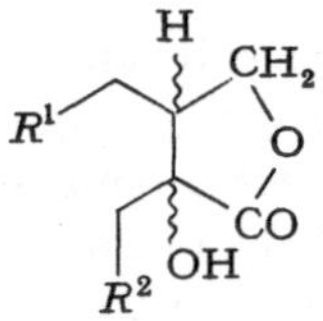

(28)
Hydroxythujaplicatin methyl ether.

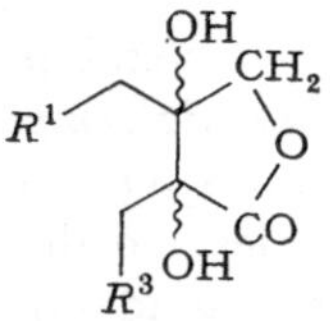

(29)
Dihydroxythujaplicatin methyl ether.

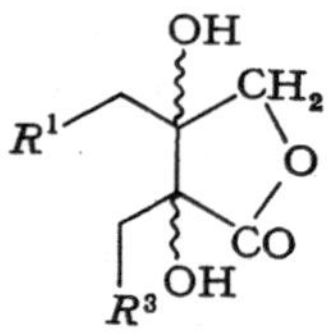

(30)
Dihydroxythujaplicatin.

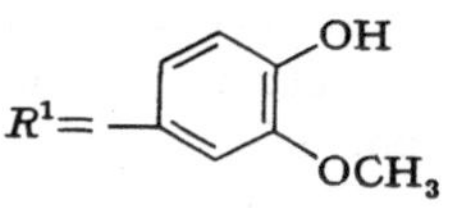

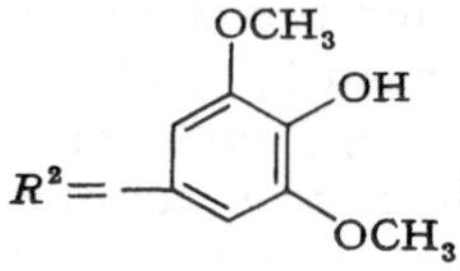

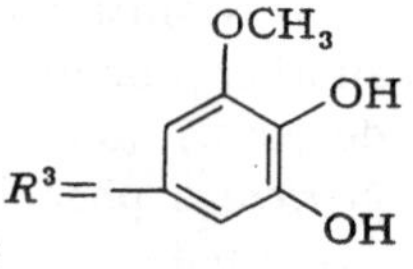

them are used, for example, in medicine for treating certain tumours. From the leaves of several Cupressales HARTWELL (*112*, cf. *147*) isolated the tumour-necrotising lignans podophyllotoxin (22) and deoxypodophyllotoxin (23) as well as the inactive savinin (24). The tumour-necrotising activity seems to be characteristic of phenyltetraline lactones possessing the arrangement of the lactone ring and the configuration of deoxypodophyllotoxin. Taiwanin A is the diphenylbutadiene analogue of savinin (24) (169).

An investigation by GARDNER and MacLEAN (100) has lead to the discovery of a series of unusual lignans in the wood of Thuja plicata, such as plicatic acid (25) and five related lignans of the butyrolactone type (26–30) (MacLEAN, private communication). Hinokinin (31) was isolated from the heartwood of hinoki (Chamaecyparis obtusa) (147) savinin (24) (147) and d-sesamin (32) from the leaves (175–177).

The distribution of the lignans in Cupressales is summarised in Table 1, p. 257; for physical constants of Cupressales lignans see Table 2, p. 258.

It is probable that in the future these and other lignans will be found in the mostly uninvestigated phenolic part of light petroleum insoluble ether or acetone extracts of the woods. So far no lignans of the diarylbutane or phenyltetraline type possessing two guaiacyl nuclei have been isolated from Cupressales but they do occur in some other orders of the Coniferae. It has been suggested that lignans are oxidative coupling products of phenols of the coniferyl alcohol type (cf. 92 and references given there).

3. Flavonoids and Biflavonyls.

Surprisingly few compounds of the flavone type have been found in the Cupressales, probably because they have not been searched for systematically. In a few instances quercetin or quercetin glycosides as well as catechin etc. were observed in barks or leaves. Fairly large quantities of taxifolin (2,3-dihydroquercetin) (33) were isolated from the heartwood of Austrocedrus chilensis (87) and Platycladus orientalis (86). The latter species also contains aromadendrin (2,3-dihydrokaempferol) (34).

The most interesting flavonoids found so far are the so-called biflavonyls. They were isolated from the leaves and are all derived from apigenin (35), obviously by oxidative dimerisation. In a few species apigenin co-occurs with biflavonyls. The biflavonyls are high-melting compounds which are sometimes difficult to obtain in a pure state. The first biflavonyl, ginkgetin, was discovered by FURUKAWA in the leaves of Ginkgo biloba. It is a dimethyl ether of the much later isolated amentoflavone (36) (from Amentotaxus formosana Li), to which it can be demethylated (145, 127). Kayaflavone, sciadopitysin and sotetsuflavone also give amentoflavone on demethylation; and on methylation all these compounds yield amentoflavone hexamethyl ether. Hinokiflavone (37) is an isomer of amentoflavone and like the latter does not contain methoxyl groups. On methylation hinokiflavone gives a pentamethyl ether and on acetylation a pentaacetate. Degradation of the methyl ethers of the members of the amentoflavone series yields the biphenyl derivative (38). Similarly, O-pentamethyl hinokiflavone gives the diphenyl ether (40).

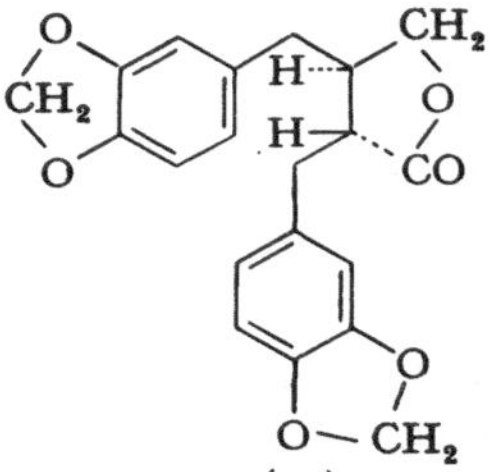

(31)
Hinokinin.

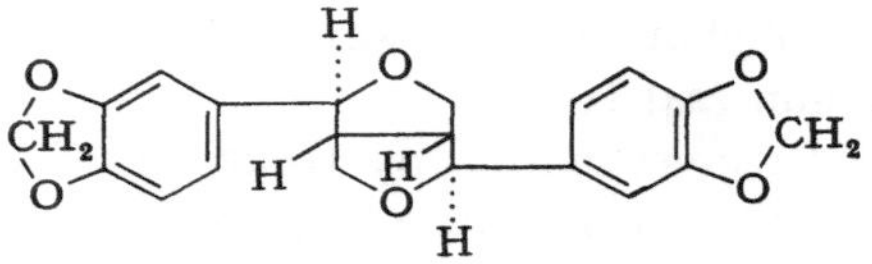

(32)
(+)-Sesamin.

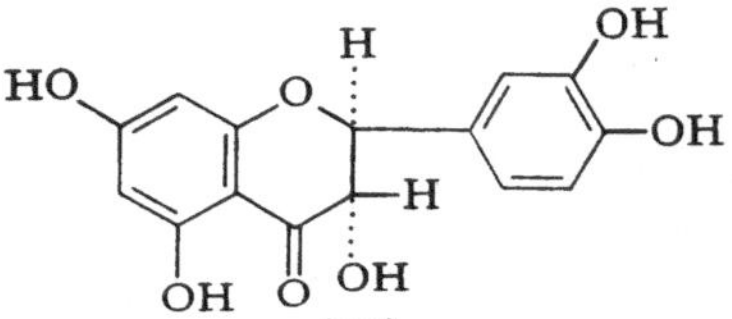

(33)
Taxifolin.

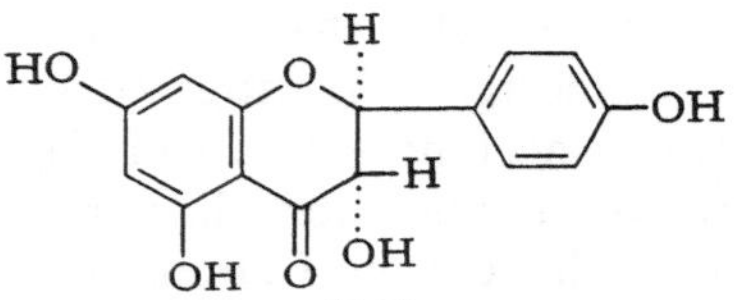

(34)
Aromadendrin.

This indicates how the apigenin molecules are joined in the pertinent biflavonyls. Those derived from amentoflavone are biphenyl derivatives, whilst hinokiflavone is a diphenyl ether. Amentoflavone hexamethyl ether has been synthesised by NAKAZAWA (cf. *15, 145*) by a mixed Ullmann coupling of 3'- and 8-iodo-apigenin trimethyl ether. The coupling product obtained from 8-iodo-apigenin trimethyl ether turned out to be identical with the hexamethyl ether of cupressuflavone recently isolated by SESHADRI (*187*) from the needles of *Cupressus sempervirens* and *C. torulosa* The relationship between apigenin (35), amentoflavone (36), cupressuflavone (39) and hinokiflavone (37) is shown below.

The structural elucidation of these unusual flavone derivatives was difficult and is intimately connected with the names of W. BAKER, T. KARIYONE, N. KAWANO and K. NAKAZAWA. [See the summaries by BAKER (*15*) and KARIYONE (*145*)]. Physical data for the Cupressales biflavonyls are given in *Table 4*, p. 260; and their botanical distribution is summarised in *Table 3*, p. 259.

4. Leaf Waxes and Other Less Investigated Constituents.

Early in this century BOUGAULT studied the leaf waxes of some conifers and found that they frequently contain polyesters (estolides) of sabinic acid, $HOCH_2(CH_2)_{10}COOH$, juniperic acid, $HOCH_2(CH_2)_{14}COOH$ and thapsic acid, $HOOC(CH_2)_{14}COOH$ (*147*). KARIYONE and his collaborators also investigated the leaf waxes and detected, in addition to the acids isolated by BOUGAULT, lauric and palmitic acids as well as a di-carboxylic acid $HOOC(CH_2)_{10}COOH$, furthermore the diols corresponding to sabinic and juniperic acids. Some of the Cupressales leaves, mainly from deciduous trees, contain little or nothing of these compounds but instead long-chain hydrocarbons, alcohols or ketones. The distribution of the estolidic and non-estolidic Cupressales is summarised in Table 3, p. 259; see also (260).

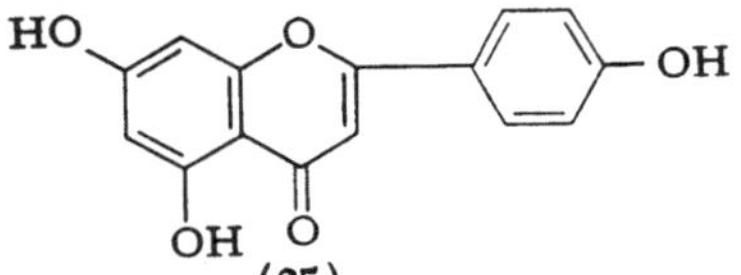

(35)
Apigenin.

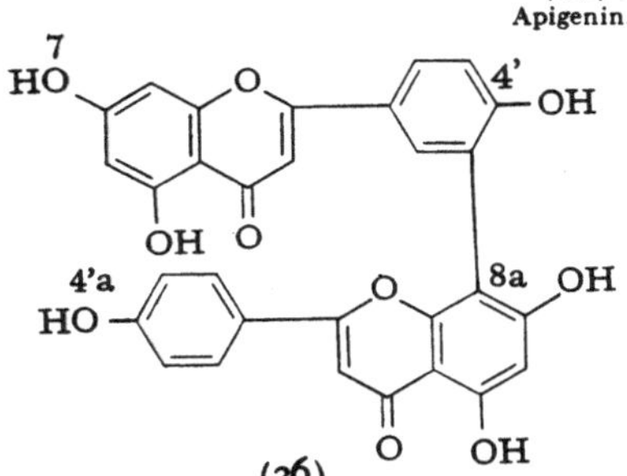

(36)
Amentoflavone.

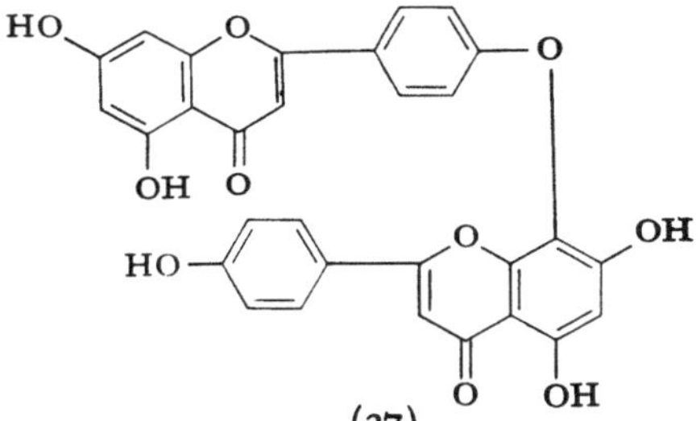

(37)
Hinokiflavone.

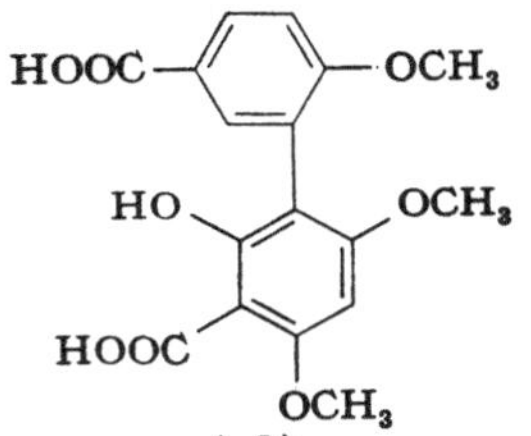

(38)

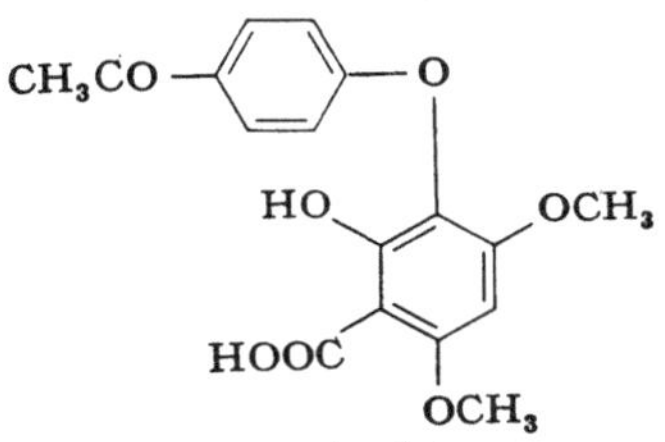

(40)

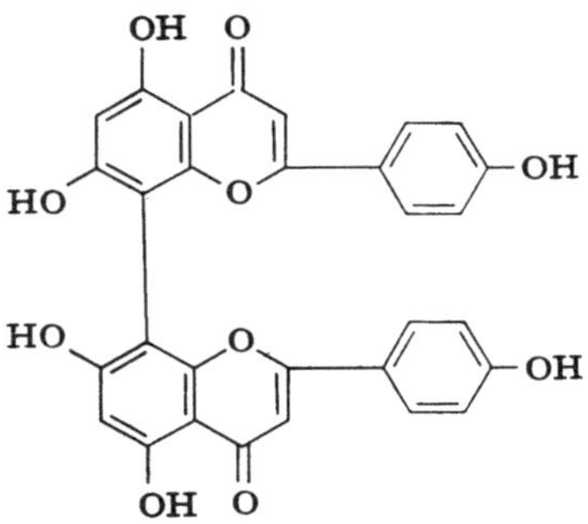

(39)
Cupressuflavone.

Several fatty acids, alcohols and sugars were isolated from the wood and leaves of the Cupressales but there has been no systematic study of their occurrence and, at present, a comparative discussion of these compounds would be of little interest.

5. Tropolones*.

All tropolones so far isolated from Cupressales exhibit distinct structural relations to the terpenes. They differ in this respect from the tropolones produced by certain moulds and *Colchicum* species (meadow saffron).

In 1933 Anderson and Sherrard (6) described a strongly fungitoxic "phenol", $C_{10}H_{12}O_2$, m. p. 82°, isolated from the heartwood of *Thuja plicata* (western red cedar). During a reinvestigation in 1947, using *Thuja plicata* wood (grown in Sweden), two isomeric compounds were isolated by Gripenberg (78, 79). One of them was identical with Anderson and Sherrard's compound; the other melted at 34°. The steam-volatile phenolic fraction of the American-grown *Thuja plicata* from which the compound melting at 82° had been recovered, slowly deposited a considerable quantity of large crystals and this material was kindly put at our disposal by Dr. Anderson. It turned out to consist of a third isomer, m. p. 52° (4, 78, 79). In the order of increasing melting points these compounds were called α-, β- and γ-thujaplicin (78, 106).

The thujaplicins exhibited puzzling properties and appeared to be of aromatic nature somewhat resembling salicylaldehyde. However, they showed no sign of carbonyl groups and were stronger acids (pK ca. 7) than normal monophenols. Furthermore, these compounds showed characteristic spectral properties (12, 13). On oxidation with permanganate α-, β- and γ-thujaplicins yielded isobutyric acid (44). Catalytic hydrogenation gave octahydro derivatives (e. g. 45) which could be oxidised, respectively, to α-, β- and γ-isopropylpimelic acids (46). This shows that the structures of α-, β- and γ-thujaplicins are (41), (42) and (43).

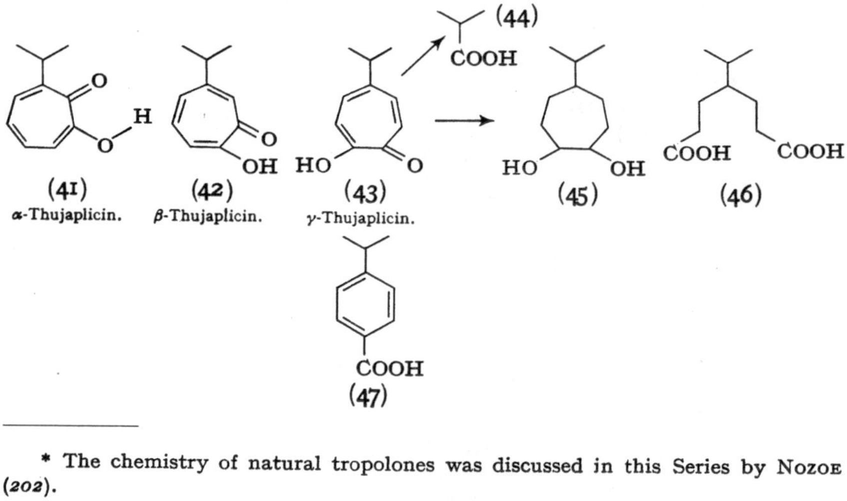

(44)

(41)   (42)   (43)   (45)   (46)

α-Thujaplicin.   β-Thujaplicin.   γ-Thujaplicin.

(47)

---

* The chemistry of natural tropolones was discussed in this Series by Nozoe (202).

References, pp. 275—287.

Although the thujaplicins were the first definitely elucidated tropolones, a few years earlier DEWAR (*53*) had suggested tropolone structures for some fungal metabolites, e. g. stipitatic acid, which when treated with strong alkali yielded a benzene derivative. γ-Thujaplicin behaves similarly and rearranges to cumic acid (47). The tropolones are very strong fungicides (*217*).

(48)
Nootkatin.

(49)

(50)

(51)

For a long time Japanese chemists have studied the chemistry of "Taiwanese hinoki" (*Chamaecyparis obtusa* Sieb. et Zucc. f. *formosana* Hayata, syn. *Ch. taiwanensis* Mas. et Suzuki). From the wood of this tree a deep red substance "hinokitin" was isolated and in 1936 NOZOE discovered that this was an iron salt of an oily compound named hinokitiol which was later obtained in crystalline form (*203*). Hinokitiol gave crystalline salts or complexes with heavy metal ions as well as with organic bases such as aniline, pyridine and quinoline (cf. *59*, *60*). The compound was often considered to have the composition $C_{10}H_{14}O_2$ instead of $C_{10}H_{12}O_2$ and was at first thought to contain a five-membered but later a seven-membered ring, being 4- or 5-isopropylcyclohept-6-ene-1,2-dione (*200*, *201*). However, tropolone structures (with H_{12}) were also discussed. Direct comparison of β-thujaplicin and hinokitiol showed identity.

Considering the theoretical, physiological and technical importance of these compounds a search for similar substances was undertaken. *Thuja occidentalis* and *Chamaecyparis nootkatensis* (Alaska yellow cedar) were first investigated. The latter was chosen because, botanically, it occupies a somewhat isolated position within the genus*.

Thuja occidentalis was found by GRIPENBERG (*108*) to contain α-thujaplicin (41), and *Ch. nootkatensis* yielded nootkatin, a new "sesquiterpene tropolone", $C_{15}H_{20}O_2$, m. p. 95° (*35*).

Nootkatin (48) reacted with performic acid to give the monoformate of a triol (49). The latter yielded acetone (50) with periodic acid, and acetone and isobutyric acid (51) on oxidation with chromic acid. Nootkatin was rearranged by strong alkali to a mixture of the benzoic acid derivatives (52) and (53) which on hydrogenation were converted into the dihydro derivative (54). This was decarboxylated and the resulting hydrocarbon (55) oxidised to phthalic acid (56) with dilute nitric acid. On similar oxidation the dihydro compound gave the lactone (57) which was

* This species was considered by ØRSTED to constitute a monotypic genus of its own, *Callitropsis nootkatensis*. It is for this reason that the New Caledonian *C. araucarioides*, which will be discussed later (p. 248) was renamed *Neocallitropsis araucarioides*.

further oxidised to trimellitic acid (58). From these degradations the complete structure of nootkatin follows (*12, 13, 58, 81*). The X-ray examination of nootkatin copper gave an identical result (*34*).

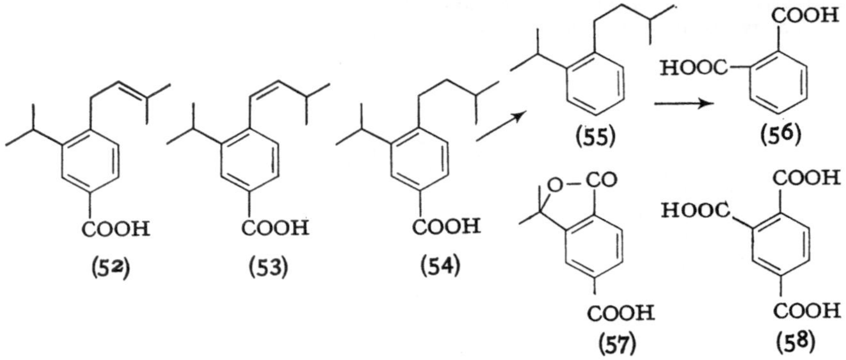

(52)   (53)   (54)   (55)   (56)   (57)   (58)

From the wood of several Cupressaceae species (*Table 5*, p. 261) quite a few tropolones of the thujaplicin or nootkatin type have since been isolated, for example β-thujaplicinol (59) which contains an extra hydroxyl group in the tropolone ring. It can be prepared from β-thujaplicin by oxidation with persulphate (*99*). β-Dolabrin (60) is a didehydro-β-thujaplicin (*98, 204*). On dehydration with formic acid nootkatinol (61) gives nootkatin and, conversely, nootkatin can be hydrated to nootkatinol by strong phosphoric or sulphuric acid (*21, 117, 118*). The hydroxylated tropolones appear to be weaker fungicides than the normal tropolones (*99*).

The structures of the known Cupressales tropolones are given below. For physical constants cf. *Table 6*, p. 263.

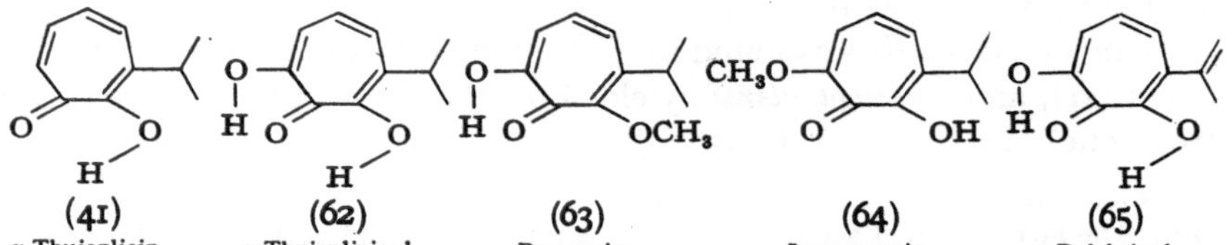

(41)   (62)   (63)   (64)   (65)
α-Thujaplicin.   α-Thujaplicinol.   Pygmaein.   Isopygmaein.   α-Dolabrinol.

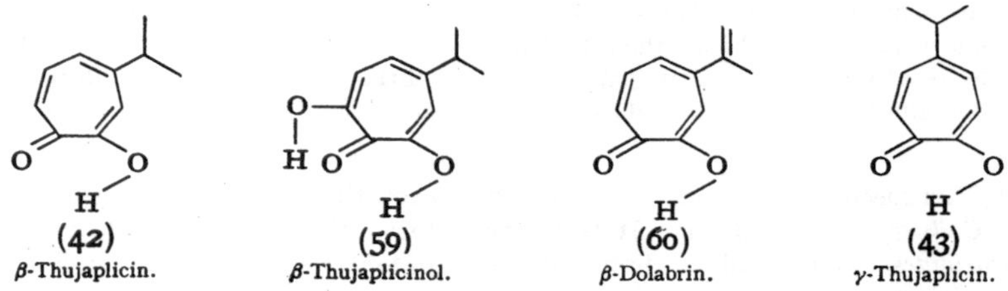

(42)   (59)   (60)   (43)
β-Thujaplicin.   β-Thujaplicinol.   β-Dolabrin.   γ-Thujaplicin.

References, pp. 275—287.

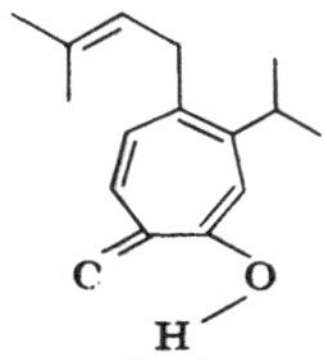

(48)	(66)	(61)	(67)
Nootkatin.	Procerin.	Nootkatinol.	Chanootin.

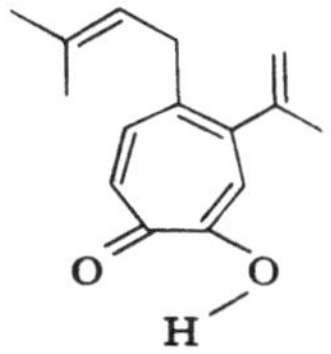

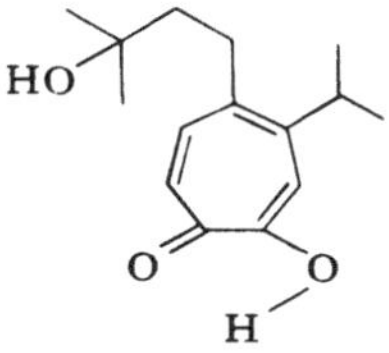

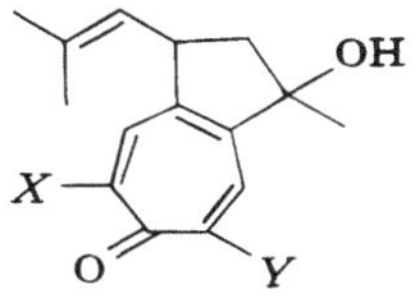

(67) Chanootin. $X = $ OH, $Y = $ H, or $X = $ H, $Y = $ OH.

Several new tropolones will most probably be found in the future; the presence of unidentified tropolones has been indicated in paper chromatographic experiments.

In some instances the tropolones have been obtained from the disintegrated wood by simple steam distillation and isolation of the "phenolic" fraction. More frequently the wood was extracted with ether or acetone and the tropolones purified by conversion to their copper complexes which then were decomposed by means of hydrogen sulphide. Salts with dicyclohexylamine, cyclohexylamine or benzylamine (275) and ion exchangers (130) have also been used for the isolation. Tropolones are fairly stable when kept in the dark; however, the copper complexes of hydroxytropolones seem to be rather unstable (274).

Paper chromatography has been very useful for the detection of the various tropolones in tropolone mixtures. ZAVARIN and ANDERSON (274, 270) recommend chromatography on paper impregnated with phosphoric acid, the mobile phase being isooctane or isooctane-toluene (cf. 171). A similar method using ethylenediaminetetraacetic acid and dimethyl sulphoxide was developed by WACHTMEISTER and WICKBERG (259).

As a rule several tropolones occur together in Cupressaceae heartwoods in quantities varying from traces to a few per cent. Many genera appear to lack tropolones and in several instances no tropolones could be detected in species belonging to a tropoloniferous genus.

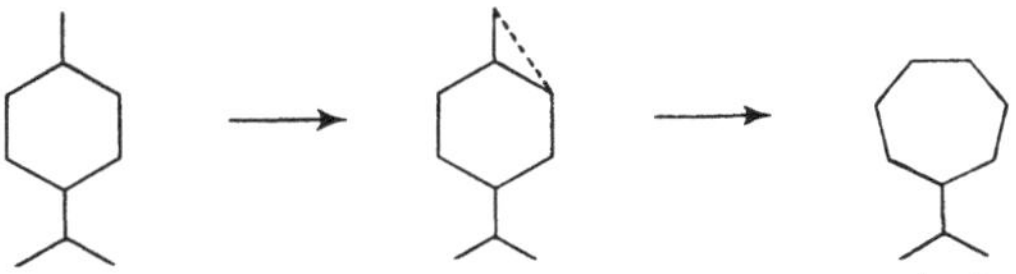

(68)   (69)   (70)

It is generally believed (71, 72, 75, 76, 96, 219) that the Cupressaceae tropolones are modified terpenes formed by ring enlargement as indicated in (68–70). There is no biosynthetic proof of this hypothesis as yet but the frequent co-occurrence of carvacrol and tropolones is a good indication of such interrelations. The co-occurrence of α-, β- and γ-thujaplicin in *Thuja plicata* and other Cupressaceae seems to indicate that they are formed via a common precursor. Nootkatin is possibly formed via a sesquiterpene intermediate of an abnormal (71) or bicyclic (73) type.

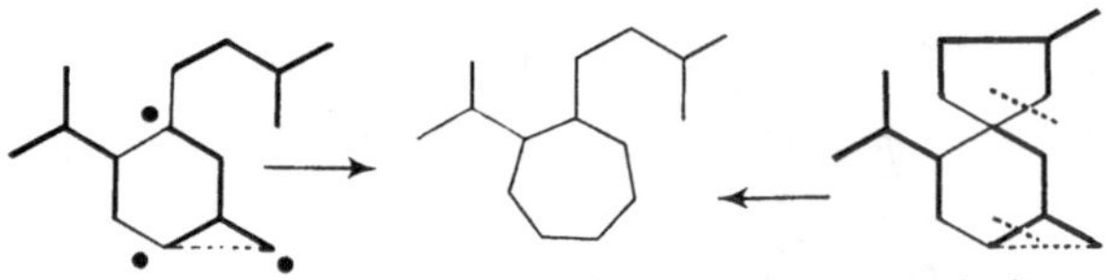

(71)   (72)   (73)

Alternatively, nootkatin may simply be a prenylation product of β-thujaplicin. It has been prepared synthetically in this way (*151*):

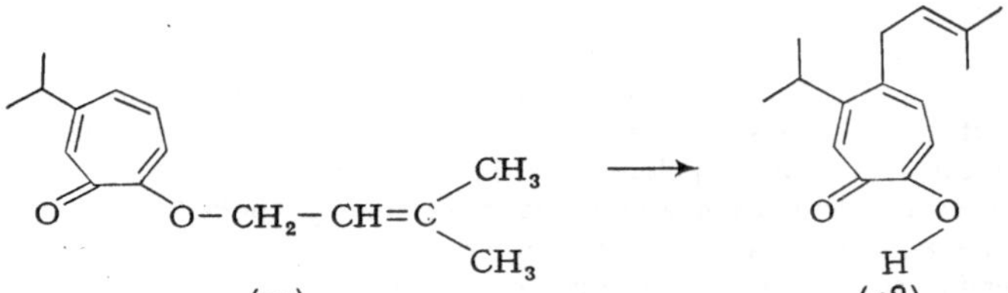

(74)   (48)

Quite recently, a bicyclic tropolone, chanootin (67), was isolated by NORIN (*194*) from the complex tropolone mixture occurring in *Chamaecyparis nootkatensis*. Chanootin was subjected to degradations closely similar to those employed for the elucidation of the nootkatin structure. Oxidation with performic acid followed by hydrolysis gave a diol which on cleavage with periodate yielded acetone. Rearrangement of chanootin with alkali gave a benzoic acid derivative which on oxidation with nitric acid furnished trimellitic acid. These results together with NMR data strongly suggest the structure (67, p. 219). Strangely enough, the substance appears to be optically inactive. Chanootin may well be derived from a monocyclic compound of the nootkatin type (75), and conversely, nootkatin may be derived from a compound of the chanootin type. The chanootin skeleton is such that it could well be derived from a normal sesquiterpene with the isoprene units arranged head-to-tail as in (76).

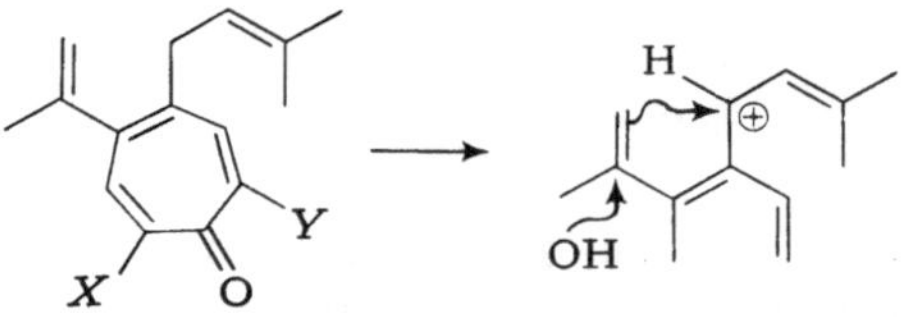

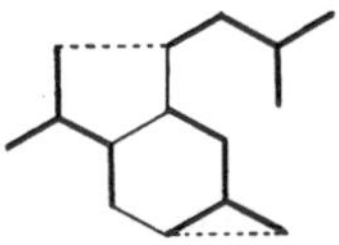

(75)   (76)

6. Terpenes.

Terpenes are the main constituents of the extractives of most woods of Cupressales. Although many species have been investigated, our knowledge of their terpenes is still limited.

References, pp. 275—287.

a. Monoterpenes.

Numerous investigations have been carried out of the steam-volatile "essential oils" obtained from the wood and leaves of many Cupressales species. Most of them were, however, conducted before the development of modern separation techniques and thus, the fractions isolated were not pure compounds. This has caused much confusion, and it is sometimes difficult to interprete the results from chemical and chemotaxonomic viewpoints. Nowadays more reliable results are obtained with the aid of modern chromatographic techniques. These results should give valuable information for future chemotaxonomic considerations. In the present review only a few monoterpenes of special interest will be discussed.

In the course of their classical work on Australian conifers BAKER and SMITH (*14*) found an acidic liquid substance in the wood of various species belonging to the genus *Callitris* ("cypress pines") including *C. (Octoclinis) macleyana*. They considered this compound to be a phenol, "callitrol". DADSWELL and DADSWELL (*47*) renamed the substance callitric acid after TRIKOJUS and WHITE (cf. *147*) had identified it as *l*-citronellic acid (*77*). The DADSWELLs found that the latter is toxic to fungi and a termite repellent (cf. also RUDMAN, *225–227*). Citronellic acid has since been isolated from the wood of *Chamaecyparis taiwanensis* (*167, 200, 201*) and *Thujopsis dolabrata* (*206*), the Japanese "hiba"-tree. The methyl ester appears to occur in the leaves of *Juniperus sabina* (*222*).

The closely related dehydrogeranic acid (*78*) is present as a geranyl ester in the steam-volatile oil from the wood of the New Caledonian conifer *Neocallitropsis araucarioides* (*147*). It was likewise isolated from the wood of *Callitris glauca* (*147*) where it probably also occurs as an ester. Geranic acid does not seem to have been encountered in any of the Cupressales.

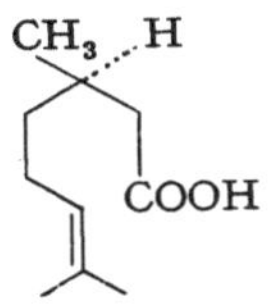

(77)
l-Citronellic acid.

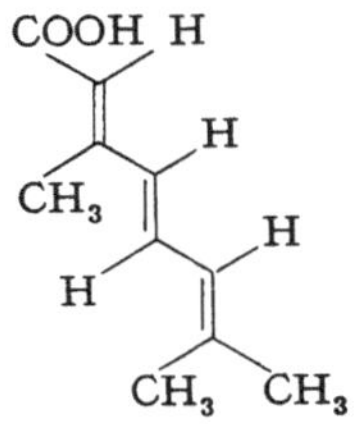

(78)
Dehydrogeranic acid.

Thujic Acid. ANDERSON and SHERRARD (*6*) have isolated from the heartwood of *Thuja plicata* a crystalline acid, $C_{10}H_{12}O_2$, which they called dehydroperillic acid since they believed it to be related to perillic acid. During our reinvestigation of this wood the same acid was obtained.

The optical properties of the compound did not agree with those expected for a compound having the structure suggested by the above authors. The acid was shown to possess the unusual cycloheptatriene structure (79) and it was renamed thujic acid (80). When briefly heated with strong hydrochloric acid, the acid rearranged to cumic acid. Oxidation with permanganate furnished dimethylmalonic acid in high yields. Catalytic hydrogenation gave a liquid hexahydro acid (amide, m. p. 147.5–148.5°) that was degraded by two different methods to β,β-dimethyladipic acid (80) and β,β-dimethylpimelic acid (81); this locates the carboxyl group (107). The alternative bicyclic structure (82) was ruled out by a NMR study which showed that thujic acid contains five vinyl protons (110). The structure (79) of thujic acid has been confirmed by total synthesis proceeding via hexahydrothujic acid and stepwise bromination and dehydrobromination (109). Thujic acid is a rather labile compound; the methyl ester is more stable and possesses an agreeable odour. Methyl thujate is also a component of the *Thuja plicata* wood (159).

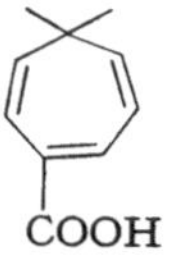

(79)
Thujic acid. (80) 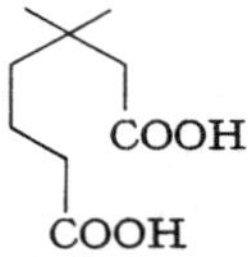 (81) 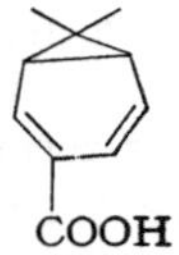 (82)

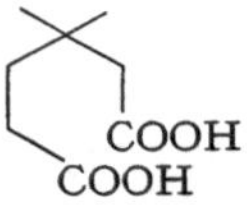

Chamic and Chaminic Acids. During the investigation of the heart-wood of *Chamaecyparis nootkatensis* two isomeric bicyclic C_{10}-acids containing two hydrogen atoms more than thujic acid were isolated (35, 82). One of them, chamic acid, was liquid and the other, chaminic acid, crystalline. Recently, the presence of small amounts of thujic acid in the crude acid mixture was ascertained by gas and thin layer chromatography (194).

Chamic acid contains one double bond which is not conjugated with the carboxyl group. Chaminic acid, however, is an α,β-unsaturated acid. When chamic acid is heated with alkali the double bond migrates into conjugation furnishing the α,β-unsaturated isochamic acid. To our surprise isochamic acid was found to be the optical antipode of chaminic acid. Catalytic hydrogenation of chamic and isochamic acids gave different dihydro derivatives and on further hydrogenation another mole of hydrogen was slowly consumed. The infrared absorption spectra of chamic, isochamic and dihydrochamic acids indicated the presence of a cyclopropane ring.

The difference between dihydrochamic and dihydroisochamic acids must therefore be due to a *cis-trans*-relationship between the carboxyl group and the cyclopropane ring. Bromination of chamic acid followed by dehydrobromination gave some *References, pp. 275— 287.*

m-isopropylbenzoic acid and a similar treatment of isochamic acid gave *m*- and *p*-isopropylbenzoic acid which shows the position of the carboxyl group. The catalytic hydrogenation of isochamic acid would be expected to give a dihydro acid possessing a *cis*-relationship between the carboxyl group and the cyclopropane ring. Thus, dihydrochamic acid can be assigned a *trans*-relationship (*82*). Chamic acid should therefore possess the structure (83) and isochamic acid the structure (84). There remains, however, the possibility that the double bond migration proceeds further with the ultimate formation of (85) in which the double bond is conjugated with both the carboxyl group and the cyclopropane ring.

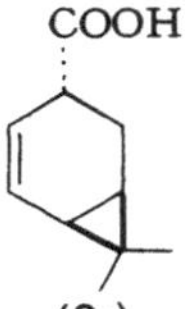

(83) 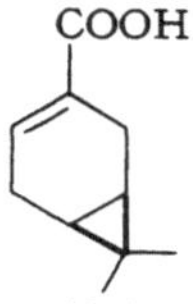   (84) 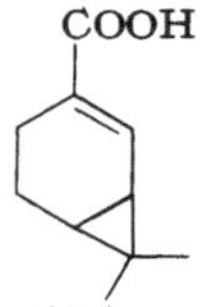   (85) 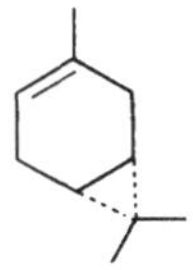   (86)

The structures for chamic and isochamic acids have recently been confirmed by an NMR-investigation (*196*). Moreover, the stepwise reduction of the carboxyl group of isochamic acid to a methyl group using conventional methods afforded the optical antipode of the naturally occurring $(+)$-Δ^3-carene (86). The absolute configurations of chamic and chaminic acids are therefore as indicated in (87) and (88). The co-occurrence of chamic and chaminic acids of "antipodal configuration" may be due to different cyclisation of a common precursor represented by the ion (89).

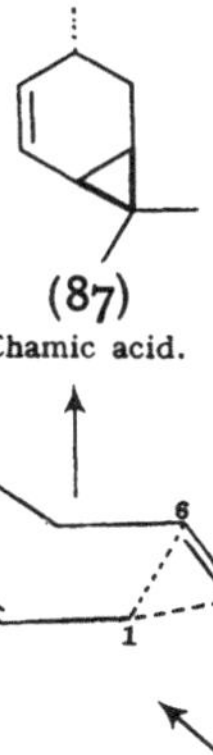

(87)
Chamic acid.

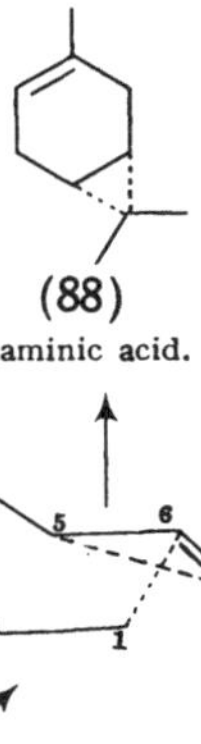

(88)
Chaminic acid.

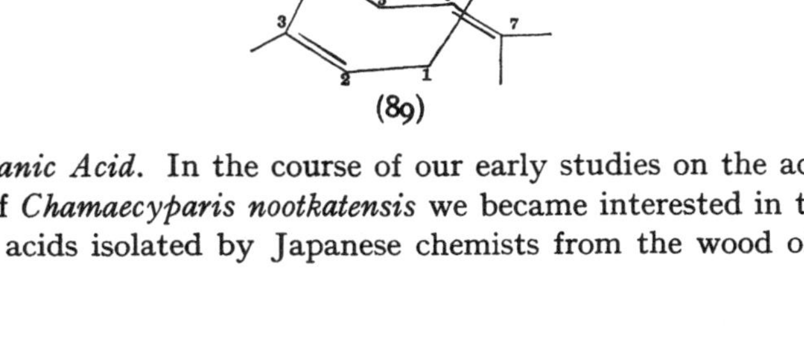

(89)

Shonanic Acid. In the course of our early studies on the acid constituents of *Chamaecyparis nootkatensis* we became interested in the nature of some acids isolated by Japanese chemists from the wood of the Taiwanese *Calocedrus formosana* (syn. *Libocedrus formosana* Florin) or "shônan boku". These acids had been extracted by means of dilute alkali and as much as 500 g was obtained from about 100 kg of wood. Vacuum distillation gave liquid as well as solid acids. One of them, the weakly levorotatory shonanic acid, $C_{10}H_{14}O_2$, m. p. 40–41°, was extensively investigated by Ichikawa (*128*) who suggested the structure (90). On catalytic hydrogenation shonanic acid yielded a liquid tetrahydro acid whose amide melted at 144–145°. This melting point coincides closely with that of hexahydrothujic acid amide. On oxidation with permanganate a high yield of dimethylmalonic acid was obtained. Assuming that tetrahydroshonanic and hexahydrothujic acids are identical, this indicates for shonanic acid the structure (91) (*71–74*). The formation of dimethylmalonic acid was confirmed using a sample of shonanic acid kindly provided by Professor Lin (Taipei). The structure (91) was corroborated by him and his co-workers who succeeded in transforming iso-shonanic acid (see below) into thujic acid by bromination and dehydrobromination (*165*).

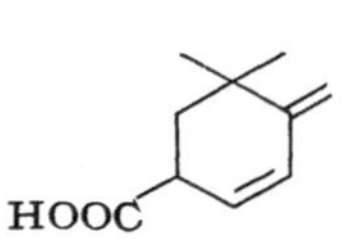

(90) 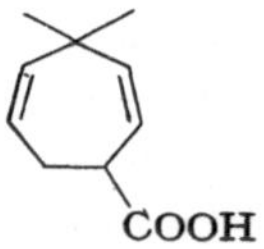   (91)   (94)
           Shonanic acid.

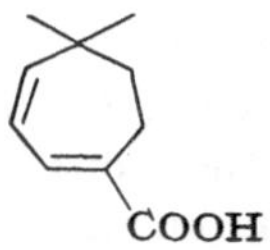

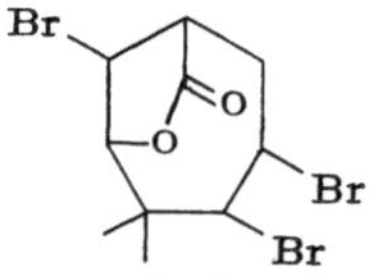

(92) 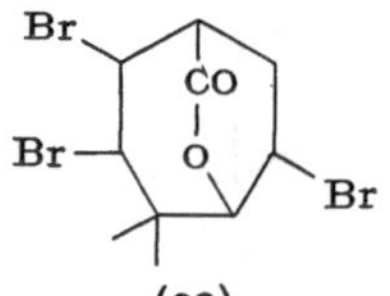   (93)

Bromination of shonanic acid gave a liquid dibromo product which on vacuum distillation underwent decomposition and yielded cumic acid. When heated with water the dibromo product gave a liquid monobromolactone which on bromination furnished a crystalline tribromolactone. On the basis of the revised structure of shonanic acid the tribromolactone may possess structure (92) or (93). A strange reaction reported by Ichikawa (*128*) is the formation of o-dinitrobenzene on heating shonanic acid with dilute nitric acid.

On treating shonanic acid with strong alkali "isoshonanic acid" is formed, m. p. 45–46°. Lo and Lin (*170*) found that this substance is not a pure compound; pure isoshonanic acid melts at 87–87.5°. Pasto (*210*) favours the structure (94) for isoshonanic acid. The double bonds in shonanic acid are very mobile and this may explain why tetrahydroshonanic acid is identical with hexahydrothujic acid. Another explanation may be that shonanic acid is partly racemised either in the wood or during the isolation process. A similar mobility of the double bonds in partially reduced thujic acid has been reported (*107, 210, 36*).

References, pp. 275—287.

Other Monoterpene Acids from Calocedrus formosana. The other acids from "shônan boku" were little investigated by ICHIKAWA (*128*). He reported that one of them melts at 78–80° which is close to the melting point of thujic acid; and another acid, an isomer of shonanic acid, melts at 103°. The optical rotation of this acid was not given but it was thought to be chaminic acid. The liquid acids from "shônan boku" could well be, or contain, chamic acid. That these assumptions were correct was proved by LIN and his co-workers (*164*). There is obviously a great similarity between the patterns of acids occurring in *Calocedrus formosana* and in *Chamaecyparis nootkatensis*. It remains to be investigated whether the latter contains shonanic acid.

b. Sesquiterpenes.

Nerolidol. The root oil of *Fokienia hodginsii* was investigated by DOLEJŠ and HEROUT (*54*). One of its main constituents is the sesquiterpene alcohol (+)-nerolidol (95), evidently identical with the "fokienol" previously isolated from the wood oil. Surprisingly, the absolute configuration of (+)-nerolidol is still unknown. So far neither nerolidol nor any other acyclic sesquiterpene has been isolated from other Cupressales species.

Bisabolanes. Sesquiterpenes of the monocyclic bisabolane type were isolated from a few species only. The leaves of *Libocedrus bidwillii* are reported to contain (+)-γ-curcumene (96) (*126*). (—)-β-Bisabolene (97) is a constituent of the stump oil of *Chamaecyparis lawsoniana* (*158*). Recently, NAGAHAMA (*188*) reported the occurrence of a ketone, cryptomerione, in the wood oil of *Cryptomeria japonica*. On catalytic hydrogenation cryptomerione furnishes a product with spectral properties similar to those of hexahydrobisabolene (cf. structure 97). The NMR spectrum of this ketone is in agreement with structure (98) or (99).

The sesquiterpene "X_2" ($C_{15}H_{24}$) isolated by RUNEBERG (*230*) from the heartwood of *Juniperus thurifera* appears to be of the bisabolane type, since it gave dihydroarcurcumene (100) on selenium dehydrogenation. The hydrocarbon "X_2" also appears to be present in the essential oil of *J. virginiana* (*229*).

Humulene and Caryophyllene. The structure of the monocyclic sesquiterpene humulene (101) containing an eleven-membered ring has been firmly established by two independent X-ray studies (*111, 178*). This compound and the related bicyclic nine-membered ring compound caryophyllene (102) were isolated from only a few Cupressales species (*Table 9*, p. 266). Recently, SUTHERLAND and co-workers (*105*) have converted humulene into (±)-caryophyllene by a series of interesting reactions.

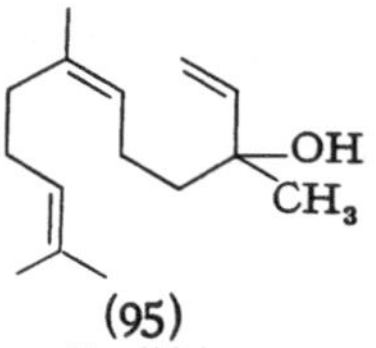

(95)
Nerolidol.

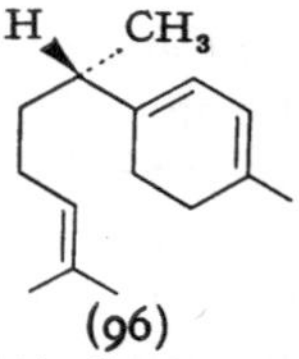

(96)
(+)-γ-Curcumene.

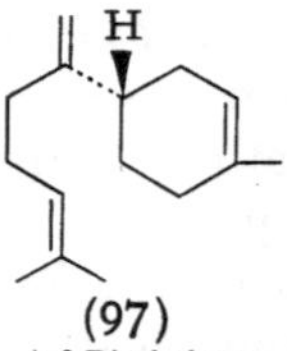

(97)
(—)-β-Bisabolene.

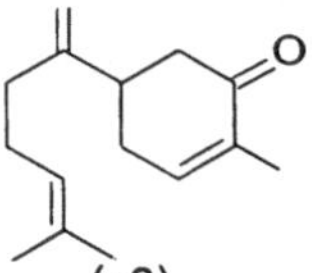

(98)
Cryptomerione (?).

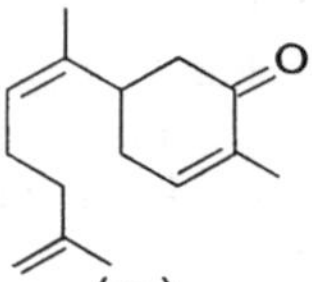

(99)

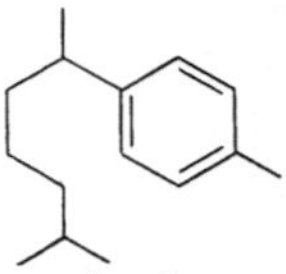

(100)

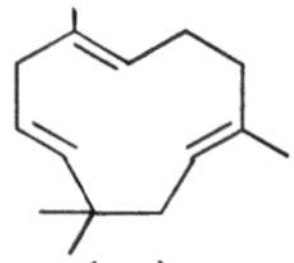

(101)
Humulene.

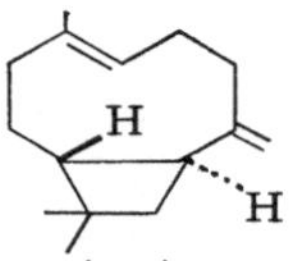

(102)
(—)-Caryophyllene.

Elemanes. A few sesquiterpenes of the elemane type were isolated from Cupressales species (*Table 7*, p. 264); thus, β-elemene (103) and elemol (104) occur in *Juniperus* species. Recently, elemenal (105) was found in the Japanese "hiba" wood *(Thujopsis dolabrata)* (*134*). Sesquicryptol (C₁₅H₂₆O, m. p. 49–51°, [α]D +22.7°), obtained from the leaf oil of *Cryptomeria japonica* (*244*), was suggested to be a mixture in which elemol is a main constituent (S. Nagahama, private communication).

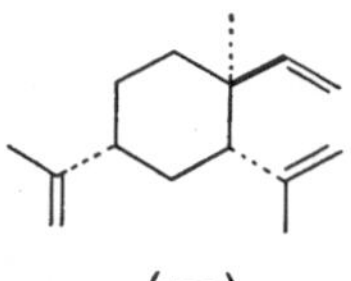

(103)
(—)-β-Elemene.

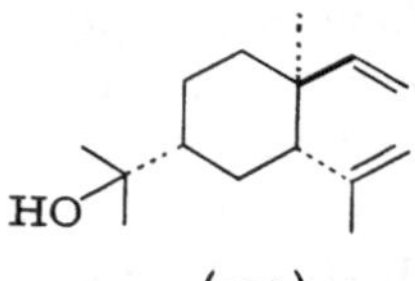

(104)
(—)-Elemol.

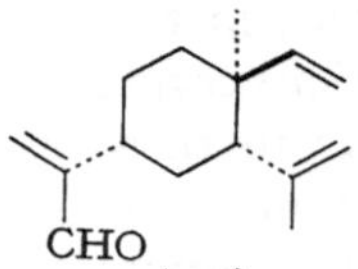

(105)
(—)-Elemenal.

Selinanes. Sesquiterpenes of the selinane type and related compounds are common Cupressales constituents (Table 7), particularly the three isomeric eudesmols (106) and the closely related cryptomeridiol. The structure of cryptomeridiol (107) has been elucidated by Sumimoto (*249, 251*), who also found that his diol was identical with the 4,11-selenadiol previously isolated from the heartwood of *Widdringtonia dracomontana* (*89*) and from the root oil of *Fokienia hodginsii* (*54*). "Cryptomeradol", an alcohol isolated from the root oil of *Cryptomeria japonica*, was shown by Gardner and Horton (*101*) to be a mixture of α- and β-eudesmol.

References, pp. 275—287.

The main alcohols of the heartwood extractives of the eastern white cedar *(Thuja occidentalis)* are occidentalol (108) and occidol (109), the latter of a rearranged selinane type. The minor constituents have recently been investigated by VON RUDLOFF and NAIR *(224)* who identified small amounts of the three isomeric eudesmols (106) and isolated a new sesquiterpene diol, occidiol, for which they tentatively propose the structure (110).

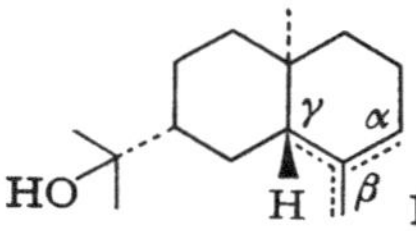

(106)*
Eudesmol.

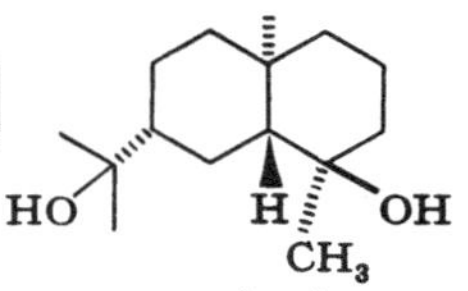

(107)
Cryptomeridiol.

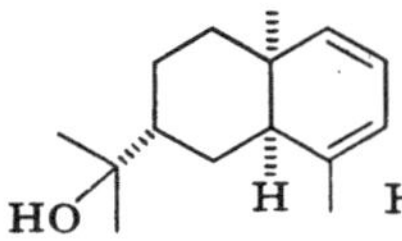

(108)
Occidentalol.

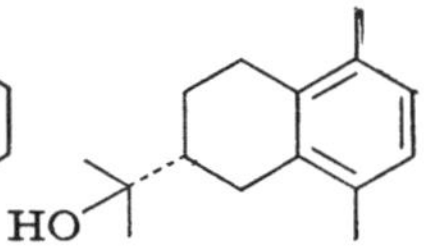

(109)
Occidol.

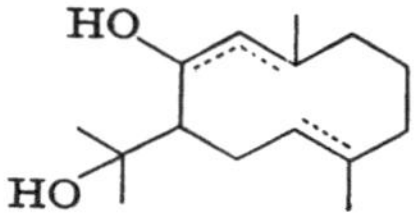

(110)
Occidiol (?).

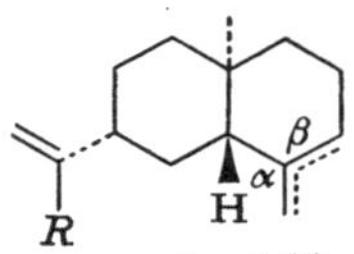

(111)**
Costol, R = CH₂OH
Costal, R = CHO.

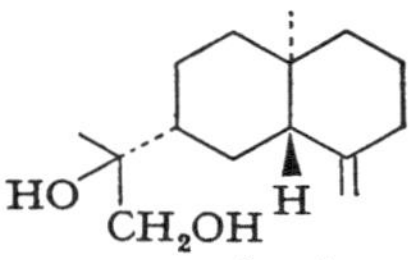

(112)
Sesquibenihidiol.

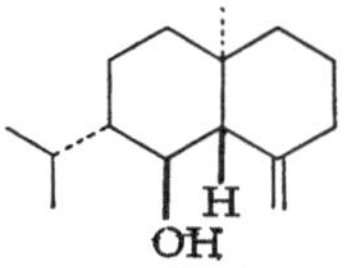

(113)
Junenol.

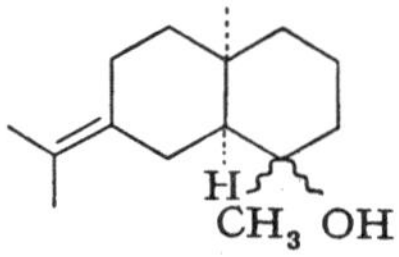

(114)
Juniper camphor.

* α, β and γ indicate double bond positions in α-, β- and γ-eudesmol, respectively.

** α and β indicate double bond positions in α- and β-costol (costal), respectively.

Sesquibenihiol is an alcohol isolated from the essential oil of *Chamaecyparis formosensis* and has been shown to be identical with costol, a constituent of costus root studied by ŠORM and co-workers *(20)*. Although the structure (111, R = CH₂OH) of costol (sesquibenihiol) had been settled, a recent investigation by BOWDEKAR and KELKAR *(23)* has shown that ŠORM's costol sample was a mixture of three components one of which was the true (+)-α-costol (111, R = CH₂OH). Sesquibenihidiol (112) is another constituent of *Ch. formosensis* and is related to costol *(148)*. The isolation of a mixture of α- and β-costal (cf. 111, R = CHO) from the wood of *Thujopsis dolabrata* was reported recently by ITO and co-workers *(134)*.

Junenol (113) and juniper camphor (114) are alcohols of the selinane type which so far have been found only in the berries of *Juniperus communis* (*182, 183*).

Eremophilanes. From the heartwood extractives of the Alaska yellow cedar *(Ch. nootkatensis)* ERDTMAN and HIROSE (*83*) isolated an α,β-unsaturated ketone, named nootkatone. Upon treatment with methyl magnesium iodide nootkatone gave an alcohol which was dehydrogenated with selenium to 1,3-dimethyl-7-isopropylnaphthalene (115). On the basis of this experiment and of a NMR investigation (with one of the early 40 Mc/s instruments) the structure (116) or an eremophilone-like structure was tentatively proposed for nootkatone. The structure of nootkatone has recently been reinvestigated by MacLEOD (*172*) who found the compound in the grapefruit peel oil. On the basis of spectral and chemical studies, he concludes that nootkatone is indeed of the eremophilane type and has the structure (117). It should be noted, however, that the proposed configuration is different from that derived from biogenetic considerations (see p. 245).

The sesquiterpene hydrocarbon nootkatene is another heartwood constituent of *Ch. nootkatensis* described by ERDTMAN and TOPLISS (*90*). On selenium dehydrogenation it gave eudalene (118) and when subjected to ozonolysis it yielded formaldehyde but no acetone. Spectroscopic studies of nootkatene have recently been carried out in our laboratory (NORIN and LEITZ, unpublished). Nootkatene shows a UV-spectrum (λ_{max} 227, 235 and 243 mμ, in ethanol; log ε 4.11, 4.15 and 4.25) characteristic of a heteroannular diene. Since the NMR spectrum exhibits signals corresponding to five olefinic protons, the selinane type of structure (119) is ruled out. The two eremophilane structures (120) and (121) are, however, in complete agreement with the spectral data. Dr. MacLEOD has kindly informed us that he has succeeded in transforming nootkatone (117) into nootkatene by a series of reactions which indicate that nootkatene is correctly represented by (121) (*83*).

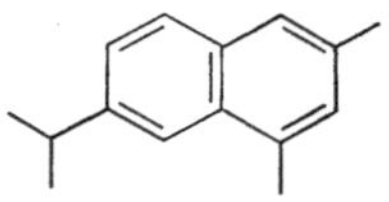

(115) 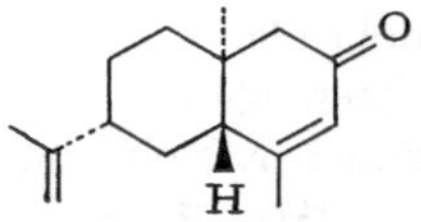   (116) 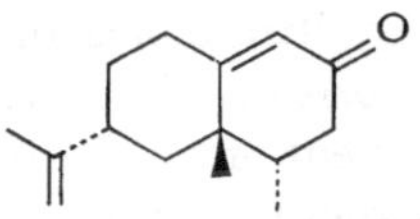   (117)
                                  Nootkatone.

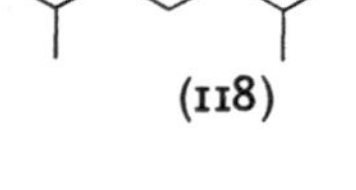

(118) 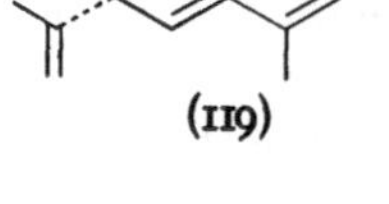   (119) 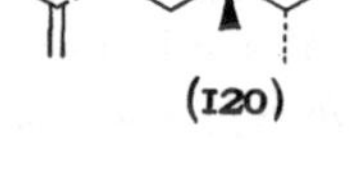   (120) 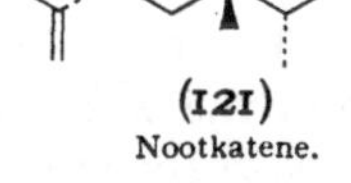   (121)
                                                   Nootkatene.

RUDMAN (*226*) has recently shown that on treatment with boron-trifluoride in methanol guaiol (**122**) yields small amounts of eudesmols (**106**). This is of interest, because the terpenes of the guaiane and selinane types are considered to be biogenetically related (see p. 244). Guaiol occurs with eudesmols and related selinanes in the heartwood extractives of various *Callitris* species and in *Neocallitropsis* (Table 7, p. 264).

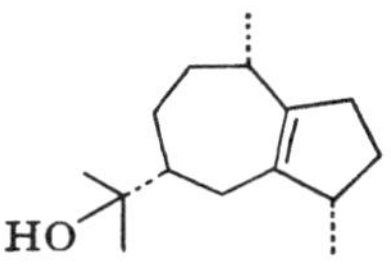

(**122**)
Guaiol.

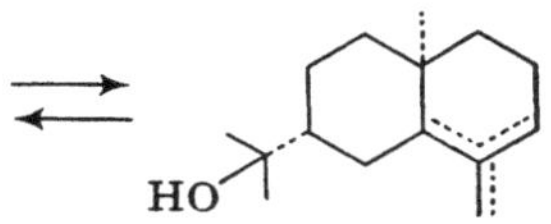

(**106**)

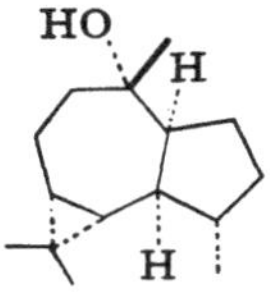

(**123**)
Viridiflorol.

The berries of *Juniperus oxycedrus* contain an alcohol, viridiflorol (**123**) of the aromadendrane type (*55, 56*). Neither this nor other compounds of this type have been reported to occur in other Cupressales species.

Cadinanes and Related Terpenes. Cadinenes, cadinols and related compounds are widely distributed in Nature and usually occur as mixtures of considerable complexity. Their presence in such mixtures can often be demonstrated by treatment with hydrogen chloride when crystalline cadinene dihydrochloride is formed. There has been much confusion in the chemistry of these compounds because of the heterogeneity of the fractions investigated. However, the cadinenes and related compounds are now being studied with the aid of modern techniques by a number of research groups and thus rapid progress can be expected.

Cadinane type terpenes (**124**–**135**) are common Cupressales constituents (*Table 8*, p. 265). The tricyclic copaene (**127**) and ylangene (**130**) have also been listed within this class of compounds. Some cadinane products of doubtful homogeneity have only been characterised by formation of cadinene dihydrochloride or by dehydrogenation to cadalene. In Table 8 such fractions are listed under "cadinene" or "cadinol".

Longifolene and Juniperol. SIMONSEN (*243*) isolated a tricyclic sesquiterpene, longifolene, from the oleoresins of various Pinales species and the structure (**136**) has been rigorously established (cf. *19, 208*). AKIYOSHI, ERDTMAN and KUBOTA (*3*) have shown that junipene, a bark constituent of *Juniperus communis*, is identical with longifolene. Although the name junipene has priority, it has been suggested that the well-known name longifolene should be retained for this compound as a nomen conservandum. Kuromatsuene isolated by Japanese workers from various pine species is also identical with longifolene.

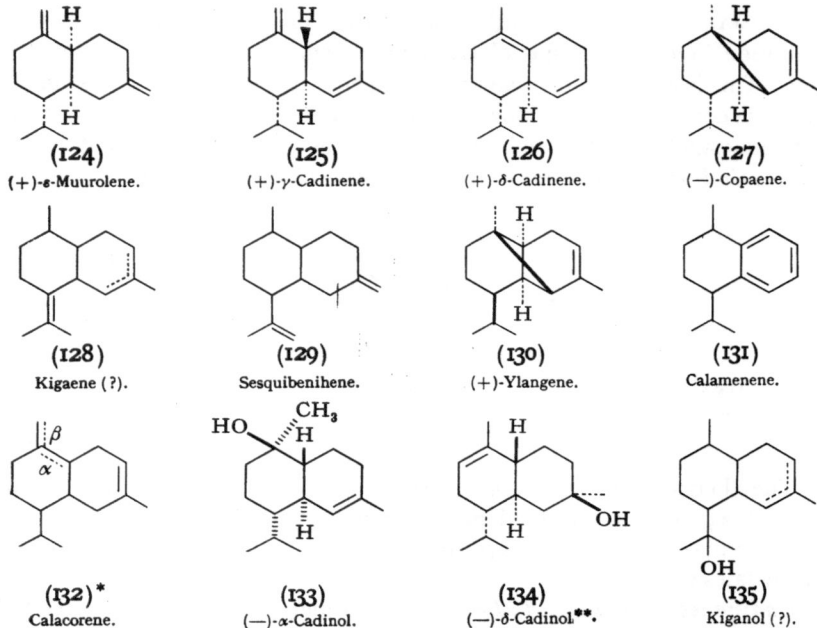

(124)
(+)-ɛ-Muurolene.

(125)
(+)-γ-Cadinene.

(126)
(+)-δ-Cadinene.

(127)
(—)-Copaene.

(128)
Kigaene (?).

(129)
Sesquibenihene.

(130)
(+)-Ylangene.

(131)
Calamenene.

(132)*
Calacorene.

(133)
(—)-α-Cadinol.

(134)
(—)-δ-Cadinol**.

(135)
Kiganol (?).

* α and β indicate double bond positions in α- and β-calacorene, respectively.
** δ-Cadinol (134) has been shown to belong to the muurolane type of sesquiter-
penes (cis-fused rings) (WESTFELT, unpublished). "δ-Cadinol" is also present in
Chamaecyparis lawsoniana (50a). (Added in Proof.)

Juniperol, another bark constituent of *J. communis*, is identical with
macrocarpol, an alcohol from the leaf oil of *Cupressus macrocarpa* (3);
it is also identical with longiborneol (138) obtained from longifolene (136)
via longifolene hydrobromide (137). The name juniperol has priority
and should be retained instead of longiborneol, macrocarpol and kuromatsuol (pinewood constituent).

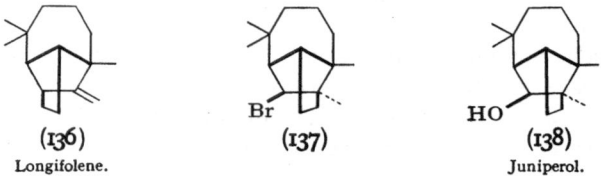

(136)
Longifolene.

(137)

(138)
Juniperol.

Many important aspects of the chemistry of longifolene and related
compounds have been discussed recently by OURISSON (208). So far
longifolene and juniperol have been isolated only from the two Cupressales species mentioned above (Table 9, p. 266).

References, pp. 275—287.

Cedranes. Cedrene (**139**) and cedrol (**140**) are the sesquiterpenes which have been most frequently encountered in Cupressales woods (Table 9) and were first isolated from "cedar wood oil" *(J. virginiana)* in 1841 by WALTER (see *147*). Careful structural studies of the compounds were performed by SEMMLER and HOFFMANN (*239*) in 1907, and continued by many other research workers. However, it was not until 1953 that the correct structures were proposed, independently by PLATTNER et al. (*214*) and by STORK and BRESLOW (*247*). STORK and CLARKE (*248*) have effected the synthesis of these compounds.

The specific rotation, $[\alpha]_D \sim -55°$, of natural α-cedrene isolated from various conifers has been reported to differ from that of samples obtained by dehydration of cedrol, $[\alpha]_D - 91°$. It is generally believed that the natural "α-cedrene" is a mixture of α- and β-cedrenes (cf. **139**). These two compounds can now be easily separated by chromatography in the presence of silver ions by the preparative method described by WICKBERG (*262*) or that of NORIN and WESTFELT (*199*). Using WICKBERG's method RUNEBERG (*230*) obtained pure α-cedrene from the heartwood extractives of *J. thurifera*.

GORYAEV and co-workers (*103*) have reported the isolation of "natural β-cedrene", $[\alpha]_D - 2.6°$, from *J. semiglobosa*. However, a sample of β-cedrene, prepared by pyrolysis of the *N,N*-dimethyl derivative of primary cedranyl amine, had a higher rotation, $[\alpha]_D + 11.3°$ (*102*).

"Pseudocedrol" is an isomer of cedrol which co-occurs with cedrene and cedrol in cedar wood oil *(J. virginiana)*. "Pseudocedrol", like cedrol, was found to be a tertiary alcohol which on dehydration yielded cedrene and was therefore concluded to be a stereoisomer (**141**) of cedrol (*240*). "Pseudocedrol" has later been found in the wood oil of *J. oxycedrus* (*19*).

Another alcohol of the cedrane type was isolated by SEMMLER and MAYER (*240*) from cedar wood oil *(J. virginiana)*. It was assumed to be the primary alcohol derived from α-cedrene and therefore called "prim-cedrenol". The properties of a sample (**142**) prepared by selenium dioxide oxidation of α-cedrene do not agree well with those of the natural product (*245*).

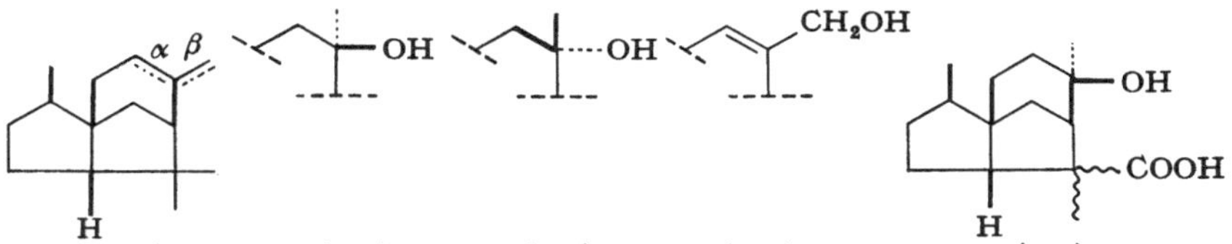

(**139**)*   (**140**)   (**141**)   (**142**)   (**143**)
Cedrene.   Cedrol.   Pseudocedrol (?).   "Prim-cedrenol".   Cedrolic acid.

* α and β indicate double bond positions in α- and β-cedrene respectively.

Cedrolic acid (143) was obtained by Runeberg (234) from the heartwood of *J. foetidissima*. Its structure was proved by conversion to (+)-cedrol but the steric arrangement of the carboxyl group remains to be elucidated (235).

Thujopsanes. Widdrene, a tricyclic sesquiterpene hydrocarbon, $C_{15}H_{24}$, was isolated by Erdtman and Thomas (89) from the heartwood extractives of various species of the South African genus *Widdringtonia*. On oxidation with selenium dioxide widdrene gave widdrenal, a crystalline α,β-unsaturated aldehyde, (cf. 145). The same aldehyde was later obtained by Hirose and Nakatsuka (120) by a similar oxidation of thujopsene, the main neutral constituent of the wood oil of "hiba" *(Thujopsis dolabrata)* which had been studied for a long time by Japanese investigators. However, the physical constants of thujopsene (e. g. $[\alpha]_D - 80°$) and widdrene ($[\alpha]_D - 109°$) were different. A careful investigation of "hiba" oil was therefore conducted by Erdtman and Norin (84) who isolated the pure hydrocarbon thujopsene. The latter is identical with widdrene; but since the name thujopsene has priority, it was suggested that this name should be retained.

The α,β-unsaturated aldehyde, "widdrenal" (now "thujopsenal") can be oxidised to the corresponding acid, "widdrenic acid", which was also isolated from the heartwood of various *Widdringtonia* species. It was noticed that there were considerable similarities between the physical properties of widdrenic acid and hinokiic acid, which had been isolated from the steam-volatile oil of the leaves of the Japanese "hinoki" tree, *Chamaecyparis obtusa*. A direct comparison of the two acids proved that they were identical (84). The name hinokiic acid has priority and has been retained.

Thujopsene and hinokiic acid are common Cupressales constituents and very often occur together (Table 9, p. 266). Several groups of investigators have studied these two compounds over a number of years and some incorrect structures were proposed. As a result of extensive chemical and spectroscopic experimentation the structures and configurations of thujopsene (144) and hinokiic acid (146) were conclusively established by Norin (95, 193, 195, see the review 197). The structure of thujopsene has been confirmed by two independent total syntheses of the optically inactive compound [Dauben and Ashcroft (48), Büchi et al. (33)].

Chetty and Dev (37) have investigated the wood oil of "mayur pankhi" *(Platycladus orientalis)* from which they isolated a number of sesquiterpene ketones. One of these, mayurone, was recently reported to be a C_{14}-ketone of the thujopsane type (147) (38, 134). It is also a wood constituent of *Thujopsis dolabrata* (134).

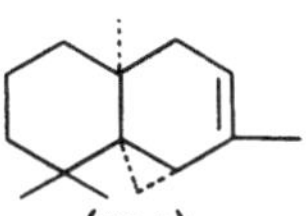

(144)
Thujopsene.

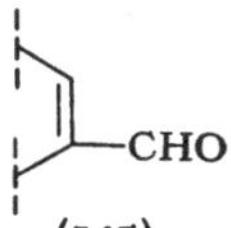

(145)

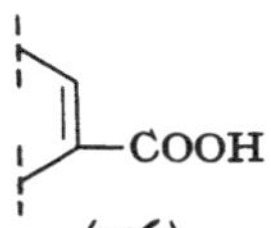

(146)
Hinokiic acid.

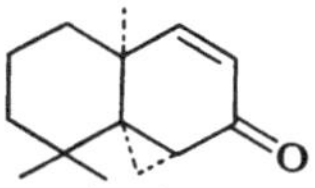

(147)
Mayurone.

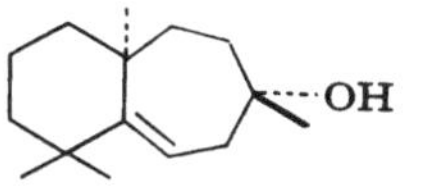

(148)
Widdrol.

Widdrol. Widdrol was first obtained from various *Widdringtonia* species by ERDTMAN and THOMAS (89). It has since been found to be a common Cupressales constituent (Table 9, p. 266), frequently co-occurring with thujopsene and hinokiic acid. ENZELL's extensive chemical and spectroscopic studies (65) have established the structure and configuration of widdrol (148), which was fully confirmed by synthesis (61). His assignment of the α-configuration to the hydroxyl group in widdrol (148) was tentative. Independently, ITO et al. (135) proposed the same basic structure for widdrol, but they favoured the β-configuration of the hydroxyl group. Later these workers reinterpreted their data and produced additional evidence for the α-configuration (136). Simultaneously with ENZELL's work NAGAHAMA (189) made the important observation that thujopsene furnishes widdrol on hydration with aqueous oxalic acid. The mechanism of this conversion has recently been studied by DAUBEN and FRIEDRICH (49). They used deuterium-labelled thujopsene (149) to show that the reaction proceeds via a "cyclopropyl carbonyl cation" as indicated in (149) etc.

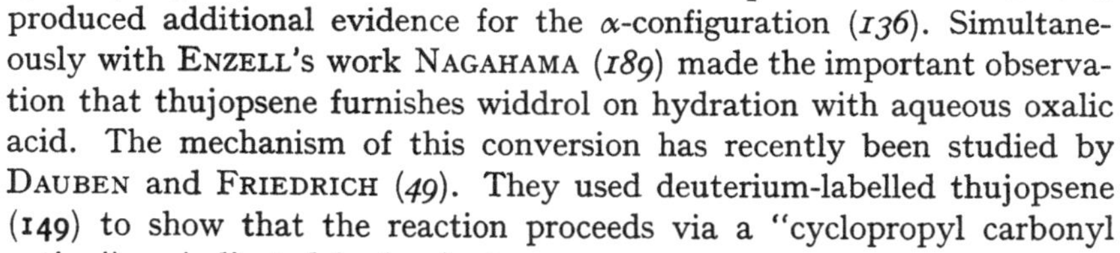

(149)

Widdrol-α-epoxide (150) was prepared in connection with the structural elucidation of widdrol (65). It is identical with an alcohol $C_{15}H_{26}O_2$ (m. p. 154°), first isolated from the heartwood of *Widdringtonia juniperoides* (89) and then from *Juniperus utahensis* (228). Since widdrol-α-epoxide is formed as a by-product on ozonisation of widdrol, there is a possibility that it is an artifact produced by the action of air. However, at least in *W. juniperoides*, widdrol-α-epoxide appears to be a true heartwood constituent, since it occurs in comparatively large amounts in the fresh wood.

Cuparanes. A high-boiling dextrarotatory sesquiterpene fraction called "widdrene II" was also isolated from the heartwood of various *Widdringtonia* species (*89*), and fractions with similar properties were obtained from *Platycladus (Biota) orientalis* heartwood (*86*).

ENZELL and ERDTMAN (*67*) found that a similar fraction from *Chamaecyparis thyoides* contained about 60% of an aromatic compound which they named cuparene. It was isolated in a pure state by a mild oxidative degradation of accompanying non-aromatic unsaturated constituents. By the same procedure cuparene was obtained from the "widdrene II" fractions. On oxidation with dilute nitric acid cuparene (**151**, $R = CH_3$) yielded terephthalic acid (**152**) whereas treatment with ozone at room temperature gave (+)-camphonanic acid (**153**) of known configuration. These experiments together with UV- and IR-data proved the cuparene structure and the absolute configuration. Cuparene has since been synthesized by NOZOE and TAKESHITA (*205*) as well as by PARKER, RAMAGE and RAPHAEL (*209*).

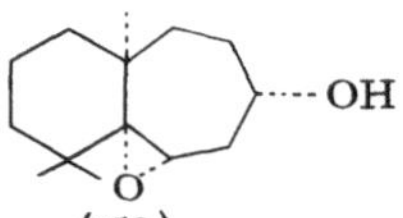

(**150**)
Widdrol-α-epoxide.

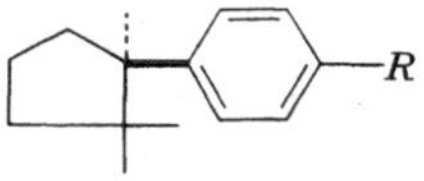

(**151**)
Cuparene, $R = CH_3$.
Cuparenic acid, $R = COOH$.
γ-Cuparenol, $R = CH_2OH$.

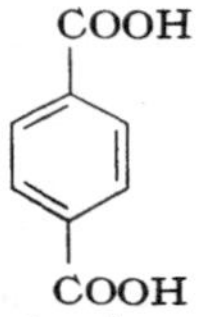

(**152**)

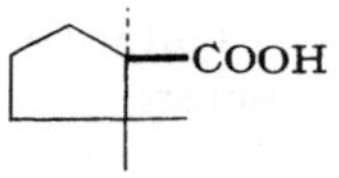

(**153**)
(+)-Camphonanic acid.

Chromic acid oxidation of cuparene (**151**, $R = CH_3$) yielded cuparenic acid (**151**, $R = COOH$) which occurs together with cuparene in the heartwood of *Ch. thyoides*. This acid was subsequently found to be identical with a *p*-substituted benzoic acid, "acid III", isolated from *Widdringtonia* species (*89*). Two other acids, "II" and "IV", were isolated from the same species. A recent chromatographic study of "acid II" revealed that it is a constant melting mixture of at least three components, two of which were identified as hinokiic acid and cuparenic acid (NORIN, unpublished). Similarly, "acid IV" has been shown to be a constant melting mixture of about equal amounts of hinokiic and cuparenic acids (*193*).

From a biogenetic point of view cuparene may be derived from a dihydrocuparene (*67*). NOZOE and TAKESHITA (*205*) have concluded from chemical and spectroscopic data that two compounds of the dihydrocuparene type, α- and β-cuprenene (**154** and **155**), are present in "hiba"-wood oil *(Thujopsis dolabrata)*; however, the two compounds have not been isolated in pure form. DAUBEN and OBERHÄNSLI (*50*) have recently reinvestigated the "hiba" oil and were unable to demonstrate the presence

References, pp. 275—287.

of β-cuprenene (155); among other compounds they isolated α-cuprenene and a new cuprenene to which the structure (156) was assigned. Reduction of cuparene with lithium in liquid ammonia under forcing conditions gave (156) as the only product. Isomerization of this non-conjugated hexadiene yielded a mixture in which α-cuprenene (154) was predominant.

Upon prolonged standing in air (50) or treatment with ozone (205) the cuprenenes are converted to cuparene. Therefore, cuparene isolated from the various Cupressales species listed in Table 9 (p. 266) may more or less be derived from cuprenenes. This is particularly true when pure cuparene is isolated by treatment of a crude hydrocarbon fraction with ozone in order to remove unsaturated impurities.

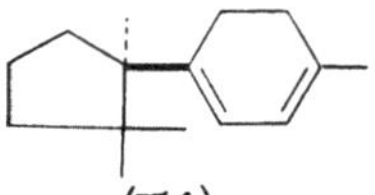

(154)
α-Cuparenene.

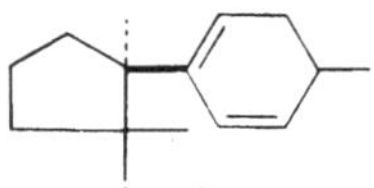

(155)
β-Cuparenene.

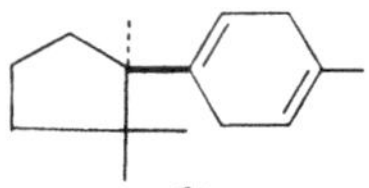

(156)

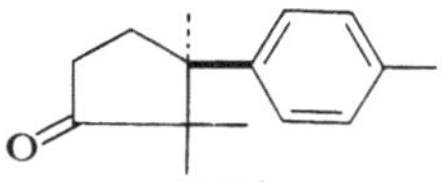

(157)
α-Cuparenone.

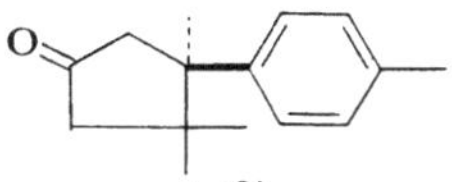

(158)
β-Cuparenone.

Two cuparene type ketones were isolated by CHETTY and DEV (37) from the wood oil of "mayur pankhi" (Platycladus orientalis). The spectral properties of one of them, α-cuparenone, were identical with those of a synthetic compound (157) previously described by PARKER, RAMAGE and RAPHAEL (209). The second ketone, β-cuparenone (158), gave cuparene on Wolff-Kishner reduction. IR- and NMR-data showed the position of the carbonyl group. A mixture of the alcohols corresponding to the cuparenones was also isolated from the same source.

The properties and leading references to recent literature on the chemistry of sesquiterpenes from Cupressales are listed in Tables 10, 11 and 12 (pp. 268, 269, 274).

c. Diterpenes.

The early investigations on Cupressales terpenes referred mainly to volatile mono- and sesquiterpenes. However, new separation techniques and methods have made rapid development possible also in the diterpene field. Thus, within the last few years several Cupressales diterpenes with novel ring systems have been reported, such as dolabradiene, hibaene and verticillol.

Labdanes. The occurrence of manool (**159**, $R = CH_3$) in *Cupressus sempervirens* was unexpected since this bicyclic diterpene had previously been found only in some *Dacrydium* species of the order Podocarpales (*66*). Manool has since been detected in the heartwood of *C. torulosa* (*17*) where it occurs together with torulosol (**159**, $R = CH_2OH$) and torulosal (**159**, $R = CHO$), two other compounds of the manool type elucidated by ENZELL (*64*). Torulosyl monoacetate (**159**, $R = CH_2OCOCH_3$) was reported by GOUGH (*104*) to occur in sandarac resin *(Tetraclinis articulata)*. This resin also contains "torulosolic acid" (*104*); no details of its properties or structure are available. Cupressic acid (**159**, $R = COOH$), isolated by MANGONI and BELARDINI (*174*) from the resin of *C. sempervirens*, might be identical with GOUGH's "torulosolic acid". Isocupressic acid (**160**) is another constituent of the *C. sempervirens* resin (*174*).

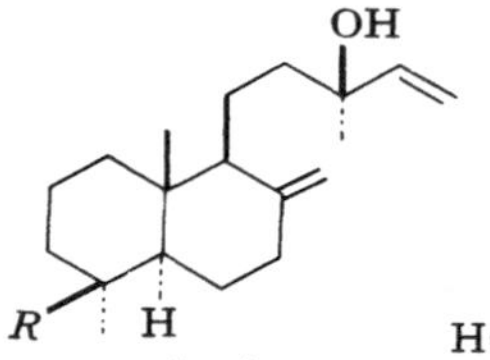

(**159**)
Manool, $R = CH_3$.
Torulosol, $R = CH_2OH$.
Torulosal, $R = CHO$.
Cupressic acid, $R = COOH$.

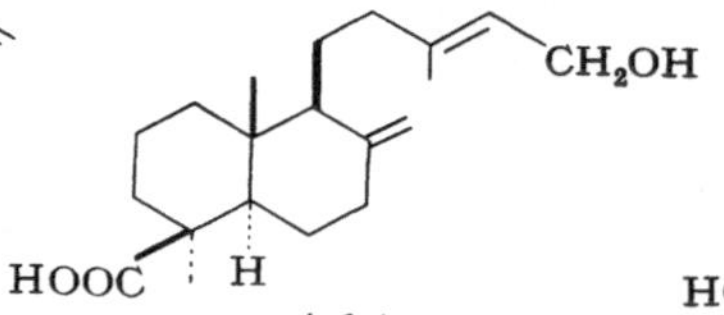

(**160**)
Isocupressic acid.

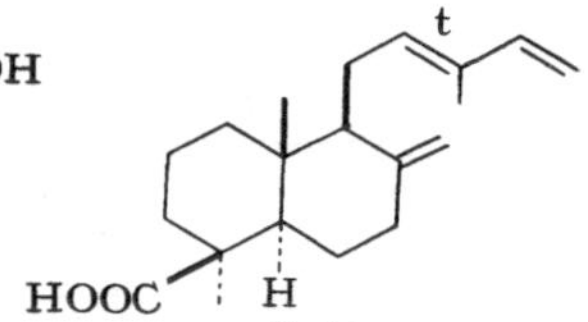

(**161**)
Communic acid
(Elliotinoic acid).

Communic acid was first isolated from the bark of common juniper, *J. communis* (*11*). It has since been found in the bark of other *Juniperus* species (*77, 137*) and also in the bark of *Cupressus arizonica* (*77*). Its occurrence in sandarac resin from *Tetraclinis articulata* (*174*), in the leaves of *J. horizontalis* (*192*) as well as in the cones of *C. sempervirens* (*2*) has also been reported. Communic acid is identical with elliotinoic acid obtained from the wood resin of *Pinus elliottii* (*137*). Except for the configuration of the trisubstituted double bond in the side-chain, the full structure of communic acid (**161**) was established by ARYA, KUBOTA and ERDTMAN (*11*). Recently, NORIN (*198*) has presented evidence for the *trans*-configuration of the side-chain double bond as shown in (**161**). Communic acid is very sensitive to acids and polymerizes on keeping. A fraction resembling that of "polycommunic acid" was obtained from the resin of *Callitris* species (*104*).

The Japanese umbrella pine *Sciadopitys verticillata* contains a number of interesting diterpenes. From its wood a new bitter principle, sciadin, was isolated and extensively studied by SUMIMOTO (*250*) who proposed structure (**162**) for this compound. He also describes the isolation of a new diterpene, methyl sciadopate, from the same source. Spectral studies and oxidative degradation to the C_{16}-keto-acid (163) have established the structure (164) (252). The keto-acid (163) had previously been obtained from agathic acid and its enantiomer from daniellic acid. Methyl sciadopate has also been related to communic acid (161) (179). Sciadinone and dimethyl sciadinonate are two other diterpenes isolated from *S. verticillata*; structures (165) and (166), respectively, have been proposed (143).

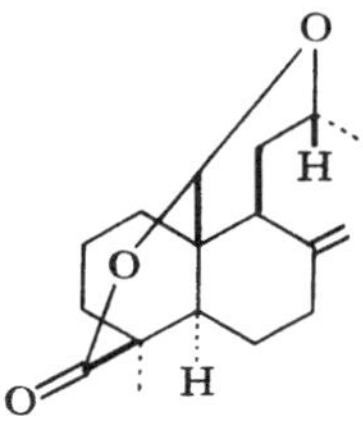

(162)
Sciadin.

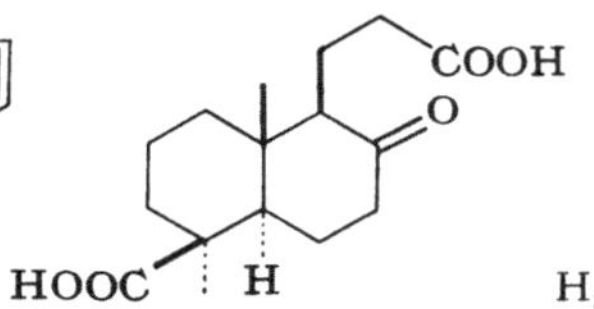

(163)

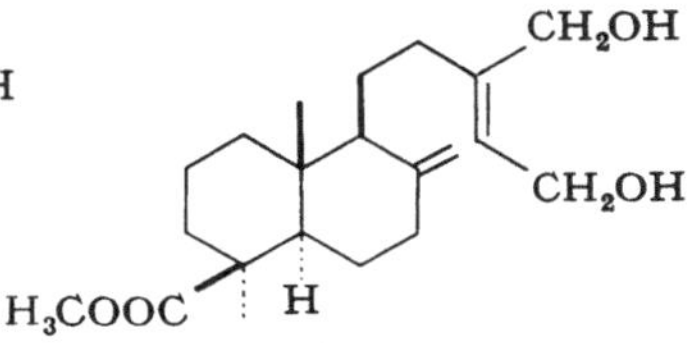

(164)
Methyl sciadopate.

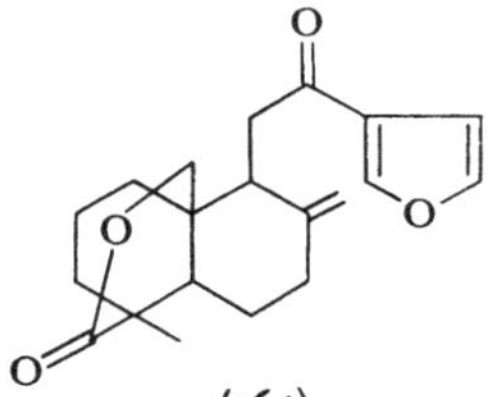

(165)
Sciadinone.

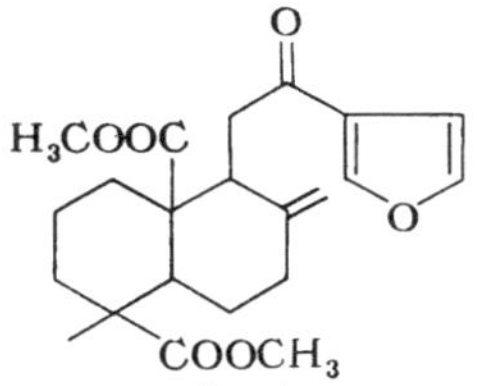

(166)
Dimethyl sciadinonate.

Pimaranes. Several pimarane type diterpenes have been found in Cupressales. The most common constituents of this type are isopimaric acid (167) and sandaracopimaric acid (168, $R^1 = COOH$, $R^2 = H$). Although the chemistry of these acids has been studied for a long time, their structures were elucidated only recently, mainly by the synthetic work of IRELAND and co-workers (43, 132a, 133).

The fruits of various *Juniperus* species (*J. japonica, J. rigida* and *J. conferta*) were investigated by UKITA et al. (257). They isolated two resin acids one of which is identical with isopimaric acid (167; "iso-dextropimaric acid"). Structure (169) was tentatively proposed for the second acid (m. p. 162.5–165°; $[\alpha]_D — 8.9°$); however, its homogeneity seems to be doubtful.

Cryptopimaric acid, from the heartwood resin of *Cryptomeria japonica*, is identical with sandaracopimaric acid and since the latter name has priority it has been retained (9, 10). Sandaracopimaric acid was also found in the resin of *Callitris* species (104). The corresponding 12β-acetoxy compound (**168**, R^1 = COOH, R^2 = OCOCH$_3$) was isolated from sandarac resin (8, *104*) and from *Callitris* species (*104*). Recently NAGAHAMA (*190*) found sandaracopimarinol (**168**, R^1 = CH$_2$OH, R^2 = H) in the wood oil of *Cr. japonica*.

Rimuene. The diterpene hydrocarbon rimuene has been known for a long time but only recently was its structure (**170**) established (*44, 46*) and confirmed by synthesis (*131*). This diterpene of rearranged pimarene type has been detected in the leaves of one Cupressales species *(Libocedrus plumosa)* by means of gas chromatography (*7*). Another diterpene of a similar rearranged type is dolabradiene, isolated by KITAHARA and YOSHIKOSHI (*152*) from *Thujopsis dolabrata* leaves. Chemical and spectral studies established its structure (**171**). ($\pm$)-Dolabradiene was synthesized by the same group of investigators (*154*).

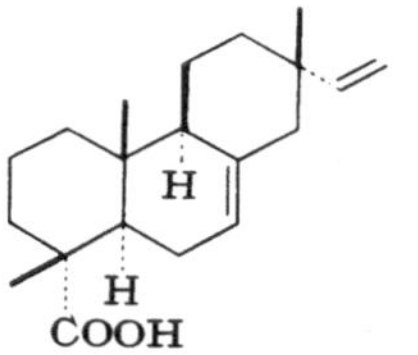

(**167**)
Isopimaric acid.

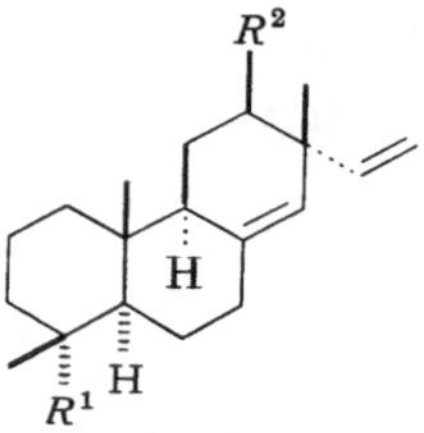

Sandaracopimaric acid,
R^1 = COOH, R^2 = H.
(**168**)

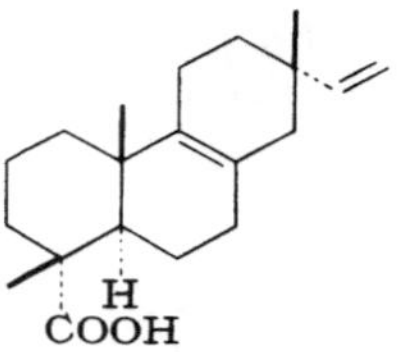

(**169**)

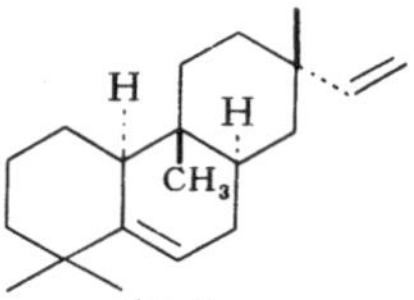

(**170**)
Rimuene.

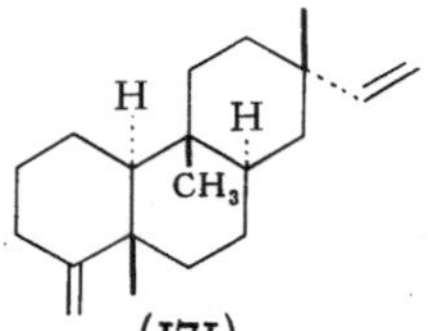

(**171**)
Dolabradiene.

Ferruginol and Related Terpenes. Ferruginol type diterpenes are common Cupressales constituents (cf. *Table 13*, p. 270) of which ferruginol (**172**) and sugiol (**173**, 7-ketoferruginol) have been known for a long time. 6,7-Dehydroferruginol (**174**) is a more recent Cupressales compound which was isolated from the wood of *Juniperus communis* (*25*). The exuded resin from *Cryptomeria japonica* contains a phenolic diterpene, crypto-japonol C$_{21}$H$_{30}$O$_3$, which appears to be of the ferruginol type. Structure (**175**) was proposed for this new phenol (*157*).

A yellow phenolic diterpene, xanthoperol C$_{20}$H$_{26}$O$_3$, was obtained from the wood of *J. communis* (*28*). Its structure (**176**) has been established by BREDENBERG (*26*). According to NMR data xanthoperol has a 5β-

hydrogen indicating that the stable configuration has *cis*-fused rings. A careful reinvestigation of the wood extract under mild conditions revealed the presence of the *trans*-compound (**177**) which easily isomerised to xanthoperol, and therefore is believed to be the genuine natural product and the precursor of xanthoperol.

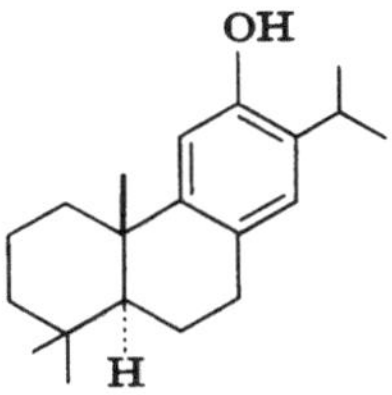

(**172**)
Ferruginol.

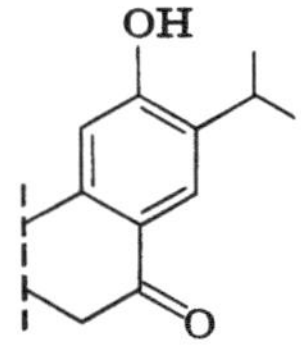

(**173**)
Sugiol.

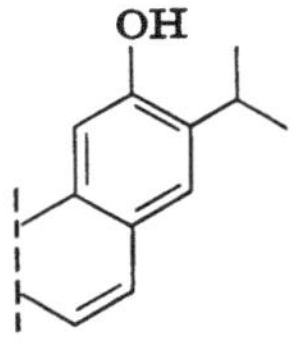

(**174**)
6,7-Dehydroferruginol.

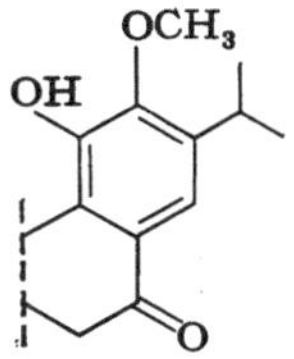

(**175**)
Cryptojaponol (?).

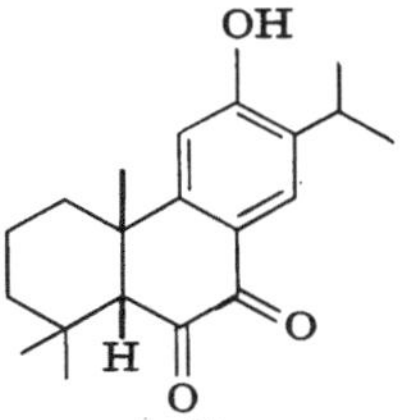

(**176**)
Xanthoperol.

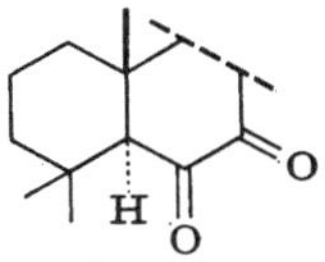

(**177**)

The phenolic diterpene alcohol hinokiol and the corresponding ketone hinokione were first isolated from "hinoki" wood *(Chamaecyparis obtusa)* (*147*). Later they were found in the heartwood of *Tetraclinis articulata*

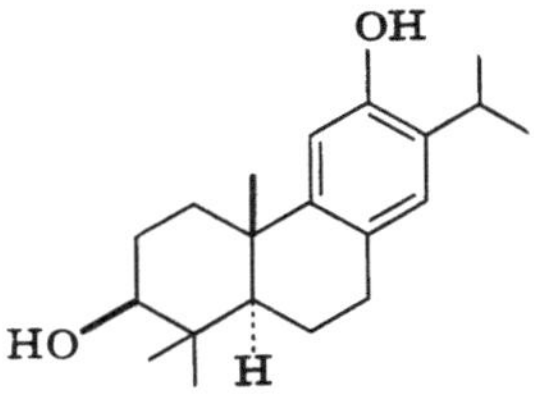

(**178**)
Hinokiol.

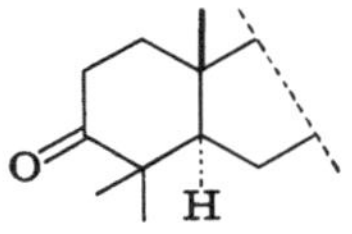

(**179**)
Hinokione.

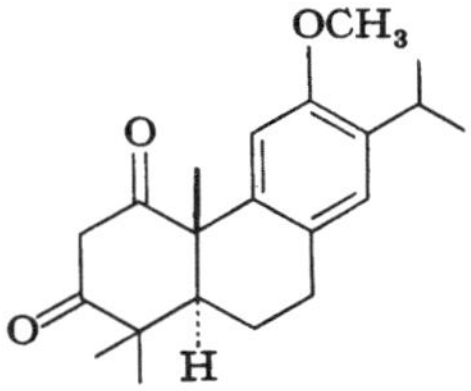

(**180**)

(*40*) and *Cupressus torulosa* (*17*). Hinokiol is also present in *Athrotaxis cupressoides, A. selaginoides* (*91*) and *Taiwania cryptomerioides* (*163*). Previous structural investigations of the two compounds had shown that they were ferruginol derivatives and the additional oxygen function was believed to be in the 2-position. However, according to Chow and Erdtman (*41*) hinokione is 3β-hydroxyferruginol (**178**) and hinokione is the corresponding 3-oxo derivative (**179**). These structures were conclusively established by Chow (*39*) who synthesized (±)-hinokione methyl ether.

Recently, Mangoni and Belardini (*173*) have isolated two diterpene 1,3-diones from the wood resin of *Cupressus sempervirens*. Chemical and spectral studies indicated that one compound ("B") was of the ferruginol type possessing structure (**180**); the second ("A") appeared to be of the totarol type (cf. **181**) and structure (**182**) was proposed for it.*

Totarol and Related Terpenes. There are several other Cupressales constituents of the totarol type. The parent compound, totarol (**181**), has been isolated from the bark of *Juniperus communis* (*77*; Arya and Erdtman, unpublished) as well as from several other Cupressales species (Table 13, p. 270). In the heartwood of *Tetraclinis articulata* (*40*) it co-occurs with totarolone (**183**) whose structure was determined by Chow and Erdtman (*42*). Totarolenone (**184**), another constituent of the above species, has only been isolated in admixture with totarolone (*40*). The leaves of *Thujopsis dolabrata* contain totarol and a mixture of 7α- and 7β-hydroxytotarol (**185**) (*125*). The latter two compounds have never been separated from each other but chromic acid oxidation of the mixture gave pure 7-oxo-totarol (**186**).

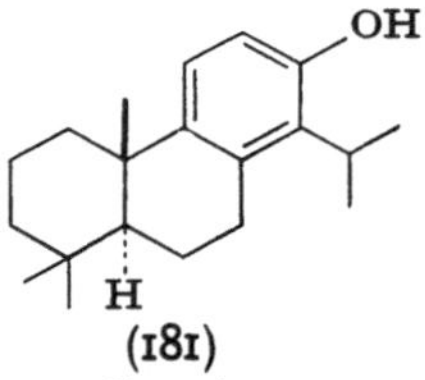

(**181**)
Totarol.

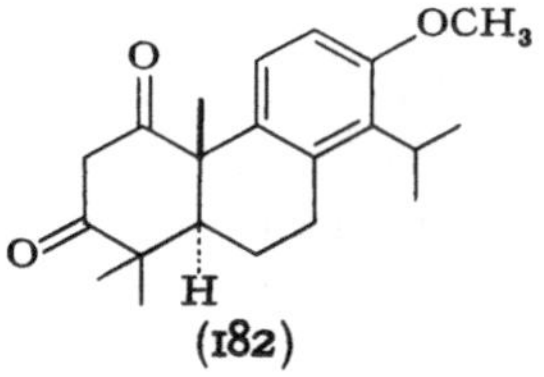

(**182**)

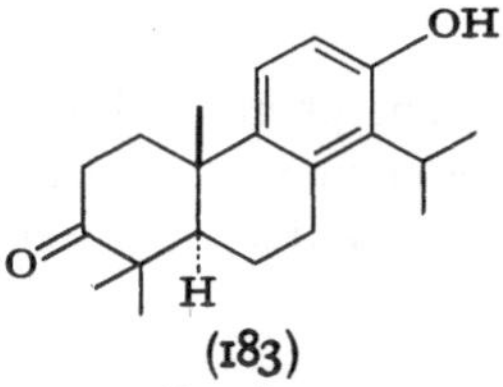

(**183**)
Totarolone.

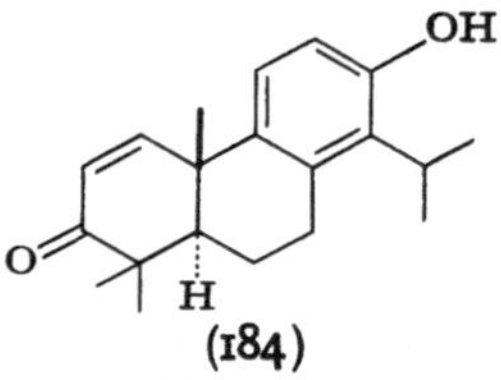

(**184**)
Totarolenone.

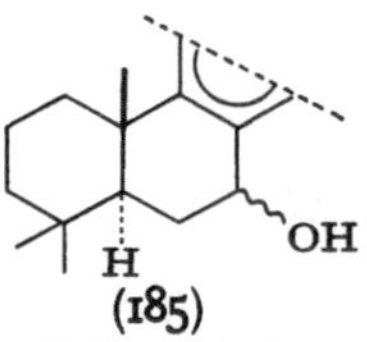

(**185**)
7-Hydroxytotarol.

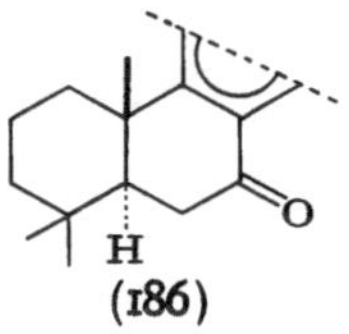

(**186**)

---

* Lin and Lin (*162 a*) have recently isolated a new diterpene phenol, shonanol ($C_{20}H_{26}O_2$; m. p. 187–188°; $[\alpha]_D$ + 3.01° in ethanol; λ_{max} 224 and 283 mμ, log ε 4.23 and 3.52), from the wood of *Calocedrus formosana* (syn. *Libocedrus formosana* Florin). Shonanol is proposed to be the ferruginol derivative corresponding to totarolenone (**184**). (Added in Proof.)

References, pp. 275—287.

All tetracyclic diterpenes isolated from Cupressales species are hydrocarbons, except the tertiary alcohol phyllocladanol, isolated from the wood resin of *Cryptomeria japonica* (*156*) where it co-occurs with (+)-isophyllocladene (**188**). On dehydration phyllocladanol yielded (+)-isophyllocladene (**188**), and structure (**189**) was therefore proposed for the alcohol. Isophyllocladene has recently been detected by gas chromatography in *Libocedrus bidwillii* and *L. plumosa* leaves (*7*).

Phyllocladene and Related Tetracyclic Diterpenes. The essential oil from the leaves and terminal branchlets of *Cupressus macrocarpa* was investigated by BRIGGS and SUTHERLAND (*32*) who isolated (+)-isophyllocladene and a diterpene hydrocarbon, $C_{20}H_{32}$ (m. p. 74–75°), which they named cupressene. It was shown later to be a mixture of isophyllocladene (32%), phyllocladene (trace) and a main component (67%) to which the name cupressene was transferred (*7*). Chemical and spectroscopic studies of an enriched cupressene fraction (85% purity) were carried out recently by BRIGGS and co-workers (*31*) and structure (**190**) was proposed for the pure hydrocarbon. During BRIGG's cupressene work KITAHARA and YOSHIKOSHI (*153*) reported the isolation from *Thujopsis dolabrata* leaves and structure (**190**) of a diterpene hydrocarbon, $C_{20}H_{32}$ (m. p. 29.5–30°), named hibaene. Hibaene and cupressene should therefore be identical; and this was supported by the observation that the IR and NMR spectra of hibaene and of the enriched cupressene sample were practically coincident. Also the gas chromatographic properties of hibaene and cupressene were identical. Recently, (±)-hibaene has been synthesised (*132*).

The literature concerning the diterpene hydrocarbon content of *Sciadopitys verticillata* is confusing. However, a pertinent critical examination has been conducted by APLIN, CAMBIE and RUTLEDGE (*7*) who conclude that (—)-kaurene (**191**) and (—)-isophyllocladene (enantiomer of **188**) are true constituents of the leaves of this species. Their presence was also confirmed by gas chromatography. It is interesting to note that these tetracyclic diterpenes have an A,B-ring system of the type antipodal to that of the bicyclic wood constituents sciadin (**162**) and methyl sciadopate (**164**, p. 237).

α-Cryptomerene is a diterpene hydrocarbon that was isolated from *Cryptomeria japonica* leaves (*256*). It was later shown to be identical with (—)-isokaurene (**192**) (*30*) and is thus of the type antipodal to that of the structurally related (+)-isophyllocladene (**188**) which is a wood constituent of the same species.

Kaurene and isokaurene of undefined abs. configurations have been detected by gas chromatography in the leaves of *Libocedrus bidwillii*, *L. plumosa* and in *Cupressus macrocarpa* (*7*).

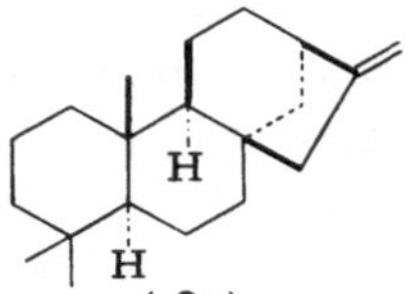

(187)
(+)-Phyllocladene.

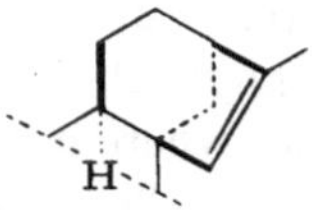

(188)
(+)-Isophyllocladene.

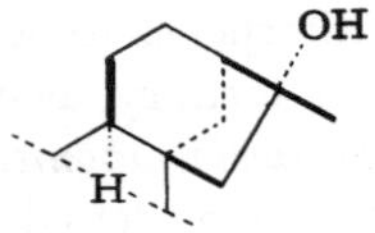

(189)
(+)-Phyllocladanol.

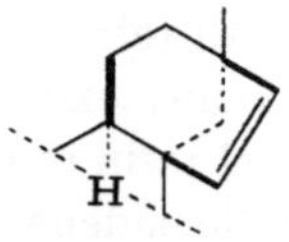

(190)
(—)-Hibaene
(Cupressene).

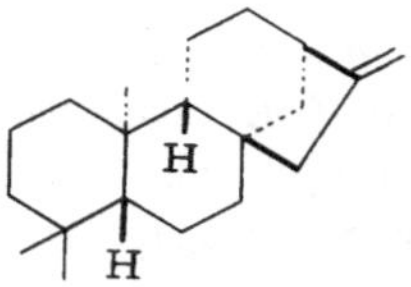

(191)
(—)-Kaurene.

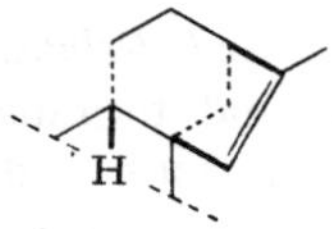

(192)
(—)-Isokaurene.

Verticillol. In the course of our work on the chemistry of the order Cupressales some diterpenes of still unknown structures have been isolated. One of them, verticillol, from the wood of *Sciadopitys verti-*

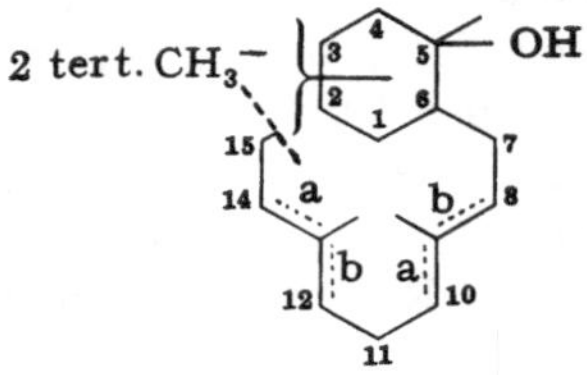

(193)

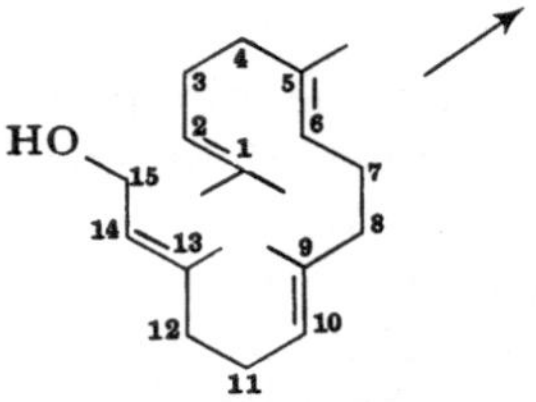

Geranylgeraniol.

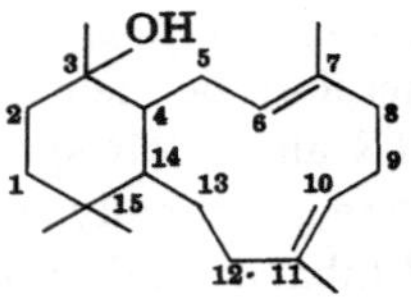

(197)

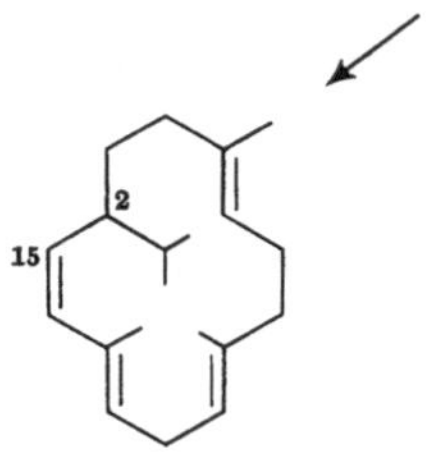

(194)
Thunbergene.

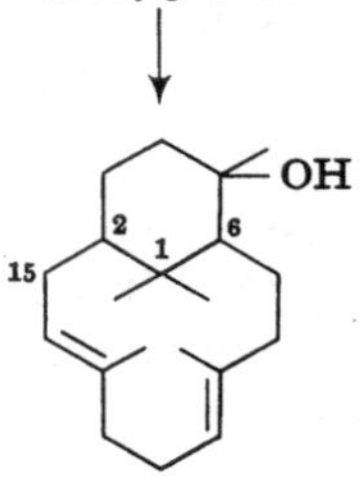

(195)

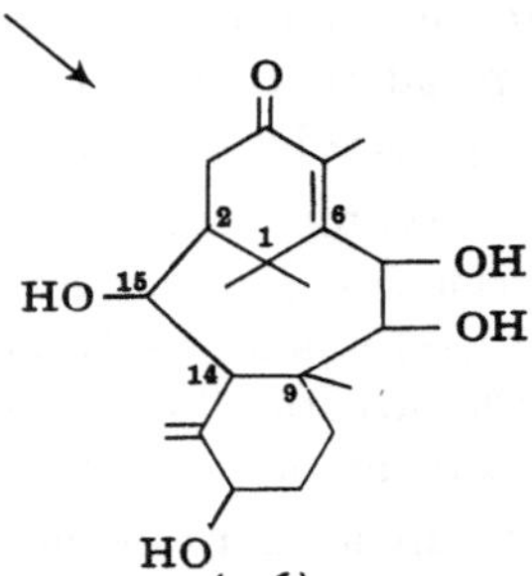

(196)
Taxicin II.

References, pp. 275—287.

cillata, has been investigated in some detail (*85*). It is a bicyclic tertiary alcohol, $C_{20}H_{32}O$, with two trisubstituted double bonds. Preliminary spectral and chemical studies led to the assignment of the partial structure (193) for this new diterpene. Based on similar evidence, this partial structure was also proposed by KANEKO et al. (*142*). The two double bonds can be either in position "a" or "b". It is possible that verticillol is related to the novel macrocyclic diterpenes of the thunbergane (cembrane, duvane) type, e. g. thunbergene (194) which is a gymnosperm constituent. Consequently, structure (195) was tentatively proposed for verticillol which would also relate it to the terpenes of the taxane type, such as taxicin II (196) (*85*). KANEKO and co-workers (*142*) favour structure (197). Further studies on this interesting diterpene are in progress.

d. Triterpenes.

A diol (m. p. 243.5–244°, $[\alpha]_D + 14°$; acetate: m. p. 127–128°), for which the analysis indicated the composition $C_{30}H_{34}O_2$ was isolated from the heartwoods of *Juniperus thurifera* (*230*) and *J. procera* (*211*). It also appears in the heartwood of *J. californica* (*211*). This diol might be the first triterpene obtained from a Cupressales species.

e. Biogenetic Aspects.

During the last few years the important "isoprene rule" formulated by RUZICKA (cf. *237*) has received experimental foundation, and the basic unit in the biosynthesis of some representative terpenes and steroids was shown to be isopentenyl pyrophosphate. However, very few experimental studies have been conducted on the biosynthesis of conifer terpenes. There has been much speculation in this field and many important problems have still to be solved, such as the cyclisation mechanisms which are considered to be of cationic types. However, there are some recent indications of more complex mechanisms. Thus, in a preliminary investigation of the biosynthesis of thujone (198) in *Thuja occidentalis*, which is the sole experimental study on a Cupressales terpene, 2-C^{14}-mevalonate was shown to be incorporated in the manner indicated below (*238*). This labelling pattern is not in agreement with a simple cationic cyclization as indicated below (— OPP indicates a pyrophosphate unit)*:

---

* In a recent communication BANTHORPE and TURNBULL (*16a*) have presented experimental results on the biosynthesis of thujone and related monoterpenes in *T. occidentalis*, *T. plicata* and *Juniperus sabina*. These results are in agreement with a cationic cyclisation and thus contrary to those reported by SANDERMANN and SCHWEERS (*238*). (Added in Proof.)

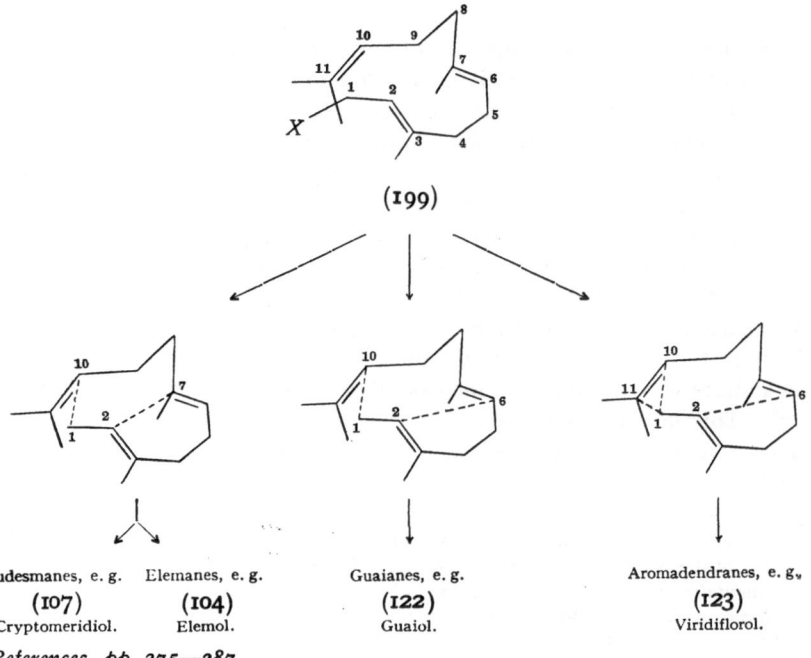

(198)
Thujone.

Some hypothetical considerations and experimental studies on the biosynthesis of terpenes have been discussed extensively by Richards and Hendrickson (*218*). Therefore, we shall only briefly expose some aspects of the biogenesis of Cupressales sesquiterpenes. The original farnesol hypothesis in sesquiterpene biogenesis as developed by Ruzicka (cf. *237*) was extended by Hendrickson (*115*) who discussed configurational and conformational implications in the cyclisation of the 2,3-*cis*- and 2,3-*trans*-farnesyl cation. Some sesquiterpenes are derived from a *cis*-farnesyl precursor and others from a *trans*-precursor. The terpenes of eudesmane, elemane, guaiane and aromadendrane types are of the *trans*-farnesol origin and they may all be derived from the common properly oriented precursor (**199**) as indicated below:

(**199**)

Eudesmanes, e. g.	Elemanes, e. g.	Guaianes, e. g.	Aromadendranes, e. g.
(**107**)	(**104**)	(**122**)	(**123**)
Cryptomeridiol.	Elemol.	Guaiol.	Viridiflorol.

References, pp. 275—287.

Occidentalol (108), which occurs together with the eudesmols in *Thuja occidentalis*, possesses a *cis*-ring juncture (223). VON RUDLOFF and NAIR (224) have suggested that these compounds are derived from the same precursor (199 → 108).

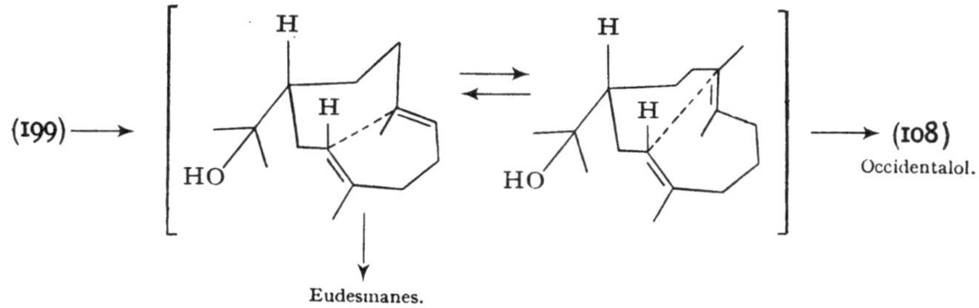

Eudesmanes.

The eremophilane type of sesquiterpenes is considered to be derived from an intermediate of the eudesmane type by methyl group migration (cf. 200–201). The angular methyl group of nootkatene and nootkatone would therefore be expected to be α-oriented like the isopropenyl group in formula (201). It is therefore surprising that the proposed configurations for these two compounds (121 and 117, p. 228) are not in agreement with those derived from the biogenetic considerations mentioned.

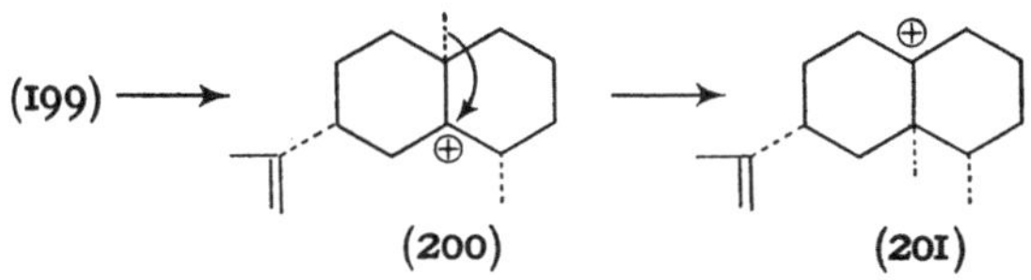

Humulene (101) and caryophyllene (102) may be derived from a common, properly oriented *cis*-farnesyl precursor (202) (cf. *115*). Juniperol (138), longifolene (136) and the cadinanes may also be derived from the same precursor as outlined on p. 246.

The thujopsane, widdrane, cuparane and cedrane groups of sesquiterpenes are all common heartwood constituents which often co-occur. Biogenetically these terpenes may be mutually related and derived from a common properly oriented precursor (203) of the *cis*-farnesyl type (*62, 197*) as indicated in the formulae (next page).

(—)-β-Bisabolene (97, p. 226) which has only been found in *Chamaecyparis lawsoniana* so far, may also be derived from the same hypothetical precursor (203). Similarly, juniperol (138) and the cadinanes may be derived

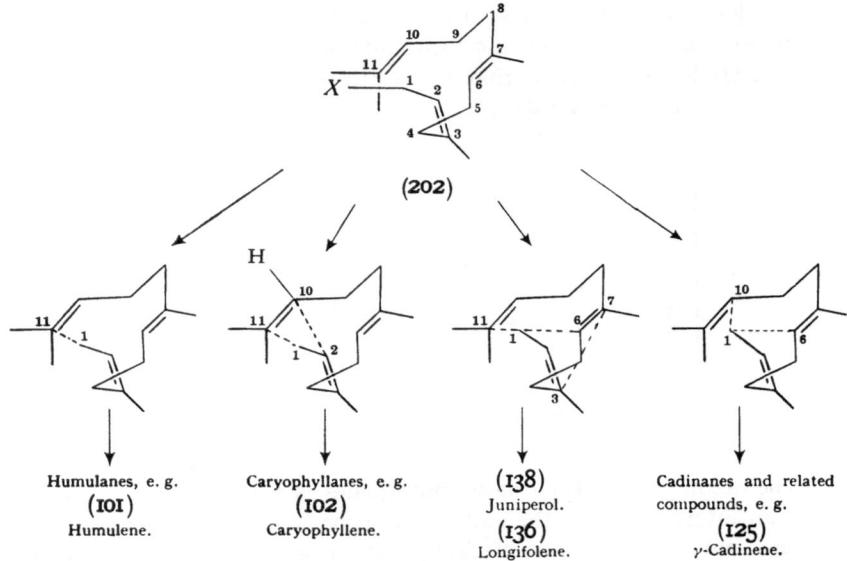

(202)

Humulanes, e. g.
(**101**)
Humulene.

Caryophyllanes, e. g.
(**102**)
Caryophyllene.

(**138**)
Juniperol.
(**136**)
Longifolene.

Cadinanes and related
compounds, e. g.
(**125**)
γ-Cadinene.

from this precursor thus, providing an additional possible biogenetic route for these terpenes (p. 245). In connection with the biogenesis of juniperol and longifolene it should be noted, however, that some recent studies on the biosynthesis of the structurally related fungal metabolite

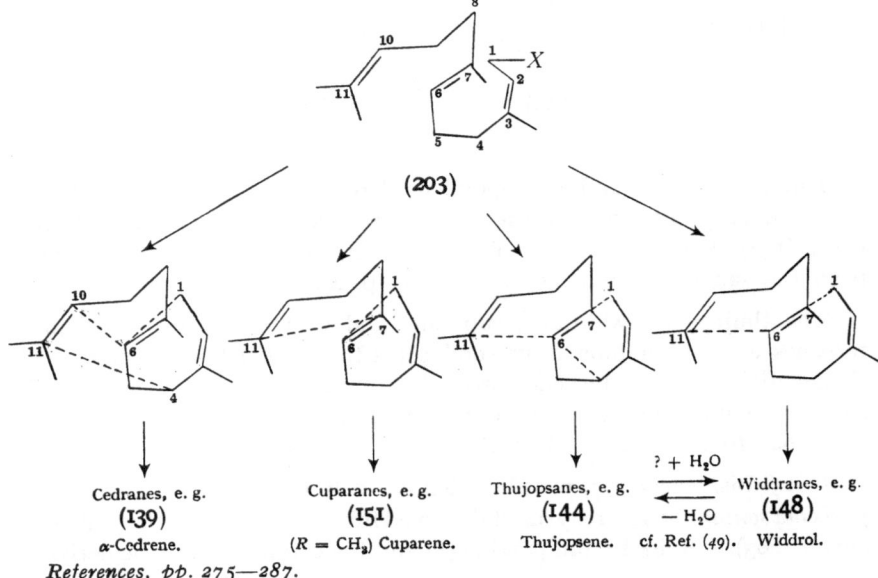

(203)

Cedranes, e. g.
(**139**)
α-Cedrene.

Cuparanes, e. g.
(**151**)
(R = CH₃) Cuparene.

Thujopsanes, e. g.
(**144**)
Thujopsene.

? + H₂O
─────────→
── H₂O
cf. Ref. (49).

Widdranes, e. g.
(**148**)
Widdrol.

helminthosporal (**204**) have shown that when $2\text{-}C^{14}$-mevalonate was incorporated, one third of the labelling was located in the α,β-unsaturated aldehyde group (**51**). This supports an initial cyclisation involving the $C_{(1)}$-atoms via the precursor (**202**):

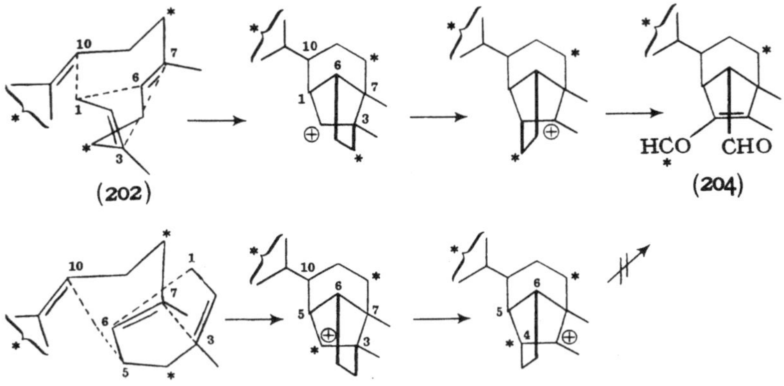

(202)   (204)

(203)

IV. Some Chemotaxonomic Aspects.

Before discussing the chemistry of the genera some remarks should be made about the classification of the Cupressales. The family Taxodiaceae (p. 208) is very complex; it embraces the scattered remnants of old groups of only distantly related plants. *Sciadopitys* differs in so many respects from the other Taxodiaceae that it is frequently separated into a family of its own. The family Cupressaceae (p. 208) is often divided into the subfamilies Cupressoideae, Juniperoideae and Thujoideae, the third subfamily being very heterogeneous. Some botanists prefer to divide Cupressaceae into two subfamilies, Cupressoideae and Callitroideae. The essentially northern Cupressoideae is further divided into three tribes: Cupresseae, Junipereae and Thujopsideae (Thujoideae, group "a", p. 208). Callitroideae is essentially southern but includes the northern monotypic *Tetraclinis*.

Phenolic Compounds. Many simple phenolic compounds have been found in Cupressales and they are mostly related to the terpenes. The new group of C_{17}-phenols, such as sugiresinol (**18**) and hinokiresinol (**19**), is perhaps more interesting from a taxonomic point of view. So far they have been found with certainty only in Cupressales, but at least one so far ill-defined phenol, agatharesinol, from *Agathis australis*, Araucariales, appears to belong to this group (**70**). Agatharesinol has been described as a C_{18}-phenol but, according to an unpublished mass spectrometric investigation by ENZELL, it has only seventeen carbon atoms. Sugiresinol and related phenols were isolated from several genera of the Taxodiaceae (*Sequoia, Cryptomeria* and *Athrotaxis*) as well as from one *Chamaecyparis* species. These or similar compounds will possibly be found to be more wide-spread in Cupressales, since they are easily overlooked and are sometimes difficult to isolate in pure state. A phenol (m. p. 217°) isolated from *Pilgerodendron* (*88*) might also belong to this group.

The peculiar occurrence of podophyllotoxin in the leaves of many genera of Cupressales is interesting and should be further investigated.

The biflavones are not restricted to Cupressales. They occur in non-coniferous plants such as *Ginkgo, Taxus, Torreya, Cephalotaxus*, seed ferns *(Cycas)*, a cryptogam *(Selaginella)* and even in angiosperms *(Casuarina, Viburnum)*. It is interesting to note, however, that biflavonyls do occur in all orders of Coniferae with the notable exception of Pinales, of which over forty species from *Abies, Keteleeria, Pseudotsuga, Tsuga, Picea, Larix, Pseudolarix* and *Cedrus* have been investigated, all with negative results. In Cupressales C—C-coupled biflavonyls are common and frequently co-occur with hinokiflavone, which is the dominant biflavonyl in this order.

Tropolones. Among the conifers tropolones have been found in Cupressales and they are all of the modified terpene type. In no case have tropolones been encountered in the family Taxodiaceae. They are widespread in Cupressaceae but for some unknown reason they seem not to be present in certain species of otherwise tropoloniferous genera, such as *Chamaecyparis* and *Juniperus*. It is, of course, possible that the aberrant species contain tropolones in such small amounts that they escape observation, but it seems to be more probable that they are absent as a result of some genetic changes. Many genera of Cupressales not containing tropolones are small or monotypic, and it is possible that earlier members of these genera have contained tropolones, the recent species being exceptions. No conclusions can therefore be based solely on the presence or absence of tropolones in Cupressales.

Tetraclinis with a single species (*T. articulata* Vahl. Masters) grows in southern Spain and north-western Africa. It does contain tropolones. Botanically it is generally considered as related to the southern Thujoideae (Callitroideae) and it was for some time included with the Australian *Callitris* (under the name of *C. quadrivalvis* Ventenet). According to this classification *Tetraclinis* would stand close to such genera as *Callitris, Neocallitropsis* and *Widdringtonia*, none of which contains tropolones.

References, pp. 275—287.

Callitris is a fairly large genus and *Widdringtonia* has five species. It would seem improbable that all recent species of these genera would constitute non-tropoloniferous remnants of larger, partly tropoloniferous genera.

Only a few southern genera contain tropolones. The heartwood of *Austrocedrus chilensis* is rich in β-thujaplicin and at least one of the three *Papuacedrus* species (New Guinea) contains tropolones. It is, of course, tempting to believe that the southern tropoloniferous genera are of northern origin; but according to FLORIN (94) there does not seem to be any paleophytogeographic support for such an assumption.

Among the southern genera, *Libocedrus** (sensu stricto) is confined to New Zealand. In none of the five species have tropolones been found. *Libocedrus* (sensu lato) included *Austrocedrus*, the northern *Calocedrus*, *Papuacedrus* and the Chilean *Pilgerodendron*. The chemistry of these genera is in accordance with the view that *Libocedrus* (sensu lato) is very heterogeneous.

The Tasmanian *Diselma archeri* was briefly investigated in our laboartory and no tropolones were found. (It is a small tree or erect shrub and it is uncertain whether the wood samples tested contained any heartwood.)

No tropolones have been reported from the northern *Fokienia*.

Almost all *Chamaecyparis* species have now been studied and it has been found that whilst some contain tropolones others apparently do not. The latter group includes the Japanese *Ch. obtusa* (Sieb. et Zucc.) Endl., "hinoki", the Formosan *Ch. formosensis* Mats., "beniki", and the Japanese *Ch. pisifera* (Sieb. et Zucc.) Endl., "sawara". The Formosan *Ch. taiwanensis* Mas. et Suzuki, "Arisan hinoki", has been considered to be only a variety of *Ch. obtusa* (*Ch. obtusa* Sieb. et Zucc. var. *formosana* Hayata) but it is chemically quite different from "hinoki" and "beniki" and contains tropolones. *Ch. lawsoniana* from Oregon is very poor in tropolone content. The wood of *Ch. thyoides* (L) B. S. P. has been investigated in some detail by ENZELL (63) in this laboratory and was found not to contain tropolones, whereas other authors state that it contains thujaplicins (270). Quite recently, this species, on which Linnaeus established the genus *Chamaecyparis*, has been split by LI (161) into two species of which *Ch. henryae* Li is confined to south-eastern and *Ch. thyoides* to north-eastern U. S. A. It is therefore possible that the

---

* The genus *Libocedrus* (sensu lato) has been split into several genera. The old *Libocedrus decurrens* (incense cedar) was unfortunately called *Heyderia decurrens*. The name *Heyderia*, however, had already been used for quite another genus; and for this reason *Libocedrus decurrens* is now called *Calocedrus decurrens*. (Botanists are for good reasons very particular about priority questions.)

Floridan wood which Enzell studied was actually of *Ch. henryae*. Through the good offices of Mrs. M. G. Henry, who noted the difference between these species, we were able to investigate an authentic sample of *Ch. henryae*. The results were practically identical with those obtained by Enzell (unpublished work by Lindström and Norin). Li makes the comment that the general characteristics of *Ch. henryae* indicate a somewhat closer relationship with the western *Ch. nootkatensis*, but this is certainly not in agreement with its chemistry.

As mentioned earlier, *Ch. nootkatensis* is in several respects a unique species and it is the only *Chamaecyparis* species from which nootkatin has been isolated. A careful reconsideration of the systematic position of this species is indicated.

Only about half of the genus *Cupressus* has been studied and almost all of these species have been found to contain C_{10}- as well as C_{15}-tropolones*.

The genus *Juniperus* is large. The tropolones investigated show a notable similarity to those from *Cupressus*. No distinct chemical difference between the subgenera *Oxycedrus* and *Sabina* has yet been found.

Terpenes. The simple monoterpene acids of the thujic-chamic acid types occur in some northern genera. Possibly, these compounds will be found to be more widely distributed in Cupressaceae than they appear to be at present; they should be systematically searched for.

The sesquiterpenes so far isolated from *Cupressus* are of the "cis-farnesyl type" but the closely related *Chamaecyparis* species sometimes produce sesquiterpenes of both the *cis*- and *trans*-farnesyl types. In *Ch. nootkatensis* only *trans* types have been found.

The genus *Juniperus* appears to be more similar to *Cupressus* than to *Chamaecyparis*. Thus, for example, only sesquiterpenes of the *cis*-farnesyl type were found in the heartwood of *Juniperus* but "*trans*-farnesyls" are present in the fruits. It would be of interest to know whether they occur in the fleshy part of the berries or in the seeds.

Cedrol occurs in several genera of Cupressaceae and is particularly common in *Juniperus*. In Taxodiaceae it seems to be restricted to *Cunninghamia* and *Sciadopitys*. Sesquiterpenes of the cedrane, thujopsane, widdrane and cuparane types have not been found in any conifers except Cupressales.

The subfamily Thujoideae (Thujopsideae and Callitroideae) appears to be very complex from both the botanical and chemical viewpoints, and there are great differences within the genera as can be seen by

---

* *C. bakeri* Jepson is considered to be merely a variety of *C. macnabiana* Murr. and *C. goveniana* Gord. = *C. sargentii* Jeps.

References, pp. 275—287.

comparing *Calocedrus decurrens* (*Libocedrus decurrens*, incense cedar) with *C. formosana* (*L. formosana*). An investigation of *C. macrolepis* (*L. macrolepis*) originating from southern China and Viet-Nam would be desirable; it might be more similar to the Formosan *C. formosana* than to the incense cedar occurring on the opposite side of the Pacific.

There are also considerable chemical differences between the *Thuja* species.

In southern Thujoideae (Callitroideae), the wood of the South African *Widdringtonia* species contains sesquiterpenes, such as thujopsene, widdrol, cuparene, cedrol etc., which occur in many Cupressales genera (*Cupressus, Chamaecyparis, Juniperus, Thujopsis, Platycladus*; cedrol also in *Tetraclinis*). The *Widdringtonia* species differ considerably from the Australian—New Caledonian *Callitris* and *Neocallitropsis* which contain eudesmol and guaiol, although some *Widdringtonia* species contain selinane derivatives. The New Caledonian *Callitris sulcata* also produces guaiol and this is of some interest, since Australia and New Caledonia have been separated for many millions of years. Unfortunately, the Australian *Actinostrobus* has not yet been investigated; it would be of considerable interest to know whether or not it is a guaiol producing conifer. So far guaiol has not been found in any conifer except those mentioned above. In view of the biogenetic relationship that exists between eudesmol and guaiol, the latter ought to be looked for in conifers known to produce eudesmol and related compounds.

Nothing is known about the chemistry of the monotypic genus *Fitzroya* from northern Patagonia and Chile.

Diterpenes are widely distributed in Cupressales, in Taxodiaceae as well as in Cupressaceae. *Athrotaxis* and *Taiwania* are considered to be botanically related and contain the sesquiterpene cadinol and the diterpene hinokiol. The chemistry of the diterpenes of *Sciadopitys* is very unusual; it is sufficient to mention the peculiar diterpene verticillol. The leaves of *Sciadopitys* contain the tetracyclic diterpenes (—)-kaurene (191) and (—)-isophyllocladene (the antipode of 188) both possessing an *A,B*-ring system antipodal to that of the bicyclic wood constituents of the same species, e. g. sciadin (162) and methyl sciadopate (164) with the "normal" conifer *A,B*-configuration. Diterpenes of the same antipodal type as (—)-kaurene, e. g. (—)-isokaurene (192), occur in *Cryptomeria* leaves. Strangely enough the wood of *Cryptomeria* contains the tetracyclic (+)-isophyllocladene (188) with "normal" Cupressales *A,B*-configuration. In no other Cupressales species have diterpenes of the antipodal type been found. It is interesting to note that (—)-kaurene and some sterically related compounds also occur in the leaves of certain species of the orders Araucariales and Podocarpales.

Eurasian species of *Cupressus (C. sempervirens, C. torulosa)* contain large amounts of bicyclic diterpenes such as manool and its oxidation products. They have not been found in any American species. It would be of great interest to learn whether or not this represents a general difference. The two groups of *Cupressus* species must have been geographically separated for a very long time and may have developed in different directions. *Cupressus*-like fossil remains from the Miocene period have been found in southern USA and from the middle Jurassic period in Asia minor [Florin (94)].

Resin acids of the abietane and pimarane types are very common in Pinales. However, in Cupressales only a few acids of the pimarane type have been found, namely isopimaric acid (167) and sandaracopimaric acid (168, $R^1 = COOH$, $R^2 = H$); they occur in *Cryptomeria* and *Juniperus*. Strangely enough no acids of the abietane type have yet been encountered in Cupressales although several compounds possessing the same skeleton were isolated, such as hinokiol and ferruginol (totarol from *Juniperus*).

The acids of the labdane group, e. g. communic acid (161) and cupressic acid (159, $R = COOH$) possess axial $C_{(4)}$-carboxyl groups. In Pinales, however, acids of both configurations at $C_{(4)}$ are known. Pinifolic acid from the leaves of *Pinus silvestris* is of the equatorial type (69) whereas communic acid (= elliotinoic acid) (161) from the oleoresin of *P. elliottii* has an axial configuration.

General Remarks. A fairly large number of species of the order Cupressales covering almost all genera have now been subjected to at least some chemical investigation. In many instances these studies have been superficial and only one or a few compounds were detected in, or isolated from, each species; sometimes wood from only a single tree was investigated. A few species have been studied more systematically with the aim of finding some characteristic patterns of chemical constituents that might prove to be of taxonomic importance. These studies suffer from the weakness that only a single organ, generally the wood, was tested. A detailed knowledge of the constituents of the other organs is just as desirable and thus much remains to be done on the chemistry of the barks, leaves, seeds etc. before it will be possible to claim that a reasonably satisfactory chemical description of a single species has been obtained. Hence, we must admit that our chemical knowledge about the order Cupressales is meager, at least when compared with the great number of established botanical facts. It is hoped, however, that even at the present stage of research the botanical taxonomists will be able to include at least some of the chemical characteristics of the Cupressales species in their discussions. Obviously, there exist certain chemical

References, pp. 275—287.

features that are characteristic of this order and others that would indicate relationships between Cupressales and other orders of the conifers. The latter may be of interest when the problem of the mono- or polyphyletic origin of the conifers is debated.

V. Chemical Constituents of Cupressales Species.

T. = Table; sp. = species; ? indicates uncertain identification.

Taxodiaceae.

Sequoia (1 sp.). *S. sempervirens:* pyrogallol, phloroglucinol, gallic acid, protocatechuic acid, catechol (*147*); sugiresinol (sequirin A), sequirin B, sequirin C (*16*); estolide, kayaflavone, hinokiflavone (T. 3).

Sequoiadendron (1 sp.). *S. giganteum:* pyrogallol, phloroglucinol, sequoyitol (*147*).

Metasequoia (1 sp.). *M. glyptostroboides:* sequoyitol (*146*), hinokiflavone (T. 3).

Taxodium (3 sp.). *T. distichum:* hinokiflavone (T. 3): eudesmol (T. 7).

Glyptostrobus (1 sp.). *G. pensilis:* hinokiflavone, sotetsuflavone, apigenin (T. 3).

Cryptomeria (1 sp.). *C. japonica:* sugiresinol, hydroxysugiresinol (*97*); estolide, kayaflavone, sciadopitysin, hinokiflavone (T. 3); cryptomerione (*188*); elemol (?), β-eudesmol, cryptomeridiol (T. 7); copaene, "kigaene", calamenene, calacorene, δ-cadinol, "kiganol" (T. 8); (—)-isokaurene, (+)-isophyllocladene, sandaracopimarinol, (+)-phyllocladanol, sugiol, cryptojaponol, ferruginol acetate, sandaracopimaric acid (T. 13).

Cunninghamia (2 sp.). *C. konishii:* estolide, kayaflavone, hinokiflavone (T. 3); cedrol (T. 9). — *C. lanceolata (= C. sinensis):* estolide, kayaflavone, sotetsuflavone, hinokiflavone (T. 3); cedrol (T. 9).

Sciadopitys (1 sp.). *S. verticillata:* eugenol methyl ether (ERDTMAN, NORIN and SUMIMOTO, unpubl.); sciadopitysin (T. 3); cedrol (T. 9); (—)-kaurene, (—)-isophyllocladene, verticillol, sciadin, sciadinone, methyl sciadopate, dimethyl sciadinonate (T. 13).

Athrotaxis (3 sp.). *A. cupressoides:* "cadinene", x-cadinol, δ-cadinol, x-cadinol (T. 8); α-cedrene (T. 9). — *A. selaginoides:* athrotaxine (*91*); sugiresinol (see p. 211); "cadinene", α-cadinol, δ-cadinol, x-cadinol (T. 8).

Taiwania (1 sp.). *T. cryptomerioides:* savinin, taiwanin A (T. 1); estolide (?), hinokiflavone (T. 3); δ-cadinol (T. 8); caryophyllene, humulene (T. 9); hinokiol (T. 13).

Cupressaceae.

Cupressus (20 sp.). *C. arizonica:* hinokiflavone (T. 3), carvacrol, β-, γ-thujaplicin, β-thujaplicinol, nootkatin (T. 5); humulene, x-cedrene, thujopsene, widdrol, cuparene (T. 9). — *C. abramsiana:* β-thujaplicin, β-dolabrin, α-thujaplicinol and/or α-dolabrinol, nootkatin (T. 5). — *C. bakeri:* carvacrol, nootkatin (T. 5). — *C. funebris:* hinokiflavone (T. 3). — *C. goweniana:* β-thujaplicin, β-dolabrin, α-thujaplicinol and/or α-dolabrinol, pygmaein, nootkatin (T. 5). — *C. lindleyi:* nootkatinol (T. 5). — *C. macnabiana:* carvacrol, α-, β-, γ-thujaplicin, nootkatin (T. 5). — *C. macrocarpa:* carvacrol, α-, β-thujaplicin, nootkatin (T. 5); juniperol (T. 9); (—)-hibaene, (+)-isophyllocladene (T. 13). — *C. pygmaea:* β-thujaplicin, β-dolabrin, pygmaein, α-dolabrinol, α-thujaplicinol, nootkatin (T. 5). — *C. sargentii:* β-thujaplicin, β-dolabrin, β-thujaplicinol, nootkatin (T. 5). — *C. sempervirens:* cupressuflavone (T. 3) carvacrol, β-thujaplicin, nootkatin (T. 5); cedrol (T. 9); manool, "compound-A" ($C_{21}H_{28}O_3$), "compound-B" ($C_{21}H_{28}O_3$), communic acid, cupressic acid, isocupressic acid (T. 13). — *C. torulosa:* cupressuflavone (T. 3); carvacrol, α-, β-thujaplicin, β-thujaplicinol, nootkatin (T. 5); humulene, thujopsene, cuparene (T. 9); torulosal, manool, torulosol, hinokione, ferruginol, hinokiol (T. 13).

Chamaecyparis (7 sp.). *Ch. formosensis:* estolide, hinokiflavone (T. 3); β-thujaplicin (?) (T. 5); cryptomeridiol (T. 7); "cadinene", "cadinol", sesquibenihene (T. 8); humulene (T. 9). — *Ch. henryae:* α-cedrene, cedrol, thujopsene, hinokiic acid, widdrol (?), cuparene (?), cuparenenes, cuparenic acid (T. 9). — *Ch. lawsoniana:* deoxypodophyllotoxin (T. 1); β-thujaplicin (T. 5); β-bisabolene (*158*); α-cadinol (T. 8). — *Ch. nootkatensis:* carvacrol, nootkatin, chanootin (T. 5); thujic acid (*194*); chamic acid (*35*); chaminic acid (*35, 82*); nootkatene, nootkatone (T. 7). — *Ch. obtusa:* hinokiresinol (*124*); hinokinin, D-sesamin (T. 1); estolide, hinokiflavone (T. 3); β-thujaplicin (?) (T. 5); cadinol (?) (T. 8); humulene, hinokiic acid (T. 9); xanthoperol, hinokiol, hinokione, sugiol (T. 13). — *Ch. pisifera:* sawarinin (*129*); savinin (T. 1); estolide, hinokiflavone (T. 3); "cadinene", α-cadinol, δ-cadinol, x-cadinol (T. 8). — *Ch. taiwanensis:* carvacrol, α-, β-thujaplicin, chamaecin (T. 5); citronellic acid (*167, 200, 201*); "cadinene", "cadinol" (T. 8); α-cedrene, thujopsene, hinokiic acid (T. 9). — *Ch. thyoides:* carvacrol, α-, β-, γ-thujaplicin (T. 5, *274*); α-cedrene, cedrol, thujopsene, hinokiic acid, widdrol, cuparene, cuparenic acid (T. 9).

Juniperus (70 sp.). *J. arizonica:* communic acid (T. 13). — *J. californica:* α-cedrene, cedrol, thujopsene, hinokiic acid, widdrol, cuparene (T. 9), triterpene $C_{30}H_{54}O_2$ (?) (*211*). — *J. cedrus:* carvacrol, β-thujaplicin, nootkatin (T. 5); δ-cadinol (T. 8); α-cedrene, cedrol, thujopsene, cuparene (T. 9). — *J. chinensis:* estolide, kayaflavone, hinokiflavone (T. 3); hydrothymoquinone, 3-hydroxythymoquinone, 3,6-dihydroxythymoquinone (*213*); carvacrol, α-, β-thujaplicin, nootkatin (T. 5); α-cedrene, cedrol, thujopsene, hinokiic acid, widdrol, cuparene (T. 9). — *J. communis:* savinin (T. 1); α-thujaplicin, pygmaein, nootkatin (T. 5); β-elemene, junenol, juniper camphor (T. 7); δ-cadinene, ε-muurolene, α-cadinol, δ-cadinol (T. 8); humulene, longifolene, juniperol, cedrol, thujopsene, widdrol, cuparene (T. 9); xanthoperol, 6,7-dehydroferruginol, sugiol, ferruginol, totarol, communic acid (T. 13). — *J. conferta:* kayaflavone, hinokiflavone (T. 3); isopimaric acid (T. 13). — *J. deppeana:* β-thujaplicin, β-dolabrin, nootkatin (T. 5). — *J. excelsa:* cedrol (T. 9). — *J. foetidissima:* deoxypodophyllotoxin (T. 1); calamenene (T. 8); α-cedrene, cedrol, widdrol (T. 9). — *J. formosana:* estolide, kayaflavone, hinokiflavone, amentoflavone, apigenin (T. 3). — *J. horizontalis:* α-cedrene, cedrol, thujopsene, widdrol, cuparene (T. 9); communic acid (T. 13). — *J. japonica:* isopimaric acid (T. 13). — *J. lucayana:* podophyllotoxin (T. 1). — *J. mexicana:* α-cedrene, cedrol (T. 9); sugiol (T. 13). — *J. monosperma:* β-thujaplicin, β-dolabrin, pygmaein, nootkatin (T. 5). — *J. occidentalis:* cedrol (T. 9). — *J. osteosperma:* β-thujaplicin, nootkatin (T. 5). — *J. oxycedrus:* nootkatin (?) (T. 5); viridiflorol (T. 7); γ-cadinene, ε-muurolene, ylangene, calamenene (T. 8); caryophyllene, humulene (T. 9). — *J. phoenicea:* carvacrol, β-thujaplicin, nootkatin (T. 5); thujopsene, hinokiic acid, widdrol, cuparene (T. 9). —

References, pp. 275—287.

J. procera: carvacrol, procerin (T. 5); α-cedrene, cedrol, cuparene (T. 9); triterpene $C_{30}H_{54}O_2$ (?) (*211*). — *J. procumbens:* estolide, kayaflavone, hinokiflavone (T. 3). — *J. pseudosabina:* podophyllotoxin (T. 1). — *J. rigida:* nootkatin, nootkatinol (T. 5); isopimaric acid (T. 13). — *J. sabina:* podophyllotoxin, savinin (T. 1); citronellic acid methyl ester (?) (*222*). — *J. sargentii:* estolide, kayaflavone, hinokiflavone (T. 3). — *J. scopulorum:* podophyllotoxin (T. 1). — *J. semiglobosa:* β-cedrene, cedrol (T. 9). — *J. seravschanica:* podophyllotoxin (T. 1). — *J. silicicola:* podophyllotoxin, deoxypodophyllotoxin (T. 1). — *J. squamata* var. *morrisonicola:* estolide (?), kayaflavone, amentoflavone, sciadopitysin, hinokiflavone, apigenin (T. 3). — *J. thurifera:* savinin (T. 1); carvacrol, β-thujaplicin, nootkatin (T. 5); sesquiterpene "X_2" ($C_{15}H_{24}$) (*230*); α-cedrene, cedrol, thujopsene, hinokiic acid, cuparene (T. 9); triterpene $C_{30}H_{54}O_2$ (?) (*230*). — *J. utahensis:* carvacrol, β-thujaplicin, nootkatin (T. 5); α-cedrene, cedrolic acid, thujopsene, hinokiic acid, widdrol, widdrol-α-epoxide, cuparene (T. 9); utahin $C_{20}H_{20}O_5$ (*228*). — *J. utilis:* estolide, kayaflavone, hinokiflavone (T. 3). — *J. virginiana:* podophyllotoxin (T. 1); kayaflavone, hinokiflavone (T. 3); sesquiterpene "X_2" (?) ($C_{15}H_{24}$) (*229*); α-cedrene, cedrol, pseudocedrol, prim-cedrenol, thujopsene, widdrol, cuparene (T. 9).

Calocedrus (3 sp.). *C. decurrens:* carvacrol, hydrothymoquinone, *p*-methoxycarvacrol, *p*-methoxythymol, libocedrol, heyderiol, 3-libocedroxythymoquinone (*267–269*); hinokiflavone (T. 3); β-, γ-thujaplicin (T. 5). — *C. formosana:* estolide, hinokiflavone (T. 3); β-, γ-thujaplicin, β-thujaplicinol (T. 5); thujic acid, chamic acid, chaminic acid (*164*); shonanic acid (*128, 165*).

Fokienia (3 sp.). *F. hodginsii:* nerolidol (*54*); eudesmol, cryptomeridiol (T. 7).

Microbiota (1 sp.). *M. decussata:* —.

Platycladus (Biota) (1 sp.). *P. orientalis:* estolide, hinokiflavone (T. 3); taxifoline, aromadendrin (*86*); α-, β-, γ-thujaplicin (T. 5); thujic acid (*120*); thujopsene, mayurone, widdrol, cuparene, cuparenones, cuparenols (T. 9).

Thuja (5 sp.). *T. occidentalis:* hinokiflavone (T. 3); α-, β-, γ-thujaplicin (T. 5); thujic acid (*108, 119*); α-, β-, γ-eudesmol, occidentalol, occidol, occidiol (T. 7). — *T. plicata:* plicatic acid and four thujaplicatins (T. 1), carvacrol, α-, β-, γ-thujaplicin, β-dolabrin, β-thujaplicinol (T. 5); thujic acid (*80, 107*); methyl thujate (*159*). — *T. standishii:* estolide, hinokiflavone (T. 3); carvacrol, α-, β-thujaplicin (T. 5).

Thujopsis (1 sp.). *Th. dolabrata:* estolide, sciadopitysin, sotetsuflavone, hinokiflavone (T. 3); carvacrol, α-,β-thujaplicin, β-dolabrin (T. 5); citronellic acid (*206*); elemol, elemenal, costol, costal, sesquibenihidiol (T. 7); α-cedrene, cedrol, thujopsene, mayurone, hinokiic acid, widdrol, cuparene, cuparenenes, γ-cuparenol (T. 9); dolabradiene, hibaene, totarol, 7-hydroxytotarol (T. 13).

Actinostrobus (1 sp.). *A. pyramidalis:* —.

Austrocedrus (1 sp.). *A. chilensis:* deoxypodophyllotoxin (T. 1); taxifolin (*87*), carvacrol, α-(?), β-thujaplicin (T. 5).

Callitris (16 sp.). *C. calcarata:* citronellic acid, guaiol (*14*). — *C. drummondi:* podophyllotoxin (T. 1). — *C. glauca (C. columellaris):* hinokiflavone (T. 3); citronellic acid, dehydrogeranic acid (*147*); guaiol, α-, β-, γ-eudesmol, cryptomeridiol (*14, 225–227*). — *C. intratropica:* citronellic acid, guaiol (*14*). — *C. macleyana:* citronellic acid, guaiol (*14*). — *C. morrisonii:* guaiol (*73*). — *C. muelleri:* estolide, hinokiflavone (T. 3). — *C. preissii:* guaiol (*73*). — *C. propinqua:* guaiol (*73*). — *C. rhoei:* eudesmol (*73*). — *C. rhomboidalis:* guaiol (*73*). — *C. robusta:* estolide, hinokiflavone (T. 3). — *C. sulcata:* guaiol (ERDTMAN and McLEAN, unpubl.). — *C. verrucosa:* guaiol (*73*).

Diselma (1 sp.). *D. archeri:* No tropolones found in the wood.

Fitzroya (1 sp.). *F. cupressoides:* —

Libocedrus (5 sp.). *L. bidwillii:* humulene (T. 9); (+)-phyllocladene (T. 13). — *L. plumosa:* (+)-phyllocladene, (+)-rimuene (T. 13).

Pilgerodendron (1 sp.). *P. uviferum:* phenol m. p. 217° (*88*), "cadinene", "cadinol", δ-cadinol (pilgerol) (T. 8).

Papuacedrus (3 sp.). *P. torricellensis:* isopygmaein (T. 5).

Neocallitropsis (1 sp.). *N. araucarioides:* dehydrogeranic acid (*147*), α-, β-, γ-eudesmol, guaiol (T. 7).

Widdringtonia (5 sp.). *W. cupressoides:* eudesmols, cryptomeridiol (T. 7). — *W. dracomontana:* eudesmols, cryptomeridiol (T. 7); cedrol, thujopsene (widdrene), hinokiic acid (widdrenic acid), widdrol, cuparene (widdrene II), cuparenic acid (T. 9); "acid III" (*89*). — *W. juniperoides, W. schwarzii* and *W. whytei:* cedrol, thujopsene, hinokiic acid, widdrol, cuparene, cuparenic acid (T. 9); "acid III" (*89*). — *W. juniperoides:* Widdrol-α-epoxide (T. 9).

Tetraclinis (1 sp.). *T. articulata:* carvacrol, hydrothymoquinone, thymoquinone (*147*); β-, γ-thujaplicin (T. 5); cedrol (T. 9); torulosal, manool, torulosol, torulosyl acetate, totarolenone, hinokione, totarolone, totarol, hinokiol, communic acid, sandaracopimaric acid, 12 β-acetoxysandaracopimaric acid (T. 13).

VI. Tables.

Table 1. Distribution of Lignans in Cupressales.

Species	Organ	Lignans	References
Taiwania cryptomerioides .	Wood	Savinin ("Taiwanin B"), "Taiwanin A"	(*163, 166, 169*)
Chamaecyparis lawsoniana	Leaves	Deoxypodophyllotoxin	(*93*)
Ch. obtusa	Wood	Hinokinin	(*147*)
Ch. obtusa	Leaves	Hinokinin, *D*-Sesamin	(*175–177*)
Ch. pisifera	Leaves	Savinin	(*147*)
Juniperus communis	Wood	Savinin	(*29*)
J. foetidissima	Leaves	Deoxypodophyllotoxin	(*241*)
J. lucayana	Leaves	Podophyllotoxin	(*93, 112*)
J. pseudosabina	Leaves	Podophyllotoxin	(*241*)
J. sabina	Leaves, Berries	Podophyllotoxin, Savinin	(*93, 112, 241*)
J. scopulorum	Leaves	Podophyllotoxin	(*93, 112*)
J. seravschanica	Leaves	Podophyllotoxin	(*241*)
J. silicicola	Leaves	Podophyllotoxin, Deoxy-podophyllotoxin	(*147, 93, 112*)
J. thurifera	Wood	Savinin	(*29*)
J. virginiana............	Leaves	Podophyllotoxin	(*93, 112*)
Thuja plicata	Wood	Plicatic acid, Thuja-plicatins	(*100*)
Austrocedrus chilensis	Leaves	Deoxypodophyllotoxin	(*93*)
Callitris drummondi......	Leaves	Podophyllotoxin	(*150*)

Table 2. Physical Constants of Lignans from Cupressales.

Lignan	Structure	Physical constants, Derivatives
(+)-Sesamin	(32)	M. p. 123–124°, $[\alpha]_D$ + 69° ($CHCl_3$)
Podophyllotoxin	(22)	M. p. 183–184°, $[\alpha]_D$ — 133° ($CHCl_3$)
Deoxypodophyllotoxin	(23)	M. p. 167–168°, $[\alpha]_D$ — 116° ($CHCl_3$)
Savinin	(24)	M. p. 146–148°, $[\alpha]_D$ — 87° ($CHCl_3$)
Taiwanin A.............		M. p. 200–201°
Hinokinin	(31)	M. p. 64–65°, $[\alpha]_D$ — 34° ($CHCl_3$)
Plicatic acid............	(25)	Amorphous (Lactone = Plicatin) $[\alpha]_D$ — 9.9° (H_2O) Trimethyl ether, m. p. 210–218° (foaming, conversion into lactone, m. p. 257°) Trimethyl ether methyl ester, m. p. 204–205°, $[\alpha]_D$ — 29.9° ($CHCl_3$)
Thujaplicatin methyl ether	(26)	M. p. 167–167.5°, $[\alpha]_D$ — 48.7° (acetone) Dimethyl ether, m. p. 102–103°, $[\alpha]_D$ — 37.4° ($CHCl_3$)
Thujaplicatin	(27)	Amorphous. Trimethyl ether = dimethyl ether of (26)
Hydroxythujaplicatin methyl ether	(28)	M. p. 128.5–129.5°, $[\alpha]_D$ — 54.8° ($CHCl_3$). Dimethyl ether, m. p. 113.5–114.5°, $[\alpha]_D$ — 50° (acetone)
Dihydroxythujaplicatin methyl ether	(29)	M. p. 95–97° (CH_3OH-solvate), $[\alpha]_D$ — 97.2° ($CHCl_3$) Dimethyl ether, m. p. 163.5–164°, $[\alpha]_D$ — 105.1° ($CHCl_3$)
Dihydroxythujaplicatin	(30)	Amorphous. Trimethyl ether = dimethyl ether of (29)

Table 3. Distribution of Biflavonyls and Composition of Leaf Waxes in Cupressales.

Essentially according to KARIYONE (*145*). Statements within parentheses by HSÜ (*127*)

(A = amentoflavone, Ap = apigenin, H = hinokiflavone, K = kayaflavone, Sc = = sciadopitysin, So = sotetsuflavone, Cu = cupressuflavone, E = leaf wax of estolide type, N = leaf wax of nonestolide type.)

Taxodiaceae.

Sequoia sempervirens	E	K, H
Metasequoia glyptostroboides	N	H
Taxodium distichum	N	H
Glyptostrobus pensilis	N	H (So, Ap)
Cryptomeria japonica	E	K, Sc, H
Cunninghamia lanceolata	E	K, So, H
C. konishii	E	K, H
Sciadopitys verticillata	N	Sc
Taiwania cryptomerioides	E (N)	H

Cupressaceae

Cupressus arizonica		H
C. funebris		H
C. sempervirens		Cu
C. torulosa		Cu
Chamaecyparis formosensis	E	H
Ch. obtusa	E	H
Ch. pisifera	E	H
Juniperus chinensis	E	K, H
J. conferta		K, H
J. formosana	E	H (A, K, H, Ap)
J. procumbens	E	K, H
J. sargentii	E	K, H
J. squamata var. *morrisonicola*	E (N)	K, H (A, Sc, Ap)
J. utilis	E	K, H
J. virginiana		K, H
Platycladus orientalis	E	H
Thuja occidentalis		H
Th. standishii	E	H
Thujopsis dolabrata	E	Sc, So, H
Callitris glauca	E*	H
Calocedrus decurrens		H
C. formosana	E	H

* = *C. muelleri* and *C. robusta* (*C. glauca* not investigated for estolides).

Table 4. Physical Constants of Biflavonyls from Cupressales.

Compound	Structure	Physical constants, Derivatives
Cupressuflavone	(39)	Hexaacetate, m. p. 251–254°. Hexamethyl ether, m. p. 295–297°
Hinokiflavone ..	(37)	M. p. 355°. Pentaacetate, m. p. 222–223° Pentamethyl ether, m. p. 260–261°
Amentoflavone	(36)	M. p. 360°. Hexaacetate, m. p. 240°. Hexamethyl ether, m. p. 225–226°
Sotetsuflavone..	(36) OCH₃ instead of OH at 7 a	M. p. 263–264°. Pentaacetate, m. p. 233–234°
Kayaflavone....	(36) OCH₃ instead of OH at 7 a, 4′ and 4′a	M. p. 314–315°. Triacetate, m. p. 190–191°
Sciadopitysin ..	(36) OCH₃ instead of OH at 7, 4′ and 4′a	M. p. 286–287°. Triacetate, m. p. 264°

Table 5. Distribution of Tropolones in Cupressales.

	Carvacrol and/or methyl ether	α-Thujaplicin	β-Thujaplicin	γ-Thujaplicin	β-Dolabrin	α-Thujaplicinol	β-Thujaplicinol	Pygmaein	Isopygmaein	α-Dolabrinol	Nootkatin	Nootkatinol	Procerin	Chanootin	Chamaecin	Number of unknown Tropolones	References
.XODIACEAE all investigated species																	
PRESSACEAE *Chamaecyparis:*																	
formosensis ..		+?															
lawsoniana ...		+															(270)
nootkatensis ..	+	+									+	+					(57, 35, 194)
obtusa		+?															
pisifera																	
taiwanensis ...	+	+	+											+			(201, 167, 168)
thyoides	+	+	+	+													(63, 274)
Cupressus:																	
abramsiana ...			+		+	+?	+?				+						(274)
arizonica	+		+	+					+		+					2	(68, 274)
bakeri	+										+						HEARON (priv. comm.)
goweniana....			+		+	+		+		+	+						(274)
lindleyi												+					(21)
macnabiana ..	+	+	+	+							+						(72, 274)
macrocarpa ..	+	+	+								+						(45, 62)
pygmaea			+		+	+		+		+	+					4	(273, 271)
sargentii			+		+			+			+						(274)
sempervirens..	+		+								+						(180, 66, 274)
torulosa	+	+	+					+			+						(1, 17)
Juniperus:																	
cedrus	+	+									+						(231)
communis		+							+		+					1	(271, 27, 274)
oxycedrus											+?						LINDSTRÖM, NORIN, (unpubl.)
rigida........											+	+					(117, 118)
californica ...																	(211)
chinensis	+	+	+								+						(213)
deppeana		+		+							+						(274)
foetidissima...																	(234)
mexicana.....																	(274)
monosperma ..		+		+					+		+						(274)
occidentalis ...																	(274, 270)

(Table 5, continued.)

	Carvacrol and/or methyl ether	α-Thujaplicin	β-Thujaplicin	γ-Thujaplicin	β-Dolabrin	α-Thujaplicinol	β-Thujaplicinol	Pygmaein	Isopygmaein	α-Dolabrinol	Nootkatin	Nootkatinol	Procerin	Chanootin	Chamaecin	Number of unknown Tropolones	References
CUPRESSOIDEAE *Juniperus:*																	
osteosperma ..			+								+?						(274)
phoenicea	+		+								+						(232)
procera	+												+			3	(211)
thurifera	+		+								+						(230)
utahensis	+		+								+						(228)
virginiana																	(274, 229)
Northern THUJOIDEAE *Calocedrus:*																	
decurrens	+		+	+													(272, 267)
formosana ...			+	+		+											(162)
Platycladus: orientalis		+	+	+													(120)
Thuja: occidentalis ..		+	+	+													(108, 109)
plicata	+*	+	+	+	+	+											(78, 79)*, (9... MacLean (priv. con...
standishii	+	+	+													1	(123)
Thujopsis: dolabrata	+	+	+		+												(206, 201, 2...
Southern THUJOIDEAE *Tetraclinis:* articulata.....	+		+	+													(40)
Austrocedrus: chilensis	+**	+?	+														(87)** (unpu...
Callitris: all investigated species																	(14, 47, 225—227)
Libocedrus: bidwillii......																	(45)
Neocallitropsis: araucarioides																	
Papuacedrus: torricellensis ..							+										(274)
Pilgerodendron: uviferum																	(88)
Widdringtonia: all five species																	(89)

References, *pp. 275—287.*

Table 6. Physical Constants of Tropolones from Cupressales.

Tropolone	Structure	Physical constants, Derivatives	References
α-Thujaplicin	(41)	M. p. 34°; Cu-complex (1 : 2), m. p. 235°	(78)
β-Thujaplicin	(42)	M. p. 52–52.5°; Cu-complex (1 : 2), m. p. 178°	(78)
γ-Thujaplicin	(43)	M. p. 82°; Cu-complex (1 : 2), m. p. 259–260°	(78)
β-Dolabrin	(60)	M. p. 58–59°; Cu-complex (1 : 2), m. p. 196–198°	(204)
α-Thujaplicinol ...	(62)	Oil. Dicyclohexylamine salt, m. p. 130–130.5°; Cu-complex (1 : 1), m. p. 303.5–304.5°	(273)
β-Thujaplicinol ...	(59)	M. p. 57–58°; Cu-complex (1 : 1), m. p. 237–238°	(99)
Pygmaein	(63)	M. p. 38–40°; Dicyclohexylamine salt, m. p. 80–87°; Cu-complex (1 : 2), m. p. 211–212°	(271)
Isopygmaein	(64)	M. p. 111–112°; Cu-complex (1 : 2), m. p. 272.5–273.5°	(266)
α-Dolabrinol	(65)	Oil; Dicyclohexylamine salt, m. p. 141.5–142.5°; Benzylamine salt, m. p. 97–98°	(266)
Nootkatin	(48)	M. p. 95°; Cu-complex (1 : 2), m. p. 234–235°	(35)
Nootkatinol	(61)	M. p. 102–103°; Cu-complex (1 : 2), m. p. > 265°	(21, 117)
Procerin	(66)	M. p. 71–72°; Cu-complex (1 : 2), m. p. 203–204° (decomp.)	(233)
Chanootin	(67)	M. p. 172–173°	(194)
Chamaecin		Oil; Cu-complex, m. p. 248°	(168)
"Juniperus tropolone II" ...		M. p. 133–134°	(211)
"Juniperus tropolone IV" ..		M. p. 72–73°	(211)
"Cupressus tropolone"		M. p. 174–176°	(68)

Table 7. Distribution of Selinanes, Guaianes and Related Sesquiterpenes in Cupressales.

	β-Elemene	Elemol	Elemenal	Eudesmols	Cryptomeridiol	Occidentalol	Occidol	Occidiol	Costol	Costal	Sesquibenihidiol	Junenol	Juniper camphor	Nootkatene	Nootkatone	Guaiol	Viridiflorol	References
TAXODIACEAE																		
Taxodium:																		
distichum				+														*
Cryptomeria		+ (?)		+	+													** (188, 251)
CUPRESSACEAE																		
Chamaecyparis:																		
formosensis	+				+													(149)
nootkatensis														+	+			(83, 90)
Juniperus:																		
communis												+	+				+	(138, 182, 183)
oxycedrus																		(55, 56)
Fokienia:																		
hodginsii				+	+													(54)
Thujopsis		+	+						+	+	+							(134)
Thuja:																		
occidentalis						+	+	+										(121, 224)
Callitris sp.				+												+		(147, 226)
Neocallitropsis				+	+											+		(18, 220, 221)
Widdringtonia:																		
cupressoides				+	+													(89)
dracomontana				+	+													(89)

* HERGERT, private communication.   ** NAGAHAMA, private communication.

References, pp. 275—287.

.ble 8. Distribution of Cadinanes and Related Sesquiterpenes in Cupressales.

	"Cadinene"	"Cadinol"	γ-Cadinene	δ-Cadinene	ε-Muurolene	Copaene	Ylangene	Sesquibenihene	"Kigaene"	Calamenene	Calacorene	α-Cadinol	δ-Cadinol	"Kiganol"	"X-Cadinol"	References
AXODIACEAE																
Cryptomeria ...						+		+	+	+		+	+			(147, 188)
Athrotaxis:																
cupressoides ..	+											+	+		+	(91)
selaginoides ..	+											+	+		+	(91)
Taiwania													+			(138, 163)
CUPRESSACEAE																
Chamaecyparis:																
formosensis ..	+	+						+								(147, 149)
taiwanensis...	+	+														(149)
pisifera	+											+	+		+	(91)
obtusa		+?														(147)
lawsoniona ...												+				(158)
Juniperus:																
cedrus													+			(231)
communis				+	+							+	+			(116, 182, 185)
oxycedrus			+	+		+				+						(181, 186)
foetidissima...										+						(234)
Pilgerodendron ..	+	+											+			(88, 185)

Table 9. Distribution of Sesquiterpenes in Cupressales: Caryophyllane, Humulane, Longifolane, Cedrane, Thujopsane and Cuparane Types.

	References	Cuparenic acid	γ-Cuparenol	"Cuparenols"	Cuparenones	Cuparenes	Cuparene	Widdrol-α-epoxide	Widdrol	Hinokiic acid	Mayurone	Thujopsene	Cedrolic acid	Prim-cedrenol	Pseudocedrol	Cedrol	β-Cedrene	α-Cedrene	Juniperol	Longifolene	Humulene	Caryophyllene
TAXODIACEAE																						
Cunninghamia:																						
konishii......	(207)															+						
sinensis.......	(207)															+						
Sciadopitys......	(85)															+						
Athrotaxis:																						
cupressoides..	(91)																	+				
Taiwania.......	(163)																				+	+
CUPRESSACEAE																						
Cupressus:																						
arizonica......	(68)						+		+			+						+			+	
macrocarpa..	(32)																		+			
sempervivens..	(66)						+					+				+					+	
torulosa......	(17)																					
Chamaecyparis:																						
formosensis..	(19)																				+	
henryae.......	*					+	+?		+?	+		+				+		+			+	
obtusa.......	(19, 255, 147)	+								+												
taiwanensis...	(139, 140, 62, 63)									+		+						+				
thyoides.......	(63)	+					+		+	+		+				+		+				

References, pp. 275—287.

	(231)	(3, 27, 182)	(181)	(211)	(213)	(236)	(192)	(234)	(263)	(160)	(232)	(211)	(103)	(230)	(228)	(229, 240, 245)	(37, 38, 86)	** (50, 134)	(19)	(89)	(89)	(89)	(89)	(40)
Juniperus:																								
cedrus																				+	+	+	+	
communis																		+						
oxycedrus																	+							
californica																	+							
chinensis																		+						
excelsa	++	++	+					++	+++							+	+			+	+	+	+	
horizontalis								+										+						
foetidissima	+	++	++		+			++	+							+	+			+	+	+	+	
mexicana		++			+			++								+				+	+	+	+	
occidentalis																+	+							
phoenicea	++	++	+		+											+++	+	+		+	+	+	+	
procera							+																	
semiglobosa							+																	
thurifera							+																	
utahensis	++	+++	+++	+++			+++									+		+		+	+	+	+	+
virginiana									+															
Platycladus	+	++	+++					+								+++		+						
Thujopsis	+																							
+																								
Libocedrus:																								
bidwillii	++																	+						
Widdringtonia:																								
dracomontana	+																							
juniperoides																								
schwarzii																								
whytei																								
Tetraclinis																								

* Unpublished results (LINDSTRÖM and NORIN).   ** Unpublished results (NAGAHAMA and NORIN).

Table 10. Physical Constants of Sesquiterpene Hydrocarbons from Cupressales.

Composition	Compound	Structure	Physical Constants[a]	References[b]	Occurrence
$C_{15}H_{20}$	α-Calacorene	cf. (132)	n_D^{20} 1.5271; d^{20} 0.9399; $[\alpha]_D$ + 52.1° (l)	(215)	Table 8
	β-Calacorene		n_D^{20} 1.5317; d^{20} 0.9364; $[\alpha]_D$ + 49.9° (l)	(215)	Table 8
$C_{15}H_{22}$	(−)-Calamenene	(131)	n_D^{28} 1.5155; d^{28} 0.9276; $[\alpha]_D$ − 67°	(215)	Table 8
	(+)-Cuparene	(151), R = CH₃	n_D^{25} 1.5202; d^{21} 0.9374; $[\alpha]_D$ + 65°	(67)	Table 9
	(−)-Nootkatene	(121)	n_D^{24} 1.5222; d^{24} 0.933; $[\alpha]_D$ − 177°	(90, 172)	Table 7
$C_{15}H_{24}$	(−)-β-Bisabolene	(97)	n_D^{25} 1.4893; $[\alpha]_D$ − 67°	(19)	*Chamaecyparis lawsoniana* (158)
	(+)-γ-Cadinene	(125)	n_D^{20} 1.5074; d^{20} 0.9104; $[\alpha]_D$ + 147.9° (l)	(19)	Table 8
	(+)-δ-Cadinene	(126)	n_D^{20} 1.5090; d^{20} 0.9195; $[\alpha]_D$ + 89° (l)	(19)	Table 8
	(−)-Caryophyllene	(102)	n_D^{30} 1.4996; d^{20} 0.9019; $[\alpha]_D$ − 9.1°	(19, 105)	Table 9
	(−)-α-Cedrene	(139)	n_D^{20} 1.4982; d^{20} 0.9342; $[\alpha]_D$ − 91°	(19)	Table 9
	(+)-β-Cedrene	cf. (139)	n_D^{20} 1.5047; d^{20} 0.9350; $[\alpha]_D$ + 11.3° [c]	(19)	Table 9
	(−)-Copaene	(127)	n_D^{26} 1.4885; d^{26} 0.9055; $[\alpha]_D$ − 6.5°	(52, 144)	Table 8
	"α-Cuprenene"	(154)	λ_{max}^{EtOH} 271 mμ (ε 6350)	(50)	Table 9
	"Cuprenene"	(156)	$[\alpha]_D$ + 50°	(50)	
	(+)-γ-Curcumene	(96)	n_D^{25} 1.4975; d^{20} 0.8810; $[\alpha]_D$ + 31.8°	(19, 126)	*Libocedrus bidwillii* (147)
	(−)-β-Elemene	(103)	n_D^{20} 1.4935; d^{20} 0.8749; $[\alpha]_D$ − 20°	(19)	Table 7
	Humulene	(101)	n_D^{20} 1.5022; d^{20} 0.8922	(19, 111, 178)	Table 9
	"Kigaene"	(128) (?)	$[\alpha]_D$ − 22.4° [d]	(147)	Table 8
	(+)-Longifolene	(136)	n_D^{30} 1.4950; d^{30} 0.9284; $[\alpha]_D$ + 42.7°	(19)	Table 9
	(+)-ε-Muurolene = (+)-ε-Cadinene	(124)	n_D^{22} 1.5049; $[\alpha]_D$ + 50.7°	(19, 261)	Table 8
	Sesquibenihene	(129)	n_D^{29} 1.5055; d^{29} 0.9182; $[\alpha]_D$ + 25.2° [d]	(19)	Table 7
	(−)-Thujopsene	(144)	n_D^{25} 1.5031; d^{24} 0.932; $[\alpha]_D$ − 110°	(19, 193)	Table 9
	(+)-Ylangene	(130)	n_D^{20} 1.4905; $[\alpha]_D$ + 50.5°	(184)	Table 8

a Rotations in chloroform unless otherwise stated (l = liquid).   b Recent literature on the chemistry of the compounds.
c Prepared from α-cedrene.   d Homogeneity doubtful.

Table 11. Physical Constants of Sesquiterpene Alcohols from Cupressales.

Composition	Compound	Structure	Physical Constants[a]	References[b]	Occurrence
$C_{15}H_{22}O$	"Cuparenols"	d	not pure	(37)	Table 9
	(+)-γ-Cuparenol	(151) R = CH_2OH	[α]D + 92°; p-Nitrobenzoate m. p. 65°	(134)	Table 9
$C_{15}H_{24}O$	(+)-Occidol	(109)	M. p. 69–70°; [α]D + 163.7°	(121, 122, 191)	Table 7
	(+)-α-Costol = Sesquibenihiol	(111) R = CH_2OH	n_D^{26} 1.5180; [α]D + 34.3°	(20, 23)	Table 7
	(+)-Occidentalol	(108)	M. p. 95.5–98°; [α]D + 363.2°	(223, 276)	Table 7
	"Prim-cedrenol"	(142)	n_D^{26} 1.5212; d^{20} 1.0083[c]	(240, 245)	Table 9
$C_{15}H_{26}O$	(−)-α-Cadinol	(133)	M. p. 74.5°; [α]D − 47°	(19)	Table 8
	(−)-δ-Cadinol = Pilgerol = Albicaulol = (−)-Isocrypto-meriol (?)	(134)	M. p. 139–140°; [α]D − 107°	(19, 246)	Table 8
	(−)-"X-Cadinol"	unknown	M. p. 78°; [α]D − 86.2°	(91)	Table 8
	(−)-Elemol	(104)	M. p. 52°; [α]D − 3°	(19)	Table 7
	(+)-α-Eudesmol	cf. (106)	M. p. 75°; [α]D + 28.6°	(19)	Table 7
	(+)-β-Eudesmol		M. p. 80.5–81°; [α]D + 68.3°	(19)	
	(+)-γ-Eudesmol		n_D^{20} 1.5087; [α]D + 62.5°	(19)	
	(−)-Guaiol	(122)	M. p. 93°; [α]D − 29.8° (EtOH)	(19)	Table 7
	(+)-Junenol	(113)	M. p. 62.5–63°; [α]D + 59.6° (EtOH)	(242, 253)	Table 7
	Juniper camphor	(114)	M. p. 166.5°; [α]D 0°	(183, 258)	Table 7
	(+)-Juniperol	(138)	M. p. 110°; [α]D + 16°	(3, 19)	Table 9
	"Kiganol"	(135) (?)	[α]D − 6.34°[c]	(147)	Table 8
	(+)-Nerolidol	(95)	M. p. 37°; [α]D + 12.5°	(19)	Fokienia hodginsii (54)
	"Pseudocedrol"	(141) (?)	n_D^{20} 1.5080; d_4^{20} 1.0062[c]	(19)	Table 8
	(+)-Widdrol	(148)	M. p. 98°; [α]D + 105°	(65, 89)	Table 9
	(+)-Viridiflorol	(123)	M. p. 75°; [α]D + 4°	(19)	Table 7
$C_{15}H_{26}O_2$	(−)-Occidiol	(110) (?)	[α]D − 125.5	(224)	Table 7
	(−)-Sesquibenihidiol	(112)	M. p. 128–130°; [α]D − 36°	(148)	Table 7
	(−)-Widdrol-α-epoxide	(150)	M. p. 156–157°; [α]D − 6°	(65)	Table 9
$C_{15}H_{28}O_2$	(−)-Cryptomeridiol	(107)	M. p. 134.5–135.5°; [α]D − 33.3°	(249, 251)	Table 7

a Rotations in chloroform unless otherwise stated. b Recent literature on the chemistry of the compounds. c Homogeneity doubtful. d Mixture of isomeric alcohols related to ketones (157) and (158).

Table 13*. Physical Constants and Distribution of Diterpenes in Cupressales.

Composition	Compound	Structure	Physical Constants[a]	References	Occurrence and References
Hydrocarbons					
$C_{20}H_{32}$	(–)-Dolabradiene	(171)	n_D^{20} 1.5240; $[\alpha]_D^{25}$ – 70°	(152, 154)	Thujopsis dolabrata (152)
	(–)-Hibaene = Cupressene	(190)	M. p. 29.5–30°; $[\alpha]_D$ – 49.9°	(31, 153)	Cupressus macrocarpa (32) Thujopsis dolabrata (153) Sciacopitys verti-cillata (152)
	(–)-Kaurene-(16) = (–)-Kaurene	(191)	M. p. 50–51°; $[\alpha]_D$ – 75.1°	(19, 30)	
	(–)-Kaurene-(15) = (–)-Isokaurene (= α-Crypto-merene)	(192)	M. p. 64°; $[\alpha]_D$ – 25.9°	(19, 30)	Cryptomeria japonica (256)
	(+)-Phyllocladene-(15) = (+)-Isophyllo-cladene	(188)	M. p. 112°; $[\alpha]_D$ + 23.7°	(19)	Cr. japonica (156) Cupressus macrocarpa (32) Libocedrus bidwillii (7) L. plumosa (7) Sciadopitys verti-cillata (7)
	(–)-Phyllocladene-(15) = (–)-Isophyllo-cladene	enantiomer of (188)	M. p. 111–112°; $[\alpha]_D$ – 24.5°	(19)	
	(+)-Rimuene	(170)	M. p. 55.5°; $[\alpha]_D$ + 53.7°	(19)	Libocedrus plumosa (7)
Alcohols and Aldehydes					
$C_{20}H_{32}O_2$	(–)-Sandaracopi-marinol	(168), $R^1 = CH_2OH$ $R^2 = H$	M. p. 63–65°; $[\alpha]_D$ – 11°	(190)	Cryptomeria japonica (190)
	(+)-Torulosal	(159), R = CHO	n_D^{25} 1.521; $[\alpha]_D$ + 29°; Semicarbazone, m. p. 195–197° (decomp.)	(64)	Cupressus torulosa (17)
	(+)-Manool	(159), $R = CH_3$	M. p. 53°; $[\alpha]_D$ + 30.4° (EtOH)	cf. (64)	Tetraclinis articulata (104) Cupressus semper-virens (66)

* Table 12 see p. 274.

Formula	Compound		M.p., Rotation[a]		Source	
	(+)-Phyllocladanol	(189)	M. p. 182–183°; $[\alpha]_D$ + 14.5° (MeOH)	(156)	C. torulosa	(17)
					Tetraclinis articulata	(104)
					Cryptomeria japonica	(156)
	Verticillol	cf. (193)	M. p. 104–105°; $[\alpha]_D$ + 168°	(85)	Sciadopitys verti-cillata	(85)
$C_{20}H_{34}O_2$	(+)-Torulosol	(159), $R = CH_2OH$	M. p. 110–111°; $[\alpha]_D$ + 31°	(64)	Cupressus torulosa	(17)
					Tetraclinis articulata	(104)
$C_{22}H_{34}O_3$	Torulosyl acetate	(159), $R = CH_2OAc$	not reported	(104)	T. articulata	(104)
Phenolic and Other Aromatic Diterpenes						
$C_{20}H_{28}O_2$	Totarolenone	(184)	mixture with totarolone (183)	(42)	T. articulata	(40)
$C_{20}H_{26}O_3$	(+)-Xanthoperol	(176)	M. p. 255–270° (decomp.); $[\alpha]_D$ + 132.5 (EtOH)	(26)	Chamaecyparis obtusa	(113)
	"Xanthoperol pre-cursor"	(177)	not pure	(26)	Juniperus communis	(28)
$C_{20}H_{28}O$	6,7-Dehydroferruginol	(174)	$[\alpha]_D$ − 60° (EtOH)	(25)	J. communis	(25)
$C_{20}H_{28}O_2$	(+)-Hinokione	(178)	M. p. 191–192°; $[\alpha]_D$ + 111.9° (EtOH)	(39, 41)	Cupressus torulosa	(17)
					Chamaecyparis obtusa	(147)
					Tetraclinis articulata	(40)
	(+)-Sugiol	(173)	M. p. 292–294°; $[\alpha]_D$ + 26°	cf. (28)	Cryptomeria japonica	(147)
					Chamaecyparis obtusa	(113)
					Juniperus communis	(28)
					J. mexicana	(77)
	(+)-Totarolone	(183)	M. p. 188–189.5°; $[\alpha]_D$ + 101.5° (EtOH)	(42)	Tetraclinis articulata	(40)
$C_{20}H_{30}O$	(+)-Ferruginol	(172)	$[\alpha]_D$ + 40.6° (EtOH); acetate: m. p. 81–82°; $[\alpha]_D$ + 60.3°	(24)	Cupressus torulosa	(17)
					Juniperus communis	(25)

a Rotations in chloroform unless otherwise stated.

(Table 13, continued.)

Composition	Compound	Structure	Physical Constants[a]	References	Occurrence and References
$C_{20}H_{30}O_2$	(+)-Totarol	(181)	M. p. 127–128° (EtOH); $[\alpha]_D + 42.5°$	(42)	J. communis (77); Thujopsis dolabrata (125); Tetraclinis articulata (40)
	(+)-Hinokiol	(178)	M. p. 240–42°; $[\alpha]_D + 74°$ (EtOH)	(39, 41)	Taiwania crypto-merioides (166); ("Taiwanin D" = Hinokiol) *
$C_{21}H_{28}O_3$	7-Hydroxytotarol	(185)	not pure		Cupressus torulosa (17); Chamaecyparis obtusa (147); Tetraclinis articulata (40, 104)
	Compound "A"	(182)	M. p. 191–192°; $[\alpha]_D + 225°$	(125)	Thujopsis dolabrata (125)
	Compound "B"	(180)	M. p. 176–177°; $[\alpha]_D + 220°$	(173)	Cupressus semper-virens (173)
	Cryptojaponol	(175) (?)	M. p. 204–205°; $[\alpha]_D + 25.3°$	(173)	C. sempervirens (173); Cryptomeria japonica (157)
	Ferruginol acetate	(172) (OAc instead of OH)	cf. Ferruginol above	(157) (155)	Cr. japonica (155)

Acids, Methyl Esters and Lactones

Composition	Compound	Structure	Physical Constants[a]	References	Occurrence and References
$C_{20}H_{24}O_4$	Sciadin	(162)	M. p. 160°; $[\alpha]_D + 10.3°$	(250)	Sciadopitys verti-cillata (250)
$C_{20}H_{28}O_2$	Sciadinone	(165)	M. p. 207°; $[\alpha]_D - 59.9°$	(143)	S. verticillata (143)
	Communic acid	(161)	Methyl ester; m. p. 105–106°; $[\alpha]_D + 48°$	(11, 198)	Cupressus semper-virens (2); Juniperus arizonica (77); J. communis (11); J. horizontalis (192); Tetraclinis articulata (174)

Name	Formula	No.	Properties	Ref.	Species	Ref.
Isopimaric acid	$C_{20}H_{30}O_3$	(167)	M. p. 162–164°; $[\alpha]_D$ 0°	(132a)	Juniperus conferta, J. japonica, J. rigida	(257)
					Cryptomeria japonica	(9, 10)
Sandaracopimaric acid = Cryptopimaric acid		(168), $R^1 = COOH$ $R^2 = H$	M. p. 172°, $[\alpha]_D − 17°$	(43, 133)	Tetraclinis articulata	(10)
Cupressic acid		(159)	Methyl ester, m. p. 70–71°, $[\alpha]_D + 52.5°$	(174)	Cupressus sempervirens	(174)
Isocupressic acid		(160)	Methyl ester (oil), $[\alpha]_D + 50°$; Methyl ester p-nitrobenzoate, m. p. 102–103°, $[\alpha]_D + 39.5°$	(174)	C. sempervirens	(174)
Methyl sciadopate	$C_{21}H_{34}O_4$	(164)	M. p. 108.5°, $[\alpha]_D − 0.7°$	(252)	Sciadopitys verticillata	(252)
Dimethyl sciadinonate	$C_{22}H_{28}O_6$	(166)	M. p. 122°, $[\alpha]_D − 45°$	(143)	Sc. verticillata	(143)
12 β-Acetoxysandaracopimaric acid	$C_{22}H_{32}O_3$	(168), $R^1 = COOH$ $R^2 = OAc$	M. p. 140.5–141.5° and 156–158.5° (dimorphous), $[\alpha]_D − 50°$ (EtOH)	(8, 104)	Tetraclinis articulata	(8, 104)

a Rotations in chloroform unless otherwise stated. * Lo, private communication.

Table 12. Physical Constants of Sesquiterpene Aldehydes, Ketones and Acids from Cupressales.

Composition	Compound	Structure	Physical Constants[a]	References[b]	Occurrence
Aldehydes					
$C_{15}H_{22}O$	α-Costal	(111), R = CHO	Semicarbazone, m. p. 218° (decomp.)	(134)	Table 7
	β-Costal	cf. (111)	not pure	(134)	Table 7
	(—)-Elemenal	(105)	$[α]_{600}$ — 11°; 2,4-Dinitrophenylhydrazone, m. p. 180°	(134)	Table 7
Ketones					
$C_{14}H_{20}O$	(+)-Mayurone	(147)	M. p. 70°; $[α]_D$ + 259°	(37, 38)	Table 9
$C_{15}H_{22}O$	Cryptomerione	(98) or (99)	$[α]_D$ — 38°; $λ_{max}^{EtOH}$ 236 mμ (ε 9600) and 304 mμ (ε 110); 2,4-Dinitrophenylhydrazone, m. p. 168—171°	(188)	*Cryptomeria japonica* (188)
$C_{15}H_{20}O$	(+)-Nootkatone	(117)	M. p. 36—37°; $[α]_D$ + 195.5°	(83, 172)	Table 7
	(+)-α-Cuparenone	(157)	M. p. 52—53°; $[α]_D$ + 177.1°	(37)	Table 9
	(+)-β-Cuparenone	(158)	n_D^{30} 1.5292; $[α]_D$ + 48.0°	(37)	Table 9
Acids					
$C_{15}H_{20}O_2$	(+)-Cuparenic Acid	(151), R = COOH	M. p. 158—160°; $[α]_D$ + 63°	(67)	Table 9
$C_{15}H_{22}O_2$	(—)-Hinokiic Acid	(146)	M. p. 169—170°; $[α]_D$ — 86°	(193)	Table 9
$C_{15}H_{24}O_3$	(—)-Cedrolic Acid	(143)	M. p. 181—187°; $[α]_D$ — 47°	(235)	Table 9

a Rotations in chloroform.   b Recent literature on the chemistry of the compounds.

References.

1. AHLUWALIA, V. K. and T. R. SESHADRI: Nootkatin from *Cupressus torulosa*. Current Sci. (India) 23, 154 (1954).
2. AHOND, A., P. CARNERO and B. GASTAMBIDE: Isolement de l'acide communique des cônes d'une variété naine de cyprès. Bull. soc. chim. France 1964, 348.
3. AKIYOSHI, S., H. ERDTMAN and T. KUBOTA: The Chemistry of the Order Cupressales. 26. The Identity of Junipene, Kuromatsuene and Longifolene and of Juniperol, Kuromatsuol. Macrocarpol and Longiborneol. Tetrahedron 9, 237 (1960).
4. ANDERSON, A. B. and J. GRIPENBERG: Antibiotic Substances from the Heartwood of *Thuja plicata* (Don). IV. The Constitution of β-Thujaplicin. Acta Chem. Scand. 2, 644 (1948).
5. ANDERSON, A. B., T. SCHEFFER and C. DUNCAN: On the Chemistry of Heartwood Decay on Ageing in Incense Cedar. Chem. and Ind. 1962, 1289.
6. ANDERSON, A. B. and E. C. SHERRARD: Dehydroperillic Acid, an Acid from Western Red Cedar (*Thuja plicata* Don) J. Amer. Chem. Soc. 55, 3813 (1933).
7. APLIN, R. T., R. C. CAMBIE and P. S. RUTLEDGE: The Taxonomic Distribution of Some Diterpene Hydrocarbons. Phytochemistry 2, 205 (1963).
8. APSIMON, J. W. and O. E. EDWARDS: Sandarac Acids: 6β-Hydroxysandaracopimaric Acid. Canad. J. Chem. 39, 2543 (1961).
9. APSIMON, J. W., B. GREEN and W. B. WHALLEY: Sandaraco- and Cryptopimaric Acid. J. Chem. Soc. (London) 1961, 753.
10. ARYA, V. P., C. ENZELL, H. ERDTMAN and R. RYHAGE: The Identity of Cryptopimaric Acid and Sandaracopimaric Acid. Acta Chem. Scand. 15, 682 (1961).
11. ARYA, V. P., H. ERDTMAN and T. KUBOTA: The Chemistry of the Order Cupressales. 41. The Structure and Stereochemistry of Communic Acid. Tetrahedron 16, 255 (1961).
12. AULIN-ERDTMAN, G.: Studies in the Tropolone Series. I. Thujaplicins and Nootkatin. Acta Chem. Scand. 4, 1031 (1950).
13. AULIN-ERDTMAN, G. and H. THEORELL: Studies in the Tropolone Series. III. Infrared Spectra. Acta Chem. Scand. 4, 1490 (1950).
14. BAKER, R. T. and H. C. SMITH: A Research on the Pines of Australia. Sydney: Govt. Printer. 1910.
15. BAKER, W. and W. D. OLLIS: Biflavonyls. In: W. D. Ollis (Edit.), Chemistry of Natural Phenolic Compounds, p. 152. New York: Pergamon Press. 1961.
16. BALOGH, B. and A. B. ANDERSON: Chemistry of Sequirins, New Phenolic Compounds from the Coast Redwood (*Sequoia sempervirens*). Phytochemistry 4, 569 (1965).
16a. BANTHORPE, D. V. and K. W. TURNBULL: The Biosynthesis of Thujane Derivatives in Higher Plants. Chem. Communs. 1966, 177.
17. BARRETO, H. S. and C. ENZELL: The Chemistry of the Order Cupressales. 39. Heartwood Constituents of *Cupressus torulosa* Don. Acta Chem. Scand. 15, 1313 (1961).
18. BATES, R. B. and E. K. HENDRICKSON: γ-Eudesmol from *Callitropsis araucarioides*. Chem. and Ind. 1962, 1759.
19. BEILSTEIN: Handbuch der organischen Chemie, drittes Ergänzungswerk. Berlin: Springer (1958—). Gives references to recent literature.
20. BENEŠOVÁ, V., V. SÝKORA, V. HEROUT and F. ŠORM: The Absolute Configuration of Costol (Sesquibenihiol) and Alantolactone. Chem. and Ind. 1958, 363.
21. BICHO, J. G., E. ZAVARIN and N. S. BHACCA: On the Occurrence of Hydronootkatinol in the Heartwood of *Cupressus Lindleyi* Klotsch. J. Organ. Chem. (USA) 28, 2927 (1963).

22. BILLEK, G.: Stilbene im Pflanzenreich. Fortschr. Chem. organ. Naturstoffe **22**, 115 (1964).
23. BOWDEKAR, A. S. and G. R. KELKAR: Structure and Absolute Configuration of Costic Acid — A New Sesquiterpenic Acid from Costus Root Oil. Tetrahedron **21**, 1521 (1965).
24. BRANDT, C. W. and L. G. NEUBAUER: Miro Resin. Part I. Ferruginol. J. Chem. Soc. (London) **1939**, 1031.
25. BREDENBERG, J. B.: Ferruginol and Δ^9-Dehydroferruginol. Acta Chem. Scand. **11**, 932 (1957).
26. — Investigations on Xanthoperol and its Precursor. Acta Chem. Scand. **14**, 385 (1960).
27. — The Chemistry of the Order Cupressales. 36. The Ethereal Oil of the Wood of *Juniperus communis* L. Acta Chem. Scand. **15**, 961 (1961).
28. BREDENBERG, J. B. and J. GRIPENBERG: Constituents of the Wood of *Juniperus communis* L. Acta Chem. Scand. **10**, 1511 (1956).
29. BREDENBERG, J. B. and J. RUNEBERG: The Presence of Savinin in the Wood of *Juniperus communis* L. and *Juniperus thurifera* L. Acta Chem. Scand. **15**, 455 (1961).
30. BRIGGS, L. H., B. F. CAIN, R. C. CAMBIE, B. R. DAVIES, P. S. RUTLEDGE and J. K. WILMSHURST: Kaurene. J. Chem. Soc. (London) **1963**, 1345.
31. BRIGGS, L. H., R. C. CAMBIE, P. S. RUTLEDGE and D. W. STANTON: The Structure of Cupressene. Tetrahedron Letters **1964**, 2223.
32. BRIGGS, L. H. and M. D. SUTHERLAND: The Essential Oil of *Cupressus macrocarpa*. J. Organ. Chem. (USA) **7**, 397 (1942).
33. BÜCHI, G. and J. D. WHITE: Photochemical Reactions. XIII. A Total Synthesis of ($\pm$)-Thujopsene. J. Amer. Chem. Soc. **86**, 2884 (1964).
34. CAMPBELL, R. B. and J. M. ROBERTSON: The Structure of Nootkatin. An X-Ray Determination. Chem. and Ind. **1952**, 1266.
35. CARLSSON, B., H. ERDTMAN, A. FRANK and W. E. HARVEY: The Chemistry of the Order Cupressales. 8. Heartwood Constituents of *Chamaecyparis nootkatensis* — Carvacrol, Nootkatin and Chamic Acid. Acta Chem. Scand. **6**, 690 (1952).
36. CHENG, Y.-S., T.-B. LO, L.-H. CHANG and Y.-T. LIN: Study of the Extractive Constituents from the Wood of *Libocedrus formosana* Florin. IV. Partial Hydrogenation of Thujic Acid. J. Chin. Chem. Soc. (Taiwan) **8**, 103 (1961).
37. CHETTY, G. L. and S. DEV: Ketones from "Mayur Pankhi". Some New Cuparene-Based Sesquiterpenoids. Tetrahedron Letters **1964**, 73.
38. — — Mayurone, a C_{14}-Sesquiterpene Ketone. Tetrahedron Letters **1965**, 3773.
39. CHOW, Y.-L.: The Chemistry of the Order Cupressales. 44. The Synthesis of DL-Hinokione Methyl Ether. Acta Chem. Scand. **16**, 1301 (1962).
40. CHOW, Y.-L. and H. ERDTMAN: The Chemistry of the Order Cupressales. 42. Heartwood Constituents of *Tetraclinis articulata* (Vahl) Masters. Acta Chem. Scand. **16**, 1291 (1962).
41. — — The Chemistry of the Order Cupressales. 43. The Structure and Configuration of Hinokiol and Hinokione. Acta Chem. Scand. **16**, 1296 (1962).
42. — — The Chemistry of the Order Cupressales. 45. The Structure and Configuration of Totarol and Totarolone. Acta Chem. Scand. **16**, 1305 (1962).
43. CHURCH, R. F. and R. E. IRELAND: Experiments Directed toward the Total Synthesis of Terpenes. V. The Synthesis of the ($\pm$)-9-Isopimaradienes. J. Organ. Chem. (USA) **28**, 17 (1963).

44. CONNOLLY, J. D., R. McCRINDLE, R. D. H. MURRAY and K. H. OVERTON: The Constitution and Stereochemistry of Rimuene. Tetrahedron Letters 1964, 1983.
45. CORBETT, R. E. and D. E. WRIGHT: The Extractives of Libocedrus bidwillii and Cupressus macrocarpa. Chem. and Ind. 1953, 1258.
46. CORBETT, R. E. and S. G. WYLLIE: The Structure and Stereochemistry of Rimuene. Tetrahedron Letters 1964, 1903.
47. DADSWELL, I. W. and H. E. DADSWELL: The Relation between Durability and the Extractives of the Cypress Pines (Callitris spp.). J. Council Sci. Ind. Res. 4, 208 (1931).
48. DAUBEN, W. G. and A. C. ASHCROFT: The Total Synthesis of ($\pm$)-Thujopsene. J. Amer. Chem. Soc. 85, 3673 (1963).
49. DAUBEN, W. G. and L. E. FRIEDRICH: The Mechanism of the Transformation of Thujopsene to Widdrol. Tetrahedron Letters 1964, 2675.
50. DAUBEN, W. G. and P. OBERHÄNSLI: Constituents of Hiba Wood Oil. The Isolation and Synthesis of Two Isomeric Cuprenenes. J. Organ. Chem. (USA) 31, 315 (1966). We thank Dr. Dauben for an advance copy of this paper.
50a. DAUBEN, W. G., B. WEINSTEIN, P. LIM and A. B. ANDERSON: The Structure of δ-Cadinol. Tetrahedron 15, 217 (1961).
51. DE MAYO, P., J. R. ROBINSON, E. Y. SPENCER and R. W. WHITE: The Biogenesis of Helminthosporal. Experientia 18, 359 (1962).
52. DE MAYO, P., R. E. WILLIAMS, G. BÜCHI and S. H. FAIRHELLER: The Absolute Stereochemistry of Copaene. Tetrahedron 21, 619 (1965).
53. DEWAR, M. J. S.: Structure of Stipitatic Acid. Nature 155, 50 (1945).
54. DOLEJŠ, L. and V. HEROUT: On Terpenes. CXXIX. Composition of the Oil and Root Extract from Fokienia Hodginsii. Collect. Czech. Chem. Comm. 26, 2045 (1961).
55. DOLEJŠ, L., O. MOTL, M. SOUČEK, V. HEROUT and F. ŠORM: On Terpenes. CVIII. Epimeric Aromadendrenes. Stereoisomerism of Ledol, Viridiflorol and Globulol. Collect. Czech. Chem. Comm. 25, 1483 (1960).
56. DOLEJŠ, L. and F. ŠORM: On Terpenes. CXIII. Structure of Aromadendrene, Alloaromadendrene, Globulol, Ledol and Viridiflorol. Collect. Czech. Chem. Comm. 25, 1837 (1960).
57. DUFF, S. R. and H. ERDTMAN: Occurrence of Carvacrol Methyl Ether in the Heartwood of Chamaecyparis nootkatensis (Lamb.) Spach. Chem. and Ind. 1953, 747.
58. DUFF, S. R., H. ERDTMAN and W. E. HARVEY: The Chemistry of the Order Cupressales. 11. Heartwood Constituents of Chamaecyparis nootkatensis (Lamb.) Spach. Nootkatin. Acta Chem. Scand. 8, 1073 (1954).
59. DYRSSEN, D.: IPT — A New Extracting Agent for Metal Ions. Acta Chem. Scand. 15, 1614 (1961).
60. — Extraction of Metal Ions with β-Isopropyltropolone (IPT). I. Trans. Roy. Inst. Technol., Stockholm, No. 188 (1962).
61. ENZELL, C.: The Total Synthesis of ($\pm$)-Widdrol. Tetrahedron Letters 1962, 185.
62. — Studies on Conifer Terpenes. Svensk Kem. Tidskr. 74, 559 (1962).
63. — The Chemistry of the Order Cupressales. 24. Heartwood Constituents of Chamaecyparis thyoides (L) Britt. Acta Chem. Scand. 14, 81 (1960).
64. — The Chemistry of the Order Cupressales. 38. The Structures of Torulosol, Torulosal and Agatholic Acid. Acta Chem. Scand. 15, 1303 (1961).
65. — The Chemistry of the Order Cupressales. 47. The Structures and Absolute Configurations Widdrol and Widdrol-α-epoxide. Acta Chem. Scand. 16, 1553 (1962).

66. Enzell, C. and H. Erdtman: The Chemistry of the Order Cupressales. 19. The Occurrence of Manool in *Cupressus sempervirens* L. Acta Chem. Scand. **11**, 902 (1957).

67. — — The Chemistry of the Order Cupressales. 21. Cuparene and Cuparenic Acid, Two New Sesquiterpenic Compounds with a New Carbon Skeleton. Tetrahedron **4**, 361 (1958).

68. Enzell, C. and M. Krolikowska: The Chemistry of the Order Cupressales. 48. Heartwood Constituents of *Cupressus arizonica* Greene. Ark. Kemi **20**, 157 (1962).

69. Enzell, C. and O. Theander: The Constituents of Conifer Needles. 2. Pinifolic Acid, a New Diterpene Acid Isolated from *Pinus silvestris* L. Acta Chem. Scand. **16**, 607 (1962).

70. Enzell, C. R. and B. R. Thomas: The Chemistry of the Order Araucariales. 2. The Wood Resin of *Agathis australis*. Acta Chem. Scand. **19**, 913 (1965).

70a. — — The Chemistry of the Order Araucariales. 5. Agatharesinol. Tetrahedron Letters **1966**, 2395.

71. Erdtman, H.: Kemiska synpunkter på barrträdens systematik. Finska Kemistsamfundets Medd. **3–4, 58**, 55, 73 (1949).

72. — Organic Chemistry and Conifer Taxonomy. In: A. Todd (Edit.), Perspectives in Organic Chemistry, p. 453. New York and London: Interscience. 1956.

73. — Conifer Chemistry and Taxonomy of Conifers. In: K. Kratzl and G. Billek (Edits.), Biochemistry of Wood, p. 1. Proc. 4th Intern. Congr. Biochem. II, 1959.

74. — Some Aspects of Chemotaxonomy. In: T. Swain (Edit.), Chemical Plant Taxonomy, p. 89. London and New York: Academic Press. 1963.

75. — Heartwood Extractives of Conifers Their Fungicidal and Insect-Repellent Properties and Taxonomic Interest. Tappi **32**, 305 (1949).

76. — The Chemistry of Heartwood Constituents of Conifers and their Taxonomic Importance. 14th Congr. Pure Appl. Chem., Experientia Suppl. II, p. 156. Basel: Birkhäuser. 1955.

77. — Some Aspects of Chemotaxonomy. Pure Appl. Chem. **6**, 679 (1963).

78. Erdtman, H. and J. Gripenberg: Antibiotic Substances from the Heartwood of *Thuja plicata* Don. Nature **161**, 719 (1948).

79. — — Antibiotic Substances from the Heartwood of *Thuja plicata* Don. II. The Constitution of γ-Thujaplicin. Acta Chem. Scand. **2**, 625 (1948).

80. — — Structure of Thujic Acid ("Dehydroperillic Acid"). Nature **164**, 316 (1949).

81. Erdtman, H. and W. E. Harvey: The Chemistry of the Order Cupressales. 9. Nootkatin. Chem. and Ind. **1952**, 1267.

82. Erdtman, H., W. E. Harvey and J. G. Topliss: The Chemistry of the Order Cupressales. 16. Heartwood Constituents of *Chamaecyparis nootkatensis* (Lamb.) Spach. The Structure of Chamic and Chaminic Acid. Acta Chem. Scand. **10**, 1381 (1956).

83. Erdtman, H. and Y. Hirose: The Chemistry of the Order Cupressales. 46. The Structure of Nootkatone. Acta Chem. Scand. **16**, 1311 (1962).

84. Erdtman, H. and T. Norin: The Identity of Thujopsene and Widdrene and of Hinokiic and Widdrenic Acids. Acta Chem. Scand. **13**, 1124 (1959).

85. Erdtman, H., T. Norin, M. Sumimoto and A. Morrison: Verticillol, a Novel Type of Conifer Diterpene. Tetrahedron Letters **1964**, 3879.

86. Erdtman, H. und Z. Pelchowicz: Die Chemie der Ordnung Cupressales. 12. Inhaltsstoffe des Kernholzes von *Biota orientalis* Endl. Chem. Ber. **89**, 341 (1956).

87. ERDTMAN, H. and Z. PELCHOWICZ: The Chemistry of the Order Cupressales. 15. Heartwood Constituents of *Austrocedrus chilensis* (D. Don.) Florin et Boutelje (= *Libocedrus chilensis* (D. Don.) Endl.). Acta Chem. Scand. 9, 1728 (1955).

88. ERDTMAN, H., Z. PELCHOWICZ and J. G. TOPLISS: The Chemistry of the Order Cupressales. 17. Heartwood Constituents of *Pilgerodendron uviferum* (D. Don.) Florin (= *Libocedrus tetragona* Endl.). Acta Chem. Scand. 10, 1563 (1956).

89. ERDTMAN, H. and B. R. THOMAS: The Chemistry of the Order Cupressales. 20. Heartwood Constituents of the Genus *Widdringtonia*. Acta Chem. Scand. 12, 267 (1958).

90. ERDTMAN, H. and J. G. TOPLISS: The Chemistry of the Order Cupressales. 18. Nootkatene, a New Sesquiterpene Type Hydrocarbon from the Heartwood of *Chamaecyparis nootkatensis* (Lamb.) Spach. Acta Chem. Scand. 11, 1157 (1957).

91. ERDTMAN, H. und H. VORBRÜGGEN: Die Chemie der Ordnung Cupressales. 32. Über die Inhaltsstoffe des Kernholzes von *Athrotaxis selaginoides* Don., *Athrotaxis cupressoides* Don. and *Chamaecyparis pisifera* Sieb. et Zucc. Acta Chem. Scand. 14, 2161 (1960).

92. ERDTMAN, H. and C. A. WACHTMEISTER: Phenoldehydrogenation as a Biosynthetic Reaction. In: Festschrift A. Stoll, p. 144. Basel: Birkhäuser. 1957.

93. FITZGERALD, D. B., J. L. HARTWELL and J. L. LEITER: Distribution of Tumour-Damaging Lignans among Conifers. J. Natl. Cancer Inst. 18, 83 (1957).

94. FLORIN, R.: The Distribution of Conifer and Taxad Genera in Time and Space. Acta Horti Bergiani 20, 121–312 (1963).

95. FORSÉN, S. and T. NORIN: Proton Magnetic Resonance. Investigations on the Structure of Thujopsene and Hinokiic Acid. Acta Chem. Scand. 15, 592 (1961).

96. FUJITA, Y.: A Consideration on the Mode of Formation of Thujaplicins and Allied Tropolones in Plants. Bull. Osaka Ind. Res. Inst. 6, 199 (1955).

97. FUNAOKA, K., Y. KURODA, Y. KAI and T. KONDO: On the Phenolic Constituents from *Cryptomeria japonica* D. Don. Isolation and a Few Characteristics of Two Phenolic Substances. J. Japan. Wood Res. Soc. 9, 139 (1963).

98. GARDNER, J. A. F. and G. M. BARTON: Occurrence of β-Dolabrin (4-Isopropenyltropolone) in Western Red Cedar (*Thuja plicata* Don.). Canad. J. Chem. 36, 1612 (1958).

99. GARDNER, J. A. F., G. M. BARTON and H. MACLEAN: Occurrence of 2,7-Dihydroxy-4-isopropyl-2,4,6-cycloheptatrien-1-one (7-Hydroxy-4-isopropyltropolone) in Western Red Cedar (*Thuja plicata* Don.). Canad. J. Chem. 35, 1039 (1957).

100. GARDNER, J. A. F., B. F. MACDONALD and H. MACLEAN: The Polyoxyphenols of Western Red Cedar (*Thuja plicata* Don.). II. Degradation Studies on Plicatic Acid, a Possible Lignan Acid. Canad. J. Chem. 38, 2387 (1960).

101. GARDNER, P. D. and W. J. HORTON: The Identification of Sagittol as Eudesmol. J. Amer. Chem. Soc. 77, 3646 (1955).

102. GORYAEV, M. I. and G. A. TOLSTIKOV: Synthesis of β-Cedrene. Doklady Akad. Nauk. (USSR) 139, 363 (1961) [Chem. Abstr. 56, 2475 (1962)].

103. GORYAEV, M. I., G. A. TOLSTIKOV, L. A. IGNATOVA and A. D. DEMBITSKII: Natural β-Cedrene. Doklady Akad. Nauk. (USSR) 146, 1331 (1962) [Chem. Abstr. 58, 9149 (1963)].

104. GOUGH, L. J.: Conifer Resin Constituents. Chem. and Ind. 1964, 2059.

105. GREENWOOD, J. M., J. K. SUTHERLAND and A. TORRE: Conversion of Humulene into Caryophyllene. Chem. Communs. (London) 1965, 410.

106. Gripenberg, J.: Antibiotic Substances from the Heartwood of *Thuja plicata* Don. III. The Constitution of α-Thujaplicin. Acta Chem. Scand. **2**, 639 (1948).

107. — Antibiotic Substances from the Heartwood of *Thuja plicata* D. Don. VI. The Structure of Thujic Acid ("Dehydroperillic Acid"). Acta Chem. Scand. **3**, 1137 (1949).

108. — The Constituents of the Wood of *Thuja occidentalis*. Acta Chem. Scand. **3**, 782 (1949).

109. — The Chemistry of the Order Cupressales. 9. The Synthesis of Hexahydrothujic Acid. Acta Chem. Scand. **6**, 854 (1952).

110. — Confirmation of the Structure of Thujic Acid by Nuclear Magnetic Resonance. Acta Chem. Scand. **10**, 487 (1956).

111. Hartsuck, J. A. and I. C. Paul: Molecular Structure of Humulene; X-Ray Analysis of the Silver Nitrate Adduct. Chem. and Ind. **1964**, 977.

112. Hartwell, J. L., J. M. Johnson, D. B. Fitzgerald and M. Belkin: Podophyllotoxin from *Juniperus* Species; Savinin. J. Amer. Chem. Soc. **75**, 235 (1953).

113. Hata, K., M. Sogo and Y. Kamemaru: Chemical Studies on the Bark. IX. On the Extractives from Outer Bark of *Chamaecyparis obtusa* S. et Z. J. Japan Wood Res. Soc. **8**, 167 (1962).

114. Hegnauer, R.: The Taxonomic Significance of Alkaloids. In: T. Swain (Edit.) Chemical Plant Taxonomy, p. 389. London and New York: Academic Press. 1963.

115. Hendrickson, J. B.: Stereochemical Implications in Sesquiterpene Biogenesis. Tetrahedron **7**, 82 (1959).

116. Herout, V., O. Motl and F. Šorm: On Terpenes. LX. Composition of Juniper Berry Oil. Chem. Listy **48**, 589 (1954); Collect. Czech. Chem. Comm. **19**, 990 (1954).

117. Hirose, Y.: Terpenoids. VII. The Structure of a New Tropolone, Nootkatinol. Agric. Biol. Chem. (Tokyo) **27**, 795 (1963).

118. — Terpenoids. VIII. On the Occurrence of Nootkatinol (Hydronootkatinol) in the Wood of *Juniperus rigida* Sieb. et Zucc. J. Japan Wood Res. Soc. **10**, 251 (1964).

119. Hirose, Y. and T. Nakatsuka: Terpenoids. Part I. The Structure of Occidentalol, a New Sesquiterpene Alcohol from *Thuja occidentalis* L. Bull. Agric. Chem. Soc. Japan **20**, 215 (1956).

120. — — Terpenoids. III. Composition of the Essential Oil from the Wood of *Biota orientalis* Endl. (*Thuja orientalis* L.). J. Japan Wood Res. Soc. **4**, 26 (1958).

121. — — Terpenoids. IV. The Structure of Occidol, a New Sesquiterpene Alcohol from *Thuja occidentalis* L. Bull. Agric. Chem. Soc. Japan **23**, 143 (1959).

122. — — Terpenoids. V. The Synthesis of Occidol. Bull. Agric. Chem. Soc. Japan **23**, 253 (1959).

123. — — Terpenoids. IX. Composition of the Wood of *Thuja standishii* Carr. J. Japan Wood Res. Soc. **10**, 253 (1964).

124. Hirose, Y., N. Oishi, H. Nagaki and 1. Nakatsuka: The Structure of Hinokiresinol. Tetrahedron Letters **1963**, 3665.

125. Hodges, R.: Constituents of the Leaves of *Thujopsis dolabrata*. J. Chem. Soc. (London) **1961**, 4247.

126. Honwad, V. K. and A. S. Rao: Absolute Configuration of (—)-α-Curcumene. Tetrahedron **21**, 2593 (1965).

127. Hsü, H. Y.: Studies on the Chemical Constituents of the Plants of Coniferae and Allied Orders in Taiwan. Bull. Taiwan Provinc. Hyg. Lab., Taipei, August 1959, 1. Cf. also C. T. Chang, T. S. Chen, T. Ueng, S. T. Choong and F. C. Chen. J. Formosan Sci. **14**, 1 (1960).

128. ICHIKAWA, N.: Studies on the Constitution of Shonanic Acid, One of the Two Characteristic Volatile Acids from *Libocedrus formosana* Florin. I–VI. Bull. Chem. Soc. Japan **11**, 759 (1936); **12**, 233, 243, 253, 258, 267 (1937).

129. IMAMURA, H.: The Heartwood Extractives of *Chamaecyparis pisifera.* On the Constitution of Sawarinin. Bull. Govt. Forest Exp. Station (Japan) No. **138**, 1 (March 1962).

130. INONE, H. and T. NOGUCHI: Treatment of Thuja Oil by Ion Exchange Resin. Yukagaku **11**, 523 (1962) [Chem. Abstr. **59**, 2706 (1963)].

131. IRELAND, R. E. and L. N. MANDER: The Total Synthesis of ($\pm$)-Rimuene. Tetrahedron Letters **1964**, 3453.

132. — — The Total Synthesis of ($\pm$)-Hibaene. Tetrahedron Letters **1965**, 2627.

132a. IRELAND, R. E. and J. NEWBOULD: Experiments Directed toward the Total Synthesis of Terpenes. VI. The Stereochemistry of Isopimaric Acid. J. Organ. Chem. (USA) **28**, 23 (1963).

133. IRELAND, R. E. and P. W. SCHIESS: Experiments Directed toward the Total Synthesis of Terpenes. IV. The Synthesis of ($\pm$)-Sandaracopimaradiene and ($\pm$)-Pimaradiene. J. Organ. Chem. (USA) **28**, 6 (1963).

134. ITO, S., K. ENDO, H. HONMA and K. OTA: New Constituents of *Thujopsis dolabrata.* Tetrahedron Letters **1965**, 3777.

135. ITO, S., K. ENDO and T. NOZOE: Configuration of Widdrol. Chem. Pharm. Bull. (Japan) **11**, 132 (1963).

136. — — — Stereochemistry of Widdrol. Tetrahedron Letters **1964**, 3375.

137. JOYE, N. M., Jr., E. M. ROBERTS, R. V. LAWRENCE, L. J. GOUGH, M. D. SOFFER and O. KORMAN: The Structure of the Dicyclic Diterpenoids of Slash Pine. The Identity of Elliotinoic Acid and Communic Acid. J. Organ. Chem. (USA) **30**, 429 (1965).

138. KAFUKU, K. und R. KATO: Über das ätherische Öl der Taiwaniaceder. Bull Chem. Soc. Japan **6**, 65 (1931) [Chem. Zbl. **1931**, I, 3731].

139. KAFUKU, K. und T. NOZOE: Untersuchungen über die Bestandteile des flüchtigen Öles aus dem Blatt von *Chamaecyparis obtusa*, Sieb. et Zucc., *F. Formosana*, Hayata oder Arisan-Hinoki. II. Bull. Chem. Soc. Japan **6**, 111 (1931) [Chem. Zbl. **1932**, I, 83].

140. KAFUKU, K., T. NOZOE und C. HATA: Untersuchungen über die Bestandteile des flüchtigen Öles aus dem Blatt von *Chamaecyparis obtusa*, Sieb. et Zucc., *F. Formosana*, Hayata oder Arisan-Hinoki. I. Bull. Chem. Soc. Japan **6**, 40 (1931) [Chem. Zbl. **1931**, II, 3218].

141. KAI, Y.: On the Phenolic Constituents from *Cryptomeria japonica* D. Don. III. Sugiresinol. J. Japan Wood Res. Soc. **11**, 23 (1965).

142. KANEKO, C., S. HAYASHI and M. ISHIKAWA: On the Structure of Verticillol. Chem. Pharm. Bull. (Japan) **12**, 1510 (1964).

143. KANEKO, C., T. TSUCHIYA and M. ISHIKAWA: Isolation and Structure of New Diterpenes, Dimethyl Sciadinonate and Sciadinone, from *Sciadopitys verticillata.* Chem. Pharm. Bull. (Japan) **11**, 271 (1963).

144. KAPADIA, V. H., B. A. NAGASAMPAGI, V. G. NAIK and S. DEV: Studies on Sesquiterpenes. XXII. Structure of Muscatone and Copaene. Tetrahedron **21**, 607 (1965).

145. KARIYONE, T.: Studies on the Components of Plants Belonging to Coniferae and Allied Orders; Especially on the Wax and Biflavones of the Leaves. J. Pharmac. Soc. Japan **16**, 1 (1962).

146. KARIYONE, T., M. TAKAHASHI, K. ISOI and M. YOSHIKURA: Constituents of the Plant of Coniferae and Allied Orders. XX. Components of the Leaves of *Metasequoia glyptostroboides* (1). J. Pharmac. Soc. Japan **78**, 801 (1958).

147. KARRER, W.: Konstitution und Vorkommen der organischen Pflanzenstoffe. Basel: Birkhäuser. 1958.
148. KATSURA, S.: The Constituents of the Volatile Oil from the Root of *Chamaecyparis formosensis* Matsum. I. J. Chem. Soc. Japan **63**, 1460 (1942) [Chem. Abstr. **41**, 3447 (1947)].
149. — Studies on the Components of the Root Oil from *Chamaecyparis obtusa*. VI. J. Chem. Soc. Japan **63**, 1483 (1942).
150. KIER, L. B., D. B. FITZGERALD and S. BURGETT: Isolation of Podophyllotoxin from *Callitris drummondi*. J. Pharm. Sci. **52**, 502 (1963).
151. KITAHARA, Y. and M. FUNAMIZU: The Synthesis of Nootkatin, a Sesquiterpenoid Tropolone. Bull. Chem. Soc. Japan **31**, 782 (1958).
152. KITAHARA, Y. and A. YOSHIKOSHI: The Structure of Dolabradiene. Tetrahedron Letters **1964**, 1755.
153. — — The Structure of Hibaene. Tetrahedron Letters **1964**, 1771.
154. KITAHARA, Y., A. YOSHIKOSHI and S. OIDA: Total Synthesis of Dolabradiene. Tetrahedron Letters **1964**, 1763.
155. KONDO, T., H. IMAMURA and M. SUDA: Exuded Resin of *Cryptomeria japonica*. I. Phenolic and Neutral Constituents. 1. Yakugaku Zasshi **79**, 1298 (1959) [Chem. Abstr. **54**, 4524 (1960)].
156. — — — Wood Extractives. IX. A New Diterpene Alcohol from *Cryptomeria japonica*. Bull. Agric. Chem. Soc. Japan **24**, 65 (1960) [Chem. Abstr. **54**, 12185 (1960)].
157. KONDO, T., M. SUDA and M. TEJIMA: Exuded Resin of *Cryptomeria japonica*. II. Phenolic and Neutral Components. 2. A New Phenolic Diterpene, Cryptojaponol. Yakugaku Zasshi **82**, 1252 (1962) [Chem. Abstr. **59**, 1685 (1963)].
158. KRITCHEVSKY, G. and A. B. ANDERSON: Composition of the Volatile Oil from *Chamaecyparis lawsoniana* Stump Heartwood. J. Amer. Pharmac. Assoc. **44**, 535 (1955).
159. KURTH, E. F.: Methyl Ester of Dehydroperillic Acid, an Odoriferous Constituent of Western Red Cedar *(Thuja plicata)*. J. Amer. Chem. Soc. **72**, 5778 (1950).
160. KURTH, E. F. and H. B. LACKEY: The Constituents of Sierra Juniper Wood *(Juniperus occidentalis* Hooker). J. Amer. Chem. Soc. **70**, 2206 (1948).
161. LI, H.-L.: A New Species of *Chamaecyparis*. Morris Arboretum Bull. **13**, 43 (1962).
162. LIN, Y.-T. and K. T. LIN: Study of the Extractive Constituents from the Wood of *Libocedrus formosana* Florin. VI. Thin Layer Chromatography of Tropolones. J. Chin. Chem. Soc. (Taiwan) **10**, 156 (1963).
162a. — — A Study of the Extractive from the Wood of *Libocedrus formosana* Florin. VIII. The Phenolic Constituents. Shonanol, a New Diterpene Phenol. J. Chin. Chem. Soc. (Taiwan) **12**, 39 (1965).
163. LIN, Y.-T., Y.-S. LIN and T.-B. LO: Extractive Components from the Heartwood of *Taiwania cryptomerioides* Hayata. Part 2. Presence of Hinokiol, Savinin and α-Cadinol. J. Chin. Chem. Soc. (Taiwan) **10**, 163 (1963).
164. LIN, Y.-T., T.-B. LO and Y.-S. CHENG: Study of the Extractive Constituents from the Wood of *Libocedrus formosana* Florin. III. Isolation of Chamic Acid, Acid C and Acid E. J. Chin. Chem. Soc. (Taiwan) **7**, 166 (1960).
165. LIN, Y.-T., T.-B. LO and T.-H. LIN: Study of the Extractive Constituents from the Wood of *Libocedrus formosana* Florin. II. Interconversions Between Isoshonanic Acid and Thujic Acid. J. Chin. Chem. Soc. (Taiwan) **3**, 36 (1956).
166. LIN, Y.-T., T.-B. LO and E.-H. SHIN: Extractive Components from the Heartwood of *Taiwania cryptomerioides* Hayata. I. Isolation of Four Crystalline Components from the Acetone Extract. J. Chin. Chem. Soc. (Taiwan) **2**, 87 (1955).

167. LIN, Y.-T., K.-T. WANG and C.-L. CHEN: The Essential Oil of *Chamaecyparis taiwanensis* Masamune et Suzuki. I. Acidic Components. J. Chin. Chem. Soc. (Taiwan) **2**, 91 (1955).

168. — — — The Essential Oil of *Chamaecyparis taiwanensis* Masamune et Suzuki. II. Chamaecin, a New Natural Tropolonoid. J. Chin. Chem. Soc. (Taiwan) **2**, 126 (1955).

169. LIN, Y.-T., K.-T. WANG and B. WEINSTEIN: Phytochemical Studies. The Structure of Taiwanin A. Chem. Communs. (London) **1965**, 592.

170. LO, T.-B. and Y.-T. LIN: Study on the Extractive Constituents from the Wood of *Libocedrus formosana* Florin. I. J. Chin. Chem. Soc. (Taiwan) **3**, 30 (1956).

171. MACLEAN, H. and J. A. F. GARDNER: Analytical Method for Thujaplicins. Analyt. Chemistry **28**, 509 (1956).

172. MACLEOD, W. D., Jr.: The Constitution of Nootkatone, Nootkatene and Valencene. Tetrahedron Letters **1965**, 4779.

173. MANGONI, L. and M. BELARDINI: The Isolation and Structure of Two Diterpene 1,3-Diones. Tetrahedron Letters **1964**, 2643.

174. — — Components of *Cupressus sempervirens* Resin. I. Communic Acid, Cupressic Acid, and Isocupressic Acid. Gazz. chim. ital. **94**, 1108 (1964) [Chem. Abstr. **62**, 5304 (1965)].

175. MASAMURA, M.: Constituents of Young Leaves of Chyabohiba. IV. Hinokiic Acid and *d*-Sesamin. Nippon Kagaku Zasshi **76**, 1318 (1955) [Chem. Abstr. **51**, 17900 (1957)].

176. — Constituents of Young Leaves of Chyabohiba. I. Isolation of Crystalline Constituents. Nippon Kagaku Zasshi **76**, 423 (1955) [Chem. Abstr. **51**, 17900 (1957)].

177. MASAMURA, M. and F. S. OKAMURA: Identity of Hibalactone and Savinin. J. Amer. Chem. Soc. **77**, 1906 (1955).

178. MCPHAIL, A. T., R. I. REED and G. A. SIM: Stereochemistry of Humulene. Chem. and Ind. **1964**, 976.

179. MIYASAKA, T.: Inter-relation of Methyl Sciadopate with Communic Acid and Daniellic Acid. Chem. and Pharm. Bull. (Japan) **12**, 744 (1964).

180. MODICA, G. DI e P. F. ROSSI: Su sostanze estratte del durame di *"Cupressus sempervirens"* L. Nota II. Isolamento di nootkatina. Ann. chim. (Roma) **46**, 842 (1956).

181. MOTL, O., V. HEROUT and F. ŠORM: On Terpenes. CXII. The Composition of the Oil from *Juniperus oxycedrus* L. Berries. Collect. Czech. Chem. Comm. **25**, 1656 (1960).

182. — — — On Terpenes. LXXIV. Junenol, a New Sesquiterpenic Alcohol from Juniper Oil. Chem. Listy **50**, 1282 (1956); Collect. Czech. Chem. Comm. **22**, 785 (1957).

183. — — — On Terpenes. LXXXV. The Structure of "Juniper Camphor", Another Alcohol of the Selinane Type from Juniper Oil. Chem. Listy **52**, 116 (1958); Collect. Czech. Chem. Comm. **23**, 1293 (1958).

184. — — — Structure of the Sesquiterpenic Hydrocarbon Ylangene. Tetrahedron Letters **1965**, 451.

185. MOTL, O., V. SÝKORA, V. HEROUT and F. ŠORM: On Terpenes. LXXXVI. The Structure of Two Crystalline Cadinols. Chem. Listy **52**, 316 (1958); Collect. Czech. Chem. Comm. **23**, 1297 (1958).

186. MOUSSERON, M., R. GRANGER et M. RONATROUX: Sur la constitution de l'essence et de l'huile pyrogénée de *Juniperus oxycedrus* L. C. R. hebd. Séances Acad. Sci. **208**, 1411 (1939).

187. MURTI, V. V. S., P. V. RAMAN and T. R. SESHADRI: Cupressuflavone, a New Member of the Biflavonyl Group. Tetrahedron Letters 1964, 2995.
188. NAGAHAMA, S.: Terpenoids. VIII. Sesquiterpenoids from the Wood Oil of "Sugi" (*Cryptomeria japonica* D. Don.). Bull. Chem. Soc. Japan 37, 1029 (1964).
189. — Hydration of Thujopsene to Widdrol. Bull. Chem. Soc. Japan 33, 1467 (1960).
190. — Terpenoids. VII. The Isolation of Sandaracopimarinol from the Wood Oil of "Sugi". Bull. Chem. Soc. Japan 37, 886 (1964).
191. NAKAZAKI, M.: Absolute Configuration of (+)-Occidol. Chem. and Ind. 1962, 413.
192. NARASIMHACHARI, N. and E. VON RUDLOFF: The Chemical Composition of the Wood and Bark Extractives of *Juniperus horizontalis* Moench. Canad. J. Chem. 39, 2572 (1961).
193. NORIN, T.: The Chemistry of the Order Cupressales. 40. The Structure of Thujopsene and Hinokiic Acid. Acta Chem. Scand. 15, 1676 (1961).
194. — Chanootin, a Bicyclic C$_{15}$-Tropolone from the Heartwood of *Chamaecyparis nootkatensis* (Lamb.) Spach. Ark. Kemi 22, 129 (1964).
195. — The Chemistry of the Order Cupressales. 49. The Configuration of Thujopsene. Acta Chem. Scand. 17, 738 (1963).
196. — The Absolute Configuration of Chamic, Chaminic and Isochaminic Acids. Ark. Kemi 22, 123 (1964).
197. — Studies on the Chemistry of Terpenes. Svensk Kem. Tidskr. 76, 97 (1964).
198. — The Configuration of Communic Acid. Acta Chem. Scand. 19, 1020 (1965).
199. NORIN, T. and L. WESTFELT: Thin Layer, Column, and Gas Liquid Chromatography of Resin Acid Esters and Related Terpenes. Acta Chem. Scand. 17, 1828 (1963).
200. NOZOE, T.: Studies on Hinokitiol. III. General Survey. Sci. Repts. Tôhoku Univ. [1] 34, 199 (1950).
201. — Studies on Hinokitiol. Part V. On the Revision of the Hinokitiol Structure. Sci. Repts. Tôhoku Univ. [1] 36, 82 (1952).
202. — Natural Tropolones and Some Related Troponoids. Fortschr. Chem. organ. Naturstoffe 13, 232 (1956).
203. NOZOE, T. and S. KATSURA: On the Structure of Hinokitiol. J. Pharmac. Soc. Japan 64, 181 (1944).
204. NOZOE, T., K. TAKASE and M. OGATA: β-Dolabrin: A New Natural Tropolone. Chem. and Ind. 1957, 1070.
205. NOZOE, T. and H. TAKESHITA: Cuparene and Cuprenene. Tetrahedron Letters 1960, No. 23, 14.
206. NOZOE, T., A. YASUE and K. YAMANE: On the Acidic Constituents of the Essential Oil of *Thujopsis dolabrata*. Occurrence of α-Thujaplicin. Proc. Japan Acad. 27, 15 (1951).
207. OSHIMA, M.: Die Termiten Formosas und die Verfahren zur Verhütung ihrer Schäden. Philippine J. Sci. 15, 319 (1919) [Chem. Zbl. 1923, I, 218].
208. OURISSON, G.: Molecular Rearrangements of Terpenes. Proc. Chem. Soc. (London) 1964, 86.
209. PARKER, W., R. RAMAGE and R. A. RAPHAEL: The Total Synthesis of (±)-Cuparene. J. Chem. Soc. (London) 1962, 1558.
210. PASTO, D. J.: The Partial Reduction of Thujic Acid. J. Organ. Chem. (USA) 27, 2786 (1962).
211. PETTERSSON, E. and J. RUNEBERG: The Chemistry of the Order Cupressales. 34. Heartwood Constituents of *Juniperus procera* Hochst. and *Juniperus californica* Carr. Acta Chem. Scand. 15, 713 (1961).

212. PILGER, R. and H. MELCHIOR: Gymnospermae. Nacktsamer (Archispermae). In: A. Engler, Syllabus der Pflanzenfamilien, I, S. 312, spez. S. 332. Berlin: Bornträger. 1954.

213. PILO, C. and J. RUNEBERG: The Chemistry of the Order Cupressales. 25. Heartwood Constituents of *Juniperus chinensis* L. Acta Chem. Scand. 14, 353 (1960).

214. PLATTNER, PL. A., A. FÜRST, A. ESCHENMOSER, W. KELLER, H. KLÄUI, ST. MEYER und M. ROSNER: Über Sesquiterpene und Azulene. 106. Mitt. Die Konstitution des Cedrens. Helv. Chim. Acta 36, 1845 (1953).

215. PLÍVA, J., V. HEROUT, B. SCHNEIDER and F. ŠORM: On Terpenes. XLIII. Infrared Investigations of Terpenes. IV. Collect. Czech. Chem. Comm. 18, 500 (1953).

216. PLOUVIER, V.: Distribution of Aliphatic Polyols and Cyclitols. In: T. Swain (Edit.). Chemical Plant Taxonomy, p. 313. London and New York: Academic Press. 1963.

217. RENNERFELT, E.: Investigation of Thujaplicin, a Fungicidal Substance in the Heartwood of *Thuja plicata* Don. Physiol. Plantarum 1, 245 (1948).

218. RICHARDS, J. H. and J. B. HENDRICKSON: The Biosynthesis of Steroids, Terpenes and Acetogenins. New York: Benjamin. 1964.

219. ROBINSON, R.: The Structural Relations of Natural Products. Oxford: Clarendon Press 1955.

220. RUDLOFF, E. VON: The Isolation of Guaiol and α-Cadinol from the Wood Oil of *Neocallitropsis araucarioides*. Chem. and Ind. 1962, 743.

221. — Minor Components of the Oil of *Araucaria*. Chem. and Ind. 1964, 2126.

222. — Gas-Liquid Chromatography of Terpenes. Part IX. The Volatile Oil of the Leaves of *Juniperus sabina* L. Canad. J. Chem. 41, 2876 (1963).

223. RUDLOFF, E. VON and H. ERDTMAN: Stereochemistry of Occidentalol and its Hydrogenation Products. Tetrahedron 18, 1315 (1962).

224. RUDLOFF, E. VON and G. V. NAIR: The Sesquiterpene Alcohols of the Heartwood of *Thuja occidentalis* L. Canad. J. Chem. 42, 421 (1964).

225. RUDMAN, P.: The Causes of Natural Durability in Timber. XIII. Factors Influencing the Decay Resistance of Cypress Pine (*Callitris columellaris* F. Muell.). Holzforschg. 17, 183 (1963).

226. — The Causes of Natural Durability in Timber. XV. A Gas Chromatographic Investigation into the Nature and Radial Variation of the Heartwood Extractives of Cypress Pine (*Callitris columellaris* F. Muell.). Holzforschg. 18, 116 (1964).

227. RUDMAN, P. and F. J. GAY: The Causes of Natural Durability in Timber. XIV. Intra-specific Variations in Termite Resistance of Cypress Pine (*Callitris columellaris* F. Muell.). Holzforschg. 18, 113 (1964).

228. RUNEBERG, J.: The Chemistry of the Order Cupressales. 27. Heartwood Constituents of *Juniperus utahensis* Lemm. Acta Chem. Scand. 14, 797 (1960).

229. — The Chemistry of the Order Cupressales. 28. Constituents of *Juniperus virginiana* L. Acta Chem. Scand. 14, 1288 (1960).

230. — The Chemistry of the Order Cupressales. 29. Heartwood Constituents of *Juniperus thurifera* L. Acta Chem. Scand. 14, 1985 (1960).

231. — The Chemistry of the Order Cupressales. 30. Heartwood Constituents of *Juniperus cedrus* Webb. & Berth. Acta Chem. Scand. 14, 1991 (1960).

232. — The Chemistry of the Order Cupressales. 31. Heartwood Constituents of *Juniperus phoenicea* L. Acta Chem. Scand. 14, 1995 (1960).

233. — The Chemistry of the Order Cupressales. 33. The Structure of Procerin. Acta Chem. Scand. 15, 645 (1961).

234. Runeberg, J.: The Chemistry of the Order Cupressales. 35. Heartwood Constituents of *Juniperus foetidissima* Willd. Acta Chem. Scand. 15, 721 (1961).

235. — The Structure of Cedrolic Acid. Acta Chem. Scand. 15, 945 (1961).

236. Rutowski, B. und I. Winogradowa: Untersuchung der Zusammensetzung russischer ätherischer Öle. Trans. scient. chem.-pharm. Inst. (Moskau), Lief. 17, 5 [Chem. Zbl. 1927, II, 1311].

237. Ruzicka, L.: History of Isoprene Rule. Proc. Chem. Soc. (London) 1954, 341.

238. Sandermann, W. and W. Schweers: Über die Biogenese von Thujon in *Thuja occidentalis*. Tetrahedron Letters 1962, 259.

239. Semmler, F. W. und A. Hoffmann: Zur Kenntnis der Bestandteile ätherischer Öle (Untersuchungen über das Sesquiterpen Cedren). Ber. dtsch. chem. Ges. 40, 3521 (1907).

240. Semmler, F. W. and E. W. Mayer: Zur Kenntnis der Bestandteile ätherischer Öle. I. Pseudocedrol, ein physikalisch Isomeres des Cedrols; II. Notizen über einige Sesquiterpenalkohole; III. Tetrahydro-caryophyllen. Ber. dtsch. chem. Ges. 45, 1384 (1912).

241. Serebryakova, A. P., L. N. Filitis and L. M. Utkin: Lignans of the Soviet Union *Juniperus* Plants. Zhurn. Obshcheĭ Khimiĭ (USSR) 31, 1731 (1961).

242. Shaligram, A. M., A. S. Rao and S. C. Bhattacharyya: Conversion of Costunolide to Junenol. Chem. and Ind. 1961, 671.

243. Simonsen, J. L.: The Constituents of Indian Turpentine from *Pinus longifolia* Roxb. Part I. J. Chem. Soc. (London) 117, 570 (1920).

244. Simonsen, J. L. (Sir John) and D. H. R. Barton: The Terpenes, Vol. 3, p. 194. Cambridge: Univ. Press. 1952.

245. — — ibid. p. 170.

246. Smolders, R. R.: Contribution à l'étude structurelle du cédrélanol (δ-cadinol), alcool sesquiterpénique $C_{15}H_{26}O$ de l'huile essentielle de *Cedrela odorata brasiliensis*. Canad. J. Chem. 42, 2836 (1964).

247. Stork, G. and R. Breslow: The Structure of Cedrene. J. Amer. Chem. Soc. 75, 3291 (1953).

248. Stork, G. and F. H. Clarke, Jr.: Stereochemistry and Total Synthesis of Cedrol. J. Amer. Chem. Soc. 83, 3120 (1961).

249. Sumimoto, M.: The Identity of Cryptomeridiol with the Compound Called Selinan-4,7-diol. Chem. and Ind. 1963, 1356.

250. — Heartwood Constituents of *Sciadopitys verticillata* Sieb. et Zucc. I. The Constitution of Sciadin. Tetrahedron 19, 643 (1963).

251. Sumimoto, M., H. Ito, H. Hirai and K. Wada: Cryptomeridiol, the Direct Precursor of the Eudesmane Series. Chem. and Ind. 1963, 780.

252. Sumimoto, M., Y. Tanaka and K. Matsufuji: Heartwood Constituents of *Sciadopitys verticillata* Sieb. et Zucc. II. The Structure of Methyl Sciadopate. Tetrahedron 20, 1427 (1964).

253. Theobald, D. W.: The Chemistry of Some 6-Oxygenated Eudesmanes and the Absolute Configuration of Junenol. Tetrahedron 20, 2593 (1964).

254. Thomas, A. F.: Odeur et constitution. XXIII. Le méthyl-2-isopropényl-5-anisole, constituant de l'huile essentielle de "Hinoki" de Formose. Helv. Chim. Acta 48, 1057 (1965).

255. Uchida, S.: Über das ätherische Öl der Blätter des „Hinoki" (*Chamaecyparis obtusa* Endl.). J. Soc. Chem. Ind. (Japan) [Suppl.] 31, 159B [Chem. Zbl. 1928, II, 1577].

256. — The Essential Oil of Sugi *(Cryptomeria japonica)* Leaves. J. Amer. Chem. Soc. 38, 687 (1916).

257. Ukita, C., T. Tsumita and N. Utsugi: Resin Acids in the Fruits of *Juniperus japonica*. Structure of an Isomeric 7-Isodextropimaric Acid. Pharm. Bull. (Tokyo) **3**, 441 (1955) [Chem. Abstr. **50**, 14651 (1956); cf. also Chem. Abstr. **47**, 4861 (1953)].

258. Varma, K. R., T. C. Jain and S. C. Bhattacharyya: Structure and Stereochemistry of Zingiberol and Juniper Camphor. Tetrahedron **18**, 979 (1962).

259. Wachtmeister, C. A. and B. Wickberg: Chromatography of Tropolones on Paper Impregnated with Ethylenediaminetetraacetic Acid and Dimethyl Sulphoxide. Acta Chem. Scand. **12**, 1335 (1958).

260. Watanabe, H.: Plant Wax. V. Conifer Wax. J. Pharm. Soc. Japan **73**, 176 (1953).

261. Westfelt, L.: The Structure of (+)-ε-Muurolene ("ε-Cadinene"). Acta Chem. Scand. **18**, 572 (1964).

262. Wickberg, B.: Separation of Sesquiterpenes by Partition Chromatography. J. Organ. Chem. (USA) **27**, 4652 (1962).

263. Windemuth, N.: The Volatile Oil of *Juniperus mexicana* Schiede. Pharm. Arch. **16**, 17 (1945).

264. Zavarin, E.: Extractive Components from Incense Cedar Heartwood. VI. On the Occurrence of 3-Libocedroxythymoquinone. J. Organ. Chem. (USA) **23**, 1198 (1958).

265. — Extractive Components from Incense Cedar Heartwood. VII. On the Occurrence of Heyderiol. J. Organ. Chem. (USA) **23**, 1264 (1958).

266. — On the Structures of α-Dolabrinol and Isopygmaein. J. Organ. Chem. (USA) **27**, 3368 (1962).

267. Zavarin, E. and A. B. Anderson: Extractive Components from Incense Cedar Heartwood. I. Occurrence of Carvacrol, Hydrothymoquinone and Thymoquinone. J. Organ. Chem. (USA) **20**, 82 (1955).

268. — — Extractive Components from Incense Cedar Heartwood. II. Occurrence and Synthesis of *p*-Methoxythymol and *p*-Methoxycarvacrol, two new Phenolic Compounds. J. Organ. Chem. (USA) **20**, 443 (1955).

269. — — Extractive Components from Incense Cedar Heartwood. III. Occurrence of Libocedrol, a New Phenol Ether, and its *p*-Methoxythymol Addition Complex. J. Organ. Chem. (USA) **20**, 788 (1955).

270. — — Paper Chromatography of the Tropolones of Cupressaceae. J. Organ. Chem. (USA) **21**, 332 (1956).

271. — — On the Structure of Pygmaein, a New Tropolone from *Cupressus pygmaea* Heartwood. J. Organ. Chem. (USA) **26**, 1679 (1961).

272. — — Extrahierbare Bestandteile des Kernholzes der kalifornischen Flußzeder (Incense Cedar, *Libocedrus decurrens* Torrey). IV. Vorkommen und Chromatographie von Thujaplicinen. Chem. Ber. **89**, 545 (1956).

273. Zavarin, E., A. B. Anderson and R. M. Smith: On the Occurrence of α-Thujaplicinol in the Heartwood of *Cupressus pygmaea* (Lemm.) Sarg. J. Organ. Chem. (USA) **26**, 173 (1961).

274. Zavarin, E., R. M. Smith and A. B. Anderson: Paper Chromatography of the Tropolones of Cupressaceae. II. J. Organ. Chem. (USA) **24**, 1318 (1959).

275. — — — Characterization of Cupressaceae Tropolones as Dicyclohexylamine Salts. J. Organ. Chem. (USA) **24**, 1584 (1959).

276. Ziffer, H., T. J. Batterham, U. Weiss and E. von Rudloff: The Absolute Configuration of Occidentalol. Tetrahedron **20**, 67 (1964).

(Received, January 20, 1966.)

Quinone Methides in Nature.

By **A. B. Turner**, Aberdeen.

Contents.

Acknowledgement. I am indebted to Professor R. H. Thomson for his helpful comments on the manuscript.

I. Introduction.

Quinone methides occur in Nature both as products of fungal metabolism and as plant pigments. The chemistry of some of these compounds, which are also known as methylenequinones, quinone methines, or quinomethanes, has been covered in several reviews (*35, 81, 142, 153*). They are derived from quinones by replacement of one of the carbonyl

References, pp. 321—328.

oxygen atoms by a methylene or substituted methylene group and are systematically named as derivatives of methylene cyclohexadienones. In addition, evidence has accumulated in recent years for the participation of quinone methides as transient intermediates in certain biochemical processes. In the present review, emphasis is laid upon the characteristic properties of these quinonoid compounds, which differ in many respects from those of quinones themselves.

II. Stable Quinone Methides in Nature.

1. Citrinin.

Citrinin (1a), $C_{13}H_{14}O_3$, one of the first compounds with antibiotic activity isolated in RAISTRICK's classic investigations on fungal meta-bolites, was initially obtained from *Penicillium citrinum* THOM (77). It has since been isolated from other *Penicillium* species (115, 116, 134, 139), from *Aspergillus terreus* THOM (119), and from *Aspergillus* species of the *Candidus* group (141). Although citrinin proved too toxic for medical use, the discovery of its activity intensified the search which subsequently led to practical antibiotics. Its structure (1a) was established by ROBERTSON, WHALLEY and co-workers (19, 20).

The metabolite is a yellow, optically active compound, which behaves as a phenolic acid. A quinonoid structure is indicated by reduction to a colourless dihydro derivative, which readily reoxidises on exposure to air. A key reaction in the structural elucidation was the smooth degradation of citrinin, under the influence of hot dilute sulphuric acid, to formic acid, carbon dioxide and the resorcinol (2a). This latter product, which is obtained partially racemized, undergoes reverse aldol reaction on fusion with alkali giving acetic acid and 5-ethyl-4-methylresorcinol (3). The structure of the phenol (2a) was established by synthesis of its racemic dimethyl ether (2b) and of the butane (4a) obtained by reduction of the alcoholic hydroxyl group of the phenol (2a) with hydriodic acid and red phosphorus.

The locations of the carboxyl group and the carbon atom eliminated as formic acid were revealed by a study of the reactions of dihydro-citrinin (5a). This, on complete methylation, gives the ester (5b), which is readily oxidised to the lactone (6a). Boiling hydriodic acid converts this into the resorcinol (7), which can be further degraded to the alcohol (2a) by boiling alkali. Reduction of dihydrocitrinin with hydriodic acid and red phosphorus furnishes the phenol (8), the structure of which was proved by synthesis from the resorcinol (4a).

Methyl citrinin (1b) is prepared by esterification of citrinin with dimethyl sulphate and sodium hydrogen carbonate, and this ester under-goes C-methylation with methyl iodide to give the diketone (9) in low yield. The sclerotiorins (*42*), a group of mould pigments elaborated by a limited number of *Monascus* and *Penicillium* species, have a similar type of structure, and it is possible that their biogenesis involves an analogous methylation process.

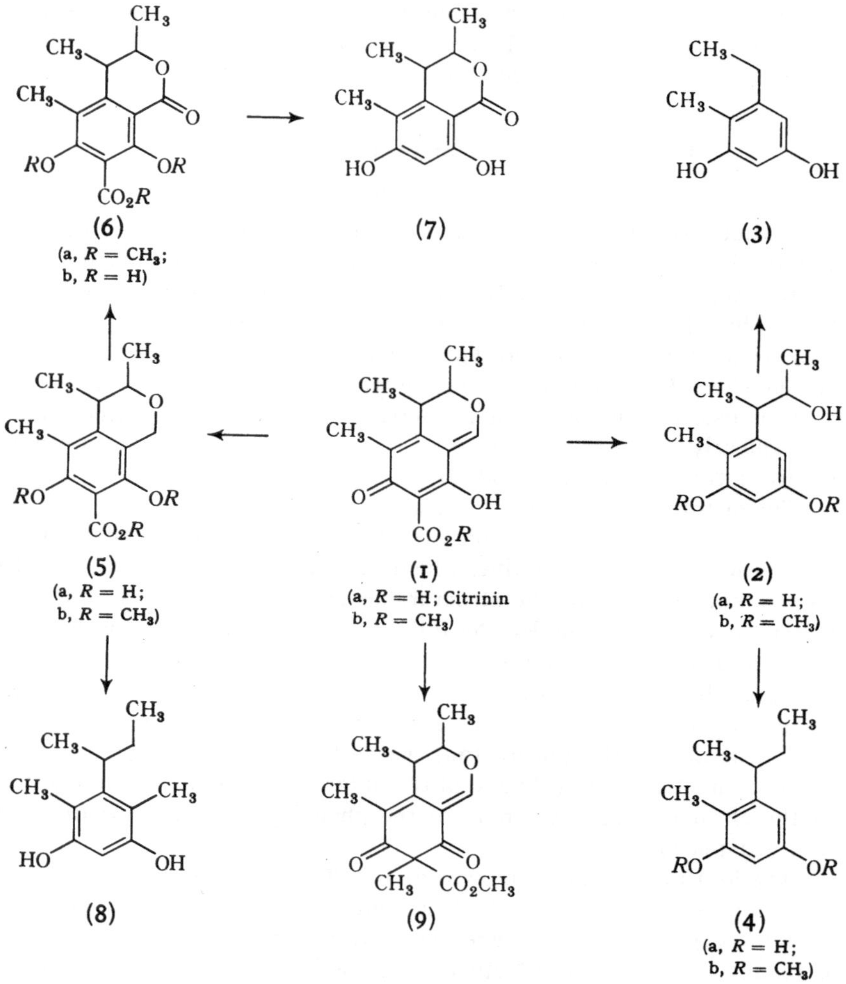

Chart 1. Reactions of Citrinin (**1a**).

The structure (**1a**) for citrinin has been confirmed by synthesis of its racemate from the racemic phenol (**2a**). Carboxylation of this gives the acid (**10a**), which is converted into citrinin by formylation followed by cyclodehydration with cold concentrated sulphuric acid. A formal total synthesis is completed by resolution through the brucine salt (*24, 84*).

References, pp. 321—328.

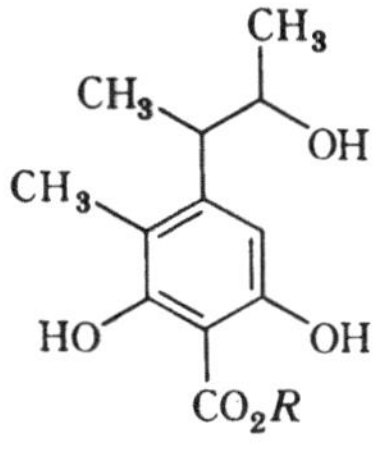

(10)
(a, R = H;
b, R = CH₃)

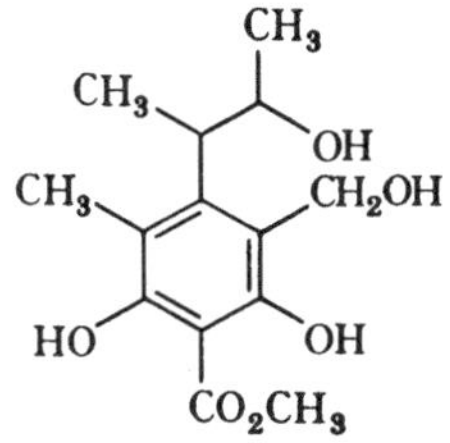

(11)

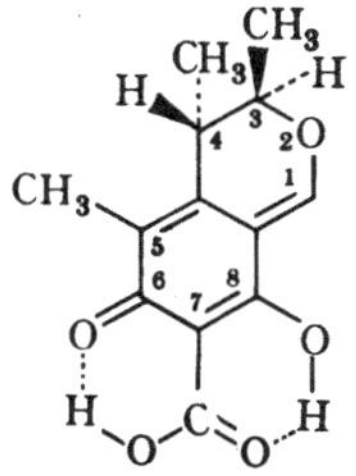

(12)
Citrinin.

Dihydrocitrinin is formed in good yield by condensation of the ester (10b) with formaldehyde in alkaline solution, the intermediate ester (11) undergoing hydrolysis as well as cyclisation (24). The final stages of these syntheses are simplified by there being only one unsubstituted position in the nucleus, and several other syntheses of citrinin and di-hydrocitrinin have been reported (65, 148, 149). As dihydrocitrinin is readily oxidised to citrinin by means of bromine (149) or mercuric oxide (150), this route has been adapted to the preparation of citrinin analogues by using various aldehydes and ketones instead of formaldehyde (150).

The acid catalysed racemization of the phenol (2a), together with its stability towards alkali, allowed CRAM (36) to make an ingenious suggestion which led to the assignment of relative configuration. He noted that if the methyl groups on the asymmetric carbon atoms of citrinin were *trans*, the observed racemization could be explained by participation of the aromatic ring to form a bridged carbonium ion intermediate. Subsequent studies (37) with model compounds confirmed this proposal, and it followed from work on the optical isomers of 3-phenyl-butan-2-ol that phenol (2a) has the *threo* configuration. The stereo-chemical work has recently been completed by assignment of the absolute configuration (12) in two laboratories. MEHTA and WHALLEY (102) applied PRELOG's atrolactic acid method to the (–)-alcohol (2b), and HILL and GARDELLA (79) proved the configurational identity of (–)-(4b) and (–)-2-phenylbutane, of known configuration, by comparison of their optical rotatory dispersion curves.

Details of hydrogen bonding in citrinin have been studied by infrared spectroscopy (88), and NMR measurements indicate the preferred con-formation to be that in which the 3- and 4-methyl groups are quasi-axial (98).

Tracer studies by BIRCH and co-workers (11) with *Aspergillus candidus* have established that citrinin is biosynthesised from five acetate units, with the remaining carbon atoms being derived from the C_1 pool, as in (13). SCHWENCK (129) independently reached similar conclusions, in particular demonstrating that methionine is a more efficient C_1-donor than formate, and that neither propionate nor bicarbonate is utilized. The two methyl groups and the carboxyl are incorporated at the same activity level from methionine, suggesting that the three C_1-fragments are all introduced as methyl groups, with the carboxyl group arising by subsequent oxidation. In consequence, the phenol (14), a known metabolite of *P. brevicompactum*, has been suggested as a precursor of citrinin (*38, 72*). The conversion of glucose-1-C^{14} to citrinin in *P. citrinum* has also been reported (*43*).

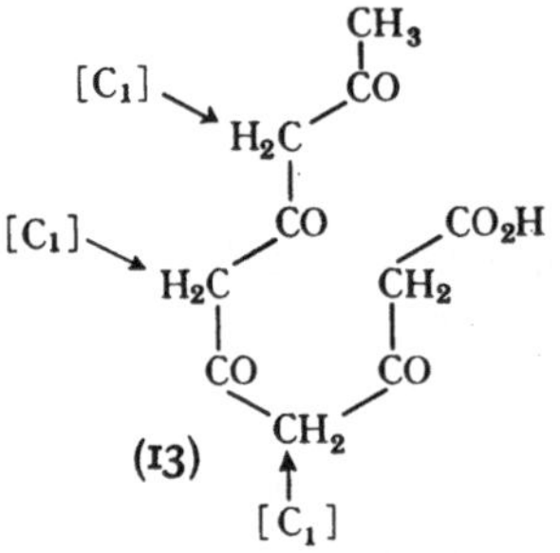

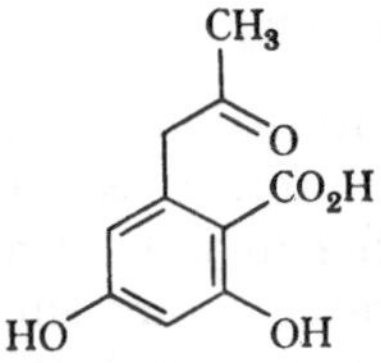

A notable feature of the biosynthesis of citrinin is the appearance of the carboxyl group of the terminal acetic acid unit in the aldehyde oxidation state. This may occur either by reduction of the carboxyl group itself, or by reduction of a ketonic function at this position in a longer precursor chain, followed by retroaldol cleavage. In this connection, it is interesting that a mutant of an organism, which no longer produces citrinin, yields instead the variant (6b) with the terminal carboxyl group appearing as a lactone function (*72*). This implies that reduction of this carbonyl group occurs at a late stage in the biosynthesis of citrinin.

2. Pulvilloric Acid.

Pulvilloric acid, $C_{15}H_{18}O_5$, is a yellow antibiotic isolated from cultures of *P. pulvillorum* TURFITT (*15*). It is non-specific in its antifungal activity, which is dependent upon pH. Its structure (15) has been established by McOMIE et al. (*101*) on the basis of degradative studies and confirmed by synthesis (*21*).

The presence of a quinone methide system was suspected from the outset, as the acid undergoes reversible addition of alcohols to give colourless 1 : 1 adducts. The metabolite was in fact purified by recrystallisation as its ethanol adduct (16c), from which it was regenerated by warming under vacuum. A second colourless adduct (16d), this time a stable one, is formed with sodium bisulphite.

Catalytic hydrogenation of pulvilloric acid yields a colourless dihydro derivative (16a), which gives a positive Gibbs reaction and couples with diazotised *p*-nitroaniline to give a mono-azo compound, thereby establishing the presence of one free position *para* to a phenolic hydroxyl group. When dihydropulvilloric acid is treated with diazomethane it gives a mono-ester (**16b**) which, on complete methylation with methyl iodide, yields the dimethyl ether (**17**). Partial oxidation of the latter with chromium trioxide furnishes the δ-lactone (**18**), whereas potassium permanganate degrades it to the phthalic acid (**19**), which readily forms a cyclic anhydride.

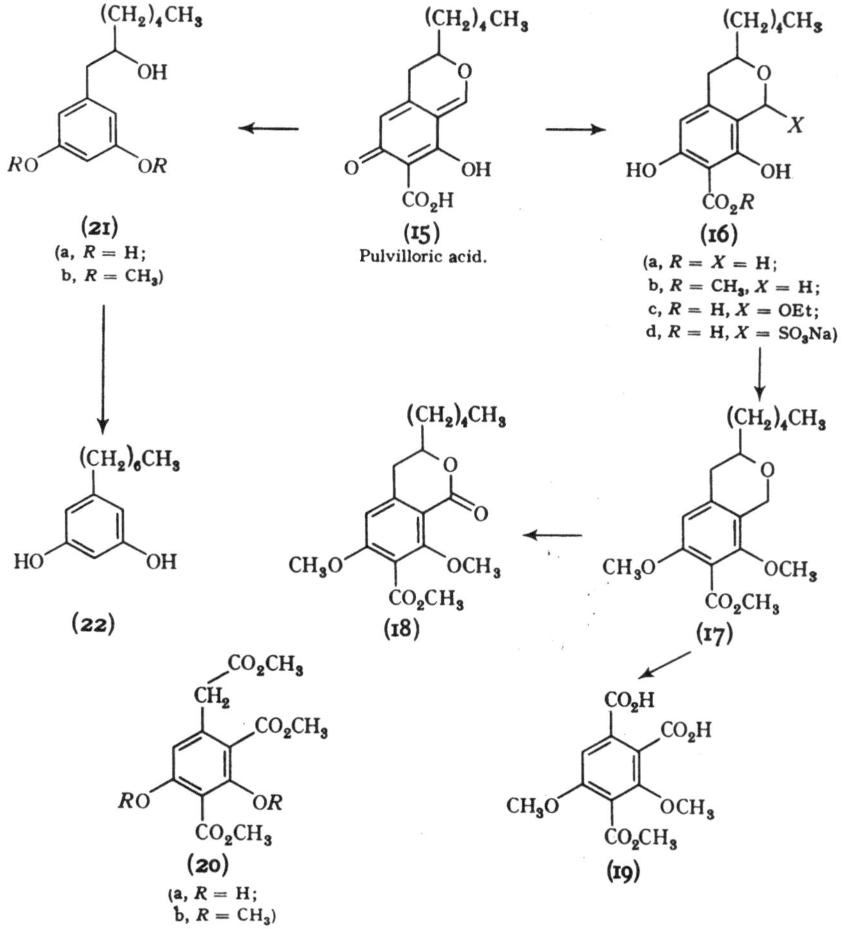

(**21**)
(a, R = H;
b, R = CH₃)

(**15**)
Pulvilloric acid.

(**16**)
(a, R = X = H;
b, R = CH₃, X = H;
c, R = H, X = OEt;
d, R = H, X = SO₃Na)

(**22**)

(**18**)

(**17**)

(**20**)
(a, R = H;
b, R = CH₃)

(**19**)

Chart 2. Reactions of Pulvilloric Acid (**15**).

For comparison, a sample of the acid (**19**) was prepared from the self-condensation product (**20a**) of methyl acetone dicarboxylate. Methylation to the dimethyl ether (**20b**), followed by permanganate oxidation, gave the desired product (**19**); during this oxidation selective hydrolysis of one ester group occurs, probably as a result of neighbouring group participation (*143*).

By this time the close resemblance between pulvilloric acid and citrinin had become apparent, and further degradative studies followed the pattern of the citrinin work. As in the case of citrinin, alkaline hydrolysis of pulvilloric acid took place with the loss of two carbon atoms, giving a phenolic alcohol, $C_{13}H_{20}O_3$. This was shown to have structure (**21a**) by reduction with hydriodic acid to the known 5-*n*-heptylresorcinol (**22**), together with evidence given above for the position of the hydroxyl group in the side-chain.

The structure of the phenol (**21a**) was confirmed by synthesis of its racemate by two routes. The first involved preparation of the dimethyl ether (**21b**) by standard methods, followed by demethylation with hydriodic acid or boron tribromide. Unfortunately, neither of these reagents gave reproducible results, owing to the instability of the secondary alcoholic group in the side-chain. This difficulty was overcome by using the corresponding ketone (**23a**), which was made by hydrolysis of the ketonitrile (**23b**). Demethylation of the ketone (**23a**) with boron tribromide, or better with pyridine hydrochloride, followed by treatment with potassium borohydride, gave the desired phenol (**21a**) as its racemic compound.

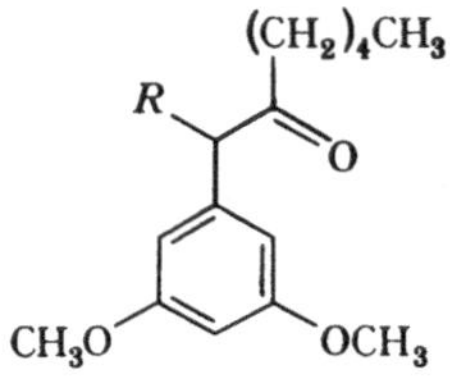

(**23**)
(a, R = H;
b, R = CN)

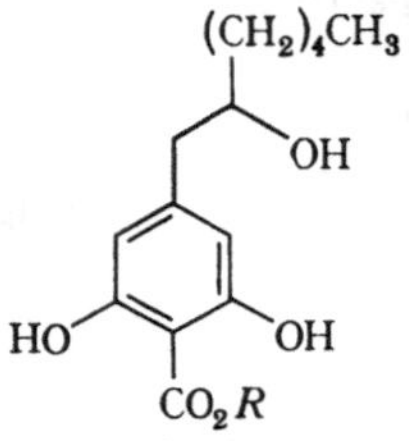

(**24**)
(a, R = H;
b, R = CH₃)

The synthesis of ($\pm$)-pulvilloric acid was completed by carboxylation of the phenol (**21a**) to give the acid (**24a**), followed by reaction with ethyl orthoformate. The ester (**24b**), obtained by the action of diazomethane on the acid (**24a**), condenses with methylal to give methyl dihydropulvillorate (**16b**). Appreciable amounts of the bridged dimer (**25**) are formed in this reaction, owing to the presence of two reactive positions in the nucleus of the ester (**24b**). Attempts to circumvent this difficulty by using the Gattermann reaction, in which the entering formyl group would deactivate the nucleus, were unsuccessful.

References, pp. 321—328.

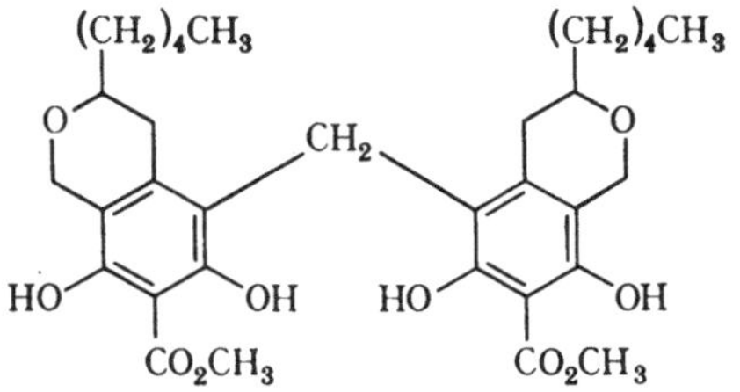

(25)

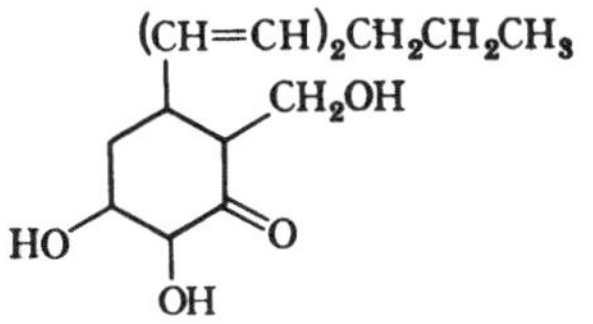

(26)
Palitantin.

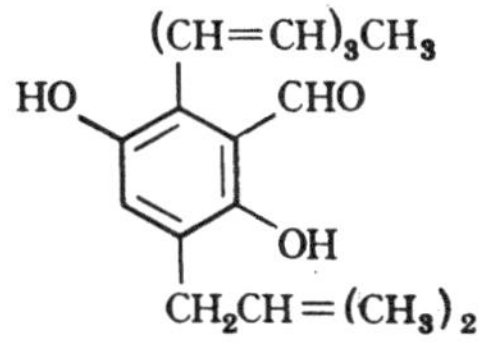

(27)
Auroglaucin.

The structure (15, p. 293) for pulvilloric acid is in accord with the acetate theory of biogenesis (10a). Thus the molecule may be built up from seven acetate units, with the introduction of an extra carbon atom, as in (28). It is probable that the primary biogenetic unit has a straight chain of fourteen carbon atoms, rather than an alternative branched chain, by analogy with the biosynthesis of citrinin and many other fungal metabolites of this order of complexity. In particular, the carbon skeleton of pulvilloric acid is the same as that in palitantin (26), apart from the extra carboxyl group. BIRCH and KOCOR (12) have shown that palitantin and auroglaucin (27) are based on a C_{14}-chain derived from a linear sequence of seven acetate units. The similarities between the biosynthetic conversions involved indicate that these mould metabolites are branch products of a common scheme arising from the precursor (29). It is likely that this compound is also an intermediate in the biosynthesis of pulvilloric acid.

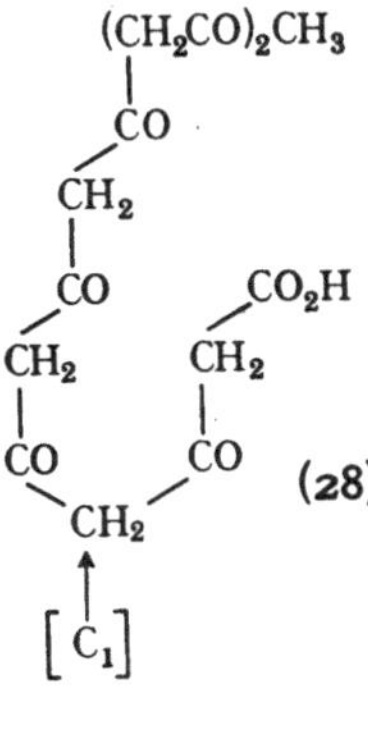

(28)

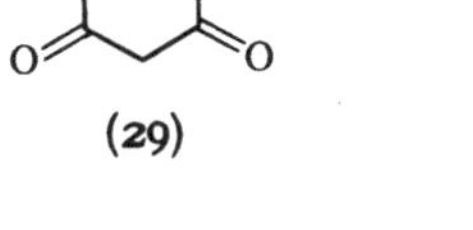

(29)

3. Ascochitine.

The yellow pigment ascochitine, $C_{15}H_{16}O_5$, first obtained from culture filtrates of *Ascochyta pisi* LIB. by BERTINI (7), was also isolated by OKU and NAKANISHI (*114*) from *Ascochyta fabae* SPEG in order to study its phytotoxic and antibiotic properties. Its structure (30) has been established by IWAI and MISHIMA (*82*). Catalytic hydrogenation of ascochitine produces a phenolic tetrahydro derivative (31a). This yields a mono-ester (31b) with diazomethane, and further methylation by means of methyl iodide leads to the dimethyl ether (32b). Reduction of the

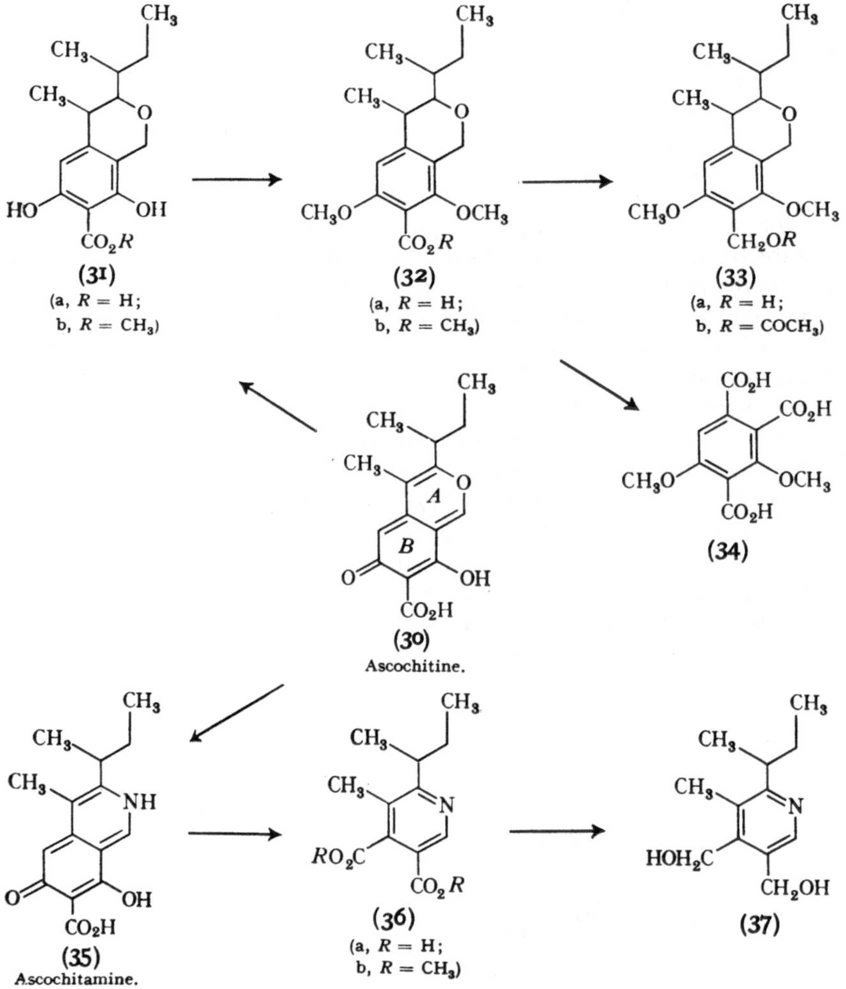

Chart 3. Reactions of Ascochitine (30).

ester group of this compound with lithium aluminium hydride gives the alcohol (33a), the acetate (33b) of which has the characteristic signal of a benzyl acetate in its NMR spectrum. This proves that the carboxyl group of (31a) is attached to the aromatic nucleus.

The acid (32a), obtained by hydrolysis of the ester (32b), is degraded by permanganate to the tricarboxylic acid (34), thus establishing the substitution pattern in ring B of ascochitine (30).

A notable feature of ascochitine is its ready conversion into amine derivatives with ammonia or primary amines, and this was put to good use in the determination of the substitution pattern in ring A. The chemistry of sclerotiorin, which also has an extended γ-pyrone structure, is likewise characterised by the ready formation of cyclic amine derivatives (42). In each case the product corresponds to the replacement of —O— by —NR—. These amine derivatives are much less sensitive than the parent compound to many reagents, thus facilitating degradative work. In view of the mild conditions under which the metabolite reacts with ammonia to form the base, ascochitamine (35), it is assumed that conversion of the pyranoid oxygen atom into an imino group takes place without molecular rearrangement. This assumption is compatible with the NMR spectra of the two compounds. On oxidation with nitric acid, ascochitamine (35) yields the basic diacid (36a), the ultraviolet spectrum of which is similar to that of pyridine-3,4-dicarboxylic acid. Its NMR spectrum, together with that of its ester (36b), revealed the presence of a methyl group at the β-position of a pyridine nucleus, an unsubstituted α-position, and a sec-butyl group. The pKa values of the diacid (36a) show that the carboxyl groups are ortho to one another. The dimethyl ester (36b) of this acid is reduced by lithium aluminium hydride to the diol (37), the structure of which was established by the following unambiguous synthesis of its racemate:

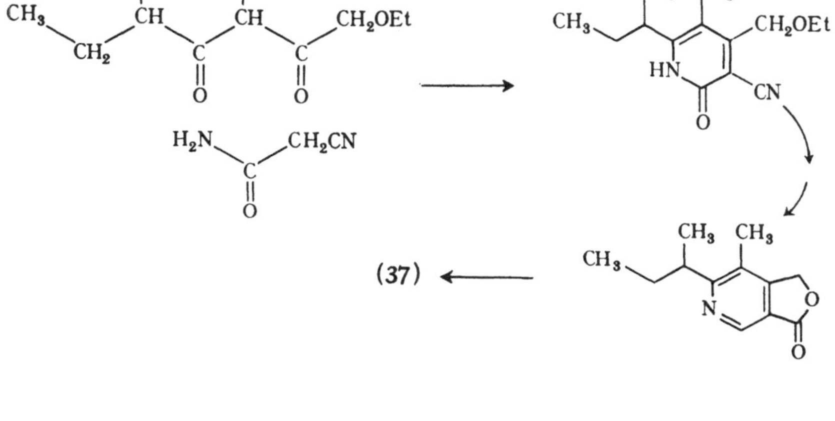

The two degradation products (34) and (36a) have six of the fifteen carbon atoms of ascochitine in common. These six carbon atoms can be superimposed in two possible ways, only one of which is compatible with the acetate theory of biogenesis and a quinonoid chromophore. This, together with the previous results, leads to the structure (30) for ascochitine (p. 296), which is therefore closely related to both citrinin (1, p. 290) and pulvilloric acid (15, p. 293). Its biosynthesis is probably based on a linear sequence of six acetic acid units, with the introduction of three C_1-fragments (cf. 38).

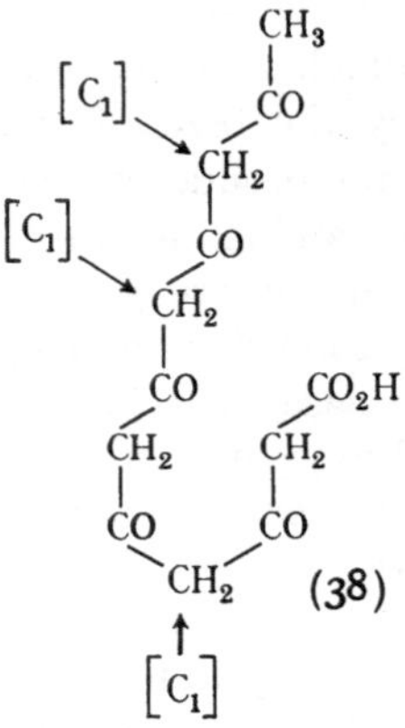

$$
\begin{array}{c}
CH_3 \\
| \\
[C_1] \searrow \quad CO \\
CH_2 \\
[C_1] \quad CO \\
\searrow \quad CH_2 \\
| \\
CO \qquad CO_2H \\
CH_2 \qquad CH_2 \\
| \qquad | \\
CO \qquad CO \\
\searrow CH_2 \swarrow \quad (38) \\
\uparrow \\
[C_1]
\end{array}
$$

4. Purpurogenone.

The mycelium of *P. purpurogenum* STOLL contains the crimson, optically active metabolite purpurogenone, $C_{14}H_{12}O_5$ (*120*). It absorbs one mole of hydrogen on catalytic hydrogenation, with the colour fading to yellow; the red colour redevelops on exposure to air, indicating a quinonoid system. No crystalline derivatives of purpurogenone could be prepared, although an amorphous triacetate was obtained.

Oxidation of purpurogenone or of its triacetate gives a phenolic acid which, by methylation and further oxidation is converted into 3-methoxybenzene-1,2,5-tricarboxylic acid (39). Furthermore, alkaline degradation of the metabolite yields formic acid together with a mixture of coloured substances, one of which was partially characterised as a derivative of 2-hydroxy-1,4-naphthoquinone on the basis of colour reactions and the ultraviolet spectrum of its amorphous acetyl derivative.

These observations leave room for both linear (40) or angular (41) fused ring structures conforming to the acetate theory of biogenesis. Of these alternative formulations for purpurogenone, the former is preferred (*120*) as being the more compatible with the ultraviolet data. However, as both (40) and (41) could exist in a number of tautomeric forms, the available evidence seems insufficient to choose between them. In addition, later work suggested a naphthopurpurin structure for the quinonoid degradation product (*40*). The linear fused-ring system in (40) is also present in the mould metabolite fusarubin, and has been discussed in terms of a branched and a linear C_{14}-polyketide chain (*104, 153*).

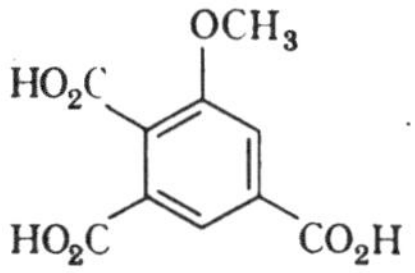

(39)

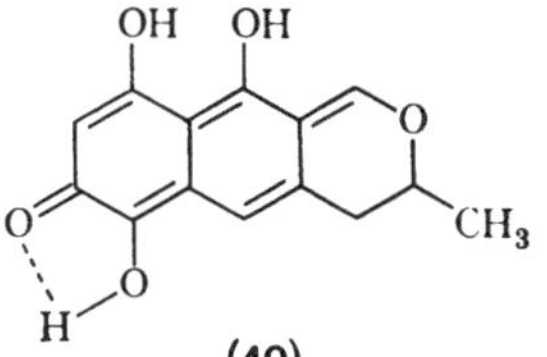

(40)
Purpurogenone (?).

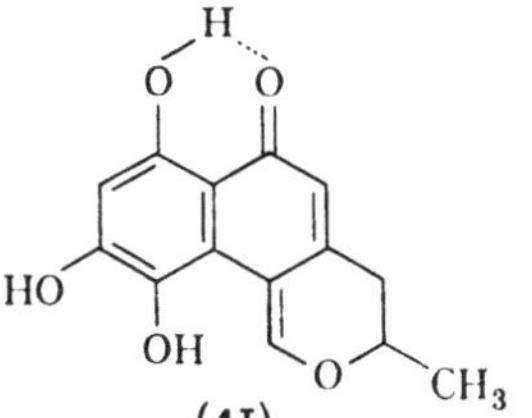

(41)
Purpurogenone (?).

5. Fuscin.

The isolation and properties of fuscin, $C_{15}H_{16}O_5$, an orange metabolite of *Oidiodendron fuscum* ROBAK, have been described by MICHAEL (*103*). Further detailed investigations of its chemistry were carried out by RAISTRICK and his colleagues (*14*), and their results were rationalised by BARTON and HENDRICKSON (*4*) and by BIRCH (*10*) in terms of the formula (42). In keeping with its quinone methide structure, fuscin forms colourless adducts with a variety of reagents, e. g. methanol, thiourea, hydrogen chloride, and acetic and thioglycollic acids. In each case 1,6-addition of one molecule of the reagent takes place across the quinone methide system, and some of the products are easily reconverted into the parent compound. Fuscin itself has one enolic hydroxyl group, and its colourless dihydro derivative behaves as a dihydric phenol, as does its methanol adduct.

Fusion of dihydrofuscin (43a) with alkali gives acetic, isovaleric, and 3,4,5-trihydroxyphenylacetic acids, and these fragments account for all the carbon atoms of the original molecule. Mild alkaline hydrolysis of dihydrofuscin leads only to the loss of acetaldehyde, by a reverse aldol reaction. The structure of the other product of this reaction, fuscinic acid (44a), was finally deduced from a study of the oxidation of its dimethyl ether (44b), together with the results of the more drastic degradation given above.

Although the molecule contains an asymmetric carbon atom, fuscin is optically inactive. This may be due to racemization by way of the tautomeric form (45), or alternatively, racemization may occur in the reduced form (43a), which contains a benzylic proton and is also present in cultures of *O. fuscum*.

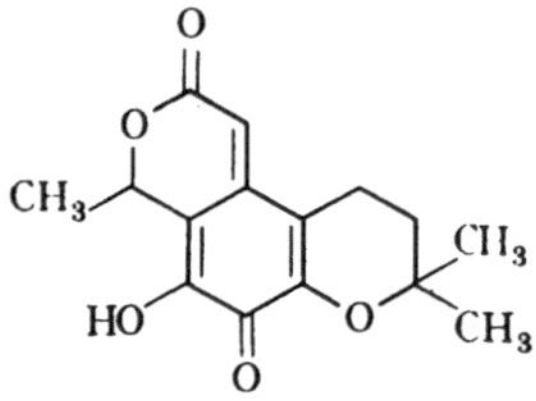

(42)
Fuscin.

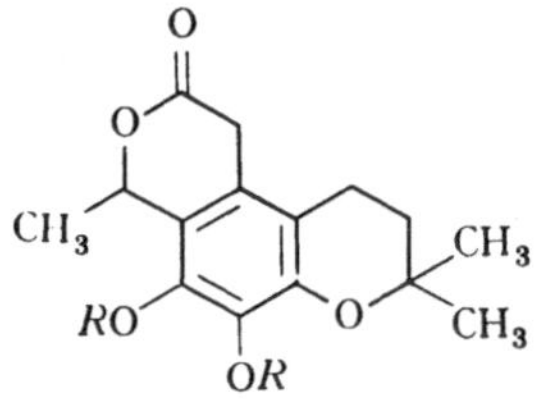

(43)
(a, R = H;
b, R = CH₃)

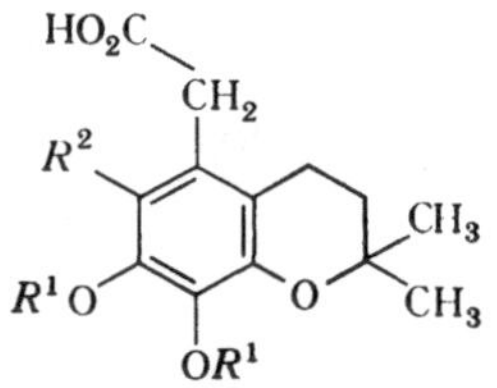

(44)
(a, R' = R² = H;
b, R' = CH₃, R² = H;
c, R' = CH₃, R² = COCH₃)

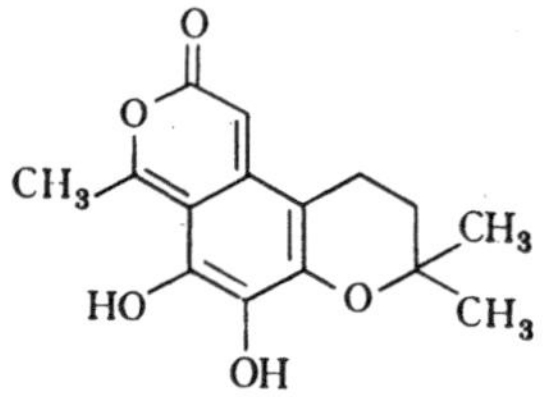

(45)

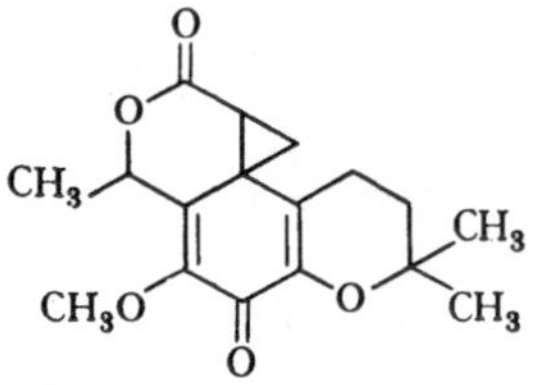

(46)

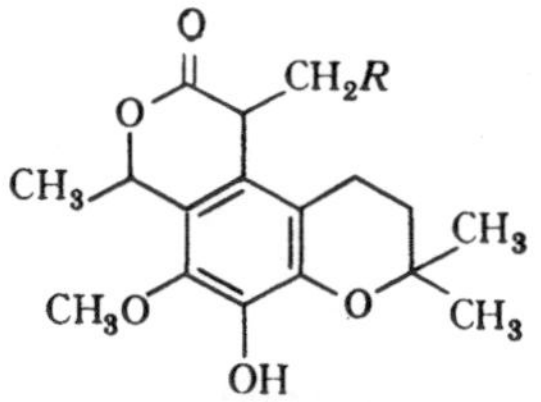

(47)
(a, R = H;
b, R = OH)

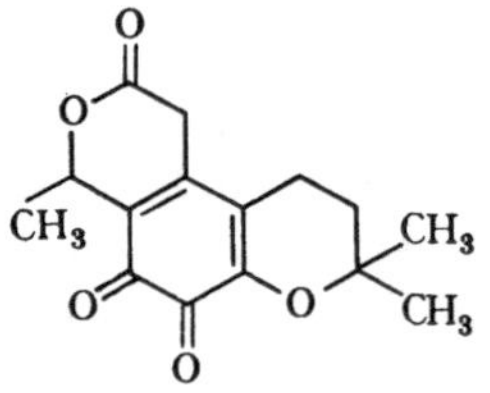

(48)

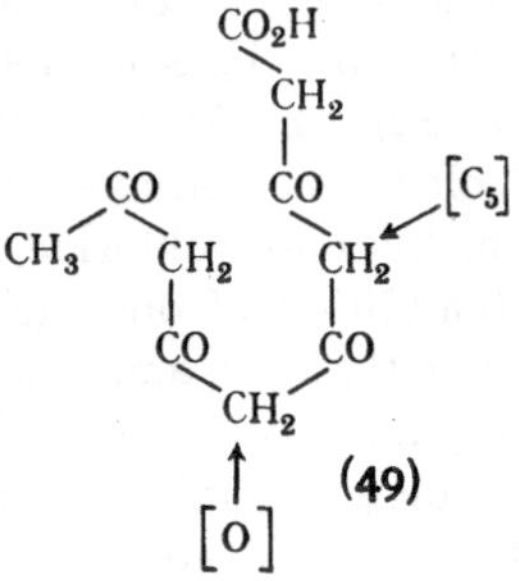

(49)

An interesting reaction of fuscin is the formation of methylhomofuscin (46) with diazomethane. This reagent normally gives benzpyrazole adducts with quinones, although loss of nitrogen can lead to cyclopropanes. The blocked cyclohexadienone system of (46) is readily aromatised, with fission of the cyclopropane ring. Thus, hydrogenation gives the phenol (47a) and titration with cold alkali yields the hydrate (47b), the cyclopropane ring reacting like a double bond in the now familiar manner. The reactions of fuscin leave no doubt that it exists in the quinone methide form (42) rather than in the tautomeric o-quinone form (48).

BARTON and HENDRICKSON (4) have confirmed the structure (42) for fuscin by a synthesis starting from methyl 3,4,5-trimethoxyphenyl-acetate. Reaction with 3-methylbut-2-enoyl chloride in the presence of aluminium chloride, followed by mild Clemmensen reduction of the resulting chromanone, leads to fuscinic acid dimethyl ether (44b). A second Friedel-Crafts condensation, this time with acetyl chloride, gives the ketoacid (44c), which is reduced by potassium borohydride to dihydro-fuscin dimethyl ether (43b). This is demethylated under controlled Zeisel conditions to dihydrofuscin (43a), which is smoothly oxidised to fuscin (42) by air in alkaline solution.

From inspection of the formula (42), two possible modes of bio-synthesis are apparent, depending on the biogenetic significance attached to the positions of the nuclear oxygen atoms. One route is from a C_6—C_3 unit derived from prephenic acid, degraded by loss of the terminal carbon atom to a C_6—C_2 unit, with the addition of one acetate and one mevalonate unit. The other (49) involves a mevalonate unit and a polyketide chain built from five acetate units. In fact, the labelling pattern of the radio-active metabolite obtained by incubating *O. fuscum* with [1—^{14}C]-acetate (*13*) was consistent with the latter pathway. Fuscin therefore arises completely from acetic acid units, but by two pathways. The extent of incorporation of label into the polyketide and terpene portions of the molecule is not the same. Furthermore, the labels in the terpene unit appear to be unequal in contrast to the situation in steroids and tri-terpenes. However, studies of the incorporation of [2—^{14}C]-mevalono-lactone show that the C_5-unit is introduced by the usual terpene route.

6. Celastrol and Pristimerin.

Pristimerin (50a) and its parent acid, celastrol (50b), are orange-red triterpenes found in the root bark of several plants of the family Celastr-aceae. Celastrol is identical with tripterine (*125*), from the roots of the thunder-god vine, *Tripterygium wilfordii* HOOK (Celastraceae). The powdered roots of this vine have been used for centuries in China as an insecticide.

Sources of celastrol: *Celastrus scandens* L. (*29, 64*), *Tripterygium wilfordii* HOOK (*125*), *Tr. regelii* SPRAGUE and TAKEDA, *Celastrus strigillosus* NAKAI (*108*). Sources of pristimerin: *Pristimera indica* (WILLD.) A. C. SMITH, *Pr. grahamii* (WIGHT) A. C. SMITH (*8*), *Celastrus dispermus* F. MUELL, *Denhamia pittosporoides* F. MUELL (*66*).

The relationship between the two pigments was established by Shah and Kulkarni (*131*, *132*). Besides its antibacterial properties (*9*), pristimerin possesses notable antitumour activity (*128*) but its high toxicity prevents its use in medicine. Celastrol and pristimerin, along with other natural and synthetic quinone methides, have been tested on the cheek-pouch tumour of the golden hamster. The tumour growth inhibiting properties of these compounds are thought to be due to their reduction to dihydro derivatives which by autoxidation give hydrogen peroxide. This substance destroys tumour cells by inhibition of their glycolysis (*128*).

Early work on these substances dealt mainly with the nature of the chromophore, as there was little evidence at first to indicate the nature of the complete carbon skeleton. As a result, a number of variants of a hydrogen-bonded, hydroxy-o-quinone system were suggested (*50*, *64*, *107*, *125*, *132*). The molecular formula $C_{30}H_{40}O_4$ for pristimerin, and therefore $C_{29}H_{38}O_4$ for celastrol, were finally established by Nakanishi and co-workers (*107*). At this stage, selenium dehydrogenation experiments leading to phenanthrene type products, together with the isolation (*107*) of a small amount of an alkyl picene from zinc dust distillation of pristimerin, indicated a triterpenoid structure for the pigments.

A detailed study of the absorption spectra and reactions of pristimerin and its derivatives led Grant and Johnson (*66*) to conclude that the pigment is not a derivative of o-benzoquinone, in spite of its being reduced to a catechol. [Pristimerol (52a), the dihydro derivative of pristimerin, behaves as a simple dihydric phenol.] Instead, they proposed that an extended quinone methide system was present, although other details of their partial structure were incorrect. The limitations of this part structure were discussed by Kulkarni et al. (*89*, *90*), and by Cooke and Thomson (*35*), with the latter authors suggesting a modified friedelane skeleton for the wood pigment in view of the occurrence of friedelane derivatives in plants of the Celastraceae family. They came close to the correct structure in putting forward (50c), which contained an ester group, hitherto thought to be an inert carbonyl group and an isolated methoxyl group. The Nottingham authors later modified their proposals in several respects (*68*) and arrived at the correct partial structure for rings A, B and C of pristimerin (50a). Evidence for the presence of the ester group was obtained by metal hydride reduction of pristimerol dimethyl ether (52b) to an alcohol with the loss of the carbonyl group along with one methoxyl group. However, this ester group was not thought to be attached to the chromophore, in spite of its low carbonyl frequency in the infrared (1725 cm^{-1}).

The full structure (50a) of pristimerin was eventually established with the aid of NMR spectroscopy by Nakanishi (*69*), who first became interested in the problem while working in Fieser's laboratory. Information on the carbon skeleton was obtained using the ester-anhydride (51), a colourless product obtained from pristimerin by permanganate oxidation of the chromophoric system (68). This compound gives a phenanthrene

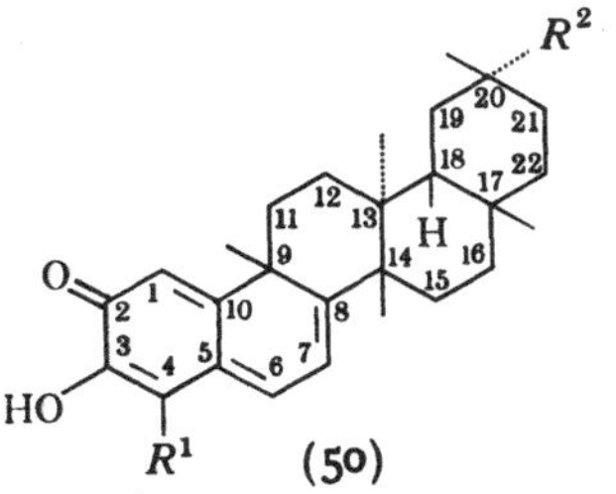

(50)

(a, $R' = CH_3$, $R^2 = CO_2CH_3$; Pristimerin.
b, $R' = CH_3$, $R^2 = CO_2H$; Celastrol.
c, $R' = CO_2CH_3$, $R^2 = CH_3$)

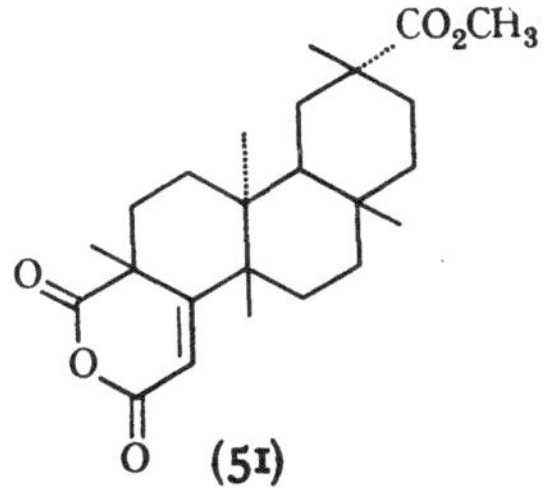

(51)

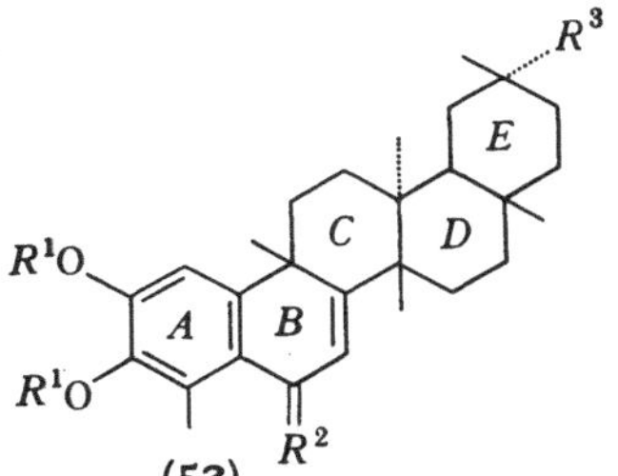

(52)

(a, $R' = H$, $R^2 = H_2$, $R^3 = CO_2CH_3$;
b, $R' = CH_3$, $R^2 = H_2$, $R^3 = CO_2CH_3$;
c, $R' = Ac$, $R^2 = H_2$, $R^3 = CH_2OAc$;
d, $R' = H$, $R^2 = H_2$, $R^3 = CH_2OH$;
e, $R' = H$, $R^2 = H$, CH_2COCH_3, $R^3 = CO_2CH_3$)

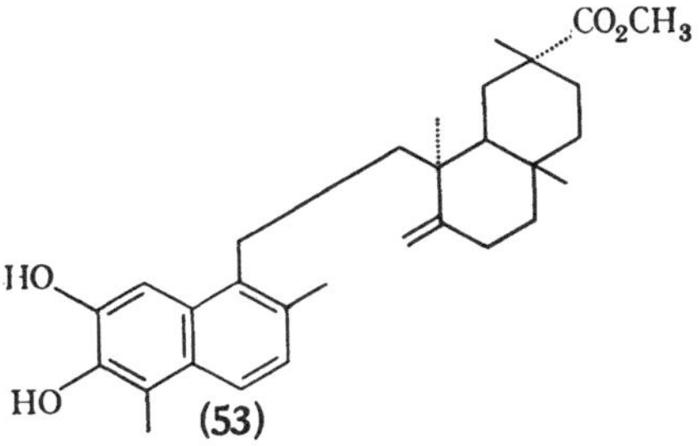

(53)

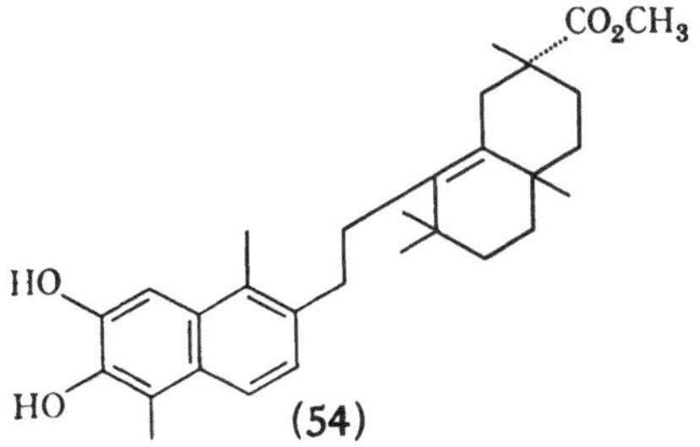

(54)

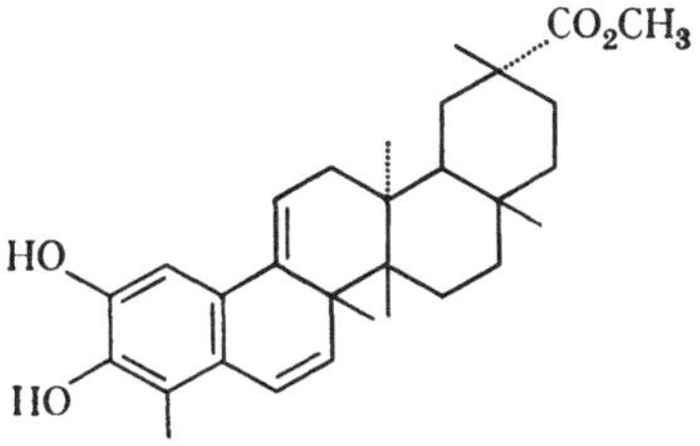

(55)

on dehydrogenation (69), thereby supporting the earlier evidence for a pentacyclic skeleton for pristimerin. The oleane-type ring E is consistent with the appearance of all the methyl groups of pristimerin and its derivatives as singlets in the NMR spectra. [Pristimerol dimethyl ether (52b), for example, shows nine methyl singlets.] In addition, NMR data on the triacetate (52c), derived from pristimerin by metal hydride reduction followed by acetylation, clearly show that the original ester group was attached to a quaternary carbon atom. Mass spectral data also support the postulated structures. The stereochemistry at the ring junctions is derived by logical biogenetic transformations from a β-amyrin type precursor, and the ester group attached to $C_{(20)}$ is regarded as being α-oriented in view of the formation of various lactones.

Independent evidence for structure (50a) for pristimerin was reported by JOHNSON et al. (83). The alcohol (52d) rearranges under the influence of phosphorus pentachloride to an amorphous product containing an ethylidene group. The presence of this grouping is shown by the formation of acetaldehyde on ozonolysis. Such a reaction sequence is common in terpene chemistry and has provided the basis for determination of the structure of many diterpene acids. In this case the result can only be rationalised if pristimerin has an ester group at position 20.

The chemistry of pristimerin shows a number of interesting features. It undergoes dienone-phenol rearrangement in the presence of acids, and three products have been isolated (83, 108). Two of them are naphthalenediols, (53) and (54), and the third is the styrene (55). Their formation is rationalised by initial migration of an angular methyl group or an alkyl residue to the terminus of the quinone methide system as set out in Chart 4.

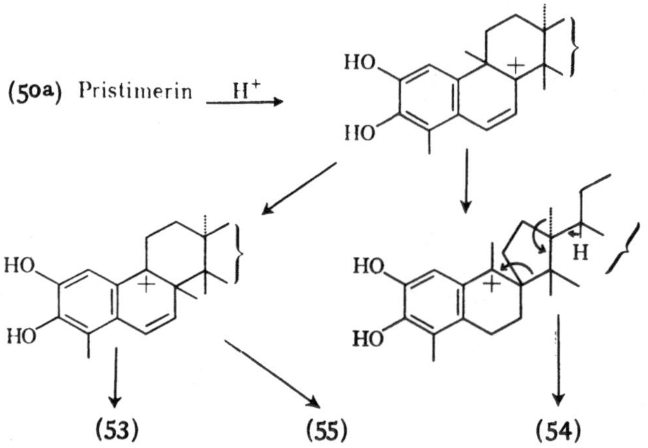

Chart 4. Skeletal Rearrangements of Pristimerin.

Nucleophilic attack takes place at position 6 of pristimerin rather than at the more hindered 8-position. Thus, acetone gives the adduct (52e) by a reaction catalysed both by acids and bases. This addition was first observed during the methylation of pristimerin with dimethyl sulphate and potassium carbonate in acetone (67). Attack of hydride ion likewise occurs at this position, giving pristimerol (52a), and the colourless bisulphite adduct no doubt has a similar structure. Unlike pulvilloric acid and fuscin, pristimerin does not form colourless adducts with alcohols. It gives a red adduct with hydrogen chloride in which the main conjugated system of the chromophore has apparently been maintained (67). Synthetic approaches to the pristimerin chromophore have been reported (78).

The diterpene **fuerstion** (fuerstiaquinone), $C_{20}H_{26}O_3$, isolated from the leaf glands of the East African plant *Fuerstia africana* (Labiatae) by KARRER and EUGSTER (85) is an unstable red pigment which absorbs one or two moles of hydrogen upon catalytic hydrogenation, depending on the catalyst used. Fuerstion has now been assigned the structure given below (84a). It has the same chromophoric system as pristimerin (50a), which it resembles in its lability towards both nucleophilic and electrophilic reagents and especially in its acid-catalysed skeletal rearrangements. (Added in Proof.)

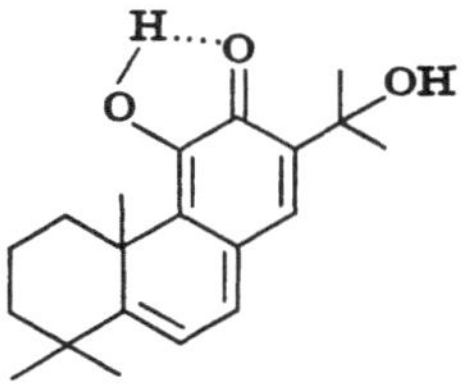

Fuerstion.

7. The Perinaphthenone Group.

a. Haemocorin.

A red glycoside, haemocorin, $C_{32}H_{34}O_{14}$, is found in the bulbous roots of the Australian plant *Haemodorum corymbosum* VAHL (33, 34). The presence of two related pigments in the same species of monocotyledon has been recently demonstrated by THOMAS (140a).

On mild acid hydrolysis, haemocorin gives cellobiose and a purple-red aglycone, $C_{20}H_{14}O_4$. Evidence for the tautomeric structure (56a ⇌ 56b) for this aglycone has been provided by COOKE et al. (32). It is likely that hydrogen bonding stabilizes the form (56a).

Methylation of the aglycone can be controlled to give two monomethyl ethers (57a) and (57b), which yield the two corresponding dimethyl ethers (58a) and (58b) upon further methylation. Permanganate oxidation of the latter products gives isomeric naphthalic anhydrides (59a) and (59b), both of which can be further oxidised to biphenyl-2,3,4-tricarboxylic acid. These reaction sequences (outlined in *Chart 5*) correspond with the behaviour of a tautomeric perinaphthenone system, and studies of model compounds indicate a close parallel with 6-hydroxyperinaphtheno-

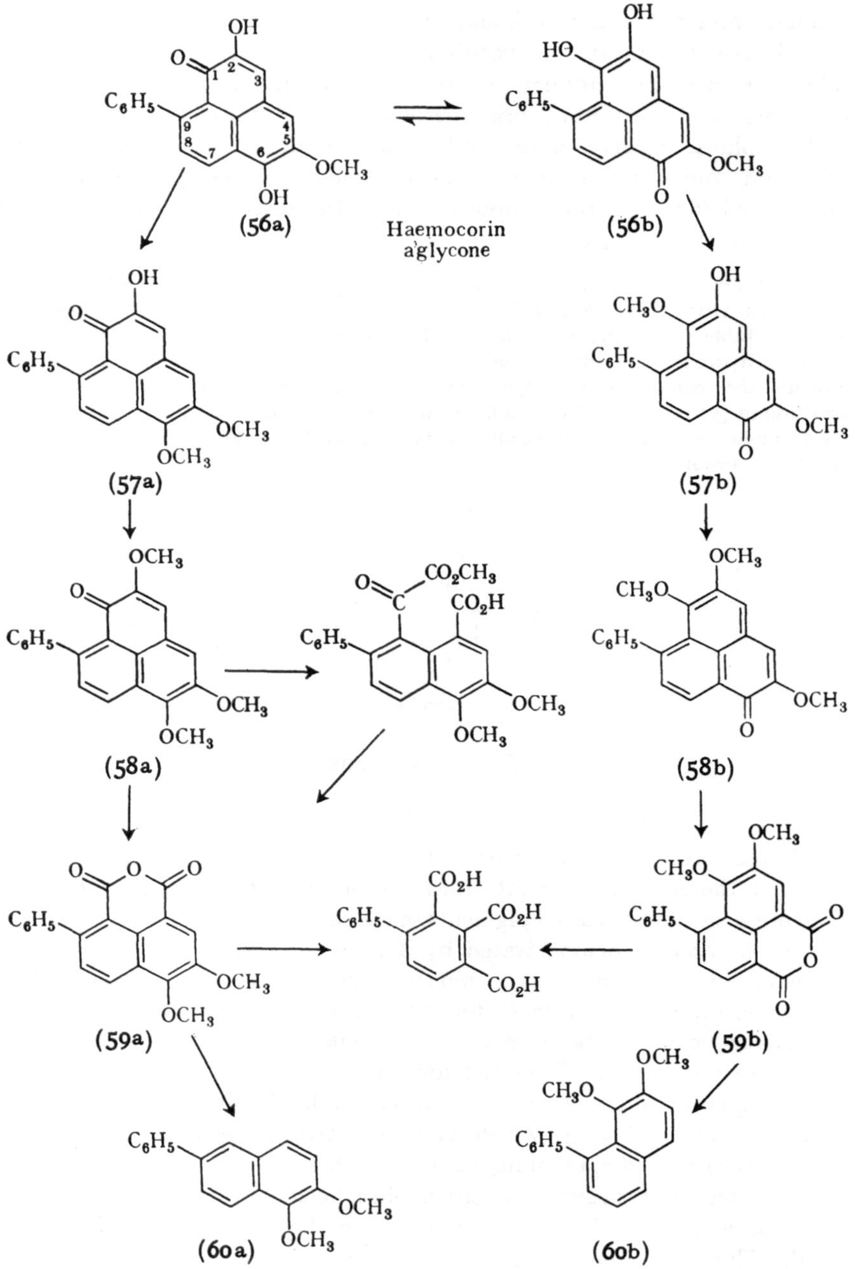

Chart 5. Reactions of Haemocorin.

nes. The orientation of substituents was finally determined by decarboxylation of the yellow anhydrides to the dimethoxyphenylnaphthalenes (60a) and (60b), whose structures were proved by synthesis.

The position and stereochemistry of the glycosidic linkage in haemocorin remain to be established. The biosynthesis of the aglycone is discussed later, together with that of the fungal perinaphthenones (p. 310).

b. Atrovenetin and Herqueinone.

The yellow metabolite atrovenetin, $C_{19}H_{18}O_6$, isolated from strains of *Penicillium atrovenetum* by NEILL and RAISTRICK (*110*), was shown to be identical to deoxynorherqueinone obtained by zinc-acetic acid reduction of norherqueinone, $C_{19}H_{18}O_7$, a red pigment from the morphologically closely related species *P. herquei*. This organism also produces herqueinone, $C_{20}H_{20}O_7$, which is a methyl ether of norherqueinone (*63, 71, 138*). All three compounds are optically active. X-Ray crystallographic studies have established the structure (61) for the ferrichloride of atrovenetin orange trimethyl ether, and hence the structure (62) for atrovenetin itself (*106, 115a*). BARTON and his colleagues (*5, 6*) had previously interpreted chemical results in terms of a 9-hydroxyperinaphthenone system, although their work suggested that the orientation of the ether ring was reversed.

Methylation of atrovenetin gives an extensive series of isomeric ethers, owing to the variety of possible tautomeric forms of the pigment. (A derivative of one of these ethers was used in the X-ray work.)

Atrovenetin is oxidised by alkaline hydrogen peroxide to the optically active naphthalic anhydride (63). This anhydride has also been isolated from cultures of *P. herquei* (*109*). The characteristic chemical and spectral properties of this product show that the anhydride grouping is flanked by the two phenolic hydroxyls.

Oxidation with nitric acid produces nitrococussic acid (64). Two of the nitro groups of this product are introduced by decarboxylative nitration, a reaction often encountered during the nitration of phenolic acids. The dilactone (65) is also formed by the action of nitric acid on atrovenetin (*5, 110*). The origin of this second product is explained by a skeletal rearrangement, probably proceeding through the intermediate (67) (*106*). An alternative structure (66) for the dilactone, differing only in the arrangement of the three methyl substituents on the lactone ring, was rejected on the basis of NMR evidence.

Herqueinone has the same carbon skeleton as atrovenetin, but contains an extra oxygen atom. Various lines of evidence suggest (*6*) that this oxygen atom is present in a tertiary hydroxyl group blocking aromaticity in one or two rings of herqueinone, as this pigment, in contrast to atrovenetin, does not show aromatic stability. This is in keeping with the

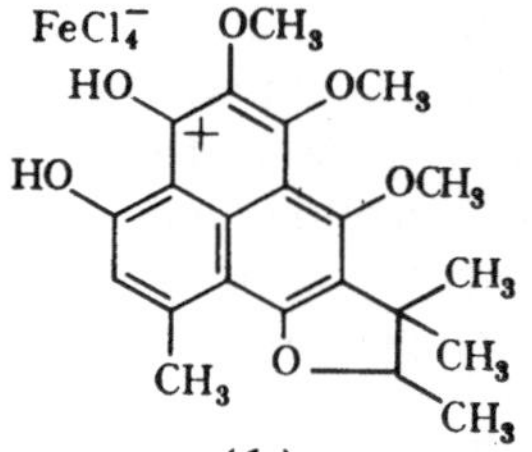

(61)

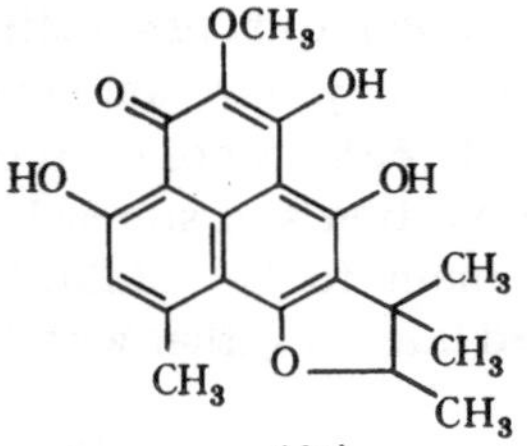

(62)
Atrovenetin.

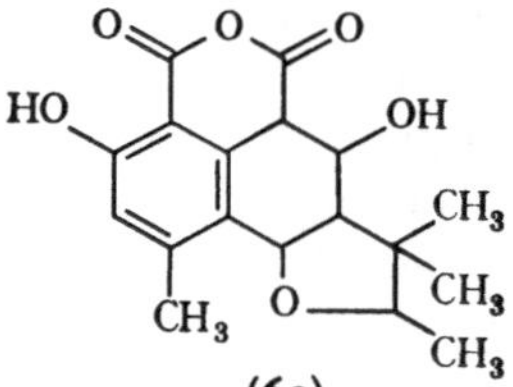

(63)

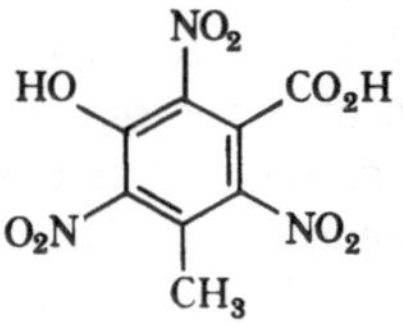

(64)
Nitrococussic acid.

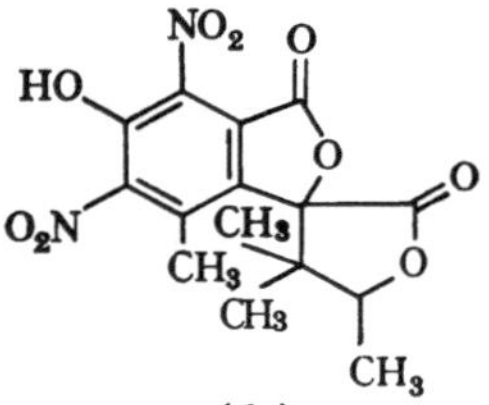

(65)

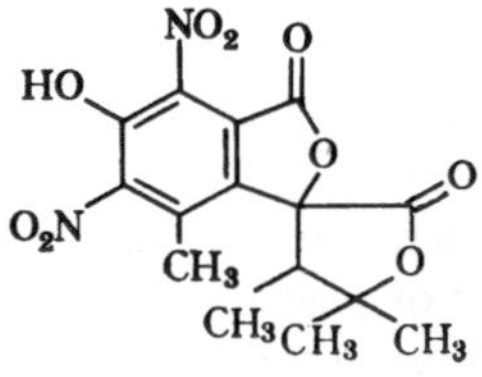

(66)

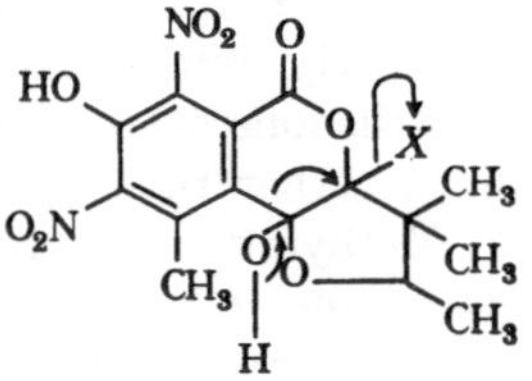

(67)

reduction of norherqueinone to atrovenetin under mild conditions. The location of the tertiary hydroxyl group is indicated by acid-catalysed cleavage of norherqueinone to isopropyl methyl ketone and optically inactive norxanthoherquein (68). Spectroscopic data identified the larger fragment as a derivative of 9-hydroxyperinaphthenone and it was assigned

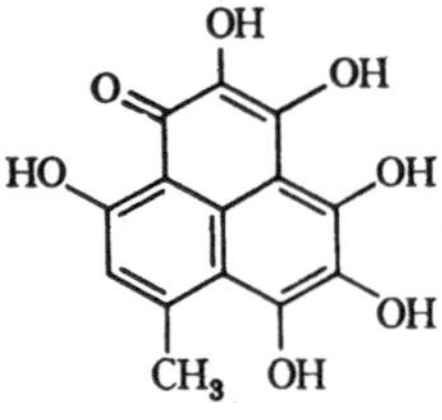

(68)
Norxanthoherquein.

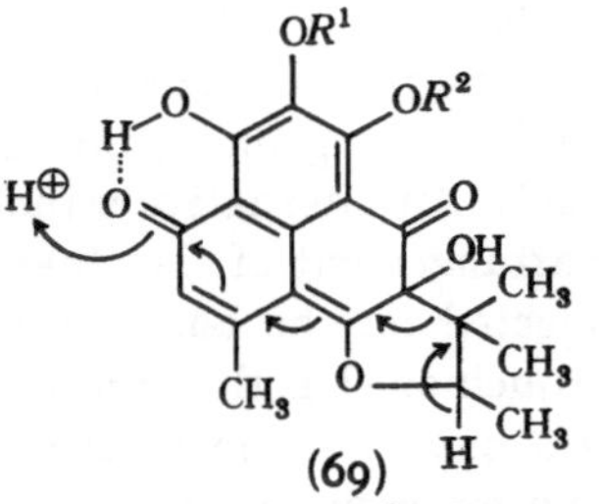

(69)
(a, $R' = R^2 =$ H;
b, $R' =$ H, $R^2 = CH_3$,
c, $R' = CH_3$, $R^2 =$ H)

References, pp. 321—328.

the structure shown to account for the seven oxygen atoms indicated by analysis and for its conversion to nitrococussic acid (64) on oxidation with nitric acid. Herqueinone is similarly hydrolysed to xanthoherquein. These transformations can be explained if norherqueinone has the tautomeric structure (69a), with herqueinone being formulated as (69b) or (69c).

Several features peculiar to the chemistry of herqueinone are explicable on the basis of its more pronounced quinone methide character when compared with atrovenetin. Thus acid hydrolysis of herqueinone to xanthoherquein and methyl isopropyl ketone, a reaction which is not parallelled by atrovenetin, may be initiated by the process depicted in (69).

The formation of trimethylherqueinone B (70), one of the trimethyl ethers of herqueinone, involves carbon migration to the terminus of the quinone methide system. In this case contraction rather than cleavage of the ether ring takes place:

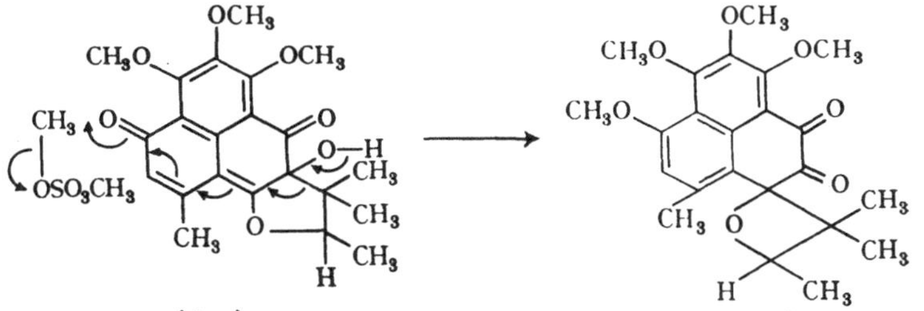

(69d)

(70)
Trimethylherqueinone B.

The structure of this rearrangement product (70) has been assigned on degradative and NMR spectral evidence (25, 95).

Catalytic hydrogenation of norherqueinone affords dihydronorherqueinone, isolated as its tetraacetate after treatment with acetic anhydride in pyridine (106). This compound contains a free hydroxyl group, apparently in a tertiary position and is probably formed by 1,6-reduction of the quinone methide chromophore. Hence the most likely structure for dihydronorherqueinone is (71a). Herqueinone is similarly reduced to herqueinic acid, which may accordingly be formulated as a methyl ether of (71a). The increased capacity for hydrogen bonding in these dihydro compounds could account for their enhanced acidity. The same explanation can be advanced for the behaviour of herqueinone as a lactone (25, 71), since the hydrated form (71b) of the pigment must be as strongly acidic as herqueinic acid. Metal hydride reduction of atrovenetin trimethyl ether, by contrast, does not yield a dihydro derivative (5). Instead, hydrogenolysis occurs with 1,2- or 1,4-addition followed by β-elimination. The quinone methide chromophore of atrovenetin is

embodied in a much more extensive aromatic system, and so the metabolite
does not exhibit characteristic quinone methide properties. This is also
true of haemocorin.

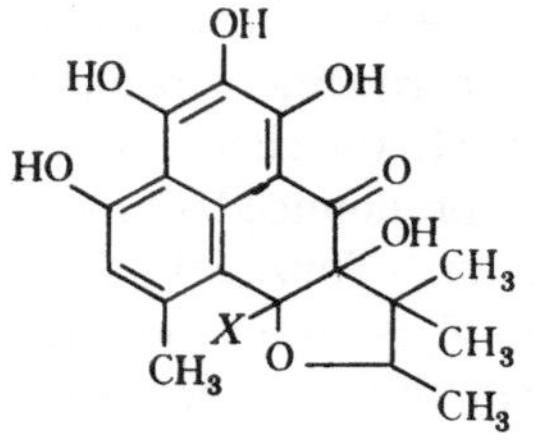

(71)
(a, X = H;
b, X = OH)

Brazilein (72a) and haematoxylein (72b) from hard-woods of the
Caesalrinia species also possess a quinone methide chromophore termi-
nated by a carbon atom bearing a tertiary hydroxyl group. However,
these pigments are artefacts formed by oxidation of the colourless precursors brazilin (73a) and haematoxylin (73b).

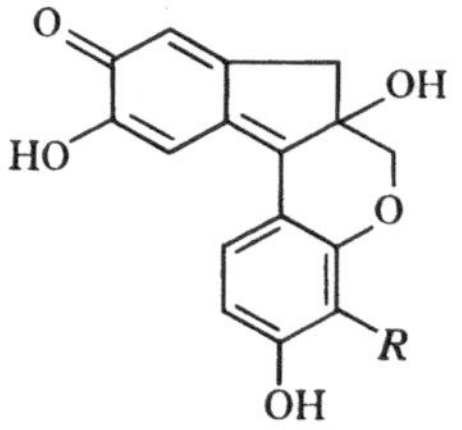

(72)   **(73)**
(a, R = H; Brazilein.   (a, R = H; Brazilin.
b, R = OH; Haematoxylein.)   b, R = OH; Haematoxylin.)

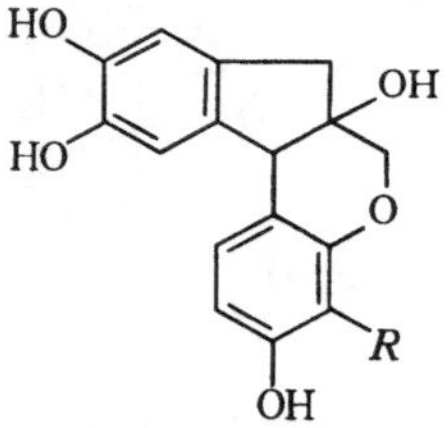

c. Biogenesis of Plant and Fungal Perinaphthenones.

Since atrovenetin, herqueinone, and norherqueinone are produced
by closely related fungal species, their common carbon skeleton can be
expected to arise by the same biosynthetic pathway. Incorporation
experiments in *P. herquei* (*140*) using sodium acetate-1-^{14}C and mevalono-
lactone-2-^{14}C are consistent with the formation of the perinaphthenone
nucleus of norherqueinone from acetate units and the side-chain from
mevalonate. Acetate is incorporated to approximately the same extent
into the aromatic nucleus and the isoprenoid ether ring. A linear C_{14}-
polyketide chain, which is also postulated in the biogenesis of the *Penicillia*
metabolites purpurogenone and pulvilloric acid, can be coiled to give
the perinaphthenones in two possible ways, (74) and (75). Although

these cannot be distinguished by [^{14}C]-tracer work, there are more analogies for cyclisation in the latter mode. The further possibility of a branched-chain C_{14}-precursor also cannot be ruled out.

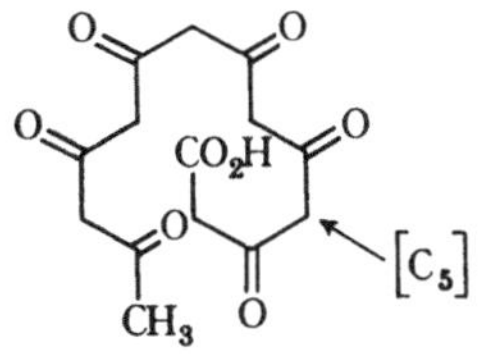

(74) 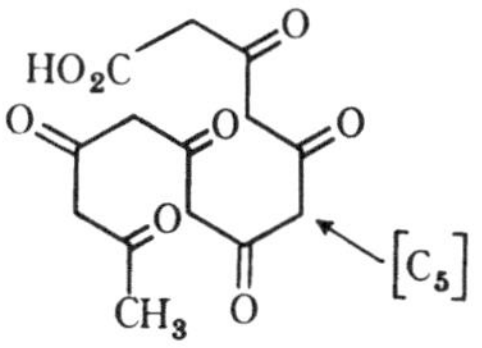   (75)

Haemocorin (56, p. 306) could arise from a polyacetic acid precursor similar to (75), but with a benzoyl terminus, although an alternative scheme has been proposed by Thomas (*140*) to account for its unusual oxygenation pattern. This involves the condensation of one acetate unit with two shikimic acid-derived C_6—C_3 units (cinnamic acid and 3,4-dihydroxyphenylpyruvic acid), followed by decarboxylation and cyclisation to give the phenylperinaphthenone nucleus of (56a).

Thomas (*140a*) has recently obtained evidence supporting his hypothesis from studies using labelled aromatic amino acids. In the biosynthesis of haemocorin by *Haemodorum corymbosum*, phenylalanine and tyrosine are utilised much more efficiently than acetate. The activity of tyrosine-2-^{14}C appears exclusively at $C_{(5)}$ of the haemocorin aglycone (56a):

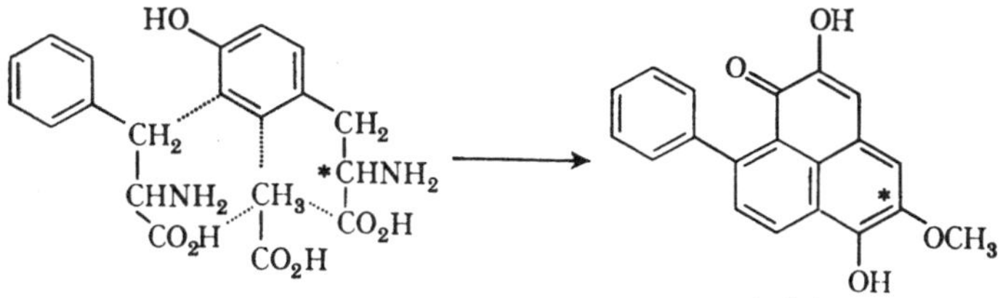

(56a)

The labelled carbon atom was located using the degradative procedure already outlined. The tyrosine-2-^{14}C-derived aglycone (56a) was converted into its dimethyl ether (58b, p. 306), which was oxidised to an inactive anhydride (59b), thereby demonstrating specific incorporation in accordance with the predicted biosynthetic pathway. The incorporation of acetate-2-^{14}C into the aglycone is also consistent with the origin of the "central" carbon atom by this route. However, experiments with acetate-1-^{14}C indicate that some general incorporation of acetate may have occurred. These results, although incomplete, strongly support the C_6—C_3

pathway for the formation of the haemocorin aglycone. Thus, the known plant and fungal perinaphthenones appear to be biogenetically unrelated, despite their structural similarities.

8. Anhydro Bases of the Flavonoid Series.

The quinone methide chromophore is present in the "anhydro-base" forms of the anthocyanins, the best known example being carajurin (**76a**), which occurs along with carajurone (**76b**) in the leaves of the South American plant *Bignonia chica* (*26, 117*).

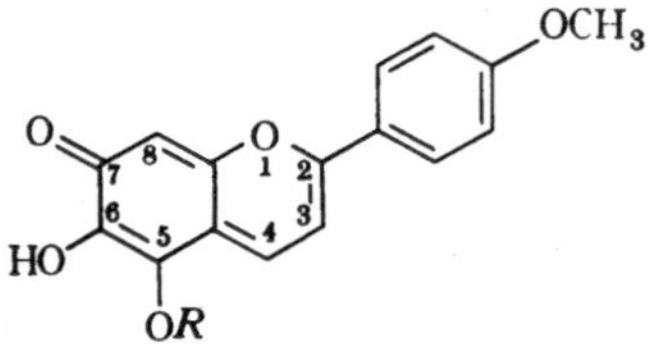

(**76**)

(a, R = CH$_3$; Carajurin.
b, R = H; Carajurone.)

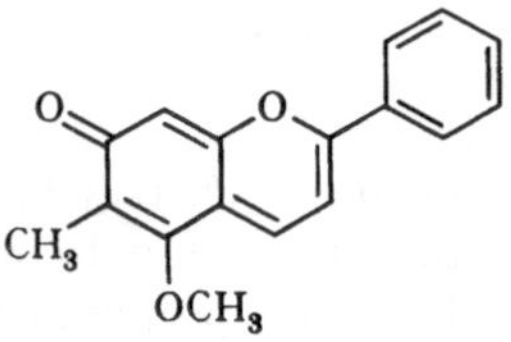

(**77**)
Dracorhodin.

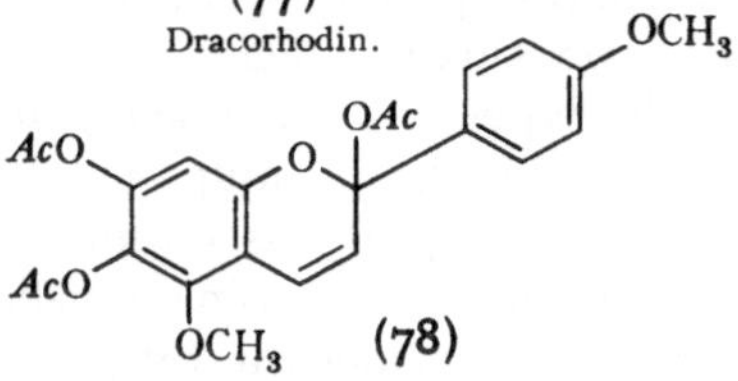

(**78**)

Dracorhodin (**77**) is one of the simpler pigments of a dark-red resin exuded by trees of the genera Dracena and Daemonorops (*76*). The resin is known as "dragon's blood" and has some use as a pigment in the varnish industry and in medicine as an astringent.

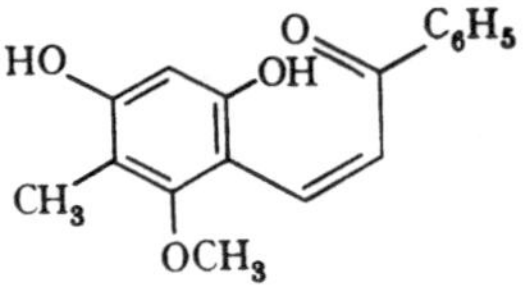

(**79**)

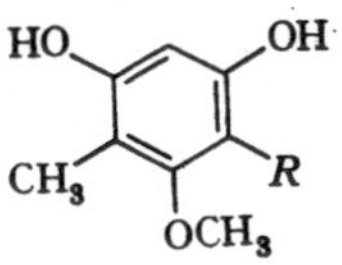

(**80**)

(a, R = CHO;
b, R = H)

Compounds of this type are stable only when position 3 is unsubstituted. On treatment with mineral acids they are converted into flavylium salts, but with other reagents their quinone methide character is more evident. Thus, with acetic anhydride in pyridine, carajurin gives a colourless triacetate (**78**) whose formation involves addition of the elements of acetic acid across the chromophore. Dracorhodin (**77**) is degraded by alkali to acetophenone and the phloroglucinol derivative (**80b**). This

References, pp. 321—328.

reaction is explained by initial attack of hydroxide ion at position 2 to form the pseudobase, followed by ring opening to the ketone (79) which can break up into the observed products via the aldehyde (80a).

Carajurin (76a) is similarly cleaved by hot aqueous alkali to *p*-methoxyacetophenone and the appropriate phloroglucinol. These anhydro bases do not yield crystalline derivatives with carbonyl reagents. Such reagents probably add to the 2-position and then induce ring fission.

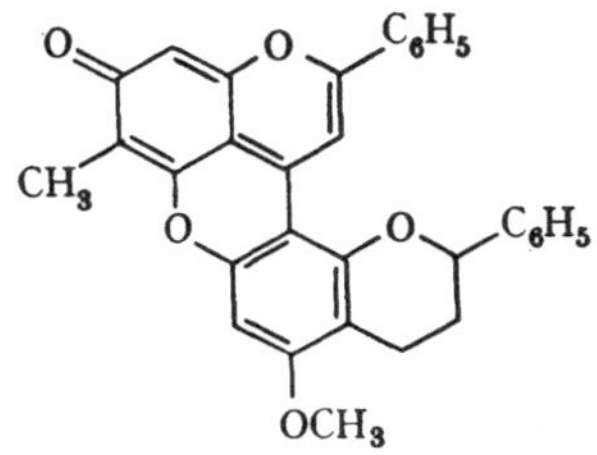

(81)
Dracorubin.

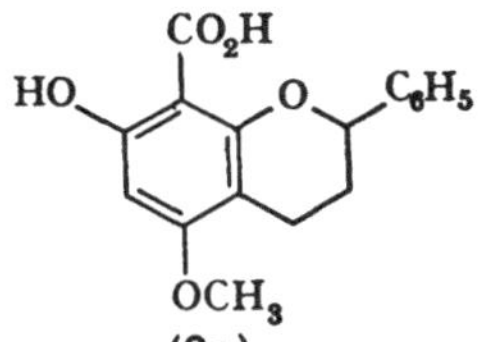

(82)
Dracoic acid.

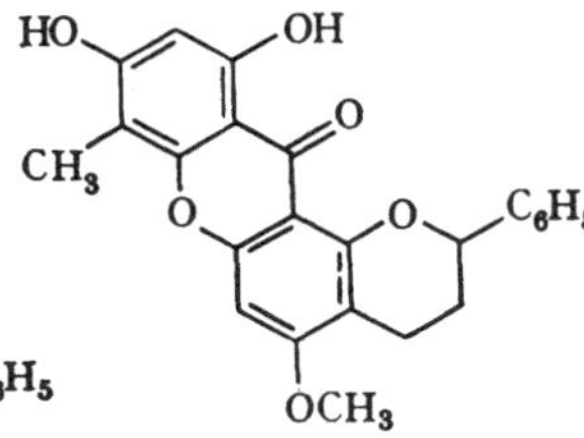

(83)

Dracorhodin (77) has been synthesised by condensation of the aldehyde (80a) with acetophenone (*17, 121*), as has carajurin by a similar process (*118*).

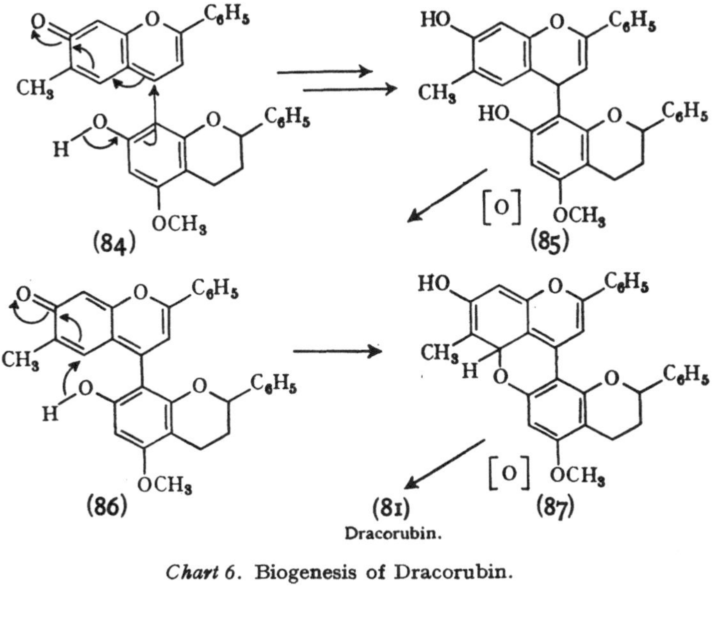

(84)

(85)

(86)

(81)
Dracorubin.

(87)

Chart 6. Biogenesis of Dracorubin.

Another pigment from dragon's blood resin, dracorubin (**81**), can also be degraded to acetophenone (**31**). It is oxidised by hydrogen peroxide (**16**) to dracoic acid (**82**), the structure of which was proved by synthesis (**123**). The precursor (**83**) of dracoic acid is also formed in the oxidation.

The probable mode of biogenesis of dracorubin, as suggested by WHALLEY (**154**), involves condensation of an anhydro base with a phenol (stages **84** to **87**) (*Chart 6*, p. 313).

Other complex anhydro bases of incompletely defined structure are present in sandalwood, barwood, and camwood (*122*). They also appear to be responsible for the redness induced in certain condensed tannins by heat or sunlight (*124*). The extent to which anthocyanins exist in their anhydro base forms in plant material has yet to be defined, as they are usually extracted as their salts. The problem has been discussed in several recent articles (cf. *41, 70*).

9. General Discussion.

a. *Structural Features Contributing to Stability.*

Simple quinone methides, lacking the symmetry of quinones themselves, are highly reactive compounds which are prone to polymerise spontaneously, in particular when the terminal methylene group is unsubstituted (*81, 142*). The ready aromatisation of the nucleus provides the driving force for these reactions. Although none of the quinone methides discussed above has an unsubstituted terminal methylene group, their stability is unusual in this class of compounds. It can be traced to a variety of factors.

A common feature of these natural products is the isolation of the quinone methide chromophore from labile hydrogen atoms, which prevents their tautomeric rearrangement to phenols (*66*). This is achieved in most cases by linking the methylene terminus of the chromophore to a cyclic ethereal oxygen atom, thereby forming a vinylogous lactone system. In the remaining compounds, carbon atoms terminating the conjugated system either lack hydrogen or form part of an aromatic system. In addition, the di-unsaturated carbonyl group in these compounds is often strongly hydrogen bonded. This factor, particularly in conjunction with the vinylogous lactone system, leads to delocalisation of the electrons of the quinonoid chromophore and confers some degree of aromatic character upon the nucleus.

b. *Tautomeric Forms.*

The mould metabolites mentioned above have been assumed to be *para*-quinone methides, although in several cases *ortho*-quinone methide structures could equally well be written. In citrinin, pulvilloric acid

References, pp. 321—328.

and ascochitine, stabilisation by hydrogen bonding is unimpaired in the *ortho* isomers (general formula **88**). Degradative evidence is equally consistent with either *ortho* or *para* formulations, and the syntheses of citrinin and pulvilloric acid are also ambiguous in this respect. The *para*-quinonoid structures have been adopted on intuitive grounds, the chromophore being considered to be more extended in these isomers. There is, however, no analogy between quinone methides and quinones themselves as far as the relative stability of their isomers is concerned. It is true that *ortho*-quinone methides appear to be more reactive than the corresponding *para* isomers, owing to their readier aromatisation by Diels-Alder type reactions, and that no such reactions have been reported for the above natural products. In any case, it seems that X-ray work will be required to settle this detail of structure.

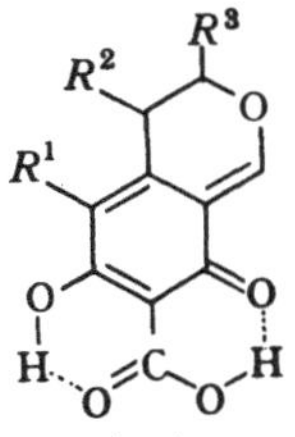

(88) 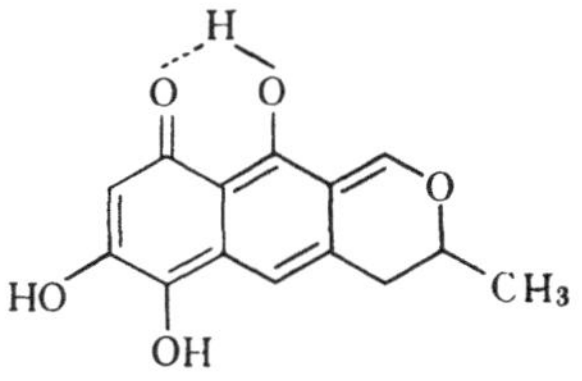　　　　　　　　(89)

The case of purpurogenone is more interesting in that the alternative tautomers should be stabilised to differing extents by hydrogen bonding. Assuming the linear structure to be correct, chelation in the six-membered ring of the *ortho* isomer (89) would be more effective than that in the five-membered ring of (40, p. 299). A recently discovered example of the stabilisation of an unexpected tautomer has been found in the case of cordeauxiaquinone (*49*), where bond lengths favour the structure (90) rather than (91) in the crystalline state.

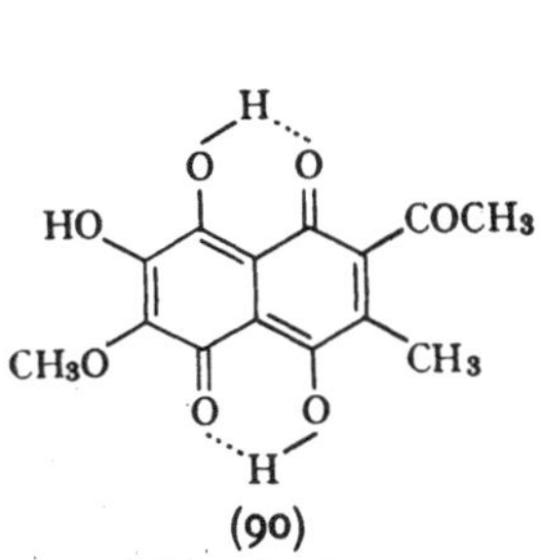

(90)　　　　　　　　(91)

Cordeauxiaquinone.

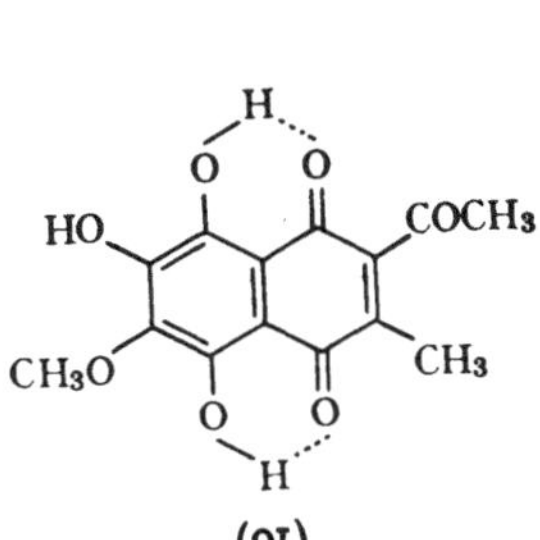

c. General Reactions.

A characteristic feature of quinone methides is their susceptibility to nucleophilic attack at the terminal methylene group with concurrent aromatisation of the nucleus. This property accounts for many of the reactions already mentioned. The most usual outcome is 1,6-addition of the reagent across the chromophore, and this type of addition is the rule even when an extended quinone methide chromophore is present (as in pristimerin):

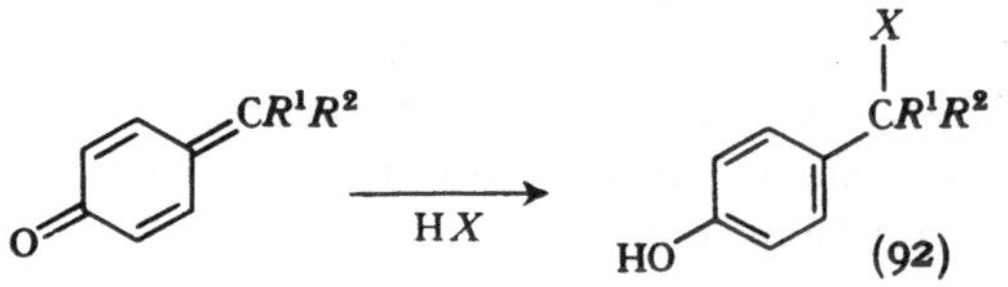

$$(92)$$

Thus, reduction of these compounds, either catalytically or with metal hydrides, leads to colourless dihydro derivatives containing one new phenolic hydroxyl group, rather than the two which appear upon hydrogenation of quinones. These various adducts of type (92) which are formed from stable quinone methides exhibit varying degrees of stability. Some of them readily revert to the parent compound, and this has been a useful guide in identifying the chromophore of these natural products. This reaction type also embraces the retro-aldol cleavage reactions catalysed by both acids and bases which involve initial addition of the elements of water to the chromophore.

Finally, the skeletal rearrangements encountered with some of these compounds can be classified as intramolecular examples of the general process depicted above.

III. Quinone Methides as Intermediates in Biochemical Processes.

1. Oxidative Phosphorylation.

Although speculation concerning the participation of quinol phosphates in oxidative phosphorylation has been rife for some time (cf. *126*), rational mechanisms for their formation from quinones have only recently appeared (*30, 46, 91, 144, 144a*). These take advantage of structural features common to members of the biologically active vitamin K, vitamin E and ubiquinone groups (general formula 93). Quinones of these series have a methyl group attached to the nucleus and a side-chain capable of participating in chroman ring formation, thereby allowing rearrangement to an isomeric quinone methide (94). The phosphate residue can be introduced through attack by phosphate anion upon the terminal methylene group of this intermediate. The resulting benzyl phosphate (95) rearranges to the phenyl phosphate (96) by way of a cyclic, six-membered transition state. This is converted into the active quinol phosphate by reduction followed by ring opening, and the cycle is completed by oxidation to the natural quinone with the release of metaphosphate. The sequence of these steps may be varied.

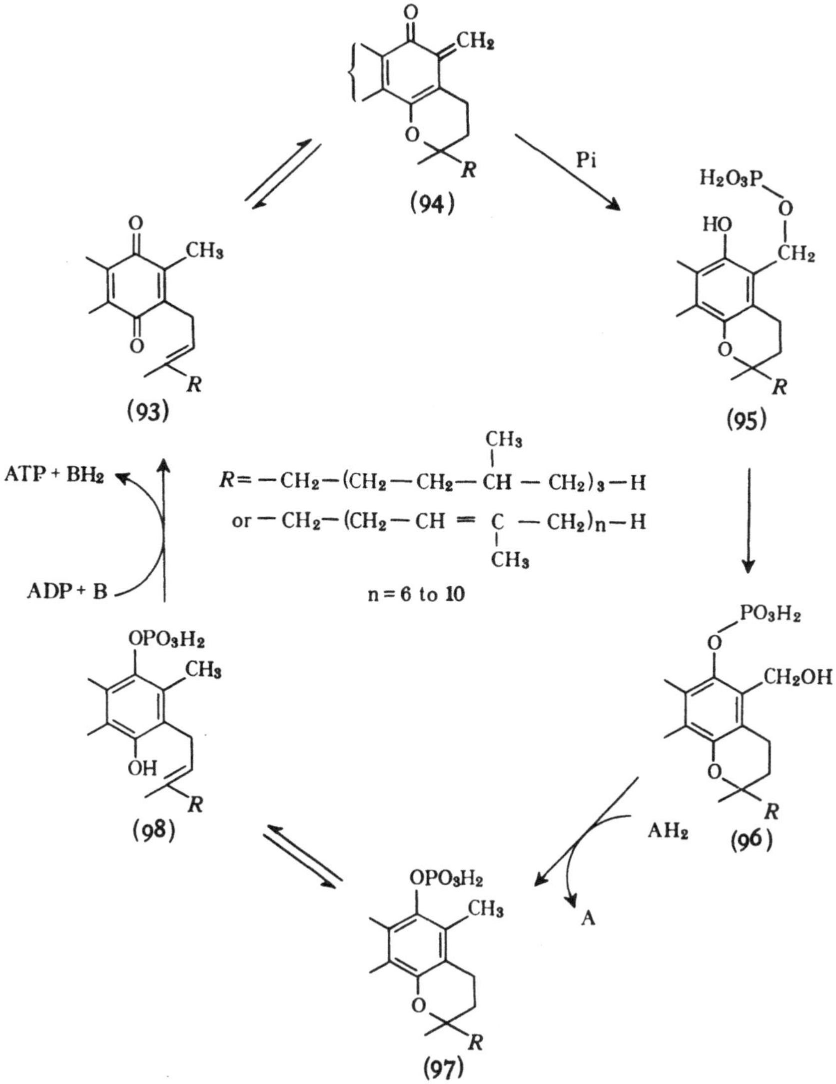
CH2
O
R
(94)
Pi
H2O3P
O
HO
CH2
R
(95)
CH3
O
R
(93)
ATP + BH2
ADP + B
R= —CH2—(CH2—CH2—CH — CH2)3—H
or — CH2—(CH2—CH = C — CH2)n—H
CH3
n = 6 to 10
PO3H2
CH2OH
R
(96)
AH2
A
OPO3H2
CH3
OH
R
(98)
OPO3H2
CH3
R
(97)

Chart 7. Vilkas-Lederer Mechanism for Oxidative Phosphorylation. (Pi = inorganic phosphate.)

This scheme, which was proposed by Vilkas and Lederer (*144, 144a*), is illustrated in *Chart 7*. Slight variants on this theme have been suggested by the same authors, and by Erickson, Wagner and Folkers (*46*), who consider that the chromanol (95) may be the active phosphorylating species, generating metaphosphate on oxidation.

The schemes (Chart 7) represent the culmination of numerous hypotheses over the years, each of which has incorporated an additional significant feature of the natural quinones (see *18, 30*). The participation of an *o*-quinone methide structure was first suggested by Chmielewska (*27, 28*), although her detailed mechanism involved 1,2-addition of phosphoric acid to this intermediate and so did not explain the lack of exchange of ^{18}O between labelled phosphate and the quinones during oxidative phosphorylation. The 1,4-addition of phosphoric acid to the intermediate (94) postulated by Vilkas and Lederer, together with the subsequent intramolecular migration of phosphate to the phenolic hydroxyl group, ensures that no oxygen atom of the original phosphate residue appears in the regenerated quinone.

Further evidence cited (*144, 144a*) in support of the participation of the quinone methide (94) in oxidative phosphorylation includes enthalpy difference calculations which demonstrate the relative ease of the isomerisation (93 → 94) in comparison with a similar example, (99 → 100), in which there is no cyclisation. This latter type of isomerisation accounts for the activity of the nuclear methyl groups of duroquinone towards a number of reagents (*22, 135–137*).

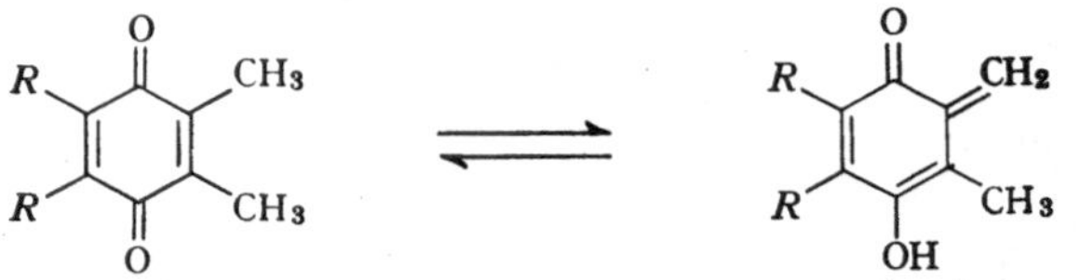

(99)   (100)

The ease of the transformation (93 → 94) is also reflected in the ready dimerisation of these compounds. Thus, in the naphthoquinone series the dimer (103) is produced by the action of sulphuric acid on vitamin K_1 (101) (*46, 96*). The same dimer is obtained by oxidation of the naphthochromanol (104) with ferricyanide *(Chart 8)*.

The formation of the dimer (103) is explained by the quinone methide (102) being an intermediate in these reactions, as it is well established that *o*-quinone methides give dimers of this type (*81, 142*). Similar dimers have been obtained in the vitamin E series (*100, 111, 127, 133*).

In contrast to these reactions, base-catalysed isomerisation of the quinone (101) leads to the alternative quinone methide tautomer (105), which cyclises to the chromenol (106) (*146*). Analogous ring-closures,

References, pp. 321—328.

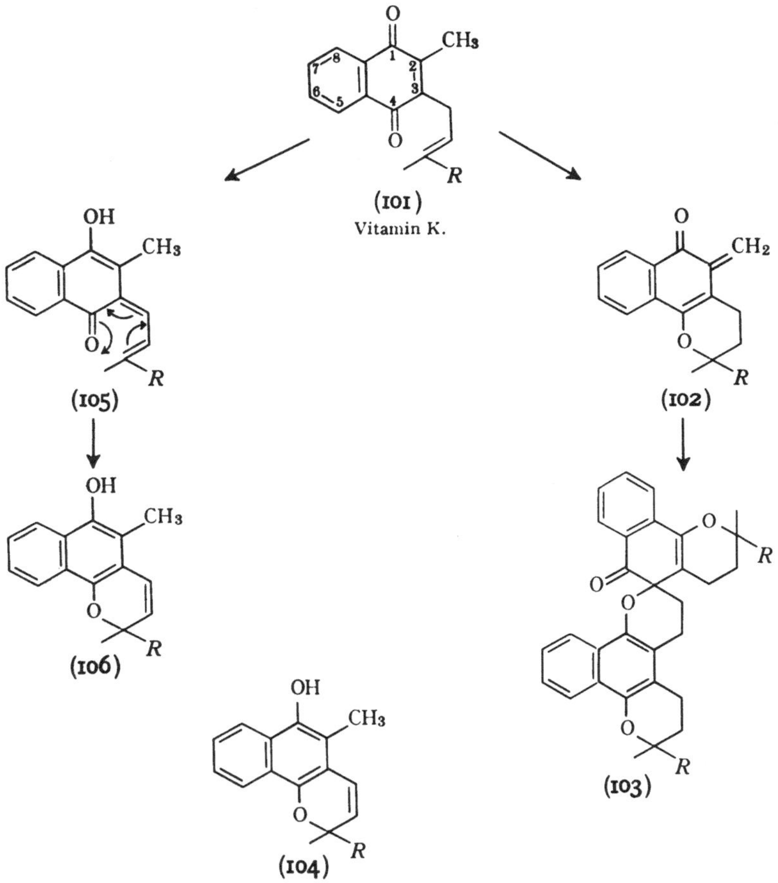

Chart 8. Reactions of Vitamin K (101).

induced by both bases and light, have been reported in the ubiquinone series (*93, 99, 105*). A recent study by McHale and Green (*100a*) has shown that simple allylbenzoquinones are converted into chromenols by heating with pyridine. The tendency for the reaction to occur increases as the quinone ring becomes more highly substituted by electron-donating groups. The co-occurrence (*47, 48*) of dalbergiones (107) and dalbergins (109) may be due to a similar cyclisation of the intermediate (108):

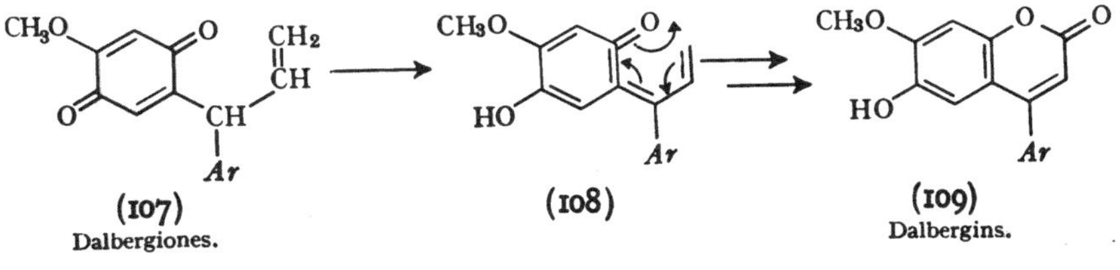

(107) Dalbergiones. (108) (109) Dalbergins.

The formation of the quinone methide (**102**) has also been demonstrated by trapping it with various reagents. Condensations of vitamin K_1-20 with styrene and 1,1-diphenylethylene in the presence of perchloric acid gives the products (**110a**) and (**110b**) respectively (*96*). Again, these Diels-Alder type additions are characteristic of *o*-quinone methides (*80*, *147*). The quinone (**101**) also reacts with acetyl chloride in perchloric acid to give the chloromethyl acetate (**111a**), and acetic anhydride likewise yields the diacetate (**111b**) (*91*, *145*).

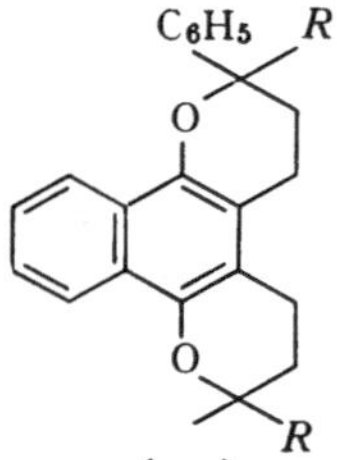

(**110**)
(a, R′= H;
b, R′ = C$_6$H$_5$)

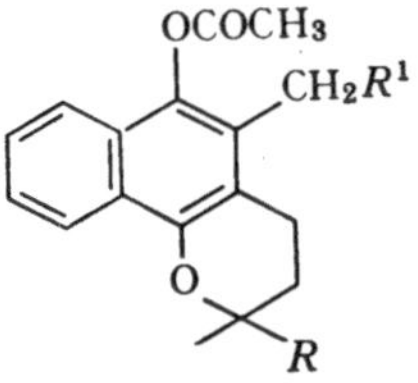

(**111**)
(a, R′ = Cl;
b, R′ = OCOCH$_3$)

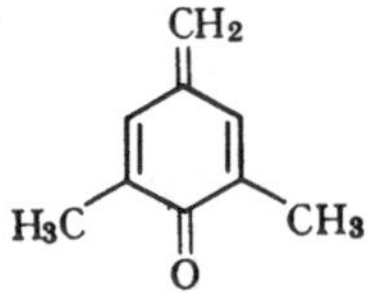

(**112**)

These results leave no doubt that vitamin K_1-20 isomerises in acid solution to the quinone methide (**102**), which subsequently undergoes 1,4-addition. The reversibility of the isomerisation (**101** → **102**) can be demonstrated by tritium labelling (*91*).

The second stage of the Vilkas-Lederer scheme has been further substantiated by the addition of phosphate anion ($PO_4^{\equiv}$) to the *p*-quinone methide (**112**) generated by oxidation of mesitol with silver oxide (*30*). Biochemical evidence, however, has not been forthcoming*.

Vitamin K also acts as a specific catalyst for prothrombin synthesis in higher animals. Here, again, the 2-methyl group and the β,γ-double bond in the 3-isoprenoid side-chain are vital structural features. It is thus expected (*91*) that a quinone methide-chroman will prove to be the active form of the catalyst. This active form could add thiol groups of enzymes or other active groups.

2. Oxidative Metabolism.

Quinone methides play a key role in the biogenesis of lignans and lignins, concerning which the reader is referred to the following articles (*44, 45, 52, 73-75, 87, 113*) and especially to FREUDENBERG's recent survey (*51*) published in this Series. The formation of lignin macromolecules from *p*-hydroxycinnamyl alcohols is initiated by oxidative coupling, leading to a variety of dimeric intermediates (see e. g., *23, 53, 59, 94, 112*), the most important of which is the quinone methide (**113a**).

* Cf. SNYDER, C. D., S. J. DI MARI and H. RAPOPORT: J. Amer. Chem. Soc. **88**, 3868 (1966). (Added in Proof.)

This is involved in further growth both by polymerisation (*62*) and by addition of the phenolic monomer units across its quinonoid chromophore (*60*, *61*). Evidence for the existence of (113a) includes ultraviolet data

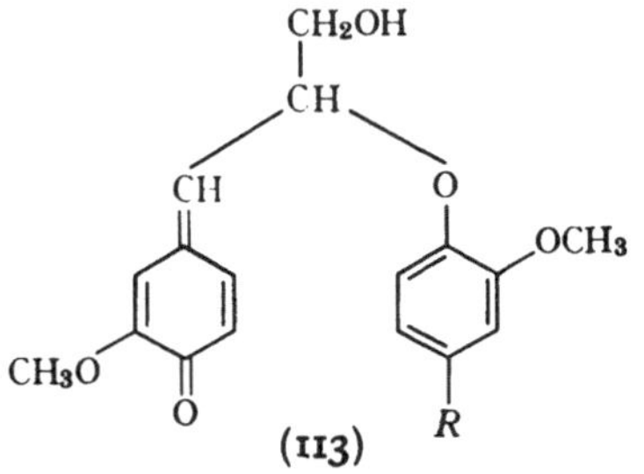

(113)

(a, R = —CH=CH—CH$_2$OH;
b, R = H)

similar to those recorded (*2*) for the model compound (113b), as well as the formation of adducts with water, methanol, carbohydrates, and simple phenols (*54–58*). Recognition of growth principles such as these has led to a schematic structural formula for lignin which is in accord with evidence from other approaches (e. g., *1*, *97*).

3. Miscellaneous Processes.

Quinone methides may also participate in other biosynthetic processes in which reaction occurs at a benzylic position *ortho* or *para* to a phenolic hydroxyl group (see e. g., *3*, *39a*, *86*, *106a*, *151*, *152*). However, these quinonoid intermediates may not be obligatory, particularly in cases where hydroxylations are involved (*39*, *92*, *130*).

References.

1. ADLER, E. and K. LUNDQUIST: Spectrochemical Estimation of Phenylcoumaran Elements in Lignin. Acta Chem. Scand. **17**, 13 (1963).
2. ADLER, E. und B. STENEMUR: Ligninchemische Modellstudien: Über Chinonmethide. Chem. Ber. **89**, 291 (1956).
3. BAROTHY, J. and H. NEUKOM: Synthesis of *p*-Hydroxybenzyl-*iso*thiocyanate and its Isolation from White Mustard Seeds. Chem. and Ind. **1965**, 308.
4. BARTON, D. H. R. and J. B. HENDRICKSON: The Constitution and Synthesis of Fuscin. J. Chem. Soc. (London) **1956**, 1028.
5. BARTON, D. H. R., P. DE MAYO, G. A. MORRISON and H. RAISTRICK: The Constitutions of Atrovenetin and of Some Related Herqueinone Derivatives. Tetrahedron **6**, 48 (1959).
6. BARTON, D. H. R., P. DE MAYO, G. A. MORRISON, W. H. SCHAEPPI and H. RAISTRICK: Some Observations on the Constitutions of Herqueinone and Related Compounds. Chem. and Ind. **1956**, 552.
7. BERTINI, S.: A Compound from *Ascochyta pisi* having Antibiotic Action. Ann. sper. agrar. (Roma) **11**, 545 (1957).

8. Bhatnagar, S. S. and P. V. Divekar: Pristimerin, the Antibacterial Principle of *Pristimera indica*. I. Isolation, Toxicity, and Antibacterial Action. J. Sci. Indust. Res. (India) 10 B, 56 (1951).

9. Bhatnagar, S. S., P. V. Divekar and N. L. Dutta: Pristimerin. Indian Patent 40970 (1951).

10. Birch, A. J.: The Structure of Fuscin. Chem. and Ind. 1955, 682.

10a. — Biosynthetic Relations of Some Natural Phenolic and Enolic Compounds. Fortschr. Chem. Organ. Naturstoffe 14, 186 (1957).

11. Birch, A. J., P. Fitton, E. Pride, A. J. Ryan, H. Smith and W. B. Whalley: Studies in Relation to Biosynthesis. Part XVII. Sclerotiorin, Citrinin, and Citromycetin. J. Chem. Soc. (London) 1958, 4576.

12. Birch, A. J. and M. Kocor: Studies in Relation to Biosynthesis. Part XXII. Palitantin and Cyclopaldic Acid. J. Chem. Soc. (London) 1960, 866.

13. Birch, A. J., A. J. Ryan, J. Schofield and H. Smith: Studies in Relation to Biosynthesis. Part XXXVII. Some Structures Derived from Acetic Acid by Two Pathways. J. Chem. Soc. (London) 1965, 1231.

14. Birkinshaw, J. H., A. Bracken, S. E. Michael and H. Raistrick: Studies in the Biochemistry of Micro-organisms. 83. Fuscin. Part 2. Derivatives and Degradation Products. Biochem. J. 48, 67 (1951).

15. Brian, P. W., P. J. Curtis, H. G. Hemming and G. L. F. Norris: Pulvilloric Acid, an Antibiotic obtained from Cultures of *Penicillium pulvillorum*. Trans. Brit. Mycological Soc. 40, 369 (1957).

16. Brockmann, H., R. Haase und E. Freiensehner: Über das Dracorubin, III. Mitt. Oxydativer Abbau zu Draconol und Dracosäure. Ber. dtsch. chem. Ges. 77, 279 (1944).

17. Brockmann, H. und H. Junge: Die Konstitution des Dracorhodins, eines neuen Farbstoffes aus dem „Drachenblut". Ber. dtsch. chem. Ges. 76, 751 (1943).

18. Brodie, A. F.: The Role of Naphthoquinones in Oxidative Metabolism. In: R. A. Morton (Edit.): Biochemistry of Quinones, p. 355. New York: Academic Press. 1965.

19. Brown, J. P., N. J. Cartwright, A. Robertson and W. B. Whalley: The Chemistry of Fungi. Part IV. The Constitution of the Phenol, $C_{11}H_{16}O_3$, from Citrinin. J. Chem. Soc. (London) 1949, 859.

20. — — — — The Chemistry of Fungi. Part V. The Constitution of Citrinin. J. Chem. Soc. (London) 1949, 867.

21. Bullimore, B. K., J. F. W. McOmie and A. B. Turner: Synthesis of Pulvilloric Acid and Methyl Dihydropulvillorate. Manuscript in preparation.

22. Cameron, D. W., P. M. Scott and Lord Todd: Side-chain Amination: A New Reaction of Nuclear-alkylated Quinones. J. Chem. Soc. (London) 1964, 42.

23. Cartwright, N. J. and R. D. Haworth: The Constituents of Natural Phenolic Resins. Part XIX. The Oxidation of Ferulic Acid. J. Chem. Soc. (London) 1944, 535.

24. Cartwright, N. J., A. Robertson and W. B. Whalley: The Chemistry of Fungi. Part VII. Synthesis of Citrinin and Dihydrocitrinin. J. Chem. Soc. (London) 1949, 1563.

25. Cason, J., J. S. Correia, R. B. Hutchison and R. F. Porter: The Structure of Trimethylherqueinone B. Tetrahedron 18, 839 (1962).

26. Chapman, E., A. G. Perkin and R. Robinson: The Colouring Matters of Carajura. J. Chem. Soc. (London) 1927, 3015.

27. Chmielewska, I.: Oxidative and Photosynthetic Phosphorylation Involving 2-Methylquinones. Biochim. Biophys. Acta 39, 170 (1960).

28. CHMIELEWSKA, I. and J. CIEŚLAK: Vitamins and Antivitamins K: Tautomerism of Dicoumarol. Tetrahedron 4, 135 (1958).
29. CHOU, T. Q. and P. F. MEI: The Principle of the Chinese Drug Lei-Kung-Teng, *Tripterygium wilfordii* HOOK. I. The Colouring Substance and the Sugars. Chinese J. Physiol. 10, 529 (1936).
30. CLARK, V. M.: The Synthesis of ADP and ATP via the Oxidation of Quinol Phosphates. In: Mechanismen enzymatischer Reaktionen, p. 276. Berlin: Springer-Verlag. 1964.
31. COLLINS, D. A., F. HAWORTH, K. ISARASENA and A. ROBERTSON: The Pigments of "Dragon's Blood" Resin. Part I. Dracorubin. J. Chem. Soc. (London) 1950, 1876.
32. COOKE, R. G., B. L. JOHNSON and W. SEGAL: Colouring Matters of Australian Plants. VI. Haemocorin: The Structure of the Aglycone. Austral. J. Chem. 11, 230 (1958).
33. COOKE, R. G. and W. SEGAL: Colouring Matters of Australian Plants. IV. Haemocorin: A Unique Glycoside from *Haemodorum corymbosum* VAHL. Austral. J. Chem. 8, 107 (1955).
34. — — Colouring Matters of Australian Plants. V. Haemocorin: The Chemistry of the Aglycone. Austral. J. Chem. 8, 413 (1955).
35. COOKE, R. G. and R. H. THOMSON: Naturally Occurring Quinone Methines and Related Compounds. Rev. Pure Appl. Chem. (Australia) 8, 85 (1958).
36. CRAM, D. J.: Mould Metabolites. III. The Structure of Citrinin. J. Amer. Chem. Soc. 70, 4244 (1948).
37. — Mould Metabolites. V. The Stereochemistry and Ultraviolet Absorption Spectrum of Citrinin. J. Amer. Chem. Soc. 72, 1001 (1950).
38. CURTIS, R. F., P. C. HARRIES and C. H. HASSALL: The Biosynthesis of Phenols. Part VIII. The Synthesis of (2-Carboxy-3,5-dihydroxyphenyl)propan-2-one. J. Chem. Soc. (London) 1964, 5382.
39. DAGLEY, S. and M. D. PATEL: Oxidation of *p*-Cresol and Related Compounds by a *Pseudomonas*. Biochem. J. 66, 227 (1957).
39a. DALY, J. W. and B. WITKOP: Recent Studies on the Centrally Active Endogenous Amines. Angew. Chem., Int. Ed. 2, 421 (1963).
40. DAVIES, J. E. and J. C. ROBERTS: Studies in Mycological Chemistry. Part V. Synthesis of 2,5-Dihydroxy-7-methyl-1,4-naphthaquinone. J. Chem. Soc. (London) 1956, 2173.
41. DEAN, F. M.: Naturally Occurring Oxygen Ring Compounds. London: Butterworths. 1963.
42. DEAN, F. M., J. STAUNTON and W. B. WHALLEY: The Chemistry of Fungi. Part XXXVI. A Revised Structure for Sclerotiorin. J. Chem. Soc. (London) 1959, 3004.
43. ELLIS, L. C.: Biogenesis of Citrinin. Dissertation Abstr. 23, 4109 (1963).
44. ERDTMAN, H.: Dehydrierung in der Coniferylreihe. I. Dehydrodi-eugenol und Dehydrodi-isoeugenol. Biochem. Z. 258, 172 (1933).
45. ERDTMAN, H. and C. A. WACHTMEISTER: Phenoldehydrogenation as a Biosynthetic Reaction. Festschrift A. Stoll, p. 144. Basel: Birkhäuser. 1957.
46. ERICKSON, R. E., A. F. WAGNER and K. FOLKERS: Coenzyme Q. XLVIII. Data on Quinone Methines as Reaction Intermediates and their Possible Role in Oxidative Phosphorylation. J. Amer. Chem. Soc. 85, 1535 (1963).
47. EYTON, W. B., W. D. OLLIS, M. FINEBERG, O. R. GOTTLIEB, I. S. GUIMARÃES and M. T. MAGALHÃES: The Neoflavonoid Group of Natural Products. II. The Examination of *Machaerium scleroxylon* and Some Biogenetic Proposals regarding the Neoflavonoids. Tetrahedron 21, 2697 (1965).

48. EYTON, W. B., W. D. OLLIS, I. O. SUTHERLAND, O. R. GOTTLIEB, M. T. MAGAL-HÃES and L. M. JACKMAN: The Neoflavanoid Group of Natural Products. I. Dalbergiones — a New Class of Quinones. Tetrahedron **21**, 2683 (1965).

49. FEHLMANN, M. und A. NIGGLI: Die Struktur des Blattfarbstoffes Cordeauxia-Chinon. Helv. Chim. Acta **48**, 305 (1965).

50. FIESER, L. F. and R. N. JONES: Celastrol. Spectrographic Characterization and Color Tests. J. Amer. Pharm. Assoc. **31**, 315 (1942).

51. FREUDENBERG, K.: Forschungen am Lignin. Fortschr. Chem. organ. Natur-stoffe **20**, 41 (1962).

52. — Lignin: Its Constitution and Formation from *p*-Hydroxycinnamyl Alcohols. Science **148**, 595 (1965).

53. FREUDENBERG, K., C.-L. CHEN, J. M. HARKIN, H. NIMZ and H. RENNER: Observations on Lignin. Chem. Communs. **1965**, 224.

54. FREUDENBERG, K. und M. FRIEDMANN: Oligomere Zwischenprodukte der Ligninbildung. Chem. Ber. **93**, 2138 (1960).

55. FREUDENBERG, K. und G. GRION: Beitrag zum Bildungsmechanismus des Lignins und der Lignin-Kohlenhydrat-Bindung. Chem. Ber. **92**, 1355 (1959).

56. FREUDENBERG, K., G. GRION und J. M. HARKIN: Nachweis von Chinonmethiden bei der enzymatischen Bildung des Lignins. Angew. Chem. **70**, 743 (1958).

57. FREUDENBERG, K. und J. M. HARKIN: Modelle für die Bindung des Lignins an die Kohlenhydrate. Chem. Ber. **93**, 2814 (1960).

58. FREUDENBERG, K., J. M. HARKIN und H.-K. WERNER: Das Vorkommen von Benzylaryläthern im Lignin. Chem. Ber. **97**, 909 (1964).

59. FREUDENBERG, K. und K.-C. RENNER: Über Biphenyle und Diaryläther unter den Vorstufen des Lignins. Chem. Ber. **98**, 1879 (1965).

60. FREUDENBERG, K. und H. TAUSEND: Weitere trimere Zwischenprodukte der Ligninbildung. Chem. Ber. **96**, 2081 (1963).

61. — — Penta- und Hexalignol. Chem. Ber. **97**, 3418 (1964).

62. FREUDENBERG, K. und H.-K. WERNER: Die Polymerisation der Chinonmethide. Chem. Ber. **97**, 579 (1964).

63. GALARRAGA, J. A., K. G. NEILL and H. RAISTRICK: The Colouring Matters of *Penicillium herquei* BANIER and SARTORY. Biochem. J. **61**, 456 (1955).

64. GISVOLD, O.: The Constitution of Celastrol. J. Amer. Pharm. Assoc. **31**, 529 (1942).

65. GORE, T. S., R. V. TALAVDEKAR and K. VENKATARAMAN: A New Partial Synthesis of Citrinin. Current Sci. (India) **19**, 20 (1950).

66. GRANT, P. K. and A. W. JOHNSON: Pristimerin. Part I. The Nature of the Chromophore. J. Chem. Soc. (London) **1957**, 4079.

67. — — Pristimerin. Part II. Further Reactions involving the Chromophore. J. Chem. Soc. (London) **1957**, 4669.

68. GRANT, P. K., A. W. JOHNSON, P. F. JUBY and T. J. KING: Pristimerin. Part III. A Modified Structure for the Chromophore. J. Chem. Soc. (London) **1960**, 549.

69. HARADA, R., H. KAKISAWA, S. KOBAYASHI, M. MUSYA, K. NAKANISHI and Y. TAKAHASHI: Structure of Pristimerin, a Quinonoid Triterpene. Tetrahedron Letters **1962**, 603.

70. HARBORNE, J. B.: Anthocyanins and their Sugar Components. Fortschr. Chem. organ. Naturstoffe **20**, 165 (1962).

71. HARMAN, R. E., J. CASON, F. H. STODOLA and A. L. ADKINS: Structural Features of Herqueinone, a Red Pigment from *Penicillium herquei.* J. Organ. Chem. (USA) **20**, 1260 (1955).

72. HASSALL, C. H. and D. W. JONES: The Biosynthesis of Phenols. Part IV. A New Metabolic Product of *Aspergillus terreus* THOM. J. Chem. Soc. (London) 1962, 4189.

73. HATHWAY, D. E.: The Lignans. In: W. E. HILLIS (Edit.), Wood Extractives, p. 159. New York: Academic Press. 1962.

74. HAWORTH, R. D.: The Chemistry of the Lignan Group of Natural Products. J. Chem. Soc. (London) 1942, 448.

75. HEARON, H. M. and W. S. MacGREGOR: The Naturally Occurring Lignans. Chem. Rev. 55, 957 (1955).

76. HESSE, G. und W. KLINGEL: Über das Drachenblut. Liebigs Ann. Chem. 524, 14 (1936).

77. HETHERINGTON, A. C. and H. RAISTRICK: Chemical Constitution of a new Yellow Colouring Matter, Citrinin, Produced from Dextrose by *Penicillium citrinum* THOM. Trans. Roy. Soc. (London) B 220, 269 (1931).

78. HILL, J. A., A. W. JOHNSON, T. J. KING, S. NATORI and S. W. TAM: Synthetical Approaches to the Pristimerin Chromophore. J. Chem. Soc. (London) 1965, 361.

79. HILL, R. K. and L. A. GARDELLA: The Absolute Configuration of Citrinin. J. Organ. Chem. (USA) 29, 766 (1964).

80. HULTZSCH, K.: Studien auf dem Gebiet der Phenol-Formaldehyd-Harze, II. Mitt. Chinonmethide als Zwischenprodukte bei der Phenolharz-Härtung. Ber. dtsch. chem. Ges. 74, 898 (1941).

81. — Chinonmethide. In: Chemie der Phenolharze, S. 63. Berlin: Springer-Verlag. 1950.

82. IWAI, I. and H. MISHIMA: Constitution of Ascochitine. Chem. and Ind. 1965, 186.

83. JOHNSON, A. W., P. F. JUBY, T. J. KING and S. W. TAM: Pristimerin. Part IV. Total Structure. J. Chem. Soc. (London) 1963, 2884.

84. JOHNSON, D. H., A. ROBERTSON and W. B. WHALLEY: The Chemistry of Fungi. Part XIII. Citrinin. J. Chem. Soc. (London) 1950, 2971.

84a. KARANATSIOS, D., J. S. SCARPA und C. H. EUGSTER: Struktur von Fuerstion. Helv. chim. Acta 49, 1151 (1966).

85. KARRER, P. und C. H. EUGSTER: Über Fuerstiachinon. Helv. Chim. Acta 35, 1139 (1952).

86. KAUFMAN, S., W. F. BRIDGERS, F. EISENBERG and S. FRIEDMAN: The Source of Oxygen in Phenylalanine Hydroxylase and Dopamine-β-hydroxylase Catalyzed Reactions. Biochem. Biophys. Res. Comm. 9, 497 (1962).

87. KING, F. E. and J. G. WILSON: The Chemistry of Extractives from Hardwoods. Part XXXVII. The Lignans of *Guaiacum officinale* L. J. Chem. Soc. (London) 1964, 4011.

88. KOVÁČ, S., P. NEMEC, V. BETINA and J. BALAN: Chemical Structure of Citrinin. Nature 190, 1104 (1961).

89. KULKARNI, A. B.: Pristimerin. Bull. Nat. Inst. Sci. (India), No. 28, p. 125 (1965).

90. KULKARNI, A. B., R. C. SHAH, S. SESHADRI and V. V. MHASKAR: Pristimerin. II. J. Sci. Indust. Res. (India) B 17, 111 (1958).

91. LEDERER, E.: Über Ursprung und Funktion einiger Methylgruppen in verzweigten Fettsäuren, in Pflanzensterinen und in Chinonen der Vitamin-K- und Ubichinongruppe. Experientia 20, 473 (1964).

92. LEVIN, E. Y. and S. KAUFMAN: Studies on the Enzyme Catalysing the Conversion of 3,4-Dihydroxyphenylethylamine to Norepinephrine. J. Biol. Chem. 236, 2043 (1961).

93. LINN, B. O., C. H. SHUNK, E. L. WONG and K. FOLKERS: Coenzyme Q. XXXVIII. Cyclization of Coenzyme Q to the Corresponding Chromenols with Sodium Hydride. J. Amer. Chem. Soc. 85, 239 (1963).

94. Lundquist, K. und G. E. Miksche: Nachweis eines neuen Verknüpfungsprinzips von Guajacylpropaneinheiten im Fichtenlignin. Tetrahedron Letters 1965, 2131.
95. Lynch, D. M.: The Structure and Synthesis of Degradation Products from the Pigment Herqueinone. Dissertation Abstr. 25, 5564 (1965).
96. Mamont, P., P. Cohen, R. Azerad et M. Vilkas: Les quinones dans l'oxydation phosphorylante. II. Dimérisation acide des vitamines K. III. Isomérisation acide des vitamines K en présence de composés insaturés. Bull. soc. chim. France 1965, 2513, 2824.
97. Marton, J. and E. Adler: Carbonyl Groups in Lignin. III. Mild Catalytic Hydrogenation of Björkman Lignin. Acta Chem. Scand. 15, 370 (1961).
98. Mathieson, D. W. and W. B. Whalley: The Conformation of Citrinin. J. Chem. Soc. (London) 1964, 4640.
99. McHale, D. and J. Green: A Dimeric Oxidation Product of γ-Tocopherol. Chem. and Ind. 1963, 982.
100. — — Potassium Ferricyanide Oxidation Product of α-Tocopherol. Chem. and Ind. 1964, 366.
100a. — — Chrom-3-en-6-ols. The Action of Pyridine on Alk-2-enylbenzoquinones. J. Chem. Soc. (London) 1965, 5060.
101. McOmie, J. F. W., A. B. Turner and M. S. Tute: The Structure of Pulvilloric Acid. Chem. and Ind. 1963, 1689; J. Chem. Soc. (London), in press (1966).
102. Mehta, P. P. and W. B. Whalley: The Absolute Configuration of Citrinin. J. Chem. Soc. (London) 1963, 3777.
103. Michael, S. E.: Fuscin. A Metabolic Product of *Oidiodendron fuscum* Robak. I. Preparation, Properties, and Antibacterial Activity. Biochem. J. 43, 528 (1948).
104. Money, T.: A Postulated Biosynthesis of Some "Anomalous" Natural Phenolic Compounds. Nature 199, 592 (1963).
105. Morimoto, H. and I. Imada: Photochemical Reactions of Ubiquinone-(35). II. Ubichromenol-(35) and Isoubiquinone-(35). Chem. Pharm. Bull. (Tokyo) 12, 739 (1964).
106. Morrison, G. A., I. C. Paul and G. A. Sim: The Structures of Atrovenetin and Herqueinone. Proc. Chem. Soc. (London) 1962, 352.
106a. Mosbach, K. and I. Ljungcrantz: On the Biosynthesis of Barnol, a new Phenolic Metabolite. Biochim. Biophys. Acta 86, 203 (1964).
107. Nakanishi, K., H. Kakisawa and Y. Hirata: The Structures of Pristimerin and Celastrol. Bull. Chem. Soc. Japan 29, 7 (1956).
108. Nakanishi, K., Y. Takahashi and H. Budzikiewicz: Pristimerin. Spectroscopic Properties of the Dienone-Phenol-Type Rearrangement Products and Other Derivatives. J. Organ. Chem. (USA) 30, 1729 (1965).
109. Narasimhachari, N. and L. C. Vining: Studies on the Pigments of *Penicillium herquei*. Canad. J. Chem. 41, 641 (1963).
110. Neill, K. G. and H. Raistrick: Metabolites of *Penicillium atrovenetum* G. Smith. I. Atrovenetin, a New Crystalline Colouring Matter. Biochem. J. 65, 166 (1957).
111. Nelan, D. R. and C. D. Robeson: The Oxidation Product from α-Tocopherol and Potassium Ferricyanide and Its Reaction with Ascorbic and Hydrochloric Acids. J. Amer. Chem. Soc. 84, 2963 (1962).
112. Nimz, H.: Isolierung von Guajacylglycerin-β-coniferyläther aus Fichtenholz. Chem. Ber. 98, 533 (1965).
113. Nord, F. F. and W. J. Schubert: The Biogenesis of Lignins. In: P. Bernfeld (Edit.), Biogenesis of Natural Compounds, p. 693. Oxford and New York: Pergamon Press. 1963.

114. Oku, H. and T. Nakanishi: A Toxic Metabolite from *Ascochyta fabae* having Antibiotic Activity. Phytopathology **53**, 1321 (1963).

115. Oxford, A. E.: The Chemistry of Antibiotic Substances other than Penicillin. Annu. Rev. Biochem. **14**, 757 (1945).

115a. Paul, I. C. and G. A. Sim: Fungal Metabolites. III. The Structure of Atrovenetin: *X*-Ray Analysis of Atrovenetin Orange Trimethyl Ether Ferrichloride. J. Chem. Soc. (London) **1965**, 1097.

116. Pollock, A. E.: Production of Citrinin by Five Species of *Penicillium*. Nature **160**, 331 (1947).

117. Ponniah, L. and T. R. Seshadri: A Survey of Anthocyanins from Indian Sources. J. Sci. Indust. Res. (India) **12 B**, 605 (1953).

118. — — Synthesis of Carajurone Hydrochloride. Proc. Indian Acad. Sci. **39 A**, 45 (1954).

119. Raistrick, H. and G. Smith: The Metabolic Products of *Aspergillus terreus* Thom. A New Mould Metabolic Product — Terrein. Biochem. J. **29**, 606 (1935).

120. Roberts, J. C. and C. W. H. Warren: Studies in Mycological Chemistry. IV. Purpurogenone, a Metabolic Product of *Penicillium purpurogenum* Stoll. J. Chem. Soc. (London) **1955**, 2992.

121. Robertson, A. and W. B. Whalley: The Pigments of "Dragon's Blood" Resin. II. A Synthesis of Dracorhodin. J. Chem. Soc. (London) **1950**, 1882.

122. — — The Chemistry of the "Insoluble Red" Woods. VI. Santalin and Santarubin. J. Chem. Soc. (London) **1954**, 2794.

123. Robertson, A., W. B. Whalley and J. Yates: The Pigments of "Dragon's Blood" Resins. III. The Constitution of Dracorubin. J. Chem. Soc. (London) **1950**, 3117.

124. Roux, D. G. and S. E. Drewes: Structural Factors Associated with the Redness induced in certain Condensed Tannins by Sunlight or Heat. Chem. and Ind. **1965**, 1442.

125. Schechter, M. S. and H. L. Haller: Identity of the Red Pigment in the Roots of *Tripterygium wilfordii* and *Celastrus scandens*. J. Amer. Chem. Soc. **64**, 182 (1942).

126. Schindler, O.: Die Ubichinone (Coenzyme Q). Fortschr. Chem. organ. Naturstoffe **20**, 73 (1962).

127. Schudel, P., H. Mayer, J. Metzger, R. Rüegg und O. Isler: Über die Chemie des Vitamins E. 2. Mitt. Die Struktur des Kaliumferricyanid-Oxidationsproduktes von α-Tocopherol. Helv. Chim. Acta **46**, 636 (1963).

128. Schwenck, E.: Tumor Action of Some Quinonoid Compounds in the Cheekpouch Test. Drug Research **12**, 1143 (1962).

129. Schwenck, E., G. J. Alexander, A. M. Gold and D. F. Stevens: Biogenesis of Citrinin. J. Biol. Chem. **233**, 1211 (1958).

130. Senoh, S., C. R. Creveling, S. Udenfriend and B. Witkop: Chemical, Enzymatic and Metabolic Studies on the Mechanism of Oxidation of Dopamine. J. Amer. Chem. Soc. **81**, 6236 (1959).

131. Shah, R. C. and A. B. Kulkarni: Structure of Pristimerin. Nature **173**, 1237 (1954).

132. Shah, R. C., A. B. Kulkarni and V. M. Thakore: Pristimerin. Part I. J. Chem. Soc. (London) **1955**, 2515.

133. Skinner, W. A. and R. M. Parkhurst: Oxidation Products of Vitamin E and its Model, 6-Hydroxy-2,2,5,7,8-pentamethylchroman. VII. Trimer Formed by Alkaline Ferricyanide Oxidation. J. Organ. Chem. (USA) **29**, 3601 (1964).

134. Smith, G.: The Effect of Adding Trace Elements to Czapek-Dox Medium. Brit. Mycol. Soc. Trans. **32**, 280 (1949).

135. Smith, L. I. and F. J. Dobrovolny: The Reaction between Duroquinone and Sodium Malonic Esters. J. Amer. Chem. Soc. 48, 1693 (1926).

136. Smith, L. I. and E. W. Kaiser: The Reaction between Quinones and Metallic Enolates. XI. Duroquinone and the Enolates of Cyanoacetic Ester and of β-Diketones. J. Amer. Chem. Soc. 62, 138 (1940).

137. Smith, L. I., R. W. H. Tess and G. E. Ullyot: The Reaction between Quinones and Metallic Enolates. XIX. The Structure of Diduroquinone. J. Amer. Chem. Soc. 66, 1320 (1944).

138. Stodola, F. H., K. B. Raper and D. I. Fennell: Pigments of *Penicillium herquei.* Nature 167, 773 (1951).

139. Terni, G. and I. Shibazaki: Studies on Citrinin and its Related Metabolites. I. Citrinin-producing Microbe and Citrinin. J. Fermentation Technol. (Japan) 26, 336 (1948).

140. Thomas, R.: Studies in the Biosynthesis of Fungal Metabolites. 3. The Biosynthesis of Fungal Perinaphthenones. Biochem. J. 78, 807 (1961).

140 a. — Private communication.

141. Timonin, M. I. and J. W. Rouatt: Production of Citrinin by *Aspergillus* Species of the *Candidus* Group. Canad. J. Public Health 35, 80 (1944).

142. Turner, A. B.: Quinone Methides. Quart Rev. (Chem. Soc. London) 18, 347 (1964).

143. Turner, A. B. and J. F. W. McOmie: The Synthesis of 3,5-Dimethoxy-4-carbomethoxyphthalic Acid. Tetrahedron 22, 31 (1966).

144. Vilkas, M. et E. Lederer: Sur un mécanisme possible de la phosphorylation oxydative. Experientia 18, 546 (1962).

144a. — — Les quinones dans l'oxydation phosphorylante. I. Généralités. Bull. soc. chim. France 1965, 2505.

145. Wagner, A. F., A. Lusi, C. H. Shunk, B. O. Linn, D. E. Wolf, C. H. Hoffman, R. E. Erickson, B. Arison, N. R. Trenner and K. Folkers: Coenzyme Q. XLVII. New 5-Phosphomethyl-6-chromanyl Derivatives from a Novel Reaction of Interest in Oxidative Phosphorylation. J. Amer. Chem. Soc. 85, 1534 (1963).

146. Wagner, A. F., P. E. Wittreich, B. Arison, N. R. Trenner and K. Folkers: Synthesis of the New 3,4-Dihydro-2 H-naphtho[1,2-b]pyran-6-yl Phosphate from Vitamin $K_{1(20)}$. J. Amer. Chem. Soc. 85, 1178 (1963).

147. Wakselman, M. et M. Vilkas: Condensations des phénols o-chlorométhylés ou o-hydroxyméthylés avec quelques hydrocarbures éthyléniques. Synthèse de chromannes. C. R. hebd. séances Acad. Sci. 258, 1526 (1964).

148. Wang, Y. and H.-S. Ting: Syntheses of d- and l-Citrinins. Science Record (China) 4, 269 (1951).

149. Warren, H. H., G. Dougherty and E. S. Wallis: The Synthesis of Dihydrocitrinin and Citrinin. J. Amer. Chem. Soc. 71, 3423 (1949).

150. — — — The Synthesis and Antibiotic Activity of Analogs of Citrinin and Dihydrocitrinin. J. Amer. Chem. Soc. 79, 3812 (1957).

151. Wenkert, E.: Biosynthesis of Hydrolyzable Tannins. Chem. and Ind. 1959, 906.

152. Wenkert, E., A. Fuchs and J. D. McChesney: Chemical Artifacts from the Family *Labiatae.* J. Organ. Chem. (USA) 30, 2931 (1965).

153. Whalley, W. B.: Oxygen Heterocyclic Fungal Metabolites. Progr. Organ. Chem. 4, 72 (1958).

154. — Some Structural and Biogenetic Relationships of Plant Phenolics. In: W. D. Ollis (Edit.), Chemistry of Natural Phenolic Compounds, p. 20. Oxford and New York: Pergamon Press. 1961.

(Received, November 2, 1965.)

The Pyrrolizidine Alkaloids. II.

By **F. L. WARREN,** Cape Town.

I. Introduction: Occurrence and Nature of the Pyrrolizidine Alkaloids.

Ten years have elapsed since the pyrrolizidine alkaloids were reviewed in this series (387), and it is a not much shorter period since they formed the subject of an excellent account by LEONARD (329) and a masterly summary by BOIT (225). The intervening years have seen considerable activity in the isolation of new alkaloids, the determination of absolute structures, the synthesis of both the "necine" bases and "necic" acids, the study of the formation of these alkaloids in the plant and the effect of structure in hepatic cirrhosis.

This review follows systematically, and must be regarded as Part II of the original communication (387). The references to the original literature and the Chart numbers follow in sequence those of the previous review. Tables which give additional information included in previous Tables carry the letter A so that Tables 1A, 2A, 4A, 4aA, 5A, 6A and 9A give additional information to that in Tables 1, 2, 4, 4a, 5, 6 and 9 in the first review article.

Table 1A. List of Additional Plant Genera Containing Pyrrolizidine Alkaloids.

[Addendum to Table 1 (387)]

Compositae	Genus	*Nardosomia (Petasites)*
Santalaceae	Genus	*Thesium*
Gramineae	Genera	*Festuca, Lolium, Thelepogon*

The pyrrolizidine alkaloids have been found in two new families (*Table 1A*), and in view of the toxicity to cattle of this type of alkaloid it is of considerable significance that the Gramineae are now included in this list. The alkaloid content of several new species and additional information of some formerly studied species are included in *Table 2 A* (p. 379). The interesting new aspect of this table is the inclusion of several unesterified pyrrolizidine bases now found in the plant.

The pyrrolizidine alkaloids were originally, and for the most part are even now, groups of alkaloids composed of hydroxylated methyl-pyrrolizidines esterified by one or two monobasic acids or by a dicarboxylic acid. The new alkaloids, and additional information on some previously investigated ones, m. p. and $[\alpha]_D$, and their hydrolysis products are shown in *Table 4 A* (p. 384). Some of the acids decompose under hydrolysis conditions and have not been isolated, as for example the acid in macrotomine, which is designated "macrotomic acid". Furthermore, the form in which the acid exists in the alkaloid is not always that form in which it is isolated after hydrolysis. For example, jacobine and jaconine both hydrolyse to retronecine and jaconecic acid. To make this clear the acids "jacobinecic" and "jaconinecic acids" are names given to the form of the acid in the alkaloid.

Some alkaloids which were originally considered to be new individuals have been found to be identical with known alkaloids or to be mixtures. These corrections are included in *Table 5 A* (p. 383). Two new alkaloids, as yet unnamed, have been isolated, and three previously unnamed alkaloids have been investigated. These addenda and corrigenda are shown in *Table 4a A* (p. 383).

II. The Free Bases and the Basic Hydrolysis Products.

1. The Absolute Configurations.

The absolute configuration of (—)-heliotridane (1) was unequivocally established by WARREN and VON KLEMPERER (*388*) by a three stage Hofmann degradation to (+)-3-methylheptane (2), which had been assigned configuration *D* on KLYNE's symbolism (*311*), *S* on that of CAHN, INGOLD and PRELOG (*236*), so that heliotridane is (1*S*,8*S*)-1-methyl-pyrrolizidine (1).

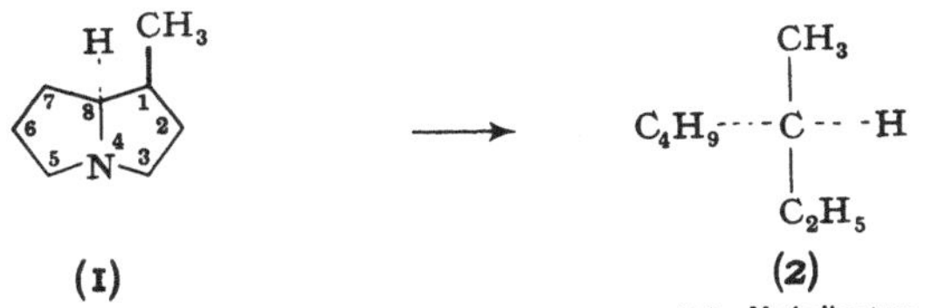

(1)   (2)
(—)-Heliotridane.   (+)-3-Methylheptane.

The absolute structure of heliotridane led (*388*) to absolute configurations of most of the "necines" as in *Chart 14*, p. 332.

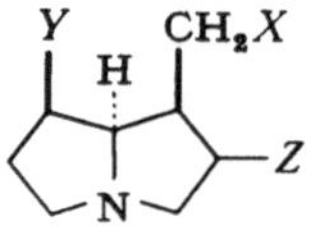

	X	Y	Z
Heliotridane:	H	H	H
Retronecanol:	H	OH	H
(—)-Isoretronecanol:	OH	H	H
Platynecine:	OH	OH	H
[Mikanecine]:	OH	OH	H
Rosmarinecine:	OH	OH	OH

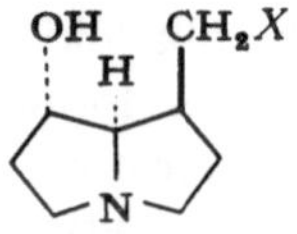

Oxyheliotridane: $X = H$
Dihydroxyheliotridane: $X = OH$

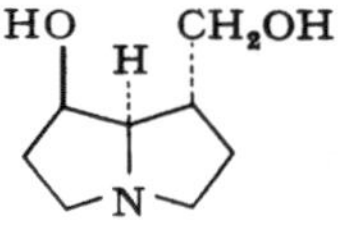

Hastanecine and/or Turniforcidine.

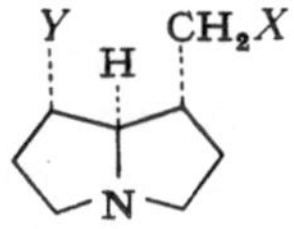

Pseudoheliotridane: $X = Y = H$
(—)-Trachelanthamidine: $X = OH$, $Y = H$
Macronecine: $X = Y = OH$
[(+)-Trachelanthamidine ⇌ Laburnine]

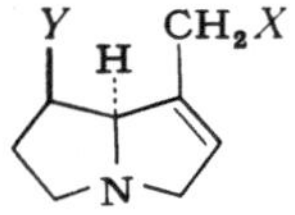

Retronecine: $X = Y = OH$
Supinidine: $X = OH$, $Y = H$
Desoxyretronecine: $X = H$, $Y = OH$

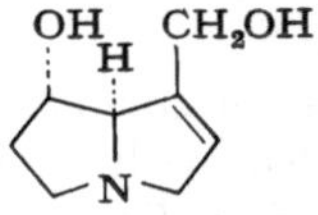

Heliotridine.

(+)-Isoretronecanol, Lindelofidine.

Chart 14. The Absolute Configuration of the Hydroxylated Methylpyrrolizidines.

Independent confirmation of this configuration was forthcoming from ADAMS and his collaborators. ADAMS and LEONARD (*13*) had previously synthesised (—)-retronecanone (3) from (—)-3-methyl-5-aminovaleric acid (4); and the correlation of this with (—)-methylsuccinic acid, known to have an S configuration, by ADAMS and FLEŠ (*197*) in a series of well chosen interconversions gave the absolute configuration of $C_{(1)}$ in (—)-retronecanone as S.

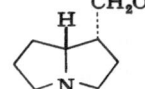

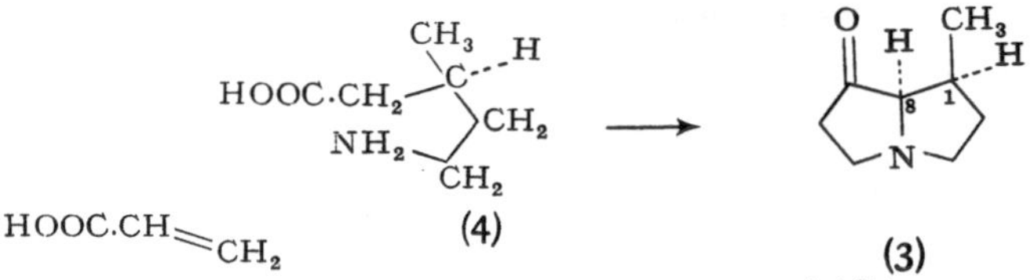

(4)

(3)
(—)-Retronecanone.

The configuration at $C_{(8)}$ was established by the same authors (*198*) by the conversion of desoxyretronecine (5, $X = OH$) to isoheliotridene (5, $X = H$), followed by ozonolysis to the ketonic acid (6), esterification and reaction with methyl magnesium iodide. The resulting product,

References, pp. 393—406.

(—)-1-(2'-hydroxy-2'-methyl-propyl)-2-(2''-hydroxy-propan-2''-yl)-pyrrolidine (7) was identical with that obtained from (S)-proline (8) and hence the configuration at $C_{(8)}$ in retronecine is R.

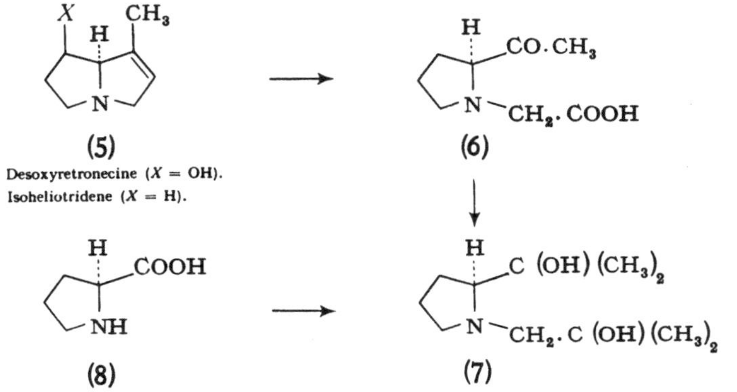

(5)

Desoxyretronecine ($X =$ OH).
Isoheliotridene ($X =$ H).

(6)

(8)

(7)

Further proof was also obtained by the X-ray studies of FRIDRICHSONS, MATHIESON and SUTOR (284) on jacobine bromohydrin (p. 353) and hence retronecine.

It is significant for future studies that the interesting comparison made by LEONARD (328) of the molecular rotation differences in the pyrrolizidine and lupinine group of alkaloids did not lead to the same absolute configuration.

1-Hydroxymethylpyrrolizidines.

The configuration of the 1-hydroxymethyl group relative to that of the 7-hydroxyl group in platynecine was previously established as *cis*. Additional evidence has been advanced by FODOR (281) for the interpretation of the action of boiling thionyl chloride on platynecine to give anhydroplatynecine (9). This conversion is envisaged as the attack at

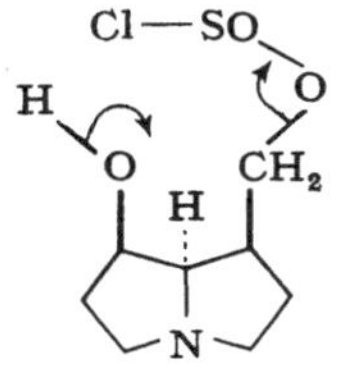

(10)

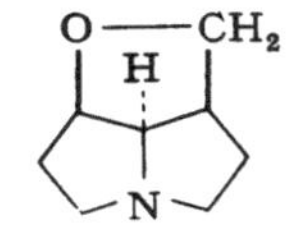

(9)

Anhydroplatynecine.

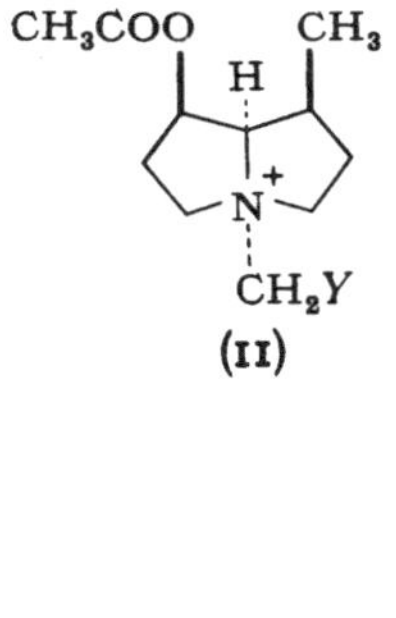

(11)

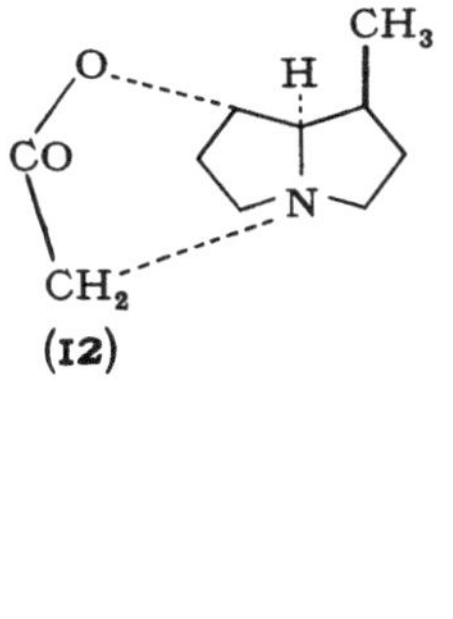

(12)

the primary hydroxyl group (**10**) followed by nucleophylic displacement to form the planar tetrahydrofuran ring, which reaction is possible only with the *cis*-configuration. A further elegant experiment from FODORs laboratory (*282*) showed that the quaternary salt (**11**, $Y = COOC_2H_5$, I⁻) formed by the addition of ethyl iodoacetate to acetylretronecanol, gave on treatment with silver oxide the betaine (**11**, $Y = COO^-$) which in turn with hydrogen iodide yielded the quaternary iodide (**11**, $Y = COOH$, I⁻). If the 7-hydroxyl group had been *cis* to the $C_{(8)}$-hydrogen, the δ-lactone (**12**) would have been formed.

The interrelation between the 1-hydroxymethylpyrrolizidines becomes clear from the careful study by LABENSKIĬ, SEROVA and MEN'SHIKOV (*325*) of the oxidation with chromic acid to give the pyrrolizidine-1-carboxylic acids. The results are summarised in *Chart 15*. The formation of one or two acids from these oxidations depends on whether the hydrogens at $C_{(1)}$ and $C_{(8)}$ are *trans* or *cis* to one another, respectively. When they are *cis* then the resulting carboxyl group is sterically hindered and partial inversion occurs to give two acids. The reduction of these acids to the corresponding 1-hydroxymethyl derivatives provides a path for the transition from heliotridane to pseudoheliotridane (*333*).

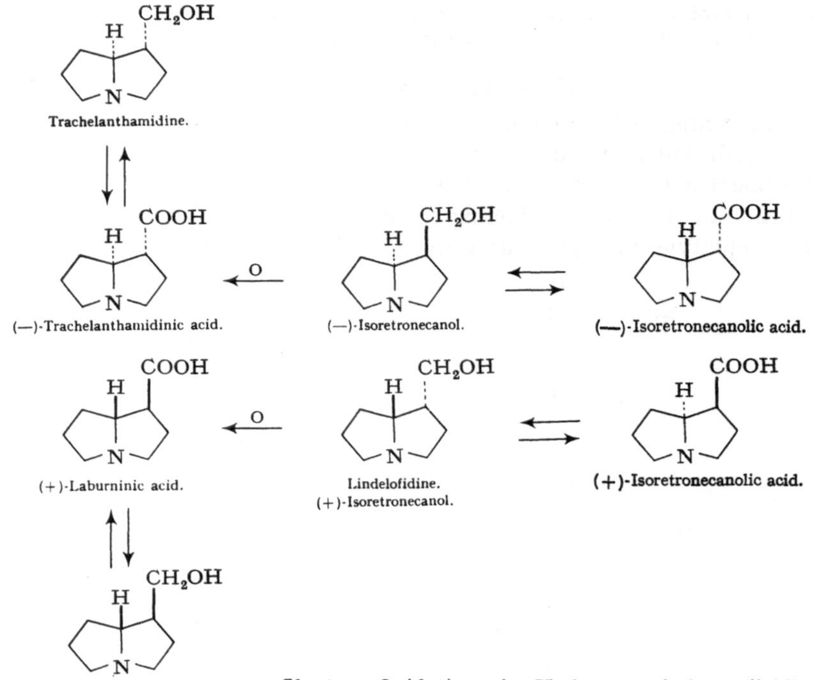

Chart 15. Oxidation of 1-Hydroxymethylpyrrolizidine.

7-Hydroxy-1-hydroxymethylpyrrolizidines.

The isomeric alkaloids, $C_8H_{15}NO_2$ (*13*), of known structure are platynecine, to which mikanecine (*120*) is identical (*204*), and dihydroxyheliotridane. CULVENOR and SMITH (*261*) have assigned structures for hastanecine (*94*) and/or turniforcidine (*139*) and the new base macronecine (*275*) on the basis of molecular rotation differences. These absolute configurations are shown in **Table 15**; but the identity of hastanecine and turniforcidine is contrary to the reported mixed melting point depression, whilst their being enantiomeric with one another is contrary to the reported signs of rotation. Retusine (*261*) is considered as identical or enantiomeric with hastanecine.

Table 15. The Configuration of 7-Hydroxy-1-hydroxymethyl-pyrrolizidines.

OH   CH₂OH

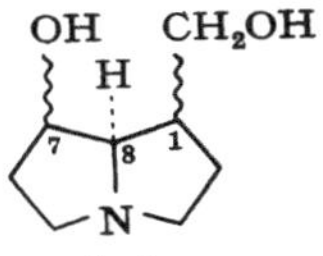

(13)

	1	7	8	$[M]_D$
Platynecine [Mikanecine]..	β	β	α	— 131°
Dihydroxyheliotridane	β	α	α	— 53°
Macronecine	α	α	α	+ 77°
Hastanecine and/or Turniforcidine	α	β	α	— 14°, — 16°

2. New Bases.

The new pyrrolizidine bases are shown in *Table 6 A* (p. 388). Two new necine bases have been found in pyrrolizidine esters: hastanecine ethochloride, $C_{10}H_{20}NO_2Cl$, occurs esterified by trachelanthic acid in the second alkaloid from *Lindelofia macrostyla* (*383*); and macronecine, $C_8H_{15}NO_2$, found as the angelic ester in macrophylline (*275*). Hastanecine and macronecine have been assigned structures as 7β- and 7α-hydroxy-1-hydroxymethyl-8α-pyrrolizidines (*261*) (see Table 15).

Several pyrrolizidine bases have been found unesterified in the plant. ARENDARUK, PROSKURNINA and KONOVALOVA (*218*) found (+)-isoretronecanol in the roots of *Thesium minkwitzianum*, whilst in the upper part of the plant, this same base as the esters, thesine and thesinine. CULVENOR and his collaborators (*261*) have reported six different unesterified bases, present frequently as the N-oxides, as for example, retronecine N-oxide in *Crotalaria retusa*.

1-Methylene-8α-pyrrolizidine (**14**) (*263*), isolated from *C. anagyroides* (*259*) and *C. damarensis*, was shown to have this structure by oxidation to formaldehyde and 1-ketopyrrolizidine and by reduction to a mixture of heliotridane (**15**) and pseudoheliotridane (**16**) from which pseudoheliotridane only was obtained pure.

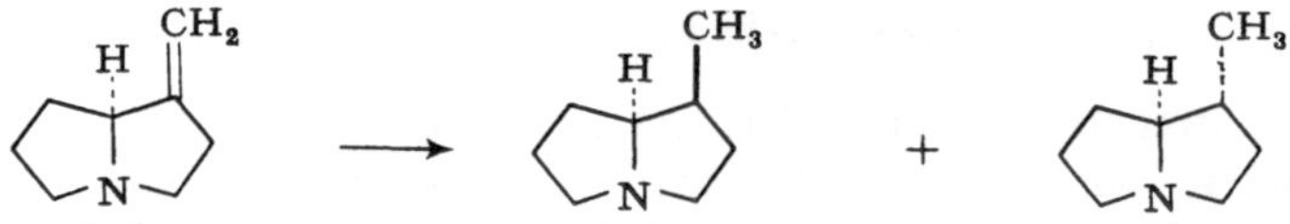

CH₂ / (14) → (15) + (16) structures

(14)
1-Methylene-8 α-pyrrolizidine.

(15)
Heliotridane.

(16)
Pseudoheliotridane.

Significant observations have been reported by CULVENOR and SMITH (263) with regard to characterisation of these two important reference compounds: the picrates are unreliable characterising derivatives, and the picrolonates are recommended. Furthermore, pseudo-heliotridane has $[\alpha]_D$ — 2.5° in contrast to $[\alpha]_D$ + 17.1° for a sample obtained from laburnine (285) and which they considered must have been impure. Moreover, they warn about ambiguity in determining rotations for liquids and make the following corrections:

	CULVENOR (263)	$[\alpha]_D$ reported previously
Heliotridene	— 159 ± 5°	— 160°, — 150° (151); + 39° (19)
Chlororetronecane	— 30°	+ 54° (19)
(—)-Pseudoheliotridane	— 2.5°	
(+)-Pseudoheliotridane		+ 17° (285)

7β-Hydroxy-1-methylene-8β-pyrrolizidine (17), and its 8α-isomer (21) isolated from *C. goreensis* (264), were similarly shown to have the methylene group. The structure of compound (17) was established as $(7R,8S)$-7-hydroxy-1-methylene-pyrrolizidine by reduction to the enantiomer (18) of hydroxyheliotridane (19) which was prepared by the reduction of heliotrine (20).

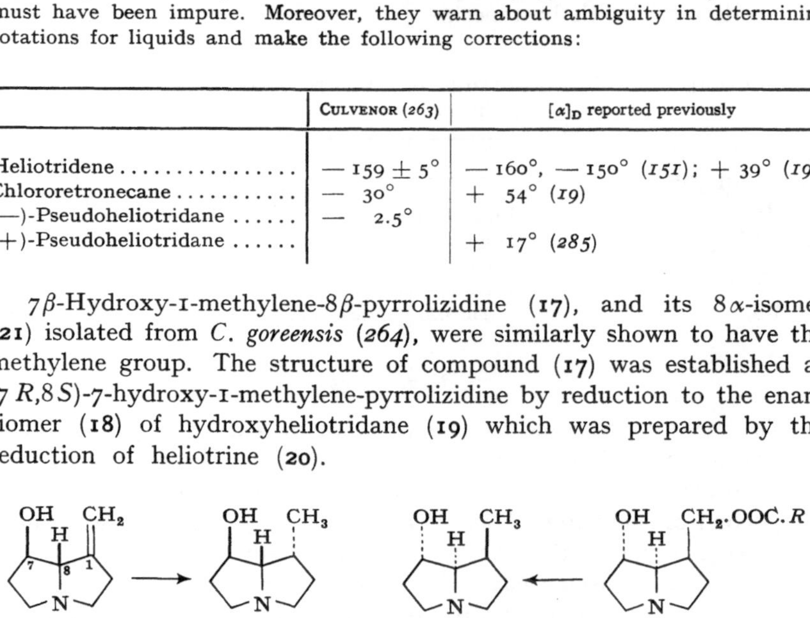

(17) → (18)   (19) ← (20) structures

(17)
7β-Hydroxy-1-
methylene-8β-pyrrolizidine.

(18)

(19)

(20)
Heliotrine.

The structure of compound (21) followed from its reduction to retronecanol. It was prepared in high yield (80%) together with desoxyretronecine (22, $X = H$) (20%) by the reduction of $(7R, 8R)$-7-hydroxy-1-chloromethyl-1,2-dehydropyrrolizidine (22, $X = Cl$) obtained from retronecine (22, $X = OH$). This interesting reduction reaction is equally applicable to other allylic pyrrolizidine alcohols, such as heliotridine and supinidine. A third alkaloid from *C. goreensis* (264), isomeric with (17) and (21), has been tentatively assigned structure 1,2-epoxy-1-methylpyrrolizidine (23).

References, pp. 393—406.

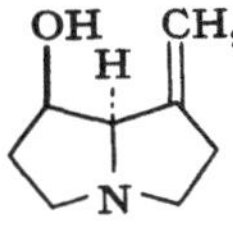

(21)
7β-Hydroxy-1-
methylene-8α-pyrrolizidine.

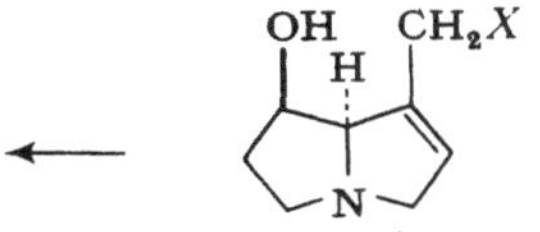

(22)
Retronecine ($X = $ OH).

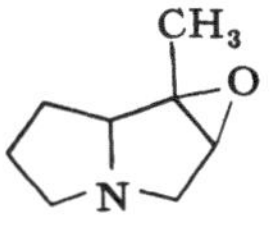

(23)

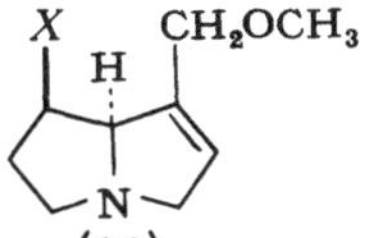

(24)

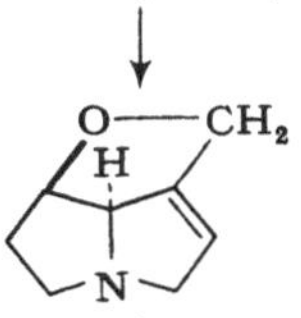

(25)
Anhydroretronecine.

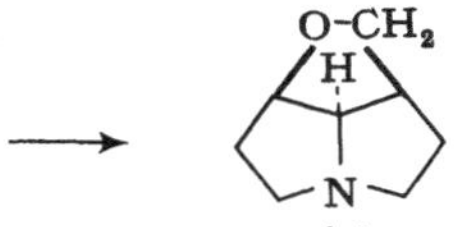

(9)
Anhydroplatynecine.

From *C. trifoliastrum* (*259*) and *C. aridicola* (*259, 265*) were obtained 1-methoxymethyl-1,2-dehydro-8α-pyrrolizidine (**24**, $X = $ H) and its 7β-hydroxy derivative (**24**, $X = $ OH), respectively the methyl ethers of supinidine and retronecine, from which they were prepared by the action of potassium tert.-butoxide and methyl iodide. Attempts to obtain the methyl ether of retronecine (**22**, $X = OCH_3$) by the action of sodium methoxide on the corresponding chloro derivative (**22**, $X = $ Cl) resulted in the strained anhydroretronecine (**25**) which was reduced to anhydroplatynecine (**9**).

Otonecine.

Otonecine, $C_9H_{15}NO_3$, first isolated as its hydrochloride by the hydrolysis of otosenine (*195*), was later obtained by DANILOVA, KORETSKAYA and UTKIN (*270*) from renardine together with anhydrootonecine, $C_9H_{13}NO_2$, isolated as its hydrochloride, m. p. 203–205°, whilst the hydrolysis of onetine (*271*) gave only the anhydro form; no basic hydrolysis product was obtained from retusamine although it was recognised as containing the base otonecine (*261*). The alkaloids containing this base are seemingly best fissioned by catalytic hydrogenolysis followed by hydrolysis to yield tetrahydrootonecine, $C_9H_{17}NO_2$, having a hydrochloride, m. p. 240–242°, by which method it has been obtained from onetine (*271*) and renardine (*270*).

The structural difficulties attendant upon handling this labile base were obviated by WUNDERLICH's X-ray diffraction analyses of a single crystal of retusamine α'-bromo-(+)-camphor-*trans*-π-sulphonate monohydrate (*389*). These crystals of the salt, $[C_{19}H_{26}NO_7]^+$, exist as the cation (**26**), but the free base retusamine, $C_{19}H_{25}NO_7$, shows a sharp medium band at 1603 cm^{-1} (in nujol), and a lower than normal carbonyl band at 1675 cm^{-1} (in CCl$_4$), indicative of a transannular interaction represented as (**27**), a type of isomerism exhaustively studied by LEONARD and his collaborators (*331, 330*).

On the basis of this structure otonecine is N-methyl-8-hydroxy-retronecine (cf. **22**) and the reactions reported by CULVENOR and SMITH (*261*) may now be interpreted. Catalytic reduction absorbed two molecules of hydrogen to give a salt-like tetrahydro derivative (**28**). Hydrolysis failed to yield a base, but the acid isolated after hydrogenation followed by hydrolysis had the same formula as that obtained by direct hydrolysis so that only the base had been hydrogenated. This interpretation explains the formation of tetrahydrootonecine from renardine (*270*) and onetine (*271*) and, as pointed out by WUNDERLICH (*389*), the formation of two oximes from tetrahydrootonecine (**29**), the structure of which has recently been studied by KORETSKAYA, DANILOVA and UTKIN (*322*).

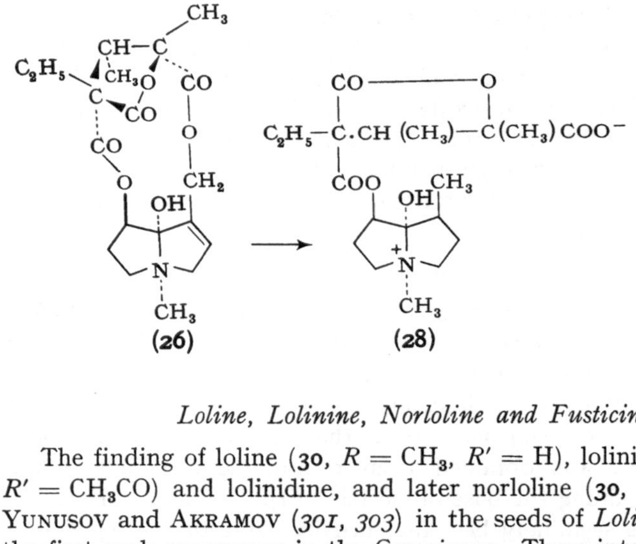

(**26**)   (**28**)   (**29**)

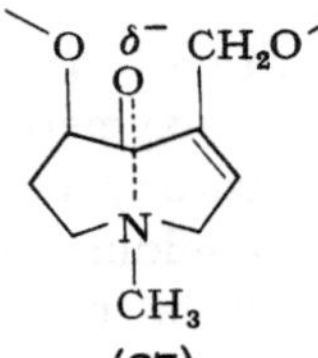

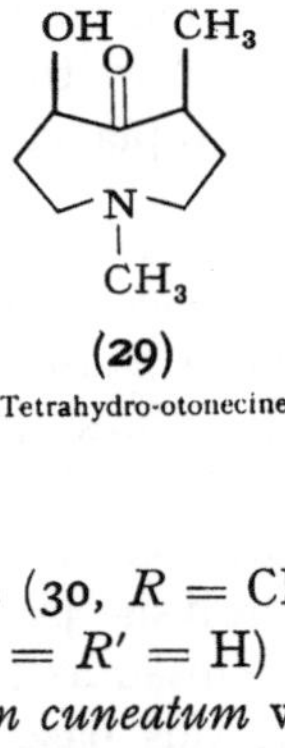

Tetrahydro-otonecine.

Loline, Lolinine, Norloline and Fusticine.

The finding of loline (**30**, $R = CH_3$, $R' = H$), lolinine (**30**, $R = CH_3$, $R' = CH_3CO$) and lolinidine, and later norloline (**30**, $R = R' = H$) by YUNUSOV and AKRAMOV (*301, 303*) in the seeds of *Lolium cuneatum* was the first such occurrence in the Gramineae. These interesting structures have been fully investigated by these same authors (*304, 305, 302*).

The reactions of loline, $C_7H_{10}NO(NHCH_3)$, set out in *Chart 16*, show clearly the presence of an ether linkage easily opened and closed, and the group $NHCH_3$ as well as a pyrrolizidine ring. The positioning of these last two groupings was established (*305*) by studying the Hofmann reaction on lolinine (acetyl-loline) in which —$NHCH_3$ of loline was deactivated to quaternisation so that only one quaternary methiodide was formed. Only two Hofmann reactions were possible, and only two moles of acetic acids (one from the acetyl group) were obtained by the Kuhn-Roth oxidation of the reduced second Hofmann product, so that one β-position to the quaternary nitrogen must carry no hydrogen.

References, pp. 393—406.

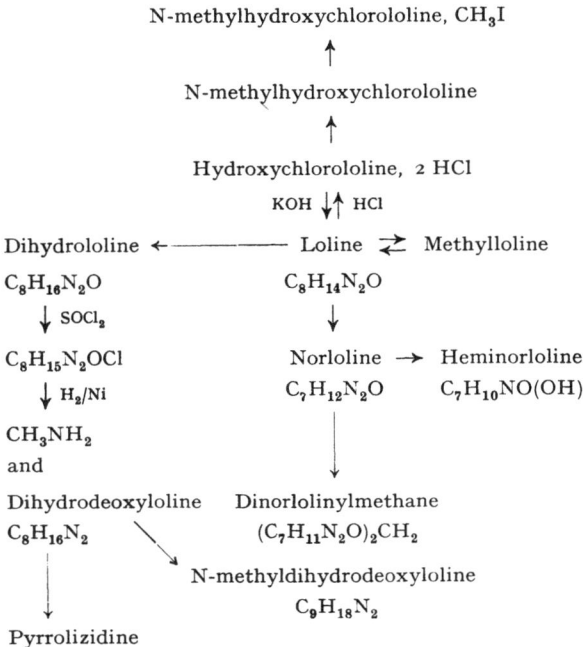

N-methylhydroxychlorololine, CH_3I

↑

N-methylhydroxychlorololine

↑

Hydroxychlorololine, 2 HCl

KOH ↓↑ HCl

Dihydrololine ← —————— Loline ⇌ Methylloline

$C_8H_{16}N_2O$ $C_8H_{14}N_2O$

↓ $SOCl_2$ ↓

$C_8H_{15}N_2OCl$ Norloline → Heminorloline

↓ H_2/Ni $C_7H_{12}N_2O$ $C_7H_{10}NO(OH)$

CH_3NH_2
and

Dihydrodeoxyloline   Dinorlolinylmethane

$C_8H_{16}N_2$ $(C_7H_{11}N_2O)_2CH_2$

            N-methyldihydrodeoxyloline

                  $C_9H_{18}N_2$

Pyrrolizidine

Chart 16. Reactions of Loline and Norloline.

Another new pyrrolizidine alkaloid is festucine, $C_8H_{14}NO$, isolated by YATES and TOOKEY (*391*), the major alkaloid of tall fescue hay, *Festuca arundinaceae*, in which it occurs with eight other bases [cf. *F. elatior* (*390*)]. Its physical properties were remarkably similar to those of loline with the exception of the melting points of the dihydrochloride salts and the optical rotation of the N-acetyl derivatives. Furthermore, festucine failed to react with 15% hydrochloric acid at 130° and required 36% acid at 160° to effect the conversion to its hydroxychloro derivative.

The reactions of the alkaloid, set out in *Chart 17*, and the NMR, IR and UV spectra were indicative of a saturated structure, and a pyrrolizidine structure was suggested. The X-ray diffraction studies of a single crystal made by McMILLAN and DICKERSON (*337*) showed structure (**31**), a positional isomer of loline (**30**, $R = CH_3$, $R' = H$) and one that recalls that of depropionyl-N-methyldecorticasine (**32**) (*217*).

(31)
Festucine.

(30)

(32)

The observation (243) that *Thelepogon elegans* contains five bases and thelepogine, $C_{20}H_{31}NO$, and thelepogidine, $C_{18}H_{29}NO_2$, is of interest in that the latter has been shown by FRIDRICHSONS and MATHIESON (283) from X-ray studies to have a structure (33) which contains a pyrrolizidine ring.

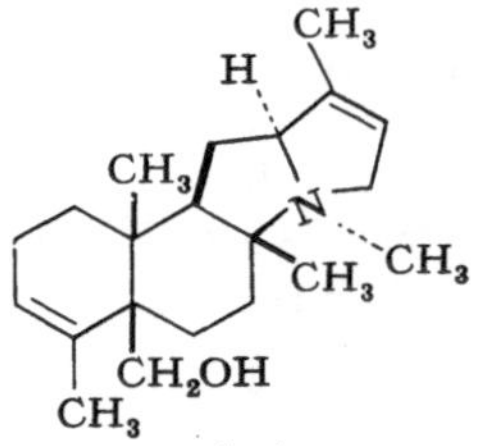

(33)

Thelepogidine.

N-Acetylfestucine HCl   N-Nitrosofestucine

$C_7H_{10}NO[N(CH_3)COCH_3] \cdot HCl$ $C_7H_{10}NO[N(CH_3)NO]$

Festucine dihydrochloride ⇄ Festucine

$C_7H_{10}NO(NHCH_3) \cdot 2\,HCl$ $C_7H_{10}NO(NHCH_3)$

Hydroxychlorofestucine dihydrochloride

$C_7H_{10}N(OH)(Cl)(NHCH_3) \cdot 2\,HCl$

$C_7H_{10}N(OOC \cdot CH_3)(Cl)[N(CH_3)(COCH_3)]HCl \;\leftarrow\; C_7H_{10}N(OH)Cl[N(CH_3)(COCH_3)] \cdot$
$\cdot\,HCl$

Chart 17. Reactions of Festucine.

3. Syntheses of the "Necine" Bases.

Pyrrolizidine.

One of the first syntheses of the pyrrolizidine ring system was that of CLEMO and his coworkers (241, 50) who prepared 7-keto-5,6,7,8-tetrahydropyrrocoline (34) but were unable to reduce it. This has now been effected by ADAMS, MIYANO and FLEŠ (208) catalytically using rhodium-on-alumina. TSUDA and SAEKI (382) treated diethyl 3-oxopentan-1,5-dicarboxylate with ammonia to give the pyrrolenone (35) which by high-pressure hydrogenation gave pyrrolizidine (36, $R = H_2$). A third synthesis was that of SEIWERTH (377) starting with 3-(tetrahydrofuran-2-yl)-propan-1-ol by the procedure shown below, cf. (39) → (40) → (41).

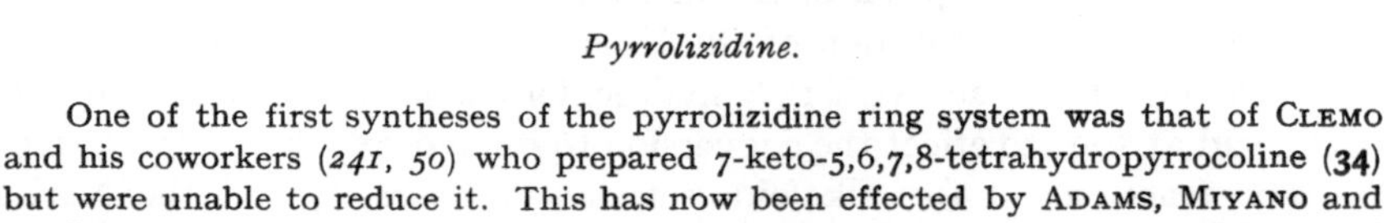

(34)   (36)   (35)

Pyrrolizidine ($R = CH_2$).

References, pp. 393—406.

1-Methylpyrrolizidine.

SEIWERTH and OREŠČANIN-MAJHOFER (*379, 350*) reacted 2-acetylfuran (**37**) with ethyl bromoacetate under Reformatsky conditions to give the ethyl 3-hydroxy-3-furanyl-butyrate (**38**) which was dehydrated and reduced to the tetrahydrofuranyl-butanol (**39**), treated with hydrobromic acid to 1,4,7-tribromo-3-methylheptane (**40**) and hence to 1-methylpyrrolizidine (**41**). LUKEŠ and JANDA (*335*) effected a novel synthesis by heating compound (**38**) with hydrochloric acid in ethanol to give diethyl 3-oxo-2-methylpentane-1,5-dicarboxylate (**42**) which by a Leuckart reaction and ring closure yielded 1-methyl-3,5-dioxopyrrolizidine (**43**) and hence (**41**).

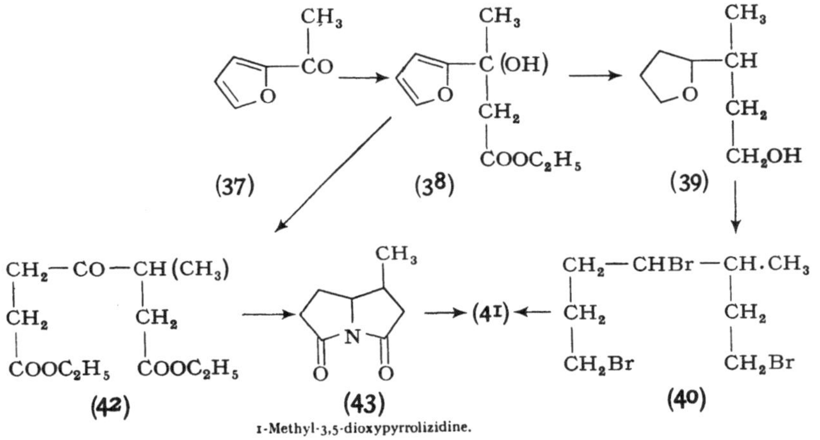

1-Methyl-3,5-dioxypyrrolizidine.

A novel stereospecific synthesis of (±)-pseudoheliotridane was reported by ČERVINKA (*238*). 1-Methyl-2-ethyl-Δ^2-pyrroline (**44**) was allowed to react with ethyl bromoacetate, and the product heated with potassium formate and sulphuric acid to give ethyl "erythro"-(±)-2-(N-methyl-2-pyrrolidyl)-butyrate (**45**, $R = = COOC_2H_5$). The reduction of this ester with lithium aluminium hydride and heating the resulting alcohol (**45**, $R = CH_2OH$) with hydrobromic acid yielded the (±)-form of pseudoheliotridane (**46**).

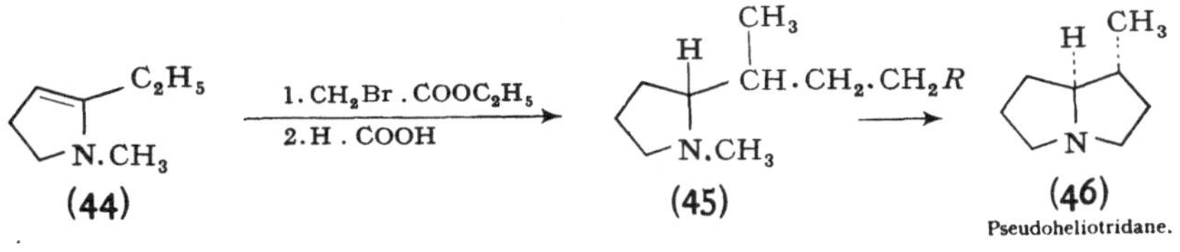

1-Methylenepyrrolizidine.

LIKHOSHERSTOV, KRITSYN and KOCHETKOV (*332*) applied the Wittig reaction for the synthesis of (±)-methylenepyrrolizidine (**36**, $R = CH_2$, p. 340) from 1-pyrrolizidine (**36**, $R = O$) and triphenylphosphenomethylene. This latter was resolved by way of its (+)-tartrate to give (—)-methylenepyrrolidene, $[\alpha]_D$ — 43.1° (*263*).

3-Methyl- and 3-Hydroxymethylpyrrolizidines.

SEIWERTH and his collaborators, employing PRELOG's (*155*) two-fold intramolecular alkylation, treated 3-(tetrahydrofuran-2-yl)-propan-1-ol (**47**, $R = H$) (*377*) and 4-(tetrahydrofuran-2-yl)-butan-2-ol (**47**, $R = CH_3$) (*350*) with hydrobromic acid to yield the 1,4,7-tribromoheptane (**48**, $R = H$) and octane (**48**, $R = CH_3$), and closed the ring with ammonia to give pyrrolizidine and 3-methylpyrrolizidine (**49**, $R = CH_3$). SEIWERTH and DJOKIĆ (*378*) employed 4-(tetrahydrofuran-2-yl)-butyric acid (**50**) to obtain pyrrolizidine-3-carboxylic acid (**49**, $R = COOH$) which with lithium aluminium hydride gave finally 3-hydroxymethylpyrrolizidine (**49**, $R = CH_2OH$).

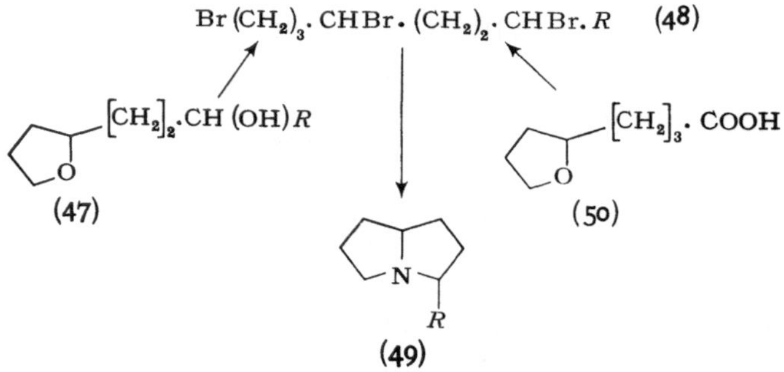
Br(CH2)3.CHBr.(CH2)2.CHBr.R (48)
[CH2]2.CH(OH)R
O
(47)
[CH2]3.COOH
O
(50)
N
R
(49)
3-Methylpyrrolizidine ($R = CH_3$).

1-Hydroxymethylpyrrolizidines.

The several syntheses of 1-hydroxymethylpyrrolizidines may be divided into two groups: those employing PRELOG's method and those proceeding by way of a pyrrolidine derivative.

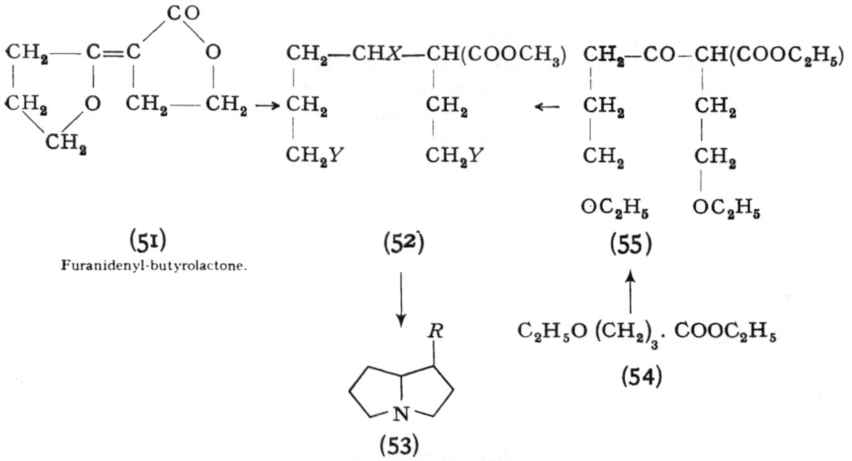
CO
CH2—C=C O
CH2 O CH2—CH2
CH2
(51)
Furanidenyl-butyrolactone.
CH2—CHX—CH(COOCH3)
CH2 CH2
CH2Y CH2Y
(52)
CH2—CO—CH(COOC2H5)
CH2 CH2
CH2 CH2
OC2H5 OC2H5
(55)
C2H5O(CH2)3.COOC2H5
(54)
R
N
(53)
Trachelantamidine ($R = CH_2OH$).

KOCHETKOV, LIKHOSHERSTOV and BUDOVSKIĬ (*312*) hydrogenated dibutolactone (furanidenyl-butyrolactone) (**51**) to give the trihydroxy acid (**52**, $X = Y = OH$) which with ammonia cyclised to methyl pyrrolizidine-1-carboxylate (**53**, $R =$

References, pp. 393—406.

= COOCH$_3$) and hence to ($\pm$)-trachelanthamidine (53, R = CH$_2$OH) (312). The tri-bromide (52) was also prepared by KOCHETKOV, LIKHOSHERSTOV and LIKHOSHERSTOV (314) starting from ethyl 4-ethoxybutyrate (54) to give the substituted acetoacetate (55) which was reduced to (52, X = OH, Y = OC$_2$H$_5$) and treated with hydrobromic acid.

The synthesis of 1-hydroxymethylpyrrolizidines by ČERVINKA, PELZ and JIRKOVSKÝ (239) can be interpreted as in *Chart 18*. 1-Benzylpyrrole (56) was condensed with acetylene-dicarboxylic acid to N-benzyl-2-pyrrolylfumaric acid (57) which was hydrogenated over a palladium-charcoal catalyst until only 1 mole of hydrogen was absorbed to give the succinic acid (58), the methyl ester of which was further hydrogenated to methyl 3-oxopyrrolizidine-carboxylate (59), and then with lithium aluminium hydride to 1-hydroxypyrrolizidine (60). The product which contained ($\pm$)-trachelanthamidine and ($\pm$)-isoretronecanol in the ratio of 9 : 1 was separated into its components by way of the picrates, and the ($\pm$)-trachelanthamidine resolved by dibenzoyltartaric acid to give (+)-laburnine.

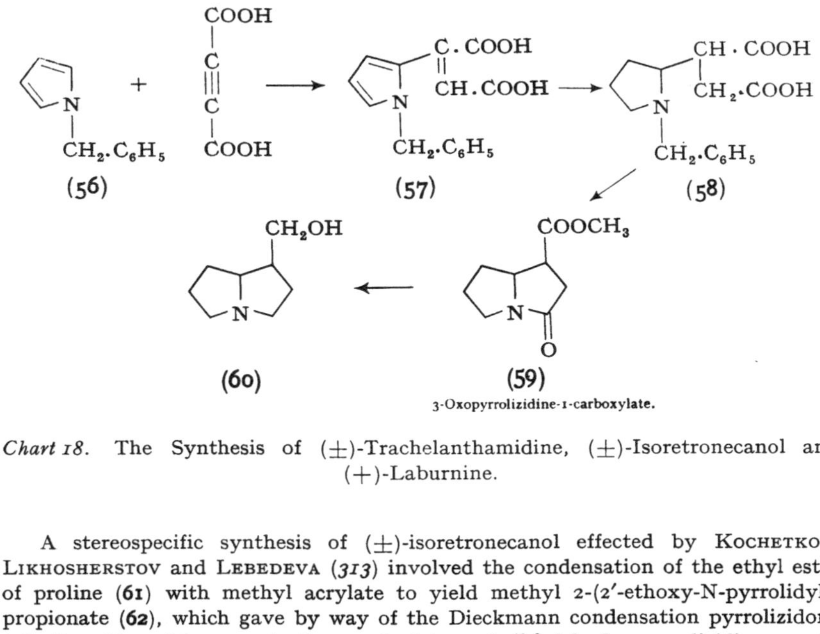

Chart 18.   The Synthesis of ($\pm$)-Trachelanthamidine, ($\pm$)-Isoretronecanol and (+)-Laburnine.

A stereospecific synthesis of ($\pm$)-isoretronecanol effected by KOCHETKOV, LIKHOSHERSTOV and LEBEDEVA (313) involved the condensation of the ethyl ester of proline (61) with methyl acrylate to yield methyl 2-(2'-ethoxy-N-pyrrolidyl)-propionate (62), which gave by way of the Dieckmann condensation pyrrolizidone (36, R = O) and hence, via the cyanhydrin and $\Delta^{1,8}$-dehydro-pyrrolizidine-1-carb-oxylate, to ($\pm$)-trachelanthamidine.

In a further study (314) the final reduction product was separated as picrates corresponding to ($\pm$)-isoretronecanol and ($\pm$)-trachelanthamidine.

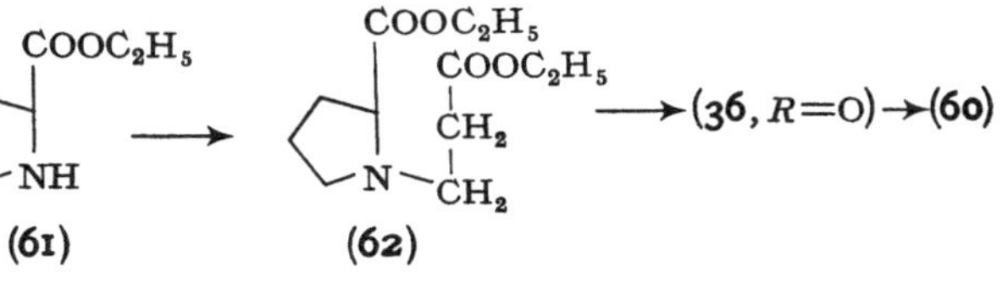

An interesting synthesis by Ježo and Kaláč (*300*) of ($\pm$)-1-hydroxymethyl-pyrrolizidine started from nitromethane and ethyl 2-methoxymethyl-acrylate (**63**) through diethyl 2,7-di-(methoxymethyl)-3-nitropimelate (**64**), 2-(1'-methoxymethyl-2'-carbethoxyethyl)-pyrrolid-5-one (**65**) and 2-(1'-methoxymethyl-3'-hydroxypropyl)-pyrrolid-5-ene (**66**).

$$CH_2{=}C(CH_2OCH_3) \cdot COOC_2H_5 \rightarrow NO_2 \cdot CH[CH_2 \cdot CH(CH_2OCH_3) \cdot COOC_2H_5]_2$$

     (63)   (64)

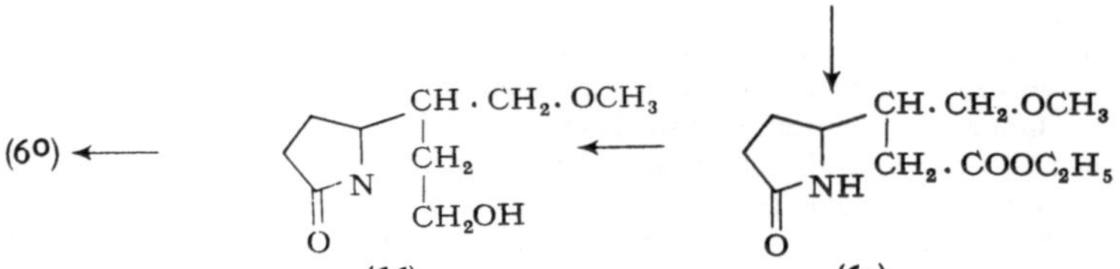

A new synthesis by Micheel and Flitsch (*346*) started with the treatment of furfurylacylic acid (**67**) with alcohol and hydrochloric acid to give diethyl 3-oxopentan-1,5-dicarboxylate (**68**) which was reduced in the presence of ammonia to the half lactam (**69**) and then the ring closed to 3,5-dioxopyrrolizidine (**70**).

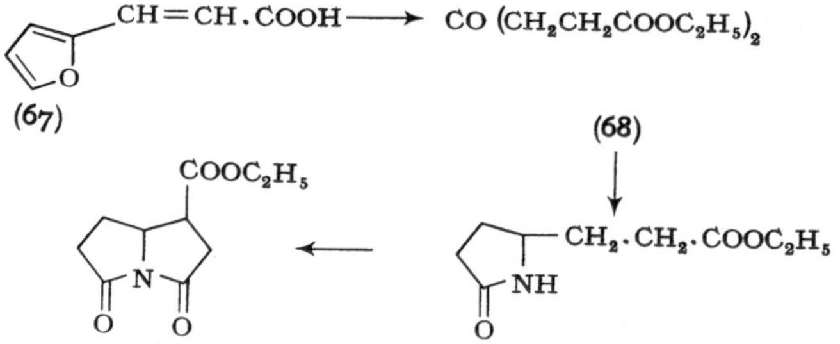

3·5-Dioxopyrrolizidine.

Dihydroxymethylpyrrolizidines.

The elegant synthesis of retronecine by Geissman and Waiss (*288*) is set out in *Chart 19*. N-Ethoxycarbonyl-3-aminopropionic ester was condensed with diethyl fumarate and the product ring closed to the pyrrolidinone (**71**). This Dieckmann condensation proceeded in the way shown, for the alternative condensation would not have given the lactone (**72**). Hydrolysis of (**72**) and treatment of the resulting secondary amine with ethyl bromoacetate gave the lactone (**73**) which by ring closure and reduction yielded ethyl 2,4-dihydroxypyrrolizidine-1-carboxylate (**74**). This ester had the required *cis* arrangement of the hydroxyl at $C_{(7)}$ and the carboxyl group at $C_{(1)}$ since it readily formed the quaternary lactone (**75**). Hydrolysis of (**74**), which was accompanied by dehydration, was followed by reduction with lithium aluminium hydride and resolution of the product as its salt with ($+$)-camphoric acid to give ($+$)-retronecine (**76**).

It has always been assumed that the dihydroxy "necine" bases were 1-hydroxy-methyl-7-hydroxypyrrolizidines. A hydroxy group, however, is found in position 2 in rosmarinecine; and Adams, Miyano and Nair (*209*) set out to synthesise the 1-hydroxymethyl-2-hydroxy-pyrrolizidine (**77**).

References, pp. 393—406.

$$C_2H_5OOC \cdot CH_2 \cdot CH_2 \cdot NHCOOC_2H_5 + C_2H_5OOC \cdot CH = CH \cdot COOC_2H_5 \rightarrow$$

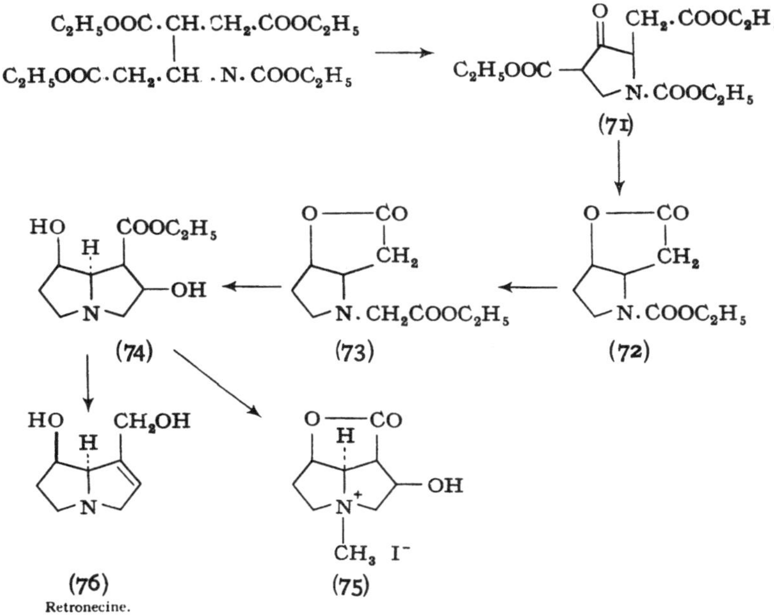

$$\begin{array}{c} C_2H_5OOC \cdot CH \cdot CH_2 \cdot COOC_2H_5 \\ | \\ C_2H_5OOC \cdot CH_2 \cdot CH \cdot N \cdot COOC_2H_5 \end{array} \longrightarrow$$

(71)

(74)   (73)   (72)

(76)   (75)

Retronecine.

Chart 19. The Geissman-Waiss Synthesis of Retronecine.

They condensed ethyl pyrrolidylacetate (78) with ethyl oxalate to give ethyl 2,3-dioxopyrrolizidine-1-carboxylate (79), the 2-keto group of which was reduced to a secondary alcohol (80, $X = OH$) in the presence of a rhodium catalyst and then with lithium aluminium hydride to yield ($\pm$)-isoretronecanol, m. p. 123–124°, definitely different from macronecine, m. p. 127–128.5°. An alternative route devised by GOLDSCHMIDT (291), was the condensation of 1-pyrroline with ethyl oxosuccinate to (79). Dehydration of the hydroxy ketone (80, $X = OH$) by way of its tosylate (80, $X = $ Tos) (347) was shown to give the 1,8-ene (81) (291).

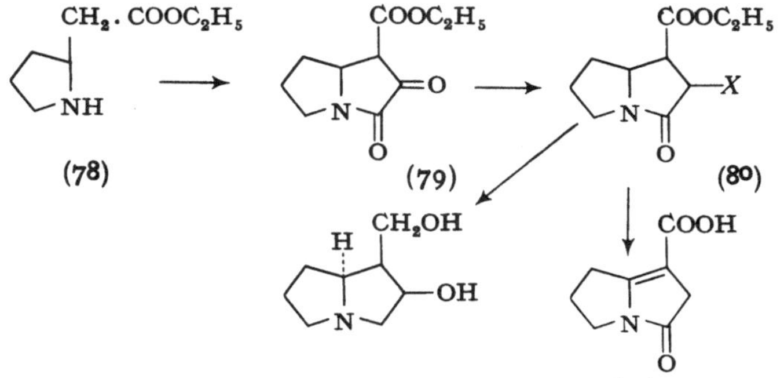

(78)   (79)   (80)

(77)   (81)

1-Hydroxymethyl-2-hydroxy-pyrrolizidine.

III. The Acids Associated with the Pyrrolizidine Alkaloids.

The new acids associated with the pyrrolizidine alkaloids and those about which additional information is now available are shown in *Table 9 A* (p. 390). New acids have been isolated and the structures of these and of many already known acids have been determined. The structures of several acids have been assigned from experiments on the native alkaloids themselves and this is particularly required where the acid is unstable. The chemistry of these acids cannot be separated from, and is accordingly treated under the section on, the total alkaloid. Where the acid itself has not been isolated in the form in which it occurs in the alkaloid, or its structure has been deduced from its degradation product, the hypothetical acid is named after the alkaloid and shown in parentheses.

1. C_{10}-Adipic Acids and Retusanecic Acid.

Hygrophyllinecic Acid.

The alkaloid $C_{18}H_{26}NO_7$, m. p. 176°, previously isolated by RICHARDSON and WARREN (*159*), has been re-isolated by SCHLOSSER and WARREN (*362*) from *Senecio hygrophylus* and named hygrophylline. It gave on acid hydrolysis platynecine and a new acid, hygrophyllinecic acid isolated as its monolactone, $C_{10}H_{14}O_5$, m. p. 180–181°, which yielded on distillation a dilactone, $C_{10}H_{12}O_4$, m. p. 103–105°. Ozonisation of the monolactone gave acetaldehyde and carbon dioxide indicative of the grouping $CH_3 \cdot CH : C(COOH)$—. The alkaloid itself showed three C-methyl groups, and on reduction with lithium aluminium hydride, followed by oxidation with periodic acid, gave formaldehyde indicative of a terminal glycol $\diagdown C(OH) \cdot CH_2OH$ from an α-hydroxy acid. These reactions together with the NMR spectrum led to the structures for the monolactone (**82**) and the dilactone (**83**). Further proof was obtained by the hydrogenolysis of hygrophyllinecic acid monolactone with the Adams catalyst activated with perchloric acid to give dihydrosenecic acid identified as its lactone (**84**), m. p. 122–124°. The hydrogenolysis by this catalyst, which is reported as reducing esters and δ-lactones to ethers, was interpreted as being indicative of the extra hydroxyl group being in the allylic position (see *Chart 20*). Final proof was obtained by the oxidation of hygrophylline (**85**) with performic acid to dihydroxydihydrohygrophylline (**86**) which was oxidised with sodium periodate to give acetaldehyde and a product (**87**) which yielded oxalic acid on hydrolysis.

Since the absolute configuration of senecic acid is known (*284*), the structure of hygrophyllinecic acid dilactone can be represented as (**88**), i. e. 2 R,3 R,4 R-2,4-dihydroxy-3-methylhept-*trans*-5-ene-2,5-dicarboxylic acid. It is impossible to form the dilactone if the $C_{(4)}$-hydroxyl group has the opposite configuration; and the *trans*-geometry* is assigned to the double bond since the absorption at λ_{max} 220 mμ has ε_{max} 4800 (*105*).

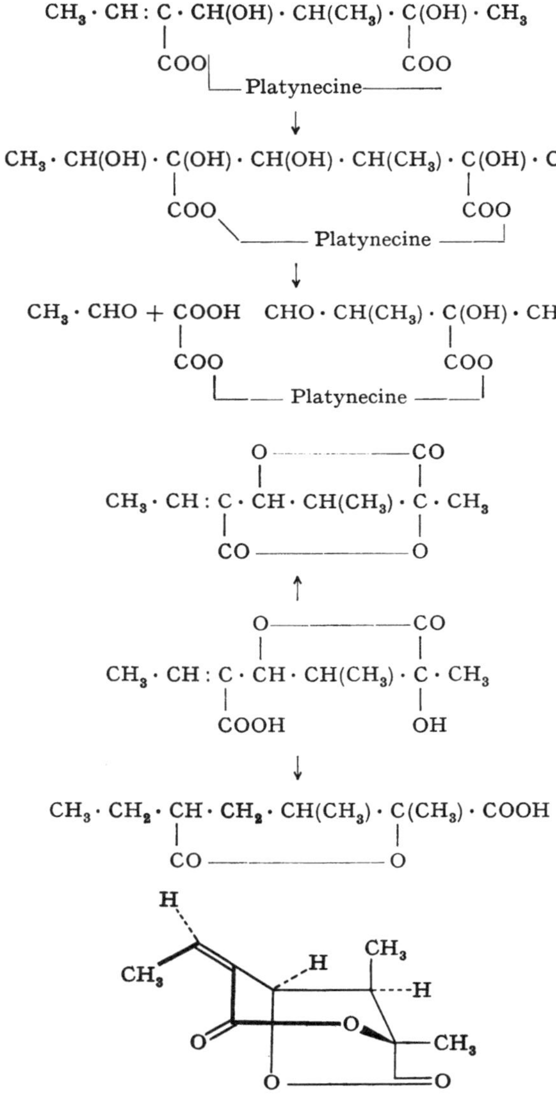

$$CH_3 \cdot CH : C \cdot CH(OH) \cdot CH(CH_3) \cdot C(OH) \cdot CH_3$$

COO COO — Platynecine —— (85)

↓

$$CH_3 \cdot CH(OH) \cdot C(OH) \cdot CH(OH) \cdot CH(CH_3) \cdot C(OH) \cdot CH_3$$

COO COO — Platynecine —— (86)

↓

$$CH_3 \cdot CHO + COOH \quad CHO \cdot CH(CH_3) \cdot C(OH) \cdot CH_3$$

COO COO — Platynecine —— (87)

O ———— CO
$$CH_3 \cdot CH : C \cdot CH \cdot CH(CH_3) \cdot C \cdot CH_3$$
CO ———— O (83)

↑

O ———— CO
$$CH_3 \cdot CH : C \cdot CH \cdot CH(CH_3) \cdot C \cdot CH_3$$
COOH OH (82)

↓

$$CH_3 \cdot CH_2 \cdot CH \cdot CH_2 \cdot CH(CH_3) \cdot C(CH_3) \cdot COOH$$
CO ———— O (84)

(88)
Hygrophyllinecic acid dilactone.

Chart 20. Some Reactions of Hygrophylline and Hygrophyllinecic Acid.

---

* Correct nomenclature IUPAC.

Seneciphyllic, Isoseneciphyllic, "Spartioidinecic" and Riddellic Acids.

ADAMS and GIANTURCO (*205*) established that spartioidine (p. 365) had a similar structure to that of seneciphylline (p. 365) and the acids present differed in geometrical isomerism and in the configuration at $C_{(2)}$. Since then the structure of isoseneciphyllic acid has been re-examined by MASAMUNE (*340*) who, from a study of the NMR spectrum and the

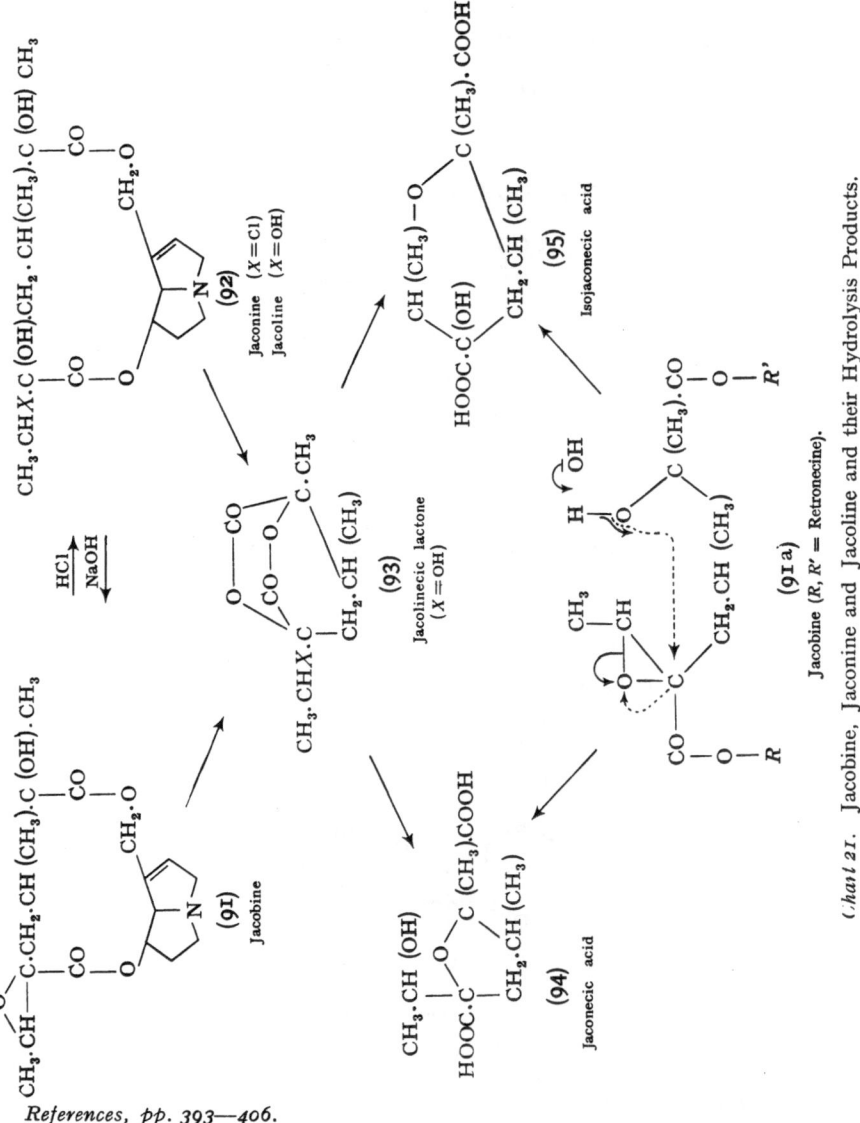

Chart 21. Jacobine, Jaconine and Jacoline and their Hydrolysis Products.

oxidation with potassium permanganate-periodate to yield formaldehyde, assigned structure (89, X = H), and for seneciphyllic acid structure (90, X = H). He concluded that "spartioidinecic" and riddellic acid had similar structures (see Table 9A, p. 390).

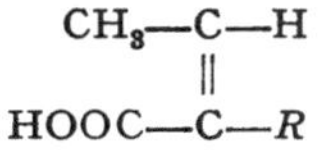

$$CH_3\text{—}C\text{—}H$$
$$\|$$
$$HOOC\text{—}C\text{—}R$$

(89)

Isoseneciphyllic acid (X = H).

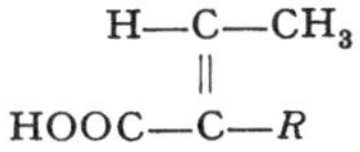

$$H\text{—}C\text{—}CH_3$$
$$\|$$
$$HOOC\text{—}C\text{—}R$$

(90)

Seneciphyllic acid (X = H).

$$R = CH_2 \cdot C(=CH_2) \cdot C(OH)(COOH) \cdot CH_2X$$

The Acids from, and the Structures of Jacobine, Jacoline and Jaconine.

A further body of valuable and painstaking experimental work on the degradation of jacobine has been accumulated by BRADBURY et al. (*41, 42, 227–232*) and so made possible the unravelling of some interesting structural changes. GEISSMAN (*287*), taking into consideration these reported reactions and the stability of the derived acids, proposed the structure of jacobine (91) and jaconine (92, X = Cl) which explained their ready interconversions (*41*), and their hydrolysis with hydrochloric acid to a chlorodilactone (93, X = Cl) (*232*) which in turn could be converted by alkaline hydrolysis to jaconecic (94) and isojaconecic acid (95) *(Chart 21)*.

GEISSMAN explained the alkaline hydrolysis of jacobine as being accompanied by alkali-induced opening of the epoxide ring which can occur in two ways [shown as dotted and solid arrows in (91a)] to give either jaconecic acid (94) or isojaconecic acid (95), respectively. Furthermore, the formula for jaconecic acid is in complete accord with the products obtained by oxidation, namely with lead tetraacetate (*232, 229*) to give acetaldehyde and carbon dioxide, 3-methyl-4-keto-pentanoic (β-methyllaevulinic) acid (96) and 3-methylbutane-2-ol-1,4-dicarboxylic acid lactone (97), possibly by way of the following conversion:

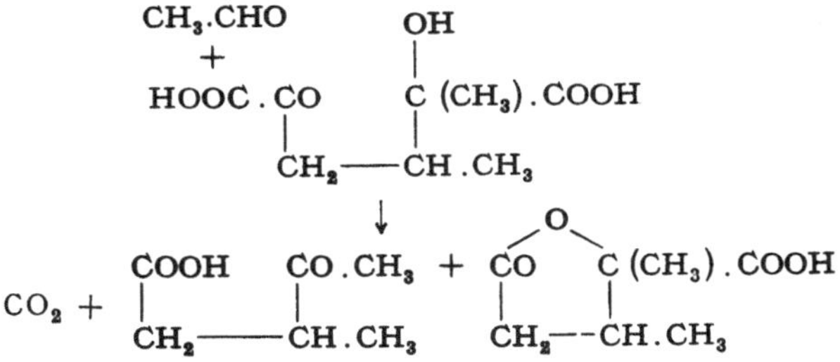

$$CH_3.CHO$$
$$+$$
$$HOOC.CO$$
$$|$$
$$CH_2\text{———}CH.CH_3$$

$$OH$$
$$|$$
$$C(CH_3).COOH$$
$$|$$
$$CH.CH_3$$

$$\downarrow$$

$$CO_2 + \quad \begin{array}{c} COOH \\ | \\ CH_2 \end{array}\text{———}\begin{array}{c} CO.CH_3 \\ | \\ CH.CH_3 \end{array} \quad + \quad \begin{array}{c} CO \\ | \\ CH_2 \end{array}\overset{O}{\diagup\diagdown}\begin{array}{c} C(CH_3).COOH \\ | \\ CH.CH_3 \end{array}$$

(96)   (97)

and with nitric acid to yield α,β-dimethylmalic acid (98) (*232*).

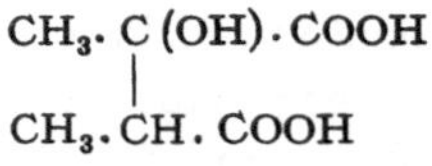

$$CH_3 \cdot \underset{\underset{CH_3 \cdot CH \cdot COOH}{|}}{C} (OH) \cdot COOH$$

(98)

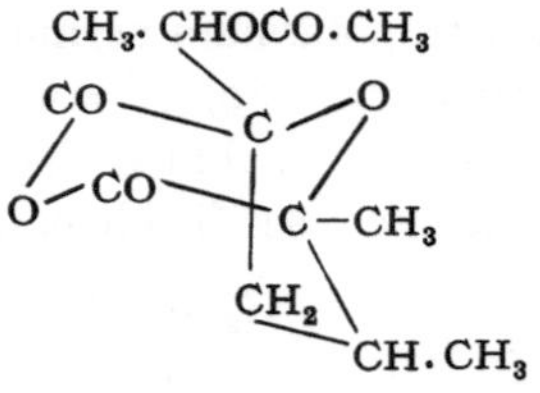

(99)

Acetyljaconecic acid anhydride.

Drastic acetylation of jaconecic acid gave acetyl jaconecic acid anhydride (99) which can be hydrolysed to acetyl jaconecic acid, also obtained from jaconecic acid and acetyl chloride.

The alkaloid jacoline, which readily consumed one mole of periodic acid to give acetaldehyde is in complete agreement with the glycol structure (92, $X = OH$). Further confirmation was obtained by the hydrolysis of jacobine with dilute sulphuric acid to a neutral dilactone (93, $X = OH$) which gave a mono-acetyl compound (93, $X = O \cdot COCH_3$).

The stereochemistry of the "necic" acid in jacobine was advanced by MASAMUNE (341) on the basis of the geometry of 3-methylbutan-2-ol-1,4-dicarboxylic acid lactone (100) and the optical activity of (—)-3-methylpentan-4-on-oic acid (101). The methyl ester of this latter acid gave by way of the Baeyer-Villiger reaction (—)-3-hydroxybutyric hydrazate (102) of known configuration, so that $C_{(3)}$ of the necic acid (103, R = retronecine base) is established (Chart 22).

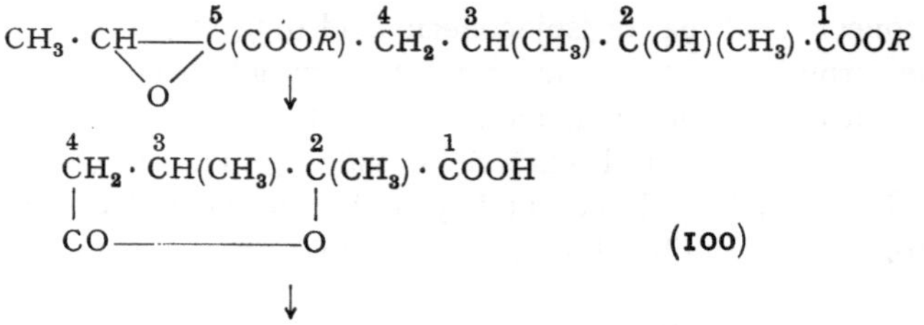

(103)

(100)

p-Bromophenacyl ester m. p. 108°, $[\alpha]_D + 13°$

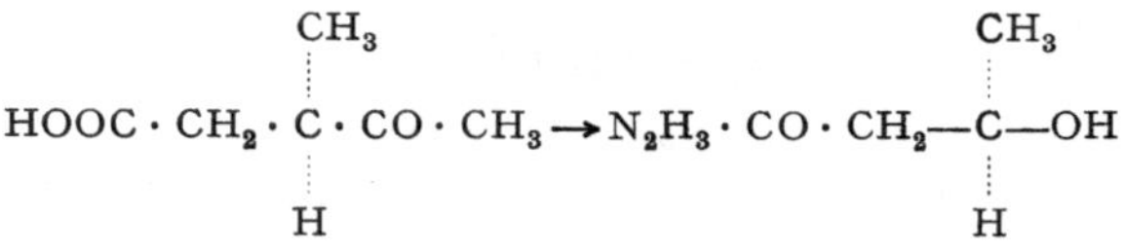

(101)　　　　　　(102)

Chart 22. The Reactions Leading to the Absolute Configuration at $C_{(2)}$ of the "Necic" Acid in Jacobine.

The absolute configuration at $C_{(3)}$ was established by identifying the (+)-3-methylbutan-2-ol-1,4-dicarboxylic acid lactone (**100**) with one of the two possible structures. For this purpose (±)-β-methyllaevulinic acid (**104**) was converted by way of its cyanhydrin to the two stereo-isomeric racemates of 3-methylbutan-2-ol-1,4-dicarboxylic acid lactone (**100**), namely A and B, shown in *Chart 23*. The lactone-B was resolved through its cinchonidine salt to an (+)-acid identical with the oxidation product. The two lactones were each in turn treated with phenyl magnesium bromide to give (±)-2,3-dimethyl-5,5-diphenyl-2-hydroxypent-4-enoic acids (**105**) which on ozonisation yielded (±)-dimethylmalic acids (**106**) that were reduced to (±)-2,3-dimethyl-1,2,3-triols (**107 a**) and (**107 b**). The geometry of the triol (**107 a**) was established by its synthesis from dimethylmaleic anhydride (**108**) by way of its lithium aluminium hydride reduction product (**109**) and the epoxide (**110**). This necessitates that the stereoisomeric triol (**107b**) has the structure shown and that the two methyl groups at $C_{(2)}$ and $C_{(3)}$ are *cis*.

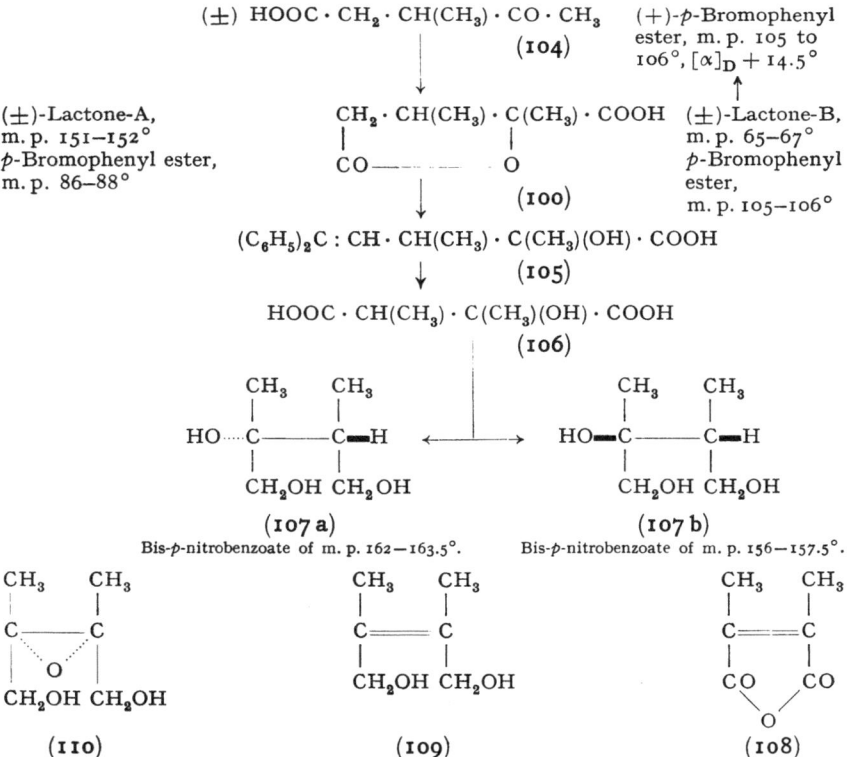

Chart 23. The Reactions Leading to the Absolute Configuration of $C_{(3)}$ of the "Necic" Acid in Jacobine.

Since the absolute configuration at $C_{(2)}$ is defined above, the absolute configuration of (+)-lactone-B is known (111). The ease of formation of the acetyljaconecic acid anhydride necessitates the two carboxy groups being *cis* and hence defines the groups around $C_{(4)}$. Finally, the action of methyl magnesium iodide on the phenylglyoxalate ester of dimethyl jaconecate gave (+)-atrolactic acid and established the absolute configuration at $C_{(5)}$ of jaconecic (112), and hence also isojaconecic (113), acids.

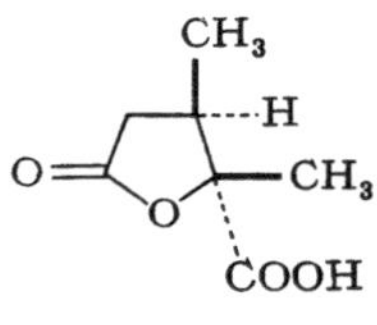

(III)	(112)	(113)
(+)-Lactone-B.	Jaconecic acid.	Isojaconecic Acid.

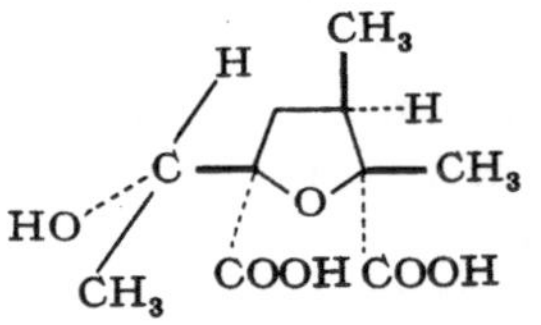

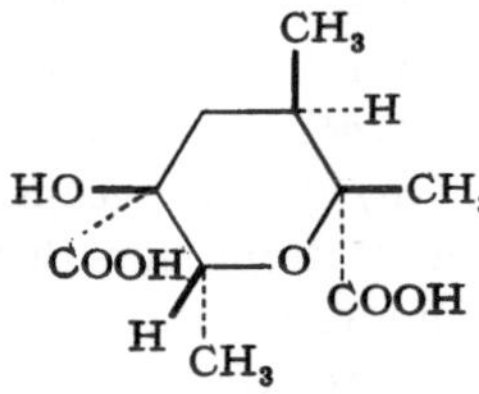

Since the absolute configuration of retronecine is known (*388*), the absolute configurations of jaconecine, jaconine and jacoline are known.

Identical structures for these acids were advanced independently by BRADBURY and MASAMUNE (*229, 230*) on reasoning similar to the above. Isojaconecic acid with lead tetraacetate gave carbon dioxide and a keto-acid (114, $R = COOH$) which yielded a bis-2,4-dinitrophenylhydrazone (115, $R = COOH$). Reduction of isojaconecic acid with lithium aluminium hydride and oxidation of the product with periodic acid gave formaldehyde and a product, presumably (114, $R = CH_2OH$), which was isolated as the bis-dinitrophenylhydrazone (115, $R = CH_2OH$). Proton magnetic resonance of the chlorodilactone, dimethyl jaconecate and dimethyl isojaconecate gave further convincing evidence of the structure.

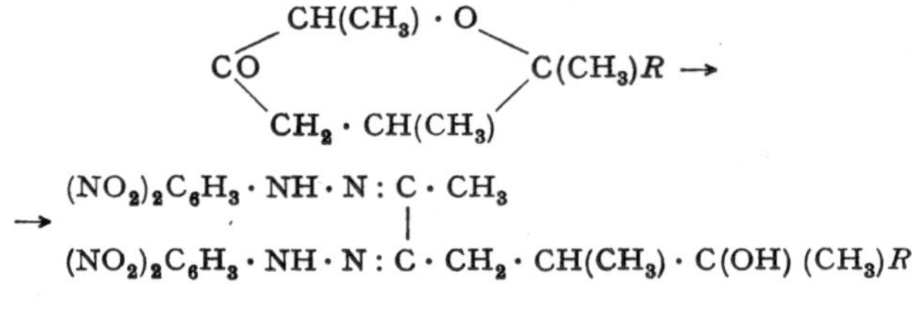

$$(114)$$

$$(115)$$

X-ray studies by FRIDRICHSONS, MATHIESON and SUTOR (*284*) on jacobine bromohydrin (116) have confirmed the above structures for jacobine as well as the absolute configuration of retronecine deduced by WARREN and VON KLEMPERER (*388*). The absolute configuration of the acid in jacobine is identical to that of senecic acid (*284*).

References, pp. 393—406.

$$CH_3 \quad OH \qquad CH_3 \quad OH$$

$$Br - \underset{\underset{H}{|}}{\overset{\overset{CH_3}{|}}{C}} - \underset{\underset{CO}{|}}{\overset{\overset{OH}{|}}{C}} - CH_2 - \underset{\underset{H}{|}}{\overset{\overset{CH_3}{|}}{C}} - \underset{\underset{CO}{|}}{\overset{\overset{OH}{|}}{C}} - CH_3$$

(116)

Jacobine bromohydrin.

The correlation of the structure of senecic acid and the acid, "jaco-binecic acid", which occurs in otosenine has been established by KORETS-KAYA, DANILOVA and UTKIN (*321*). Otosenine (117) on treatment with hydrobromic acid gave a monobromide of "jacobinecic acid", m. p. 116° (118), $[\alpha]_D$ — 30°, which was epimeric with the product (119), m. p. 113°, $[\alpha]_D$ — 56°, obtained by treatment of senecic acid lactone (120), with bromine water. Both these substances gave on reduction the dilactone (121).

$$CH_3 \cdot \overset{\overset{O}{\diagdown}}{CH} - C(COOR')R \rightarrow CH_3 \cdot CH - \overset{\overset{Br}{|}}{C}(OH)(COOH)R$$

(117)   (118)
Otosenine.

$$CH_3 \cdot CH = C(COO-)R \rightarrow CH_3 \cdot CH - \underset{\underset{Br}{|}}{C}(OH)(COOH)R$$

(120)   (119)

$$\begin{array}{c} O \rule{2cm}{0.4pt} CO \\ | \qquad\qquad | \\ C_2H_5C \cdot CH_2 \cdot CH(CH_3) \cdot C \cdot CH_3 \\ | \qquad\qquad | \\ CO \rule{2cm}{0.4pt} O \end{array}$$

(121)

Synthesis of Integerrinecic and Senecic Acids.

An elegant synthesis of integerrinecic acid and senecic acid has been effected by CULVENOR and GEISSMAN (*257*) and is set out in *Chart 24* which shows the absolute geometry of the compounds in the light of X-ray studies recently carried out by FRIDRICHSONS (*284*).

3-Acetoxy-but-1-yne (122) with nickel carbonyl gave 3-acetoxy-but-1-en-2-carboxylic acid (123) the methyl ester of which was treated with methyl methyl-acetoacetate in the presence of molar quantities of sodium ethoxide to effect a Michael condensation. Hydrolysis of the resulting product (124) gave a mixture of (±)-*cis*- and (±)-*trans*-1-ethylidene-3-methylhexan-4-en-oic acids (125) which was separated into its geometrical isomers. The *cis*-acid (126) was identical with the acid obtained by the oxidation of senecic acid (127) (*104*). The *trans*-acid (128) with sodium cyanide and then hydrolysis gave (±)-integerrinecic acid lactone, m. p. 142–143°, which was resolved by its brucine salt to the (+)-isomer, m. p. 235°, identical with the (+)-integerrinecic acid lactone (129), and the (—)-isomer (130), m. p. 221–222°. Irradiation of the integerrinecic acid, or better, of its lactone, by ultraviolet light gave 50% conversion to senecic acid (127) or senecic acid lactone (131),

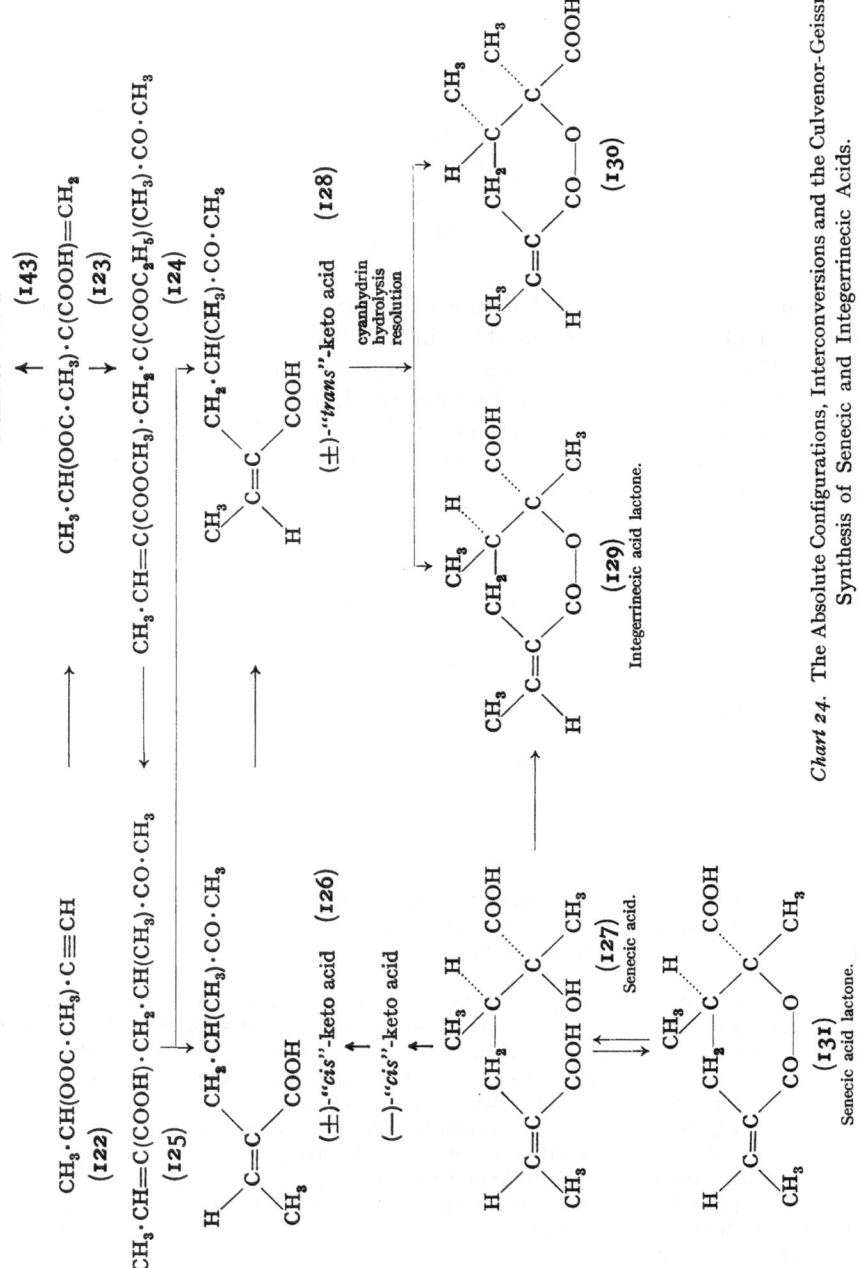

Chart 24. The Absolute Configurations, Interconversions and the Culvenor-Geissman Synthesis of Senecic and Integerrinecic Acids.

References, pp. 393—406.

respectively. This latter *cis*-lactone (**131**), unisolated previously since lactonisation gave integerrinecic acid lactone (*104*), was also prepared with partial retention of configuration by treatment of senecic acid with 60% sulphuric acid at room temperature. The synthetic route by way of the *"trans"* keto acid (**128**) was dictated by ease of racemisation of the (—)-*"cis"*-keto acid (**126**) and the certainty that the hydrolysis of the nitrile by usual procedure would be accompanied by geometrical isomerisation.

KOCHETKOV, VASIL'EV and LEVCHENKO (*317*) describe the synthesis of (±)-integerrinecic acid starting from the condensation of 3-chloromethylbutan-2-one with ethyl malonate to give the keto ester (**132**) *(Chart 25)*. The keto group was masked by the formation of the ketal with ethylene glycol, and the resulting compound condensed with 1-ethoxy-1-chloroethane to give (**133**) which, after hydrolysis, decarboxylation, and splitting of ethanol, yielded 1-ethylidene-3-methylhexane-4-en-1-oic acid (**134**); this gave by way of the cyanhydrin reaction (±)-integerrinecic acid [(±)-(**129**)].

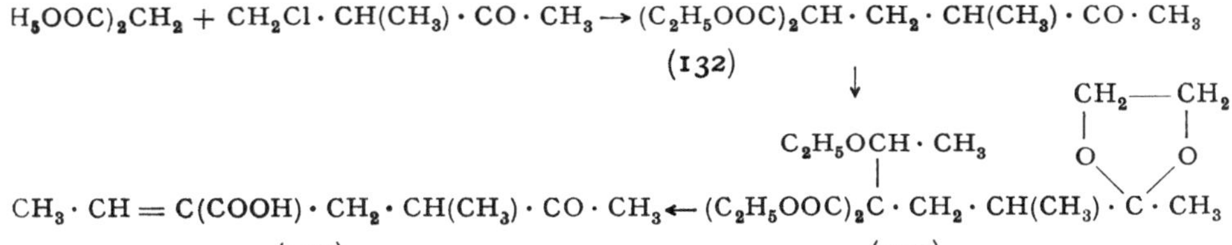

$$H_5OOC)_2CH_2 + CH_2Cl \cdot CH(CH_3) \cdot CO \cdot CH_3 \rightarrow (C_2H_5OOC)_2CH \cdot CH_2 \cdot CH(CH_3) \cdot CO \cdot CH_3$$

$$(132)$$

$$C_2H_5OCH \cdot CH_3$$

$$CH_3 \cdot CH = C(COOH) \cdot CH_2 \cdot CH(CH_3) \cdot CO \cdot CH_3 \leftarrow (C_2H_5OOC)_2C \cdot CH_2 \cdot CH(CH_3) \cdot C \cdot CH_3$$

$$(134) \qquad\qquad (133)$$

Chart 25. The Kochetkov-Vasil'ev-Levchenko Synthesis of (±)-Integerrinecic Acid.

The Synthesis of Dihydrosenecic Acid.

The synthesis of (±)-dihydrosenecic acid has been reported by KOCHETKOV and VASIL'EV (*315, 316*). The Michael addition of ethyl ethylmalonate to 2-methyl-but-1-en-3-one in the presence of sodium ethoxide gave diethyl 5-methylheptan-6-one-2,2-dicarboxylate which by hydrolysis, decarboxylation, esterification, addition of hydrocyanic acid and hydrolysis gave the (±)-dihydrosenecic acid lactone (**136**).

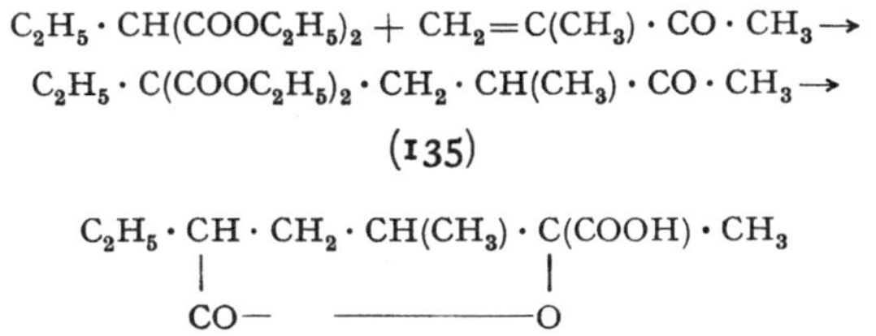

$$C_2H_5 \cdot CH(COOC_2H_5)_2 + CH_2{=}C(CH_3) \cdot CO \cdot CH_3 \rightarrow$$

$$C_2H_5 \cdot C(COOC_2H_5)_2 \cdot CH_2 \cdot CH(CH_3) \cdot CO \cdot CH_3 \rightarrow$$

$$(135)$$

$$C_2H_5 \cdot CH \cdot CH_2 \cdot CH(CH_3) \cdot C(COOH) \cdot CH_3$$

$$CO- \qquad\qquad -O$$

$$(136)$$

Sceleranecic and Sceleratinic Acids.

A new structure for sceleranecic acid has been advanced by DE WAAL, WIECHERS and WARREN (*385*), on the basis of a new method of degradation. Lithium aluminium hydride gave a glycol which was quantitatively oxidised with periodic acid to formaldehyde and 2,3-dimethyllaevulic acid (**139**), which with sodium hypobromite yielded butan-2,3-dicarboxylic acid, m. p. 132–134°, which was shown to be an optically active form from infrared data, but no rotation was assigned. Oxidation of disodium sceleranecate with lead tetraacetate gave the same acid (**139**) and the previously reported γ-lactone monocarboxylic acid (**140**), m. p. 100°. These results permit the formulation of sceleranecic acid as (**137**, $Y = CH_2OH$), the structures of the glycol (**138**), and the γ-lactone monocarboxylic acid (**140**). This new structure (**137**) readily accounts for the oxidation to a dilactone monocarboxylic acid (**137**, $Y = COOH$); the action of alkali to give an hydroxy-dicarboxylic acid (**141**, $X = OH$), which can be converted to a chloro-dicarboxylic acid (**141**, $X = Cl$) and sceleratinic acid (**137**, $Y = CH_2Cl$). These interconversions are shown in *Chart 26*.

The bicyclic dilactone structure was further substantiated by the increased carbonyl frequency of sceleranecic acid in the infrared. The ready opening of one lactone ring which was originally envisaged as due

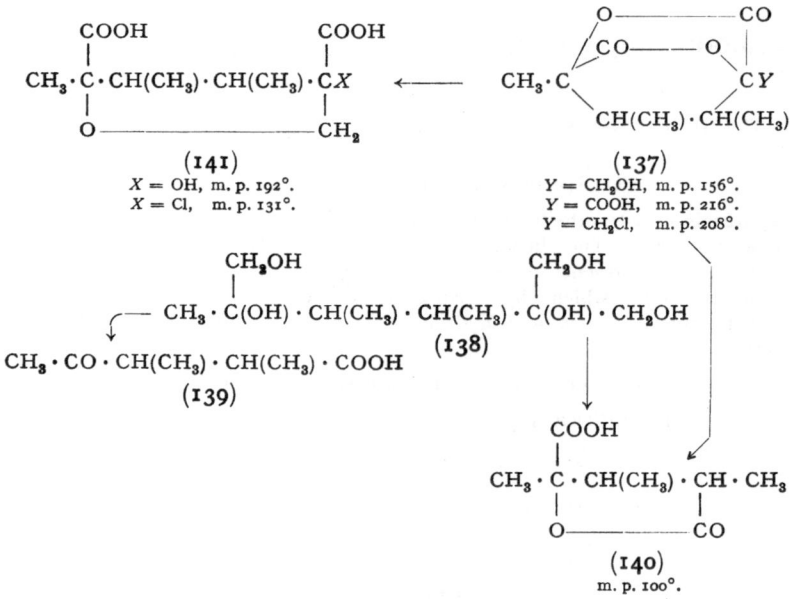

Chart 26. The Structure of Sceleranecic Acid, m. p. 156°, and Sceleratinic Acid, m. p. 208°, and their Reaction Products.

to a γ- and δ-lactone ring is interpreted as due to the instability of the "boat" six ring of the dilactone compound (**137**, $Y = CH_2OH$) to give the more stable "chair" form of the monolactone.

References, pp. 393—406.

These several structures for the C_{10}-adipic acids are coordinated in *Chart 27*. They show one main carbon skeleton predicted previously as a general formula by KROPMAN and WARREN (*105*) showing different degrees of hydroxylation and unsaturation. Sceleranecic acid has similar two C_5-units joined differently.

$$\overset{5}{CH_3} \cdot \overset{4}{CH} = \overset{3}{C(COOH)} \cdot \overset{}{CH}Y \cdot \overset{}{CH(CH_3)} \cdot \overset{2}{C(OH)(COOH)} \cdot \overset{1}{CH_2}X$$

$Y = H$	$X = H$	[2 *R*, 3 *R*]	*cis*	Senecic Acid.
		[2 *R*, 3 *R*]	*trans*	Integerrinecic Acid.
		[2 *S*, 3 *R*]	*trans*	Usaramoensinecic Acid.
$Y = H$	$X = OH$	[? *R*, 3 *R*]	*cis*	Isatinecic Acid.
		[? *R*, 3 *R*]	*trans*	Retronecic Acid.
$Y = OH$	$X = H$	[2 *R*, 3 *R*]	*cis*	Hygrophyllinecic Acid.

$$CH_3CH = C(COOH) \cdot CH_2 \cdot C(=CH_2) \cdot C(OH)(COOH) \cdot CH_2X$$

$X = H$	[2 *R*]	*cis*	Isoseneciphyllic Acid.
$X = H$	[2 *R*]	*trans*	Seneciphyllic Acid.
$X = H$	[2 *S*]	*trans*	Spartionecic Acid.
$X = OH$			Riddellic Acid.

$$CH_3 \cdot CHX — CY(COOH) \cdot CH_2 \cdot CH(CH_3) \cdot C(OH)(COOH) \cdot CH_3$$

—X—Y— = —O—	[2 *R*, 3 *R*]		"Jacobinecic Acid".
$Y = OH$	$X = OH$	[2 *R*, 3 *R*]	"Jacolinecic Acid".
$Y = OH$	$X = Cl$	[2 *R*, 3 *R*]	"Jaconinecic Acid".

$$CH_3 \cdot CH(OH) \cdot C(COOH) \cdot CH_2 \cdot CH(CH_3) \cdot C(COOH) \cdot CH_3$$
$$\underset{\text{O}}{\rule{4cm}{0.4pt}}$$

Jaconecic Acid*.

$$CH_3 \cdot CH \cdot CH(COOH) \cdot CH_2 \cdot CH(CH_3) \cdot C(COOH) \cdot CH_3$$
$$\underset{\text{O}}{\rule{4cm}{0.4pt}}$$

Isojaconecic Acid*.

$$CH_3 \cdot CH_2 \cdot C(COOH)_2 \cdot CH(CH_3) \cdot C(OH)(COOH) \cdot CH_3$$

Retusanecic Acid.

$$CH_3 \cdot C(OH)(COOH) \cdot CH(CH_3) \cdot CH(CH_3) \cdot C(OH)(COOH) \cdot CH_2X$$

Sceleranecic Acid. $X = OH$

Sceleratinic Acid. $X = Cl$

Chart 27. C_{10}-Acids from Pyrrolizidine Alkaloids.

---

* Not naturally occurring (see **Chart 21**, p. 348).

Sarracinic and Mikanecic Acids.

DANILOVA and her collaborators (*268*) have continued their studies on sarracine, whose acid hydrolysis they have shown to yield angelic and sarracinic acids. The latter acid gave with lead dioxide acetaldehyde and formaldehyde, and by hydrogenation a mixture of dihydrosarracinic acid and butan-2-carboxylic acid, which reactions can be interpreted on the basis of sarracinic acid having structure (**142 a**) or (**142 b**).

$$CH_3 \cdot CH = C(COOH) \cdot CH_2OH \rightleftharpoons CH_3 \cdot CH(OH) \cdot C(COOH) = CH_2$$

(**142a**)   (**142b**)

CULVENOR and GEISSMAN (*258*) considered that sarracinic acid in the alkaloid itself existed as (**142 a**) on the basis of the NMR spectra and the formation of only acetaldehyde on ozonolysis of the alkaloid.

The structure of mikanecic acid advanced by ADAMS and GIANTURCO (*204*) has been criticised by CULVENOR and GEISSMAN (*256, 258*) on the basis of its reported absorption at 216 mμ ($\varepsilon \pm$ 11000). They re-examined *Senecio mikanoides* and found only the alkaloid sarracine (and its N-oxide).

When sarracine was hydrolysed by alkali it gave sarracinic acid, $C_8H_8O_3$, and mikanecic acid, $C_{10}H_{12}O_4$, which was also formed by the action of alkali on sarracinic acid itself. Mikanecic acid was identified as a 4-vinylcyclohexene-1, 4-dicarboxylic acid (**143**) and its formation from sarracinic acid envisaged as proceeding by dehydration to a butadiene derivative followed by a Diels-Alder condensation.

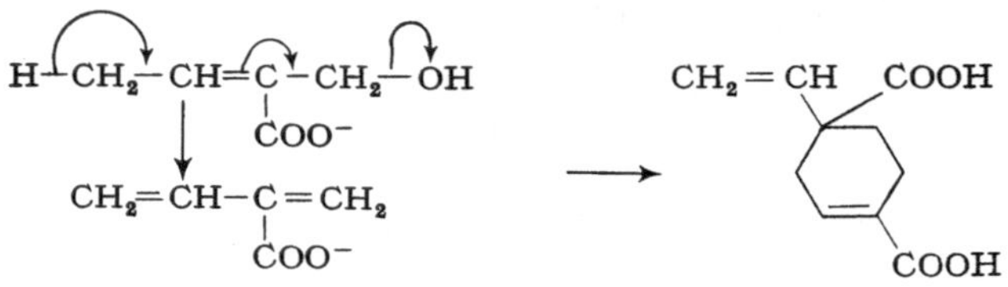

2. The Glutaric Acids.

Monocrotalic, Dicrotalic, Fulvinic and Crispatic Acids, and the Acids from Retusine.

Fulvinic acid, $C_8H_{14}O_5$, m. p. 114°, was obtained by SCHOENTAL (*369*) from fulvine isolated from the flowers of *Crotalaria fulva*. It was a di-carboxylic acid showing 2.4 methyl groups by Kuhn-Roth oxidation, furthermore, two equivalent units of the structure $>C\!-\!CH(CH_3)\!-\!C<$ from its NMR spectrum, and did not contain an α-hydroxyl by the ferric chloride test. This led to the structure 3-hydroxy-2,3,4-trimethyl-

References, pp. 393—406.

glutaric acid. A similar acid, crispatic acid, $C_8H_{14}O_5$, m. p. 133–134°, was later obtained by CULVENOR and SMITH (267) from the acid hydrolysis product of crispatine, and assigned a similar structure. Since both these acids showed no optical activity they were assigned structures (144) and (145).

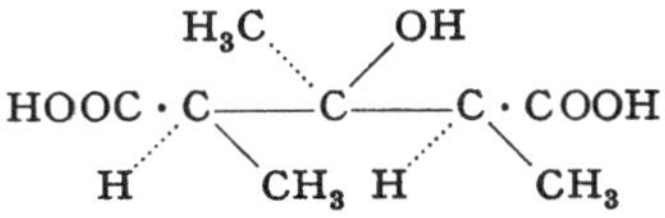

$$\underset{\text{H}}{\overset{\text{H}_3\text{C}}{\text{HOOC}\cdot\text{C}}}\overset{\text{OH}}{\underset{\text{CH}_3}{\text{—C—}}}\underset{\text{H}}{\text{—C}}\cdot\underset{\text{CH}_3}{\text{COOH}}$$

(144)
Fulvinic acid.

$$\underset{\text{H}}{\overset{\text{HO}}{\text{HOOC}\cdot\text{C}}}\overset{\text{CH}_3}{\underset{\text{CH}_3}{\text{—C—}}}\underset{\text{H}}{\text{—C}}\cdot\underset{\text{CH}_3}{\text{COOH}}$$

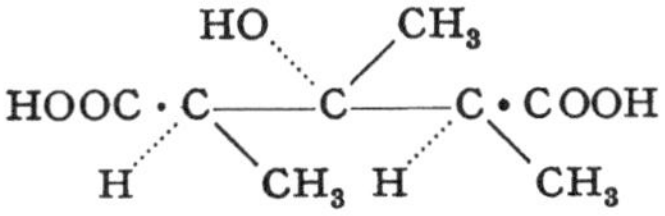

(145)
Crispatic acid.

CULVENOR and SMITH (261) obtained by the hydrolysis of retusine two isomeric acids, $C_8H_{12}O_4$, the α-acid, m. p. 130–131°, $[\alpha]_D + 3.3°$, and the β-acid, m. p. 118°, $[\alpha]_D - 60°$, which were identified by comparison with authentic specimens as epimeric dihydroanhydromonocrotalic acids (4-hydroxy-2,3,4-trimethylglutaric acid lactone) (146, $X = H$). The α-form was that present in the alkaloid. These structures

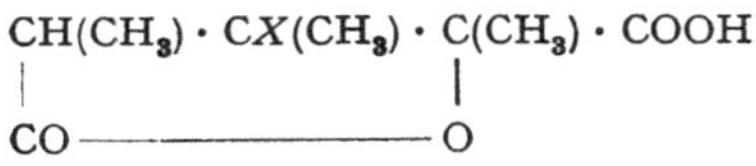

$$\underset{\text{CO}———————\text{O}}{\text{CH(CH}_3)\cdot\text{C}X\text{(CH}_3)\cdot\text{C(CH}_3)\cdot\text{COOH}}$$

(146)
Monocrotalic acid (X = OH).

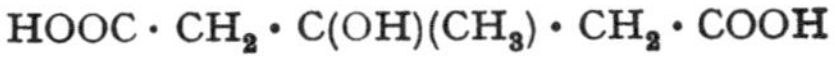

$$\text{HOOC}\cdot\text{CH}_2\cdot\text{C(OH)(CH}_3)\cdot\text{CH}_2\cdot\text{COOH}$$

(147)
Dicrotalic acid.

are of interest in relation to two other acids from *Crotalaria* spp., namely monocrotalic acid (146, $X = OH$), which also occurs as its β-acetyl derivative (146, $X = CH_3COO^-$) in spectabiline and dicrotalic acid (147).

Trichodesmic and Junceic Acids.

Trichodesmic acid, $C_{10}H_{16}O_5$, obtained from the catalytic hydrogenation of trichodesmine (145) was assigned structure (148, $X = H$, $Y = OH$) on the basis of the reactions of the alkaloid itself (p. 369) and the similarity of the alkaline fission products, methyl isobutyl ketone and ($\pm$)-lactic acid, with those from monocrotalic acid, namely, methyl ethyl ketone.

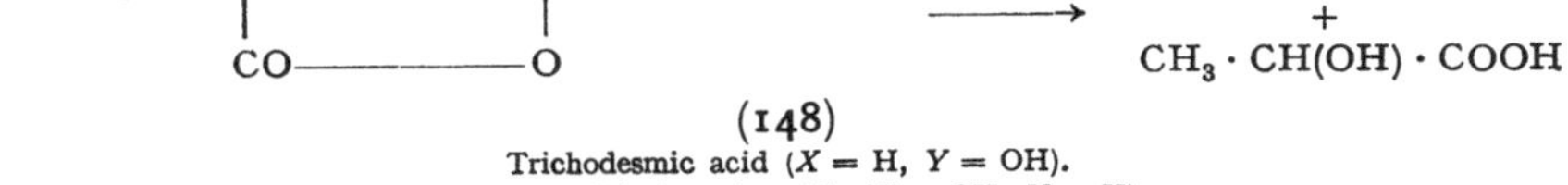

$$\underset{\text{CO}———————\text{O}}{\text{(CH}_3)_2\text{CH}\cdot\text{CH}\cdot\text{C}Y\text{(CH}_3)\cdot\text{C(COOH)}\cdot\text{CH}_2X} \longrightarrow \underset{\text{CH}_3\cdot\text{CH(OH)}\cdot\text{COOH}}{\overset{\text{(CH}_3)_2\text{CH}\cdot\text{CO}\cdot\text{CH}_3}{+}}$$

(148)
Trichodesmic acid (X = H, Y = OH).
Incanic and isoincanic acids (X = OH, Y = H).
Junceic acid (X = Y = OH).

Comparative oxidation studies on trichodesmine (p. 369) and mono-crotaline by ADAMS and GIANTURCO (200) have been interpreted as indicating a *trans* glycol group in trichodesmine, whilst YUNUSOV and PLEKHANOVA (306) assign a *cis* configuration to the two hydroxyl groups on the basis of the formation of a sulphinic ester of trichodesmine (p. 369).

Junceic acid, $C_{10}H_{16}O_6$, obtained by ADAMS and GIANTURCO (201) by the hydrogenation of junceine, was a monobasic acid containing a five-membered lactone ring. When junceine was treated with periodic acid two mole equivalents were taken up to give formaldehyde so that there were three vicinal hydroxyl groups terminating in a primary hydroxyl group. Hydrolysis of the alkaloid gave methyl isobutyl ketone and the structure assigned was (148, $X = Y = OH$).

Incanic acid, $C_{10}H_{16}O_4$, m. p. 163°, and isoincanic acid, m. p. 122–123.5°, were two readily interconvertible isomeric acids obtained from the hydrolysis of incanine. They gave on oxidation with chromic acid in sulphuric acid 2.3 mols of acetic acid and acetone indicative of one $C(CH_3)_2$ and two $C—CH_3$ groups. Reduction of the two acids gave two isomeric glycols, the oxidation of which corresponded to 2,3-dimethyl-4-isopropyl-1,2,5-pentantriol, and incanic and isoincanic acids were assigned the structure (148, $Y = H$, $X = OH$).

Grantianic Acid and "Retusaminecic" Acid.

The structure of grantianic acid as the lactone of (149, $R = COOH$) proposed by ADAMS and GIANTURCO (202) rested on the presence of a γ-lactone group, one hydroxyl group and its similarity to the acid (149, $R = CH_3$) from trichodesmine. Such a structural relationship is found between latifolic acid (161, p. 363) on the one hand, and viridifloric and trachelanthic acids on the other. WUNDERLICH (389), however, suggests the structure of grantianic acid as (150, $X = OH$) as equally possible, since it would differ by only one OH group from the acid present in retusamine (150, $X = H$) (389).

$$CH_3 \cdot CH \cdot R$$
$$|$$
$$HOOC \cdot CH \cdot C(OH)(CH_3) \cdot C(OH)(COOH) \cdot CH_3$$
$$(149)$$

$$CH_3 \cdot CH_2 \cdot C(COOH) \cdot CX(CH_3) \cdot C(COOH) \cdot CH_3$$
$$| \qquad\qquad\qquad |$$
$$CO————————————O$$
$$(150)$$

C_7-Acids: Hydroxylated 2-Methylpentane-3-carboxylic Acids.

Two new acids belonging to the 2-methylpentane-3-carboxylic acid group are echimidinic and latifolic acids. Echimidinic acid, $C_7H_{14}O_5$, obtained by CULVENOR (249) directly by hydrogenolysis of echimidine, and isolated as its brucine salt, was oxidised with periodic acid to give acetaldehyde, acetone and oxalic acid and assigned the structure 2,3,4-trihydroxy-2-methylpentane-3-carboxylic acid (**151**). This structure had previously been assigned by MEN'SHIKOV and PETROVA (*144*) to the unisolated "macrotomic acid" from macrotomine, and by DENISOVA, MEN'SHIKOV and UTKIN (*276*) to the acid from heliosupine. DENISOVA, PETROVA and MEN'SHIKOV (*277*) showed that macrotomine gave di-hydroxybutyric acid (**152**) and acetone on alkaline hydrolysis more readily than heliosupine. Furthermore, MAN'KO and BORISYUK (*338*) gave the structure (**151**) to the acid from the optically inactive alkaloid cynoglossophine.

$$\overbrace{\text{—CH}_3 \cdot \text{CH(OH)} \cdot \text{C(OH)(COOH)} \cdot \text{C(OH)(CH}_3)_2\text{—}}$$

$$(151)$$
Echimidinic acid.

$$\text{CH}_3 \cdot \text{CH(OH)} \cdot \text{CH(OH)} \cdot \text{COOH} + \text{CO(CH}_3)_2 + \text{CH}_3 \cdot \text{CHO} + (\text{COOH})_2 + \text{CO(CH}_3)_2$$

$$(152)$$

The lasiocarpic (**153**, $R = \text{CH}_3$), macrotomic (**153**, $R = \text{H}$) and echimidinic (**153**, $R = \text{H}$) acids all have tertiary hydroxyl groups in the β-position, and alkaline hydrolysis of their esters results in a retroaldol reaction to give acetone, and hydroxymethoxy or dihydroxybutyric acid. Lasiocarpic acid itself is moderately stable to alkali whilst the ester rapidly yields acetone and a substance $\text{C}_5\text{H}_{10}\text{O}_4$ (*351*). Acid hydrolysis which effected decomposition without resinification, has formed the subject of a careful study by CROWLEY and CULVENOR (*247*) and is shown in *Chart 28*. They have established that lasiocarpic acid with concentrated hydrochloric acid at 100° yields acetaldehyde, dimethylpyruvic acid (**154**) and α-keto-β,β-dimethyl-γ-valerolactone (2-keto-3,3-dimethyl-3-hydroxybutanoic acid lactone) (**157**).

The initiating step is envisaged as the elimination of the β-hydroxyl group to give the carbonium ion (**155**) which could yield 2,3-dimethyl-butan-1-on-3-ol-1-carboxylic acid (**156**), by way of dimethylpyruvic acid (**154**) and acetaldehyde, and then lactonise to (**157**). The alternative route to (**157**) or to dimethylpyruvic acid and acetaldehyde by a pinacol rearrangement to (**156**) is not favoured since (**157**) would then be expected to be optically active.

Echimidinic acid (**153**, $R = \text{H}$) was shown to undergo a similar series of reactions even more rapidly and precludes any possibility of demethyla-ting lasiocarpic acid to echimidinic acid or its diastereoisomer. This sensitivity to concentrated hydrochloric acid is in contrast to the recovery of trachelanthic (**162**) and heliotrinic (**160**) acids after treatment with concentrated hydrochloric acid at 100°.

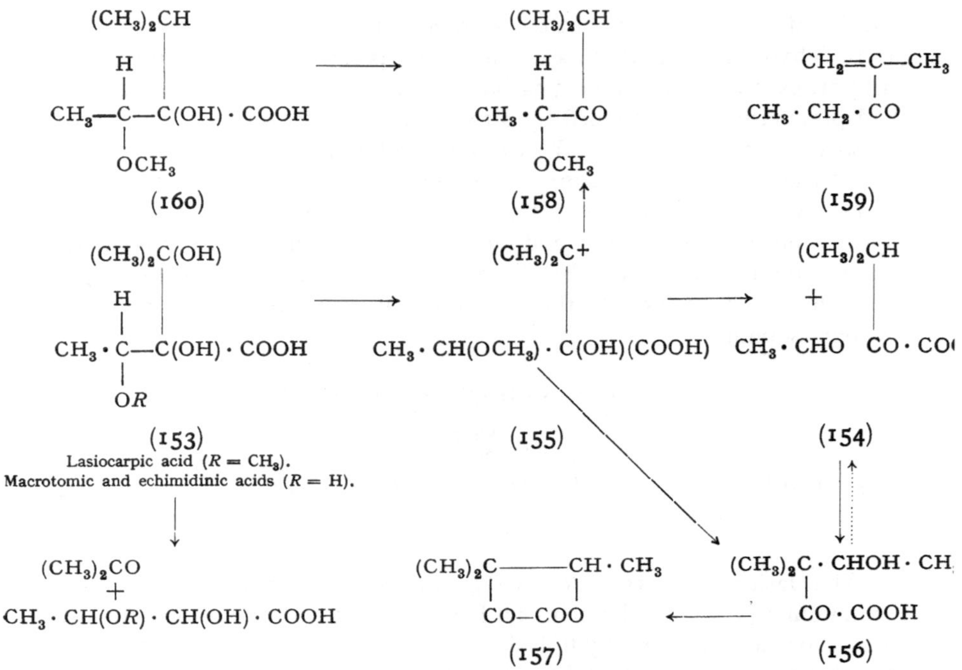

$(CH_3)_2CH$

$\underset{\underset{OCH_3}{|}}{\overset{\overset{H}{|}}{CH_3{-}C{-}C(OH)\cdot COOH}}$

(160)

$\longrightarrow$

$(CH_3)_2CH$

$\underset{\underset{OCH_3}{|}}{\overset{\overset{H}{|}}{CH_3\cdot C{-}CO}}$

(158) $\uparrow$

$\overset{\overset{CH_2{=}C{-}CH_3}{|}}{CH_3\cdot CH_2\cdot CO}$

(159)

$(CH_3)_2C(OH)$

$\underset{\underset{OR}{|}}{\overset{\overset{H}{|}}{CH_3\cdot C{-}C(OH)\cdot COOH}}$

(153)
Lasiocarpic acid $(R = CH_3)$.
Macrotomic and echimidinic acids $(R = H)$.

$\longrightarrow$

$\overset{\overset{(CH_3)_2C+}{|}}{CH_3\cdot CH(OCH_3)\cdot C(OH)(COOH)}$

(155)

$\longrightarrow$

$(CH_3)_2CH$

$+$

$CH_3\cdot CHO \quad CO\cdot CO($

(154)

$(CH_3)_2CO$
$+$
$CH_3\cdot CH(OR)\cdot CH(OH)\cdot COOH$

$\underset{\underset{CO{-}COO}{|\qquad|}}{(CH_3)_2C{-}\!\!-\!\!-\!\!-CH\cdot CH_3}$

(157)

$\underset{\underset{CO\cdot COOH}{|}}{(CH_3)_2C\cdot CHOH\cdot CH}$

(156)

Chart 28. Effect of Acid and Alkali on Lasiocarpic, Echimidinic and Macrotomic Acids.

Treatment of lasiocarpic acid (153, $R = CH_3$) with 2 N hydrochloric acid at 100° led to acetone, acetaldehyde, dimethylpyruvic acid (154), (+)-2-methoxy-4-methylpentan-3-one (158), 2-methylpent-1-en-3-one (159) and a compound A $(C_6H_{12}O_2)$. The formation of (158) from (155) is a process of decarboxylation favoured in dilute acid in which the methoxy group is retained. The isolation of (158) in the same enantiomeric form as that from heliotrinic acid (160) shows that heliotrinic and lasiocarpic acids have the same absolute configuration at the carbon atom carrying the methoxy groups.

Latifolic Acid.

Latifolic acid, $C_7H_{10}O_5$, obtained by catalytic hydrogenation of latifoline, contained a γ-lactone ring (λ_{max} 1800 cm^{-1}), and gave on Kuhn-Roth oxidation 1.9 C-methyl. On the basis of the C_7-skeleton of the acids in the Boraginaceae, as well as NMR and IR spectra, it was assigned the structure 3,4-dihydroxypentan-2,3-dicarboxylic acid lactone (161). The stereochemistry is yet unknown, and it is of interest in that the compound contains a carboxyl group in the place of a methyl group of viridifloric and tracheanthic acids (162).

References, pp. 393—406.

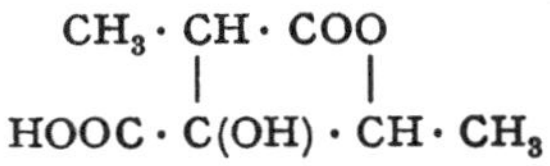

$$CH_3 \cdot CH \cdot COO$$
$$HOOC \cdot C(OH) \cdot CH \cdot CH_3$$

(161)

Latifolic acid.

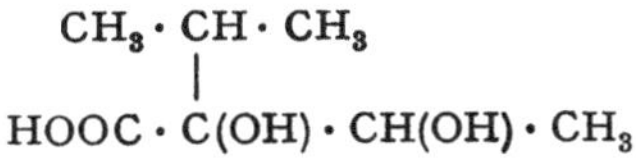

$$CH_3 \cdot CH \cdot CH_3$$
$$HOOC \cdot C(OH) \cdot CH(OH) \cdot CH_3$$

(162)

Viridifloric and trachelanthic acids.

Heliotramide.

The preparation of the *erythro* diastereoisomeric form (163, $R = OH$) of heliotric acid has already been described (66), but not that of heliotric acid itself which is known to be the *threo*-form (26). A stereo non-specific synthesis of ($\pm$)-heliotramide has now been reported by ADAMS, CULVENOR, ROBINSON and STINGL (196). Isopropylmagnesium chloride reacted with 2-methoxypropionitrile (164) to give 2-methoxy-4-methylpentan-3-one (165) which was treated with hydrocyanic acid. The resulting cyanhydrin gave on hydrolysis the two racemates, m. p. 81–82° and 145–146°, identified as ($\pm$)-*threo*-2-methoxy-4-methylpentan-2-carboxylic amide [($\pm$)-heliotramide] (166) by comparison of the IR spectrum with that of (+)-heliotramide, and the ($\pm$)-*erythro* amide (163, $R = NH_2$), respectively. The occurrence of a bonded CO in the IR spectrum of the latter was further proof of the structures assigned.

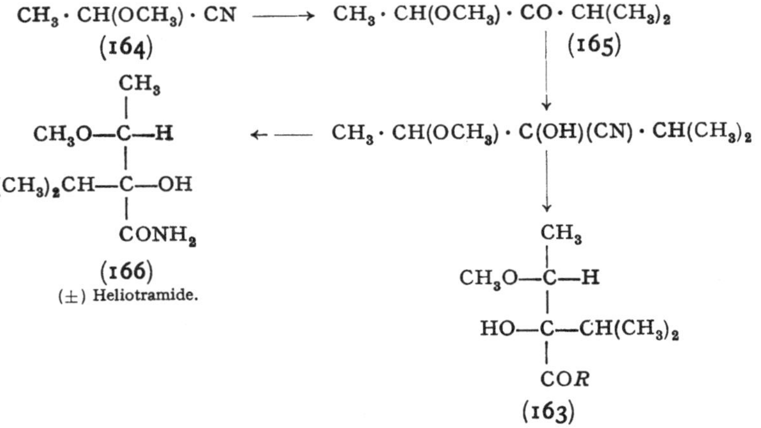

$$CH_3 \cdot CH(OCH_3) \cdot CN \longrightarrow CH_3 \cdot CH(OCH_3) \cdot CO \cdot CH(CH_3)_2$$

(164)   (165)

$$CH_3 \cdot CH(OCH_3) \cdot C(OH)(CN) \cdot CH(CH_3)_2$$

$$\begin{array}{c} CH_3 \\ | \\ CH_3O—C—H \\ | \\ (CH_3)_2CH—C—OH \\ | \\ CONH_2 \end{array}$$

(166)

($\pm$) Heliotramide.

$$\begin{array}{c} CH_3 \\ | \\ CH_3O—C—H \\ | \\ HO—C—CH(CH_3)_2 \\ | \\ COR \end{array}$$

(163)

The close relationships between the acids of this group is clearly shown in the following structures:

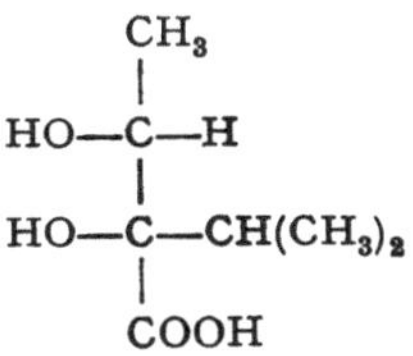

$$\begin{array}{c} CH_3 \\ | \\ HO—C—H \\ | \\ HO—C—CH(CH_3)_2 \\ | \\ COOH \end{array}$$

Viridifloric acid.

$$\begin{array}{c} CH_3 \\ | \\ RO—C—H \\ | \\ (CH_3)_2CH—C—OH \\ | \\ COOH \end{array}$$

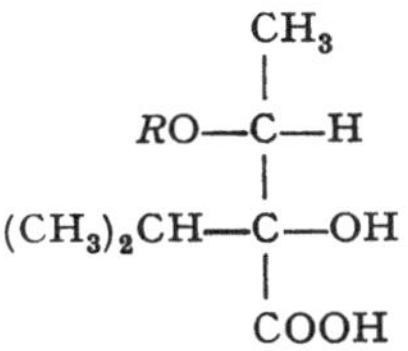

Trachelanthic acid. $R = H$.
Heliotrinic acid. $R = CH_3$.

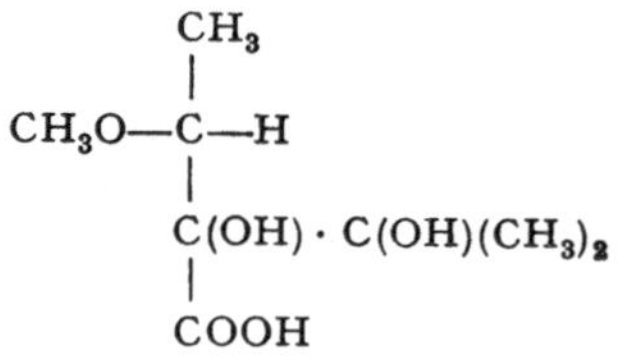

$$CH_3$$
$$|$$
$$CH_3O—C—H$$
$$|$$
$$C(OH) \cdot C(OH)(CH_3)_2$$
$$|$$
$$COOH$$

Lasiocarpic acid.

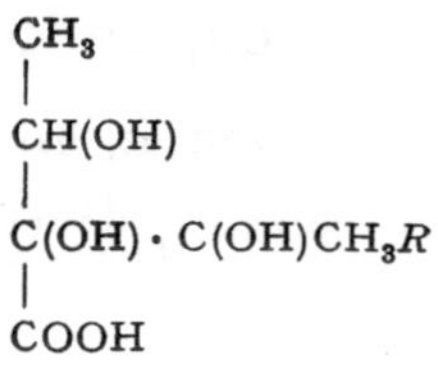

$$CH_3$$
$$|$$
$$CH(OH)$$
$$|$$
$$C(OH) \cdot C(OH)CH_3R$$
$$|$$
$$COOH$$

Macrotomic acid. $R = CH_3$.
Echimidinic acid. $R = CH_3$.
Latifolic acid. $R = COOH$.

The acid from strigosine, isolated by MATTOCKS (*343*) and shown to be identical with 2,3-dihydroxy-3-methylpentanoic acid (**167**), was synthesised by SJOLANDER, FOLKERS, ADELBERG and TATUM (*380*); it is the only acid of this particular carbon skeleton.

$$CH_3 \cdot CH_2 \cdot C(OH)(CH_3) \cdot CH(OH) \cdot COOH \qquad (\text{167})$$

IV. The Structure of the Native Alkaloids.

The structures for the pyrrolizidine alkaloids rest logically on those assigned to the acids and bases obtained on hydrolysis. In several instances, however, the acid hydrolysis product is the isomeric form of the acid in the alkaloid, e. g., jaconecic acid from jacobine, or has not been obtained owing to degradation under the conditions of hydrolysis, e. g., macronecic acid. Experiments on, and spectral data of, the alkaloid itself have permitted the determination of structural features not revealed by hydrolysis. There are, however, two possible structures for a dihydroxy-amine esterified by either two acids (**168**) or a dicarboxylic acid (**169**). A decision has been reached in these cases by studying the products of hydrogenolysis: compound (**168**) would liberate the acid, $R \cdot COOH$; and retrorsine (**169**) results in a half ester with a free hydroxyacid group (**110**). The similarity of the carbon skeletons for the different groups of acids would seemingly justify assigning analogous orientations where these have not been specifically studied.

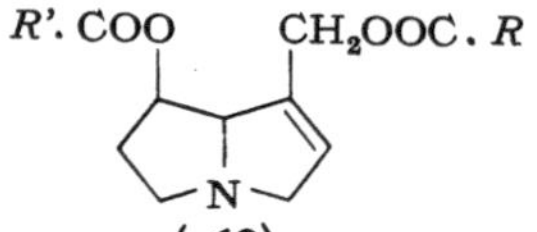

$R' \cdot COO$ $CH_2OOC \cdot R$

(168)

$CH_3 \cdot CH{:}C \cdot CH_2 \cdot CH(CH_3) \cdot C(OH) \cdot CH_2OH$
$|$
CO ... CO
$|$ ... $|$
O ... $CH_2 \cdot O$

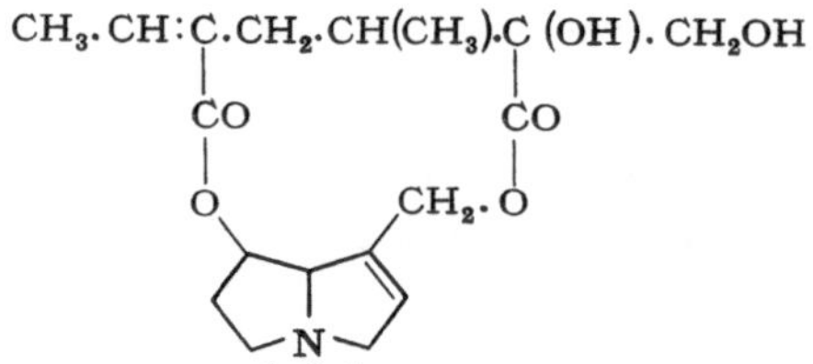

(169)

With the many structures now established an attempt is made to divide the alkaloids into groups according to the type of acid. The stereoisomerism of the different acids is not shown and reference to this must be made to Chapter III, p. 346.

References, pp. 393—406.

1. The Alkaloids Containing the C_{10}-Adipic Acids.

Seneciphylline.

DANILOVA and KORETSKAYA (269) have reduced seneciphylline with lithium aluminium hydride to obtain a triol which was characterised as a p-nitrobenzoate, m. p. 128–129°, $[\alpha]_D$ — 6.35°, identical with that obtained by a similar reduction of dimethyl cis-seneciphyllate. The trans-ester gave no solid ester and the authors conclude that seneciphyllic acid is in the cis form in the ester.

Spartioidine.

ADAMS and GIANTURCO (205) showed that spartioidine gave on hydrolysis retrorsine, and absorbed four mole equivalents of hydrogen to give a zwitterion. Since in hydrogenolysis retronecine esters give retronecanol with the absorption of two mole equivalents, the unisolated acid moiety "spartioidinecic acid", $C_{10}H_{14}O_5$, must have two double bonds.

Ultraviolet and Infrared Spectra in the Study of Geometrical and Stereo-isomerism: Seneciphylline, Spartioidine, Senecionine, Integerrimine and Usaramoensine.

The comparison of the infrared spectra of seneciphylline and spartioidine with those of senecionine, integerrimine and usaramoensine has permitted the formulation of the structure of spartioidine and certain deductions concerning the geometric configuration of the ethylidene group and the asymmetry at $C_{(2)}$. ADAMS and GIANTURCO (205) took into consideration the concept of CASON and KALM (237) that the cis form of crotonic acid with a bulky α-constituent (170) would be less hindered, that is more planar, and would have the higher extinction coefficient than the trans form. They envisage that this did not apply to the alkaloids, the more stable form of which would be trans (171), owing to the α-substituent being tied in the ring structure*.

---

* The terms cis and trans in the necic acids have been used to illustrate a similarity of structure to the angelic acid *(cis)*

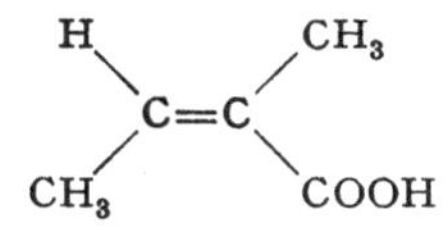

$$\underset{CH_3}{\overset{H}{\diagdown}} C = C \underset{COOH}{\overset{CH_3}{\diagup}}$$

and tiglic acid *(trans)*

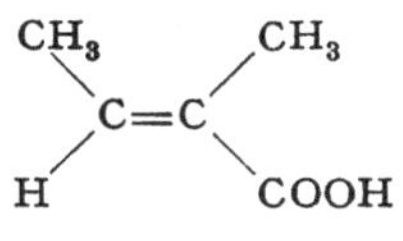

$$\underset{H}{\overset{CH_3}{\diagdown}} C = C \underset{COOH}{\overset{CH_3}{\diagup}}$$

so that the necic acid

$$\underset{CH_3}{\overset{H}{\diagdown}} C = C \underset{COOH}{\overset{CH_2\cdots}{\diagup}}$$

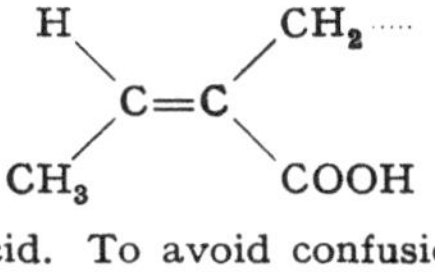

has been referred to as a cis acid. To avoid confusion in relating names used in this review to the literature names, this nomenclature has been retained but with

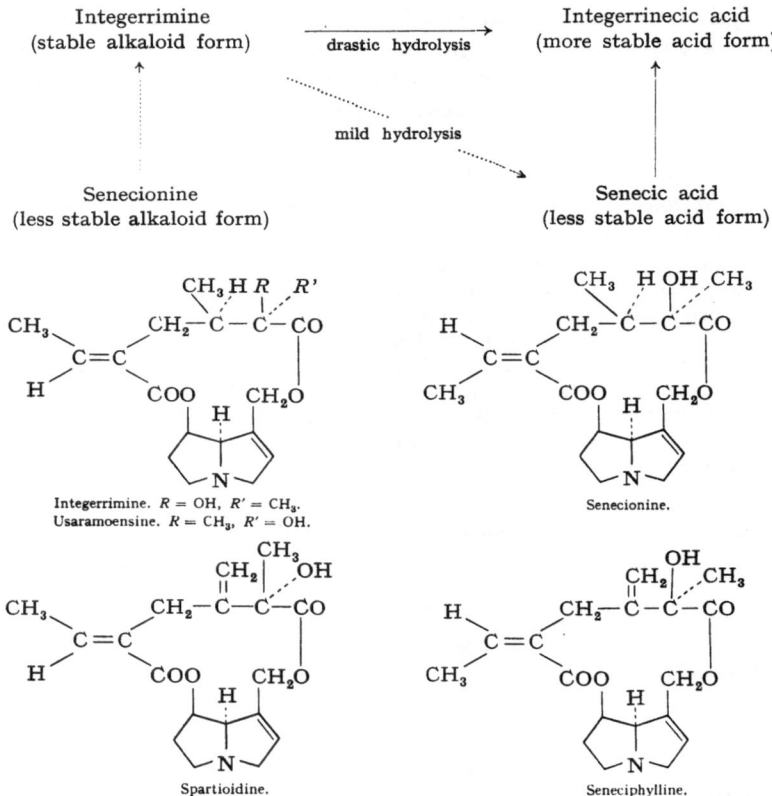

$$\begin{array}{ccc}
& C & \\
H & \vdots & \\
\diagdown & C\!-\!\!\!\!- & \\
C\!=\!C & & \\
\diagup & \diagdown & \\
CH_3 & COOH & \\
& (170) &
\end{array}$$

This study led to the conclusion that usaramoensinic acid had the same geometrical form as in the alkaloid whilst integerrinecic acid was different. This is contrary to the observation that senecic acid is the labile and integerrinecic acid the stable form (*105*). The isomerism exhibited on mild hydrolysis of senecionine is then envisaged as proceeding by way of the stable integerrimine before hydrolysis to the less stable senecic acid:

Integerrimine
(stable alkaloid form)

→ drastic hydrolysis

Integerrinecic acid
(more stable acid form)

mild hydrolysis

Senecionine
(less stable alkaloid form)

Senecic acid
(less stable acid form)

Integerrimine. *R* = OH, *R′* = CH₃.
Usaramoensine. *R* = CH₃, *R′* = OH.

Senecionine.

Spartioidine.

Seneciphylline.

Chart 29. The Absolute Structure of Integerrimine, Senecionine, Usaramoensine, Spartioidine and Seneciphylline.

the inclusion of the terms in parentheses, "*cis*" and "*trans*". It is contrary to IUPAC rules which relate *cis* and *trans* to the longest chain and the latter nomenclature must be used in defining fully the absolute structures.

References, pp. 393—406.

Hydrogen bonding with the carbonyl group of the $C_{(2)}$-hydroxyl group observed by LEISEGANG and SCHÜLER (*327*) has been successfully employed (*205*) for configuration studies: the broad carbonyl band observed in usaramoensine and spartioidine necessitates a different arrangement at $C_{(2)}$ from that for integerrimine, senecionine and seneciphylline. These findings may now be put together with absolute configuration in regard to $C_{(3)}$, defined by the isolation of ($\pm$)-methylsuccinic acid (*49*) as an oxidation product, the absolute configuration of retronecine (*388*) and X-ray crystallographic studies (*341*) to show the structure set out in *Chart 29*. A more comprehensive study covering fifteen alkaloids has been made by CULVENOR and BON (*253*).

Retusamine, Otosenine (Tomentosine), Renardine and Onetine.

The alkaloid retusamine, a minor constituent of *Crotalaria retusa* was isolated by CULVENOR and SMITH (*261*), and preliminary work suggested that it was related to otosenine (*195*) which was shown (*376*) to be identical with tomentosine (*207*). Furthermore, otosenine, renardine (*270*) and onetine (*271*) are esters of the same aminoalcohol, otonecine, $C_9H_{15}NO_3$. Using single crystal X-ray diffraction analysis, WUNDERLICH (*389*) showed that retusamine-α'-bromo-*d*-camphor-*trans*-π-sulphonate monohydrate had the structure (**172**). Furthermore, the published chemical and spectral data enabled WUNDERLICH to assign the structures otosenine (tomento-

sine) (**173,** $R = \overset{O}{\underset{/}{>}}C{-}{-}{-}CH \cdot CH_3$), renardine (**173,** $R = \overset{}{\underset{/}{>}}C{=}CH \cdot CH_3$), and onetine (**173,** $R = \overset{}{\underset{/}{>}}C(OH) \cdot CH(OH) \cdot CH_3$).

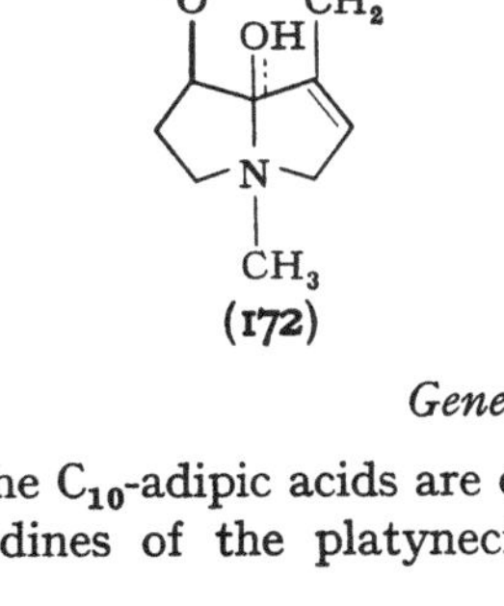

(**172**) 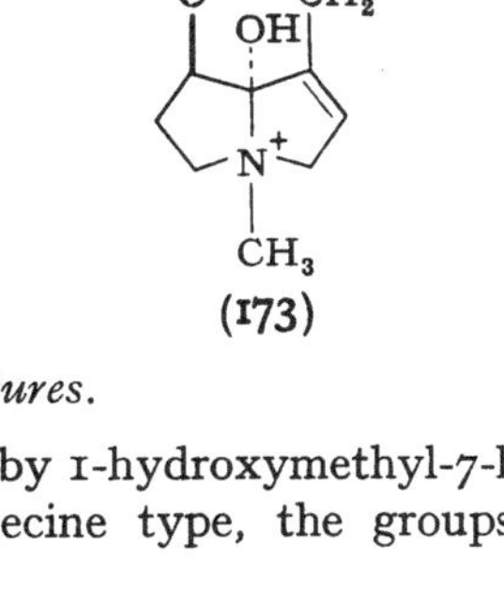   (**173**)

General Structures.

All the C_{10}-adipic acids are esterified by 1-hydroxymethyl-7-hydroxy-pyrrolizidines of the platynecine-retronecine type, the groups at $C_{(1)}$

and $C_{(7)}$ being *trans* to $C_{(8)}$-hydrogen, namely retronecine, platynecine, rosmarinecine and othonecine (see *Table 16*, p. 392).

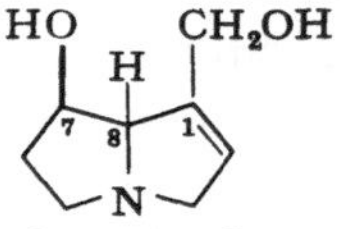

Retronecine. (R)

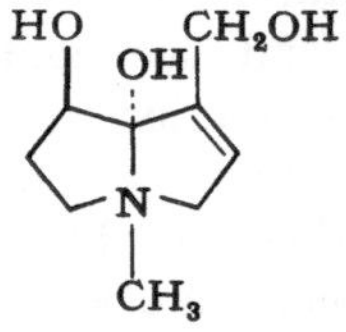

Otonecine. (O)

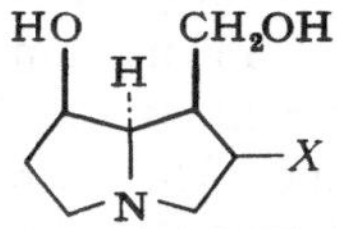

Platynecine: $(X = H)$ (P).
Rosmarinecine: $(X = OH)$ (Ros).

Sarracine.

Those alkaloids which possess an acid with the carbon skeleton $C—C—C(COOH)\cdot C—C(C)\cdot C(COOH)\cdot C$ are of interest in relation to that of sarracine (**174**) in which the sarracinic acid is placed on the primary hydroxyl group of platynecine, since this acid is more readily hydrolysed than the angelic acid (*258*).

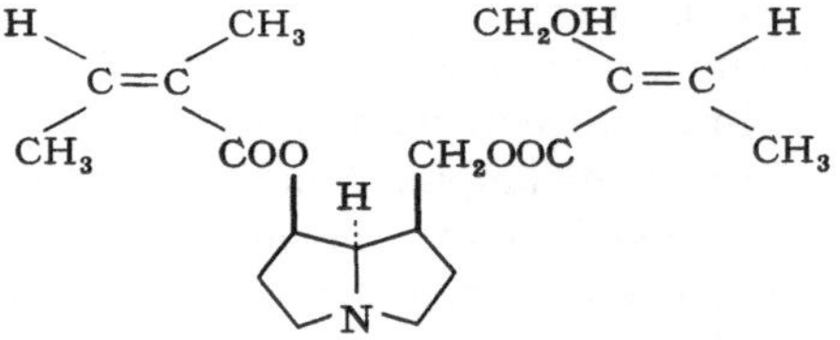

(**174**)

Sarracine.

2. Dicrotaline and the Alkaloids Containing Trimethyl Glutaric Acids.

Spectabiline (Acetyl monocrotaline).

Spectabiline, $C_{18}H_{25}NO_7$, which was found and studied by CULVENOR and SMITH (*262*) with monocrotaline in *C. spectabilis*, gave on alkaline hydrolysis retronecine and an oily acid mixture, whilst acid hydrolysis yielded monocrotalic acid or monocrotaline according to the conditions.

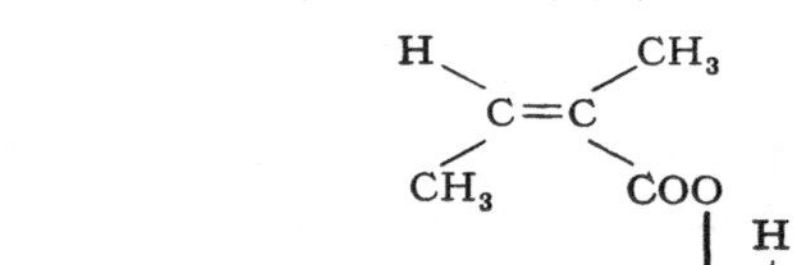

(**175**)

Acetylmonocrotalic acid $(R = CH_3CO)$.

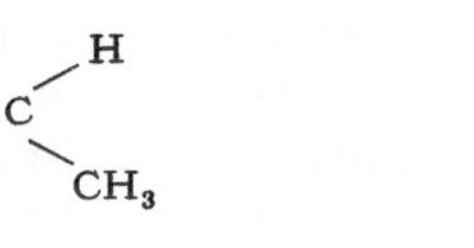

(**176**)

Anhydromonocrotalic acid.

Hydrogenation of spectabiline in very dilute acid solutions gave an amino acid which by careful hydrolysis with 2 N sulphuric acid gave retronecanol and acetylmonocrotalic acid (**175**, $R = CH_3 \cdot CO$) indicative of the acetyl group at the β-position. More drastic hydrolysis after hydrogenation gave anhydromonocrotalic acid (**176**), which is not formed from mono-crotalic acid (**175**, $R = H$) itself and was further confirmation of the structure.

The alkaloids within this group are shown in *Table 17*.

Table 17. The Pyrrolizidine Alkaloids Containing
2,3,4-Trimethylglutaric Acid Skeleton.
(R = Retronecine, Has = Hastanecine.)

Alkaloid	Base	$HOOC \cdot CH(CH_3) \cdot CX(CH_3) \cdot CY(CH_3) \cdot COOH$		
Fulvine	R	Fulvinic acid	$X = OH$	$Y = H$
Crispatine	R	Crispatic acid	$X = OH$	$Y = H$
Monocrotaline	R	Monocrotalic acid	$X = OH$	$Y = OH$
Retusine	Has	Retusinecic acid	$X = OH$	$Y = OH$
Spectabiline	R	Acetylmonocrotalic acid	$X = OOCCH_3$	$Y = OH$

Dicrotaline may conveniently be included within this group of alkaloids as it contains retronecine and 3-hydroxy-3-methylglutaric acid, $HOOC \cdot CH_2 \cdot C(OH)(CH_3) \cdot CH_2 \cdot COOH$.

3. Alkaloids Containing Acids with the 2,3-Dimethyl-4-isopropylglutaric Acid Skeleton.

Trichodesmine, Junceine, Incanine and Grantianine.

Trichodesmine, $C_{18}H_{27}NO_6$, from *Trichodesma incana* was shown (*145*) to contain two hydroxyl groups and to hydrolyse to methylisobutyl ketone, $(\pm)$-lactic acid and a base, shown to be retronecine (*101*). ADAMS and GIANTURCO (*200*) reduced trichodesmine catalytically to give a salt-like tetrahydroderivative, $C_{18}H_{31}NO_6$, m. p. 182°, which gave trichodesmic acid, $C_{10}H_{16}O_5$, m. p. 209–211°, and retronecanol. A comparative study of the action of periodic acid on retrorsine [grouping $— C(OH)(COOR) \cdot \cdot CH_2OH$], monocrotaline [grouping $— C(OH)(CH_3) \cdot C(OH)(COOR) \cdot \cdot CH_3$] and trichodesmine showed that two mole equivalents were absorbed in 3, 25 and 45 minutes, respectively. The similarity of IR spectra of trichodesmine and monocrotaline and of their respective acids, led to formulation of the structure for trichodesmine (**177**, p. 370).

The arrangement of the glycol groupings *trans* in trichodesmine as distinct from the *cis* arrangement in monocrotaline is in accordance with the slower rate of action of periodic acid and the action of thionyl chloride to give an acid sulphite ester, $C_{18}H_{27}NO_8S \cdot HCl$, whilst monocrotaline gave a neutral cyclic sulphite. On the other hand, YUNUSOV and PLEKHANOVA (*306*) treated trichodesmine with thionyl chloride to obtain a sulphinic ester, $C_{18}H_{25}NO_7S$, m. p. 151–152°, which on catalytic reduction gave retronecanol and trichodesmic acid. They assigned a structure similar to that of ADAMS but with a *cis* configuration for the two hydroxyl groups to account for the sulphinic group (**178**).

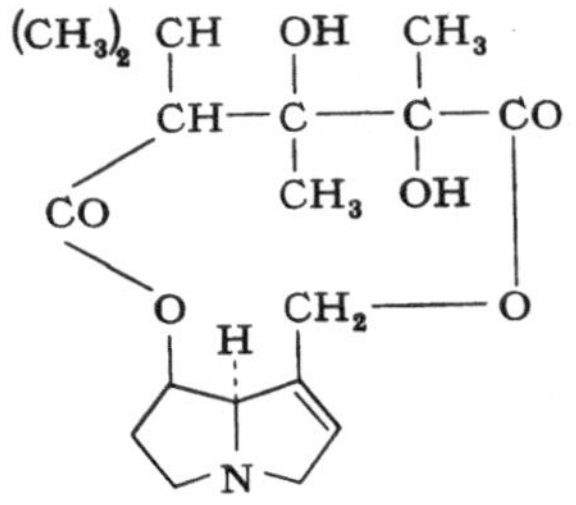

(177)
Trichodesmine.

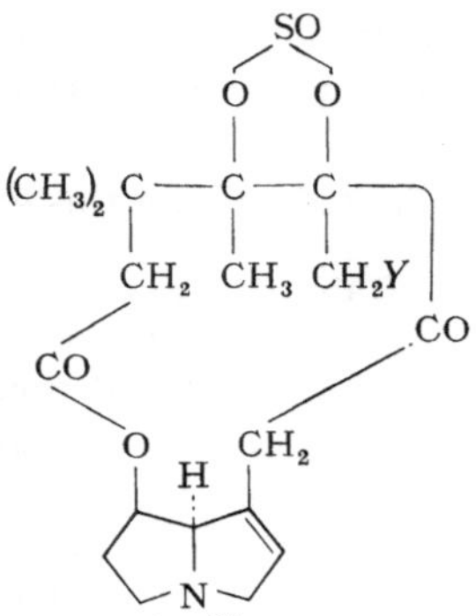

(178)

Incanine, $C_{18}H_{29}NO_5$, from *Trichodesma incanum* was studied by YUNUSOV and PLEKHANOVA who showed that it gave on hydrolysis retronecine and two isomeric acids, incanic acid, m. p. 161–163° (*306*), and isoincanic acid (*307*). Catalytic hydrogenation yielded in acid solution retronecanol and incanic acid, and in alcoholic solution a solid, m. p. 156°, soluble in ammonia. This latter substance on treatment with acid decomposed to incanic acid and retronecanol probably by intramolecular rearrangement. When incanine was acetylated (**179**, $R = CH_3 \cdot CO$, $X = OH$, $Y = H$) and catalytically reduced it gave an amino acid, m. p. 194–195°, presumably (**180**, $R = CH_3 \cdot CO$) which on hydrolysis gave acetic acid, incanic acid and retronecanol. The protection of the hydroxyl group or the reduction of the alkaloid in neutral solution seemingly precluded lactonisation and the total fission of the alkaloid.

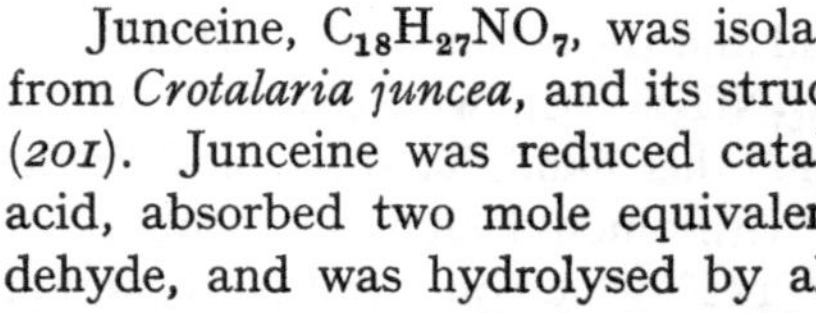

(179)

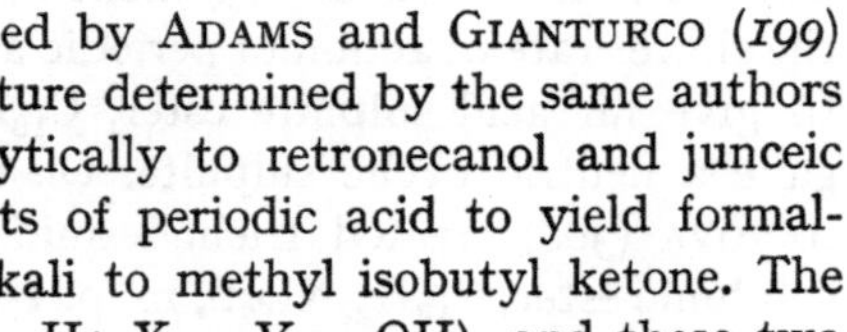

(180)

Junceine, $C_{18}H_{27}NO_7$, was isolated by ADAMS and GIANTURCO (*199*) from *Crotalaria juncea*, and its structure determined by the same authors (*201*). Junceine was reduced catalytically to retronecanol and junceic acid, absorbed two mole equivalents of periodic acid to yield formaldehyde, and was hydrolysed by alkali to methyl isobutyl ketone. The structure was accordingly (**179**, $R = H$; $X = Y = OH$), and these two alkaloids are another example of the occurrence of the grouping $C(OH)(COOH) \cdot CH_2X$, where X is H or OH.

References, pp. 393—406.

Table 18. Pyrrolizidine Alkaloids Containing 2,3-Dimethyl-4-isopropyl-glutaric Acid Skeleton.

$$\begin{array}{ccc} (CH_3)_2C & CH_3 & CH_2Z \\ | & | & | \\ HOOC \cdot CH\!-\!CX\!-\!C(OH) \cdot COOH \end{array}$$

Alkaloid	Base			
Trichodesmine	R	Trichodesmic acid	$X = OH$	$Z = H$
Junceine	R	Junceic acid	$X = OH$	$Z = OH$
Incanine	R	Incanic acid	$X = H$	$Z = OH$

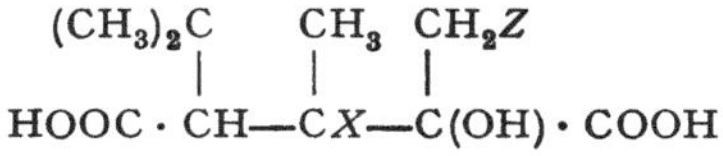

$$\begin{array}{ccc} CH_2 \quad COOH \\ \diagdown \;\; \diagup \\ C \quad CH_3 \quad CH_3 \\ | \quad\;\; | \quad\quad | \\ HOOC \cdot CH\!-\!C(OH)\!-\!C(OH) \cdot COOH \end{array}$$

| Grantianine | R | Grantianic acid | 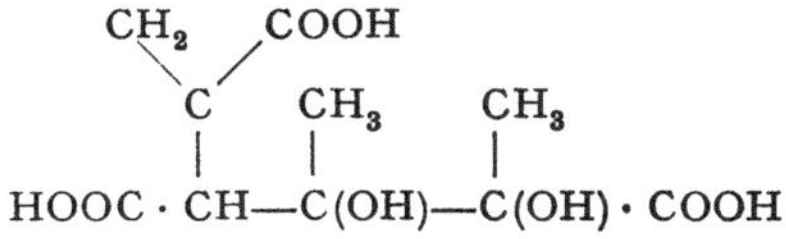 |

These three alkaloids form a structural group as shown in *Table 18*. If grantianic acid has the structure assigned by ADAMS and GIANTURCO (*202*), grantianine (**180**) could well be included here since it contains the same carbon skeleton with a methyl group of the isopropyl part replaced by a carboxyl group.

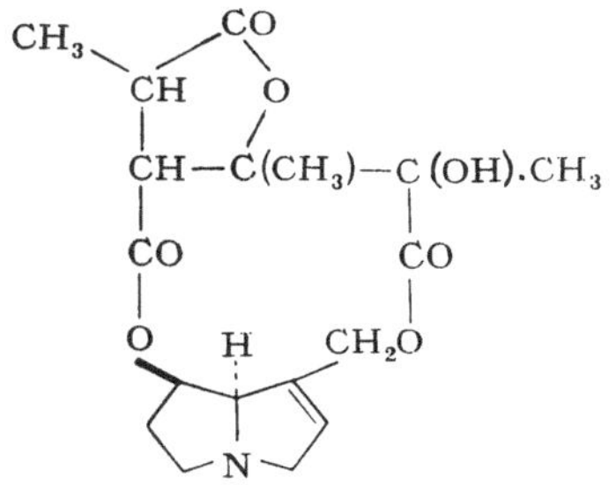

Grantianine.

4. Alkaloids Containing Hydroxylated 2-Methylpentane-3-carboxylic Acids.

Echinitine, Echiumine, Echimidine, Heliosupine, Macrotomine and Cynoglossophine.

Three new alkaloids have been added to this group: the alkaloid echinitine, isolated by MEN'SHIKOV and DENISOVA (*345*) from *Rindera echinata*, is shown to hydrolyse to viridifloric acid and heliotridine; and CULVENOR (*249*) has isolated echiumine and echimidine from *Echium plantagineum*. Echiumine, $C_{20}H_{31}NO_6$, was hydrolysed to retronecine, angelic acid and trachelanthic acid, whilst hydrogenolysis gave trachelanthic acid and retronecanyl butane-2-carboxylate, so that the trachelanthic acid is esterified by the primary hydroxyl group of retronecine. Echimidine, $C_{20}H_{31}NO_7$, on hydrolysis gave retronecine, angelic acid and acetone, whilst on catalytic hydrogenation, three moles of hydrogen were absorbed to give retronecanyl butan-2-carboxylate, and echimidinic acid, $C_7H_{14}O_5$.

Similar experimental procedure conducted by DENISOVA, MEN'SHIKOV and UTKIN (*276*) led to the structure of heliosupine from *Heliotropium supinum*, as containing heliotridine and an acid similar to, if not identical with echimidinic acid, although the acid was not isolated. A similar acid esterified to trachelanthamidine was earlier found by MEN'SHIKOV and PETROVA (*144*) in macrotomine, which as DENISOVA, PETROVA and MEN'SHIKOV (*277*) showed, gave with sodium hydroxide acetone and 2,3-dihydroxybutyric acid more readily than did heliosupine.

MAN'KO and BORISYUK (*338*) assign a similar structure to the acid from an optically inactive alkaloid, cynoglossophine, which on hydrolysis yielded angelic acid and acetone.

The alkaloids which contain this group of acids are shown in *Table 19* (p. 392) and may be classified as esters of heliotridine (**181**, $R' =$ OH or angelate), esters of trachelanthamidine (**182**), and esters of lindelofidine (**183**) where the group $R \cdot COO$ is $CH_3 \cdot CHY \cdot C(OH)[CX(CH_3)_2] \cdot COO$. The alkaloid indicine is of interest in that it is the only alkaloid in the *Crotalaria* spp. which contains the base retronecine, and the only alkaloid which contains the (—)-isomer of trachelanthic acid. There are in addition four alkaloids which are esters of angelic acid only: angelylturniforcidine (turniforcine), 7-angelylheliotridine, 7-angelylretronecine, and angelylmacronecine (macrophylline).

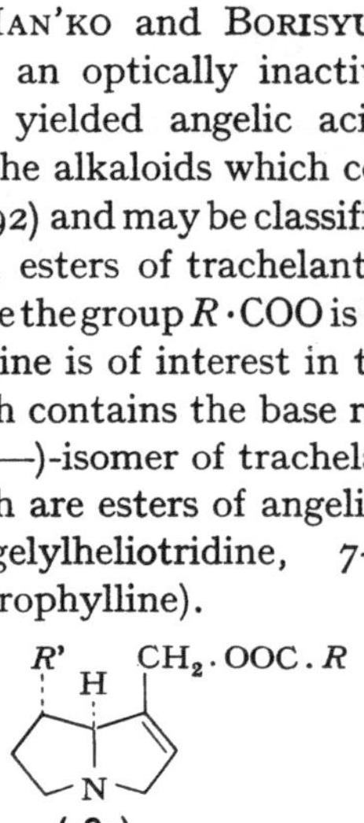

(181) 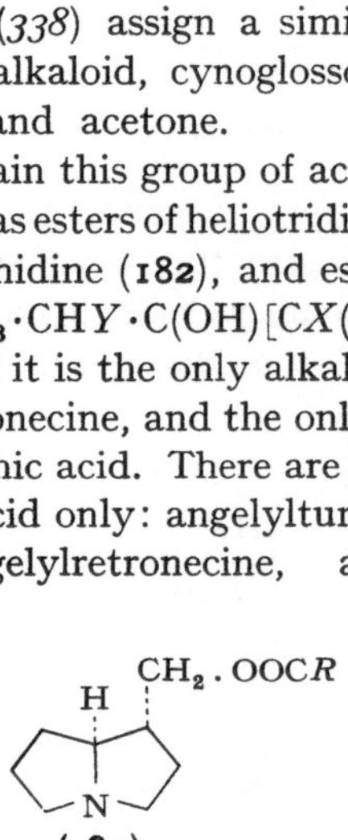   (182) 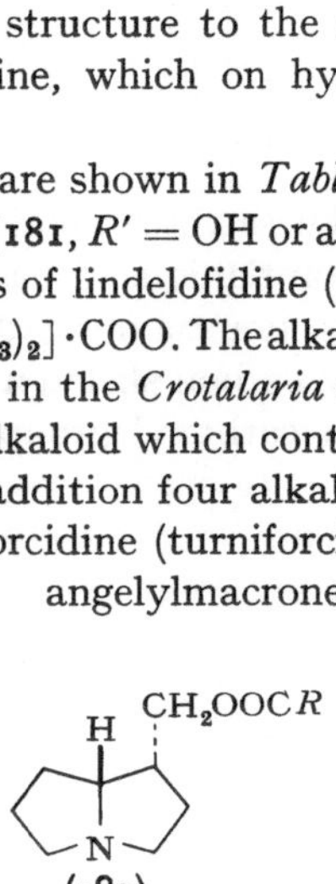   (183)

Lindelofamine is the angelyl ester of lindelofine (**184**, $R =$ H or angelyl, $R' =$ angelyl or H).

Strigosine (**185**) is the trachelanthamidine ester of 2,3-dihydroxy-3-methylpentanoic acid.

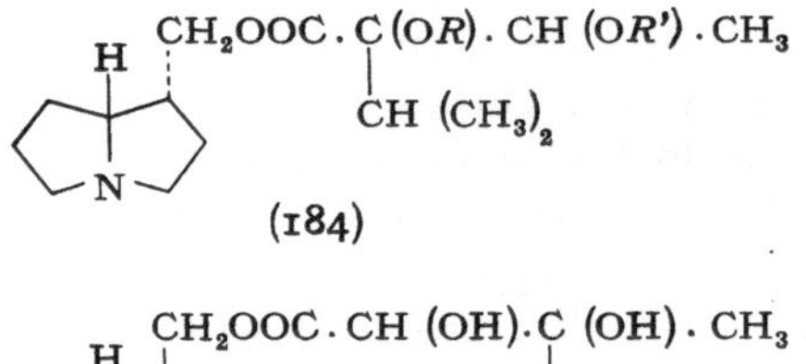

(184)

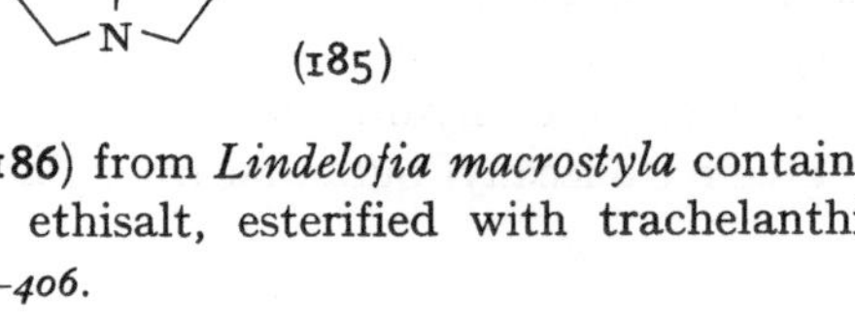

(185)

The alkaloid (**186**) from *Lindelofia macrostyla* contains the quaternary base, hastanecine ethisalt, esterified with trachelanthic acid.

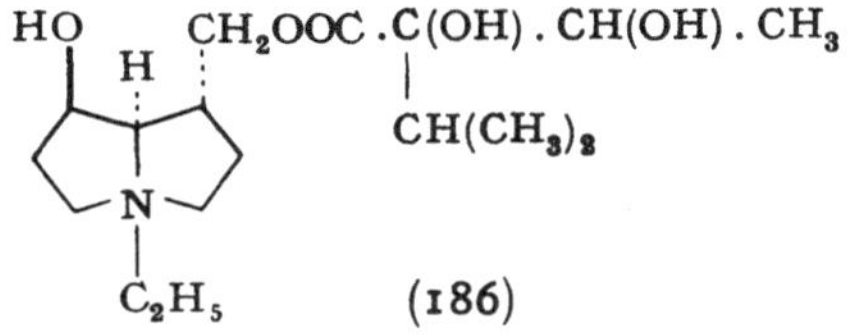

HO   CH$_2$OOC . C(OH) . CH(OH) . CH$_3$
|
CH(CH$_3$)$_2$

N
|
C$_2$H$_5$ (186)

The only syntheses of the total alkaloids have been effected in this group. CULVENOR, DANN and SMITH (255) have effected a synthesis of supinine (187, $R = R' =$ H) and heliotrine (187, $R =$ OH, $R' =$ OCH$_3$) by acting on 1-chloromethyl-1,2-dehydropyrrolizidine (188, $R =$ H) with the sodium salts of trachelanthic and heliotric acids, respectively.

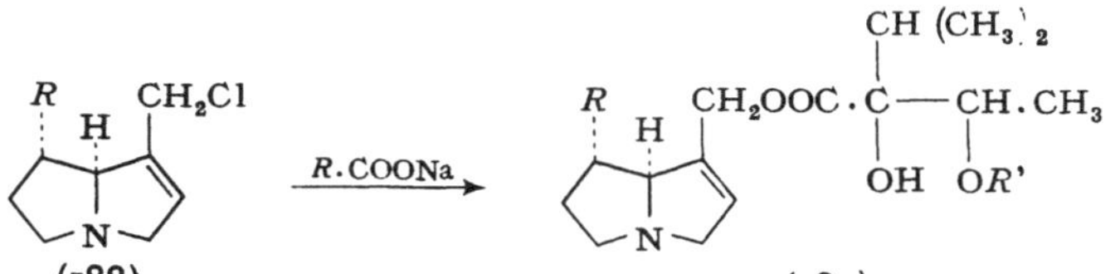

R CH$_2$Cl R CH$_2$OOC . C——CH . CH$_3$
  H   → R.COONa →   H | |
                                          OH OR'
N   N
(188)   (187)

CH (CH$_3$)$_2$

Latifoline.

Latifoline, C$_{20}$H$_{27}$NO$_7$, m. p. 103°, isolated by CROWLEY and CUL-VENOR (248) from *Cynoglossum latifolium*, may well be included in this group. It gave on alkaline hydrolysis retronecine, angelic acid and a non-crystalline mixture of at least two other acids. On catalytic hydrogenation it absorbed three moles of hydrogen to give retronecanyl butan-2-carboxylate and latifolic acid, hence latifoline must have structure (187).

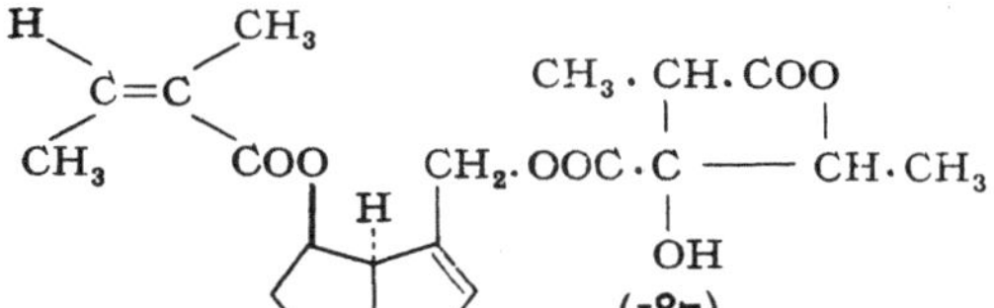

H   CH$_3$
  C=C   CH$_3$. CH . COO
CH$_3$ COO   CH$_2$. OOC . C —— CH . CH$_3$
           H |
                                   OH
N   (187)
                                 Latifoline.

This alkaloid occurs also with the 7-angelylretronecine; but the latter could be an artefact since it is found in smaller proportion in the alkaloid from the fresh, in contrast to the dried, plant. On the other hand, the presence of these two alkaloids in the same plant recalls the isolation by the same authors (246) of 7-angelylheliotridine with heliosupine from *Heliotropium supinum*. The alkaloid structure (187) is of interest in that one methyl of the isopropyl group has been replaced by a carboxyl.

Thesine and Thesinine.

The alkaloids thesine, $C_{34}H_{42}N_2O_6$, and thesinine, $C_{17}H_{21}NO_3$, were isolated by ARENDARUK, PROSKURNIA and KONOVALOVA (218) from *Thesium minkwitzianum*. The structures determined by ARENDARUK and SKOLDINOV (219, 220) are the first examples of pyrrolizidine alkaloids with aromatic acids, and thesine is the first alkaloid having a dibasic acid esterfied with two pyrrolizidine bases. Thesinine was hydrolysed to (+)-isoretronecanol and *p*-hydroxycinnamic acid and is accordingly (190, R = 189). Thesine was hydrolysed to (+)-isoretronecanol and thesinic acid, m. p. 300°, which with alkaline dimethyl sulphate gave dimethyl dimethoxythesinate, m. p. 125–126°, which in turn was hydrolysed to dimethoxythesinic acid, m. p. 250–251°. The nature of thesinic acid was established by its formation by sunlight on *p*-hydroxycinnamic acid, and by its synthesis from α-truxillic acid by nitration, reduction to the diamino acid and diazotisation to *p,p'*-dihydroxy-α-truxillic acid (191, R = H), which gave the dimethyl dimethoxy-α-truxillate identical with dimethyl dimethoxythesinate. Accordingly, thesine is di-(+)-isoretronecanyl *p,p'*-dihydroxy-α-truxillate (191. R = 189).

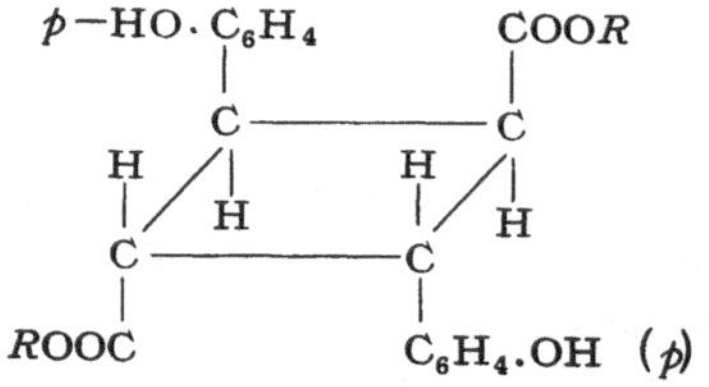

$p-HO.C_6H_4$ ──── COOR $p-(OH).C_6H_4.CH=CH.COOR$ (190)

R is

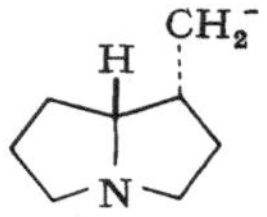

(191)   (189)

V. Biosynthesis.

Alkaloid Variation in the Plant.

Studies on the variation of the alkaloid content have shown interesting results. Varieties of Ukranian and Sakhalin Ragweed show differences ranging from 0.3–1.0% alkaloid (213). The N-oxides in *Senecio platyphyllus* (221, 215, cf. 320), namely seneciphylline and platyphylline N-oxides, are found in highest yield in the aerial portions reaching 2.95% during flowering, and 3.8% in the roots at the end of vegetation. In the resting roots in late autumn the alkaloids are exclusively in the reduced form with the N-oxide increasing again in the next growth.

Retronecine.

ROBINSON's (357) early suggestion that the pyrrolizidine ring originated from two ornithine molecules was vindicated by NOWACKI and BYERRUM (349) who fed [2-14C]ornithine to *Crotalaria spectabilis* to obtain monocrotaline which contained all the activity in the retronecine. BOTTOMLEY and GEISSMAN (*226*) have extended these studies to include the feeding of [1,4-^{14}C]putrescine, [2-^{14}C]ornithine and [5-^{14}C]ornithine to *S. douglasii*. The retronecine (**192**) was oxidised with osmium tetroxide to give (**193**) and then with metaperiodate to give formaldehyde. In all three experiments 25% of the activity was found in the formaldehyde (originating from the hydroxymethylene group), so these authors conclude that at least one of the two molecules involved in the synthesis goes through a symmetrical intermediate (**194**).

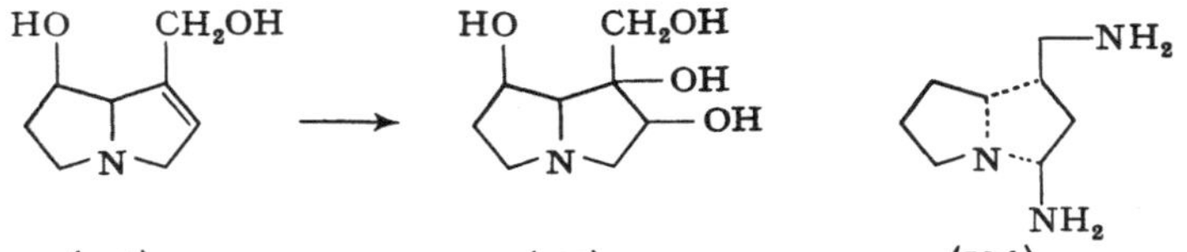

(**192**)   (**193**)   (**194**)

HUGHES, LETCHER and WARREN (*298*) fed [2-^{14}C]ornithine to *S. isatideus* and *S. sceleratus*. Heliotridane obtained from the alkaloid was stepwise degraded by Hofmann reaction. In agreement with the above authors 26% of the activity was found in the methyl group but no activity was found in $C_{(5)}$ and $C_{(3)}$. GORDON-GRAY and MORGAN (*295*) have, however, confirmed in our laboratories the observation of BOTTOMLEY and GEISSMAN (*226*) by feeding [5-^{14}C]ornithine to *S. sceleratus* and found the activity in the retronecine, 25% of which was in the —CH_2OH group. NOWACKI and BYERRUM (*349*) consider that hydroxylation occurs after the pyrrolizidine system is synthesised. They fed [6-^{14}C]4-hydroxylysine to lupin plants and found very little radioactivity in the hydroxylated alkaloid. It is of significance that ADAMS and GIANTURCO (*202*) found 3-hydroxy-N-methylnorvaline in *Crotalaria* spp.

Senecic and Retronecic Acids.

The biosynthesis pattern for the formation of the C_{10}-acids was studied by feeding [1-^{14}C] and [2-^{14}C]acetate to *S. isatideus* to give low activity retrorsine. The degradation of the retronecic acid (**195**, $X = OH$) showed that the acid could be formed from two acetoacetates substituted with a C_1-unit in the α-position. Further studies in these laboratories by GORDON-GRAY and SCHLOSSER (*295*) by feeding [2-^{14}C]acetate to *S. sceleratus*, *S. isatideus* and *S. hygrophyllus* gave retronecic acid (**195**, $X = OH$) and senecic acid (**195**, $X = H$) which showed a reproducible pattern as follows:

$$\begin{array}{cccccc} \overset{2}{CH}=\overset{3}{C} & \longrightarrow \overset{4}{CH_2}-\overset{2'}{CH}-\overset{3'}{C}(OH)-\overset{4'}{CH_2}X \\ | & | & | & | \\ CH_3 & COOH & CH_3 & COOH \\ 1 & 5 & 1' & 5' \end{array} \qquad (195)$$

C-atoms	1	2	5	1'	2'	5'	4'	3 + 4 + 3'
Activity %	8	12	13	12	20	14	4	15

Combination of Acid and Base.

[COOH-^{14}C]Integerrinecic acid was synthesised by HUGHES, GORDON-GRAY, SCHLOSSER and WARREN (297) from natural senecic acid (196) by way of the keto-acid (197) which was treated with [^{14}C]hydrocyanic acid to yield the cyanohydrin (198) and hydrolysed. This labelled acid was fed to S. adnatus; and the resulting rosmarinine had the active senecic acid labelled to the extent of 89% in the carboxyl group. Accordingly, the acid is incorporated direct into the alkaloid.

$$CH_3 \cdot CH : C(COOH) \cdot CH_2 \cdot CH(CH_3) \cdot C(OH)(COOH) \cdot CH_3 \rightarrow$$
$$(196) \qquad \uparrow \qquad CH_3 \cdot CH : C(COOH) \cdot CH_2 \cdot CH(CH_3) \cdot CO \cdot CH_3$$
$$\downarrow \quad (197)$$
$$CH_3 \cdot CH : C(COOH) \cdot CH_2 \cdot CH(CH_3) \cdot C(OH)(^{14}CN) \cdot CH_3$$
$$(198)$$

VI. Pharmacology.

(With MARJORIE E. VON KLEMPERER.)

The Hepatotoxic Activity of the Pyrrolizidine Alkaloids.

There has been considerable activity in the toxicological studies since SCHOENTAL (368) demonstrated the metastasic spread of liver tumours caused by pyrrolizidine alkaloids. SCHOENTAL and MAGEE (373) produced chronic liver lesions by administering a single dose of lasiocarpine as opposed to the malignant tumours caused by a series of sub-lethal doses. Similar results were obtained with seneciphylline, riddelliine, monocrotaline, heliotrine and lasiocarpine-N-oxide (364). The effective dose, however, varied from 20–300 mg./kg. rat and it was concluded that certain essential structural features were responsible for this hepatotoxic action. The suggestion made earlier (68) that the N-oxide forms were less toxic is not supported by SCHOENTAL (364), nor by BULL, DICK and McKENZIE (235). SCHOENTAL reports that the cyclic diesters with ten C atoms in the cycle are very toxic, while SCHOENTAL and HEAD (372) showed that monocrotaline had an additional pronounced effect on the lungs. Among the open ester structures lasiocarpine, a diester, proved more toxic than heliotrine which has the secondary alcoholic hydroxyl unesterified.

From previous reports (68) it seemed that small variations in the acidic moieties appear to exert little influence but SCHOENTAL and MATTOCKS (375) established the importance of the branched C-chain for hepatotoxic activity. They prepared esters of retronecine with various monocarboxylic acids. With these they produced liver lesions and cytological changes which, when the acid contained a branched C-chain resistant to hydrolysis by the enzymes of rat liver, were indistinguishable from those produced by the natural alkaloids. SCHOENTAL (*364*) considers that the stereochemistry of the acid moiety is of minor significance, since similar lesions are produced by lasiocarpine and the cyclic diesters.

Other factors shown to influence the toxicity are those of diet, sex and age. Rats on a low protein diet are more liable to liver damage (*363*). DICK, DANN, BULL and CULVENOR (*278*) found that *Senecio* alkaloids resisted hydrolysis in the rumen (of cattle) and were only broken down to their 1-methylenepyrrolizidine derivatives by the addition of vitamin B_{12} to the rumen liquor. Males were more susceptible than females, and young sexually immature animals produced symptoms of liver disease more rapidly than adults (*366*). SCHOENTAL (*365*) showed that lactating rats which were given lasiocarpine or retrorsine with no apparent effect on themselves or their milk production, poisoned their young which died of striking liver lesions.

Extensive studies have been made by various workers on the pathological changes in the liver and related organs as a result of feeding or injection of pyrrolizidine alkaloids (*216, 339, 355, 224, 371, 363*). Young animals which die in the first three months exhibit the characteristic enlarged parenchymal cells and gastrointestinal haemorrhage. Those which survive this first period may show a variety of changes and live until the age of 18 months. The optimum conditions for the development of hepatomas have not yet been established, but this occurs only occasionally after a single dose and more usually after intermittent small doses (*370*). A cirrhotic condition with marked fibrosis does not necessarily accompany the formation of tumours (*224*) and irradiation did not enhance the effect (*371*).

PUKHALSKAYA, PETROVA and MANKO (*355*) tested six alkaloids for their effect on the growth of inoculable tumours but, though lasiocarpine had a hepatoma inhibiting action, the most effective substances were also the most toxic.

As separate from the general pattern, SCHOENTAL (*368*) recorded deaths caused by lung lesions. These were found to develop more frequently in rats given alkaloids of *Crotalaria* spp., such as monocrotaline and fulvine. LALICH and EHRHART (*326*) also report pulmonary arteritis and haemorrhage induced by feeding monocrotaline.

Mechanism of the Toxic Action.

The irreversible and specific nature of the hepatotoxic activity of the pyrrolizidine alkaloids has aroused interest in the mechanism by which these changes are induced. It is evident that they can interact with the constituents of parenchymal liver cells to produce progressive changes. SCHOENTAL (*366*) links chemical structure with this aspect and points out that degradation of the alkaloids could yield isoprenoid units which might upset the biosynthesis of steroid hormones or of a "mitotic hormone" (*370*). The fact that sex and age differences affect susceptibility suggests that the steroid metabolism is involved. She postulates an interference in carboxylase activity and draws attention to the structural similarity between biotin and the scaffolding of the alkaloids; the basic moieties of both contain two fused 5-membered rings inclined to one another. The inactivity of the acids as such she attributes to their ready excretion. CLARK (*240*) reports irreversible cell damage which causes sterility in *Drosophila*. He finds that high mutagenic activity is a feature of the total alkaloid only, and shows that small doses may induce a degree of genetic deterioration without hepatotoxic symptoms. This work links up with the suggestion of SCHOENTAL and MAGEE (*374*) that the pyrrolizidine alkaloids may act as an interphase mitotic poison and in this connection BARNES and SCHOENTAL (*224*) draw attention to the formation of large polyploid cells comparable with those described in the livers of Kwashiorkor victims.

A further theory that the action of pyrrolizidine alkaloids on cell nuclei results from alkylation is advanced by CULVENOR, DANN and DICK (*254*). They describe a biological transformation of heliotrine involving a mechanism of alkyl-oxygen fission of the ester linkage. Heliotrine (**199**) fed orally to sheep was isolated from the rumen as 7α-hydroxyl-1-methylene-8α-pyrrolizidine (**200**) enantiomeric with the non-toxic major alkaloid of *Crotalaria goreensis*. They maintain that secondary differences in pathological effects must be expected but that the basic biochemical lesions will be of a similar type for all hepatotoxic alkaloids and an understanding of the precise nature of this is now possible. They raise the possibility of using compounds containing thiol or other nucleophilic groupings as protective agents against intoxication and cite the effectiveness of cysteine against chromosome breakage in *Allium*.

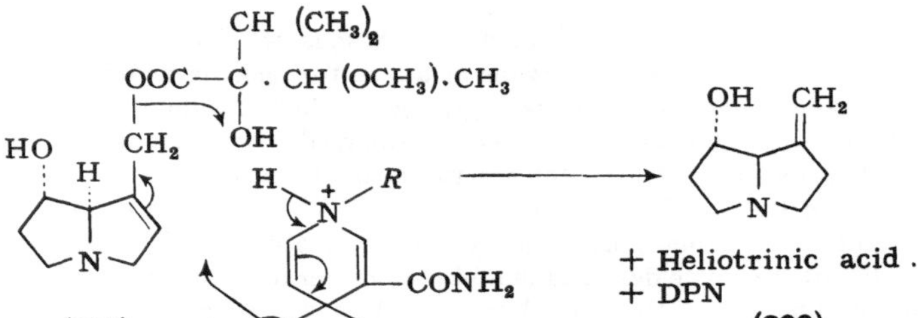

Possible Other Uses for Fully Saturated Pyrrolizidines.

An interesting series of experiments in the synthetic field have been carried out by KUZOVKOV, MASHKOVSKIĬ, DANILOVA and MEN'SHIKOV (*323, 324*). The curare-like substances based on platynecine (**201**, $R = $ ▬CH_2OH, $R' = OH$) and trachelanthamidine (**201**, $R = $ ····CH_2OH, $R' = H$) derivatives showed strong activity and the former under the name of Diplacine has been suggested as a substitute for tubocurarine chloride. Those based on pseudoheliotridane (*336*, $R = $ = ····CH_3, $R' = H$) though they blocked nervous transmission through the ganglia, showed no curare-like properties. Also unconnected with hepatotoxic activity were the studies made by McKENZIE (*336*) who investigated a series of alkaloids for their anti-cholinergic properties and ability to inhibit longitudinal tonus in rat ileum.

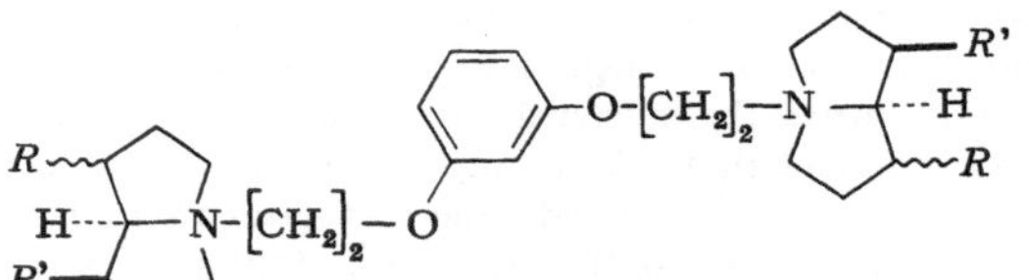

Supinine, heleurine and platyphylline were the most active antagonists of acetyl choline tested but none was as strong as atropine while senecionine was the most powerful inhibitor of longitudinal tonus.

VII. Tables.

(The short Tables I A, 15, 17 and 18 appear in the text on pp. 330, 335, 369, 371.)

Table 2 A. The Alkaloid Content of the Plant Species which have been Investigated since 1955.

[Addendum to Table 2 (387)] Where additional alkaloids have been found, or corrections made in the literature for specific plants, Table 2 A records the total alkaloid content.

Compositae	Alkaloid	Reference
[*Nardosomia laevigata* D. C.] ≡ *Petasites laevigatus* REICHB.	Platyphylline Renardine (≡ Senkirkine) Senecionine	(342)
Senecio aquaticus HILL [*S. aquaticus* HUDS.]	Seneciphylline (Jacodine) Aquaticine (? Seneciphylline)	(40, 42, 280)
S. angulatus L.	Rosmarinine Angularine	(352, 353)
[*S. ambrosioides* MART ex BAKER] ≡ *S. brasiliensis* LESS.	Retrorsine Seneciphylline Senecionine	(203)
[*S. borysthenicus* ANDRZ. ex D. C.] ≡ *S. praealtus* BERTOL.	Seneciphylline $C_{18}H_{23}NO_5$	(212, 356)
S. bupleuroides D. C.	Retrorsine	(361)
S. carthamoides GREENE	Seneciphylline Senecionine	(6, 15, 207)
S. chrysanthemoides D. C.	Seneciphylline	(386)
S. cineraria D. C. [≡ *Cineraria maritima* L.]	Senecionine Jacobine Jacodine Seneciphylline	(5, 37, 214)
S. discolor D. C.	Retrorsine Isatidine Senecionine	(296, 367)
S. douglasii D. C. [≡ *S. longilobus* BENTH.] (same alkaloids)	Retrorsine Seneciphylline Riddelliine	(6, 15, 191)
S. erraticus BERTOL. var. *barbaraeifolius* KROCK	Senecionine Otosenine ≡ [Tomentosine] Alkaloid S—C. $C_{18}H_{23}NO_6$ and other alkaloids Seneciphylline	(319, 358—360, 376)
S. erucifolius L.	Senecionine ⎫ previously Seneciphylline ⎬ $C_{18}H_{27}NO_5$ Alkaloid S—C	(40, 318)
S. fremonti TORR. and GRAY.	Senecionine Seneciphylline	(168, 203)

(Table 2A, continued.)

Compositae	Alkaloid	Reference
[*S. glabellus* D. C.] ≡ *S. campestris* D. C.	Senecionine Campestrine	(*25, 40*)
S. grandifolius LESS.	Seneciphylline Platyphylline	(*290*)
S. hygrophylus DYER and SM.	Platyphylline Rosmarinine $C_{18}H_{27}NO_6$ now called Hygrophilline	(*159, 361*)
S. jacobaea L.	Jacobine Jaconine Jacozine Jacoline Seneciphylline Senecionine	(*37, 40, 42, 118, 119, 231, 232*)
S. kirkii HOOK.	Senkirkine ≡ Renardine Renardine	(*143, 233, 59, 270*)
S. kleinia SCH. BIP.	Integerrimine	(*292, 293*)
S. longiflorus SCH. BIP.	Senecionine Seneciphylline	(*384*)
S. macrophyllus BIEB.	Macrophylline	(*273, 275*)
S. magnificus E. MUELL.	Integerrimine Senecionine	(*289, 251*)
S. mikanioides (WALP) OTTO.	Senecioic Acid ≡ Tiglic Acid Sarracine	(*37, 120, 258*)
S. othonnae BIEB.	Otosenine ≡ Tomentosine Seneciphylline Onetine	(*195, 271*)
S. palmatus PALL.	Seneciphylline	(*210*)
S. paludosus L.	Jacobine Jacodine ≡ Seneciphylline	(*40, 42, 279, 211*)
S. pampeanus CABRERA.	Senecionine	(*348*)
S. paucicalyculatus PLATT.	Retrorsine. Isatidine Paucicaline $C_{18}H_{22}NO_8$	(*354*)
S. platyphyllus D. C.	Platyphylline Seneciphylline Neoplatyphylline Sarracine (no seneciphylline or platyphylline found)	(*101, 102, 153*) (*274*) (*223*)
[*S. rivularis* D. C.] ≡ *S. crispatus* D. C.	Rivularine	(*360*)
S. tomentosus MICH.	Senecionine Otosenine (≡ Tomentosine)	(*207, 358*)
S. viminalis BREMEK.	Senecionine Retrorsine	(*384*)
S. viscosus L.	Senecionine Integerrimine Alkaloid S—F ($C_{15}H_{21}NO_3$)	(*36, 358*)

References, pp. 393—406.

(Table 2A, continued.)

Leguminosae	Alkaloid	Reference
Crotalaria anagyroides H. B. and K.	(—)-1-Methylenepyrrolizidine	(*263, 266*)
C. aridicola DOMIN.	Methyl ether of Supinidine	(*265*)
	Methyl ether of Retronecine	
C. crispata F. MUELL ex BENTH.	Monocrotaline	(*267*)
	Fulvine	
	Crispatine	
C. damarensis ENGL.	(—)-1-Methylenepyrrolizidine	(*266, 334*)
C. fulva ROXB.	Fulvine	(*369, 267*)
C. goreensis GUILL et PERR.	7β-Hydroxy-1-methylene-8β-pyrrolizidine	
	7β-Hydroxy-1-methylene-8α-pyrrolizidine	(*264*)
	Base "C" $C_8H_{13}NO$	
C. juncea L.	Junceine	(*149, 200, 201*)
	Seneciphylline	
	Riddelliine	
	Senecionine	
	Trichodesmine	
C. spartioides D. C.	Retrorsine (no experimental)	(*234*)
C. retusa L.	Monocrotaline	(*76, 18*)
	Retronecine-N-oxide	
	Retusamine	
	Retusine	(*261*)
	Base 130—132°	
[*C. spectabilis* ROTH.]	Monocrotaline	(*149, 18, 262*)
(≡ *C. sericea* RETZ.)	Spectabiline	
C. trifoliastrum WILLD.	Methyl ether of Supinidine	(*265, 259*)
	Methyl ether of Retronecine	
	Methoxymethyl-1,2-epoxy-pyrrolizidine	
[*Cytisus laburnum* L.]	Laburnine $C_8H_{15}NO$	(*72, 285*)
≡ *Laburnum laburnum* DÖRF-LER.	Alkaloid $C_{12}H_{22}N_2O$	
Boraginaceae		
Cynoglossum latifolium R. B.	Latifoline	(*248*)
	7-Angelylretronecine	
C. officinale L.	Cynoglossofine	(*338, 381*)
	Heliosupine	
	Echinatine	
Echium plantagineum L.	Echimidine	(*249*)
	Echiumine	
Heliotropium europaeum L.	Heliotrine	(*54, 53*)
	Supinine	
	Lasiocarpine	
	Europine	(*245*)
	Heleurine	

(Table 2A, continued.)

Boraginaceae	Alkaloid	Reference
H. indicum L.	Indicine Indicine N-oxide and minor unidentified alkaloids	*(344)*
H. strigosum WILLD.	Strigosine Trachelanthamidine and minor alkaloids	*(343)*
H. supinum L.	Supinine Heliosupine Echinatine	*(141, 53)*

	Angelylheliotridine
	   Trachelanthate
	Angelylheliotridine Viridi-
	   florate
		(246)

Lindelofia macrostyla BUNGE.	Lindelofine N-oxide Alkaloid $C_{17}H_{32}NO_5Cl$	*(383)*
Rindera echinata REGEL.	Echinatine	*(345)*
Trichodesma incanum BUNGE.	Trichodesmine Incanine	*(145, 129)* *(306, 309, 308,* *307, 310)*
Tournefortia sarmentosa LAM.	Supinine Supinine N-oxide	*(244)*
Santalaceae		
Thesium minkwitzianum B. FEDTSCH.	Thesine $C_{34}H_{42}N_2O_6$ Thesinine $C_{17}H_{21}NO_3$ Thesinecine $C_{10}H_{11}NO_2$	*(219)*
Gramineae		
Lolium cuneatum NEVSKI.	Loline $C_8H_{14}N_2O$ Lolinine $C_{10}H_{16}N_2O_2$ Lolinidine Norloline $C_7H_{12}N_2O$	*(301, 303)*
Thelepogon elegans ROTH. ex ROEM and SCHULT.	Thelepogine Thelepogidine	*(243)*

References, pp. 393—406.

Table 4aA. New Unnamed Pyrrolizidine Alkaloids and Reference to Previously Unnamed and Now Investigated Alkaloids.
[Addendum to Table 4a (387).]   [(e) = ethanol; (m) = methanol.]

Source	Formula	m. p.	[α]D	Ref.	Notes
[Cytisus labur-num L.] ≡ Labur-num laburnum DÖRFLER	$C_{12}H_{22}N_2O$	128–129°	+ 18.6°	(285)	—
Senecio borysthe-nicus ANDRZ. ex D. C.	$C_{18}H_{23}NO_5$	195°	—	(356)	—
S. erucifolius L.	$C_{18}H_{27}NO_5$	—	—	(318)	see Table 5 A
S. hygrophylus DYER and SM.	$C_{18}H_{27}NO_6$	176°	—67.3° (e)	(362)	see Table 4 A, p. 384
Lindelofia macro-styla BUNGE.	$C_{17}H_{32}NO_5Cl$	149–151°	+ 20.25° (m)	(383)	see Table 4 A

Table 5A. Alkaloids Originally Considered to be Individuals and now Found to be Mixtures or Identical with Previously known Forms.

Reported Name	Reference	Present Identification	Reference
Brasilinecine	(70)	Senecionine Seneciphylline Jacobine	(203)
Mikanoidine	(120)	Mixture	(258, 256)
Senkirkine*	(43)	Renardine	(233)
Tomentosine	(207)	Otosenine	(358, 376)
Squalidine	(36)	Integerrimine	(105, 293)
$C_{18}H_{27}NO_5$ (ex. S. erucifolius)	(40)	Senecionine Seneciphylline	(318)

* Original name, but was re isolated and its structure worked out later under the name renardine.

Table 4A. The Pyrrolizidine Alkaloids and their Hydrolysis Products. New Alkaloids and Additional Information on Previously Isolated Alkaloids [Addendum to Table 4 (387)]. [Solvents used for spec. rotation measurements: (c) chloroform; (e) ethanol; (m) methanol; and (w) water.]

Alkaloids	m.p.	$[\alpha]$D	Base	Acid	Reference
7-Angelylheliotridine $C_{13}H_{19}NO_3$	116–117°	—	Heliotridine	Angelic	(246)
Angelylheliotridine[a] Trachelanthate	—	—	Heliotridine	Angelic, Trachelanthic	(246)
Angelylheliotridine[a] Viridiflorate	—	—	Heliotridine	Angelic, Viridifloric	(246)
7-Angelylretronecine $C_{13}H_{19}NO_3$	76–77°	+49° (e)	Retronecine	Angelic	(248)
Angularine $C_{18}H_{26}NO_6$	200–201°	−98° (e)	Rosmarinecine	Seneciphyllic	(352, 353)
Aquaticine $C_{18}H_{25}NO_5$[b]	220°	−83° (c)	—	—	(280)
Base 4	176–178°	—	—	—	(59)
Brasilinecine (see Table 14)					(203)
Crispatine $C_{16}H_{23}NO_5$	137–138°	+41° (e)	Retronecine	Crispatic	(267)
Cynoglossofine $C_{20}H_{35}NO_8$	deliquess	0°	Cynoglossoph-idine	Angelic and $(CH_3)_2C(OH)\cdot C[CH\cdot(OR)CH_3](OH)\cdot COOH$	(338)
Echimidine $C_{20}H_{31}NO_7$	glass	13.4° (e)	Retronecine	Macrotomic	(249)
Echiumine $C_{20}H_{31}NO_6$	99–100°	14.4° (e)	Retronecine	Trachelanthic, Angelic	(249)
Echinatine $C_{15}H_{25}NO_5$	oil	Optically active	Heliotridine	Viridifloric	(246, 345, 381)
Europine $C_{16}H_{27}NO_6$ (previously Base G)	oil	+10.9° (e)	Heliotridine	Lasiocarpic	(53)
Fuchsisenecionine $C_{12}H_{21}NO_3$	225–227° (HCl salt)	—	Hydroxypyrrolizidine $C_7H_{13}NO$ m. p. 155.5°	Angelic	(148, 242)

Name / Formula	M.p.	$[\alpha]$	Necine base	Necic acid	References
Fulvine $C_{16}H_{23}NO_6$	212–213°	− 50.8° (c)	Retronecine	Fulvinic	(267, 369)
Heleurine $C_{16}H_{27}NO_4$ (previously Base C)	67–68°	− 12.0° (e)	Supinidine	Heliotrinic	(53, 245)
Heliosupine $C_{20}H_{31}NO_7$	oil	− 4.3° (e)	Heliotridine	Echimidinic, Angelic	(246, 276, 249)
Hygrophylline(d) $C_{18}H_{27}NO_6$	176°	− 67.3° (e)	Platynecine	Hygrophyllinecic	(189, 362)
Incanine $C_{18}H_{29}NO_5$	96–97°	(HI) − 58.4° (w)	Retronecine	Incaninic	(306, 309)
Indicine $C_{15}H_{25}NO_5$	97–98°	+ 23.3° (e)	Retronecine	(−)-Trachelanthic	(344)
Jacobine $C_{18}H_{25}NO_6$	226°	− 46.3° (c)	Retronecine	"Jacobinecic"(f)	(37, 42, 41, 287, 227, 232)
Jacoline $C_{18}H_{27}NO_7$	221°	+ 48° (c)	Retronecine	"Dihydroxydihydrosenecic"(f)	(42, 287)
Jaconine $C_{20}H_{32}O_7NCl$	147°	+ 52.5° (c)	Retronecine	"Jaconinecic"(f)	(37, 42, 41, 287, 227, 232)
Jacozine $C_{18}H_{25}NO_6$	228°	− 140°	Retronecine	Epoxide of Seneciphyllic	(42, 252)
Junceine $C_{18}H_{27}NO_7$	191–192°	− 3.0° (pyridine)	Retronecine	Junceic	(199, 201)
Latifoline $C_{20}H_{27}NO_7$	102–103°	+ 57.0° (e)	Retronecine	Angelic, Latifolic	(248)
Macrophylline $C_{13}H_{21}NO_3$	42–44°	+ 34.5° (e)	Macronecine	Angelic	(273, 275)
Mikanoidine (see Table 5A, p. 383)					(258, 256)
Neoplatyphylline $C_{18}H_{27}NO_5$	131–133°	+ 2.0° (c)	Platynecine	Isomers of Senecic	(274)
Onetine $C_{19}H_{29}NO_8$	192–193°	+ 73.0°	Otonecine	"Dihydroxydihydrosenecic"	(271, 389)
Otosenine $C_{19}H_{27}NO_7$ [= Tomentosine]	221°	+ 20.8° (c)	Otonecine	"Jacobinecic"(g)	(195, 389, 376, 358)
Renardine(h) $C_{19}H_{27}NO_6$ [= Senkirkine]	193–194°	0–2°	Otonecine	Senecic	(59, 233, 270)
Retusamine $C_{19}H_{25}NO_7$	174.5°	+ 13.0° (e)	Otonecine	"Retusaminecic"	(261, 389)
Retusine $C_{16}H_{25}NO_5$	174–175°	+ 16.0° (c)	Hastanecine (or enant-iomer)	α-Dihydroanhydromono-crotalic	(261)

(Table 4A, continued.)

Alkaloids	m. p.	[α]D	Base	Acid	Reference
Rivularine(i) $C_{13}H_{19}NO_3$	115–117°	−19° (c), +17° (e)	—	—	(360)
Sarricine $C_{18}H_{27}NO_5$	51–52°	−129.7°	Platynecine	Sarracinic, Angelic	(61, 272, 268)
Alkaloid SC $C_{18}H_{23}NO_6$	193°	−100° ± 2 (e)	Retronecine	$C_{10}H_{16}O_7$	(376, 318)
Alkaloid SD(t) $C_{18}H_{25}NO_3$	166–168°	−32° ± 2 (c)	—	—	(376, 360)
Senkirkine(k)	—				(233)
Alkaloid SF $C_{15}H_{21}NO_3$	228–230°	−148° (c)	—	—	(360)
Spectabiline $C_{18}H_{25}NO_7$	186°	+121° (c), +143° (e)	Retronecine	O-Acetylmonocrotalic	(262)
Strigosine $C_{14}H_{25}NO_4$	Gum	−19.3° (e)	Trachelanthamidine	α,β-Dihydroxy-β-methyl-valeric	(343)
Thesine $C_{34}H_{42}N_2O_6$	254–256°	+ 33.4°	(+)-Isoretronecanol	Thesinic	(219, 220, 218)
Thesinine $C_{17}H_{21}NO_3$	38–40°	—	(+)-Isoretronecanol	p-Hydroxycinnamic	(219)
Thesinicine $C_{10}H_{11}NO_2$ [Tomentosine] see Oto-senine	124–125°	—	—	—	(219) (207)
[Trachelanthine](k) Trachelanthamine N-oxide $C_{15}H_{27}NO_5$	167°	− 22.5° (w)	[Trachelan-thidine] Trachelanthamidine N-oxide	Trachelanthic	(135, 137, 138)
Trichodesmine $C_{15}H_{27}NO_6$	161°	+ 38° (e)	Retronecine	Trichodesmic	(145, 129, 200, 308)
$C_{17}H_{32}NO_5Cl$ un-named Alkaloid ex *Lindelofia macrostyla*	149–151°	20–25° (m)	N-Ethochloride of Hastanecine	Trachelanthic	(383)

(a) Not isolated pure but structure shown by hydrolysis.

(b) Seneciphylline has already been isolated from S. *aquaticus*, and only differs in rotation from aquaticine.

(d) Previously reported (Table 4 a, 387), now named and characterised.

(f) The hydrolysis of jacobine jacoline and jaconine, is accompanied by secondary reactions to give mixtures of acids. "Jacobinecic", "jacolinecic" and "jaconinecic" are names given to indicate the structure of the acid in combination in the alkaloid. "Jaco-

(g) Jaconecic acid was isolated, but the acid in combination has a structure similar to the acid in jacobine.

(h) Senkirkine, first isolated by BRIGGS (43), was later isolated and its structure determined as renardine.

(i) Campestrine (see Table 4, 387) has similar formula.

(j) Alkaloid SD could be integerrimine or palustrine.

(k) The N-oxides, in which form most of the alkaloids occur in the plant, are not included in this Table. Trachelanthine was previously listed as a separate alkaloid in Table 4.

Table 6A. New Bases from the Hydrolysis of Pyrrolizidine Alkaloids, together with Free Bases which have been Isolated from the Plant [Addendum to Table 6 (387)].

[Solvents used for spec. rotation measurements: (c) chloroform; (e) ethanol; (m) methanol; and (w) water.]

Base	$[\alpha]_D$	Base m.p.	Base b.p.	Picrate m.p.	Hydrochloride m.p.	Reference
$C_7H_{13}NO$ ex Fuchsisenecionine	—	155.5°	—	—	—	(242)
$C_8H_{13}N$ (8 S)-1-Methylenepyrrolizidine	− 43.1° (e)	—	120°/170 mm.	217–218°	—	(263)
$C_8H_{13}NO$ (7 R,8 S)-7-Hydroxy-1-methylene-8-pyrrolizidine	+ 36.1° (e)	34–35°	62°/0.03 mm.	173.5–174.5°	—	(264)
$C_8H_{13}NO$ (7 R,8 R)-7-Hydroxy-1-methylene-pyrrolizidine	− 150° (e)	35–36°	41°/0.1 mm.	203.5–204°	—	(264)
$C_8H_{15}NO$ Lindelofidine ≡ (+)-Iso-retronecanol	+ 79° (e)	40–41°	—	194° (picrolonate 182°)	—	(106, 218)
$C_8H_{15}NO$ Laburnin	+ 15.45° (e)	—	80–90°/0.01 mm.	—	—	(72, 285)
$C_8H_{15}NO_2$ Trachelanthidine ≡ Trachelanthamidine N-oxide	—	—	—	—	107–108°	(137)
$C_8H_{15}NO_2$ (ex Retusine) Hastanecine or Turniforcidine or enantiomer	—	—	—	—	116°	(261)
$C_8H_{15}NO_2$ Macronecine	+ 49.29° (e)	126–8°	—	—	152–153°	(275)

C₉H₁₅NO Methyl ether of Supinidine	$-24°$ (e)	—	100°/10 mm.	155-156°	—	(265)
C₉H₁₅NO₂ Methyl ether of Retronecine	$+38°$ (e)	35-40°	77°/0.4 mm.	185-186°	159-160°	(265)
C₉H₁₅NO₂ Methoxymethyl-1,2-epoxy-pyrrolizidine	$-63°$ (e)	—	53°/0.1 mm.	166-168° (picrolonate 166-168°)	—	(265)
C₉H₁₅NO₃ Otonecine	—	143°	—	amorphous 170-180° (d)	146-148°	(195, 270, 389)
C₁₀H₂₀NC₂Cl N-Ethochloride of Hastanecine	—	218.5-220°	—	173-174°	—	(383)
Cynoglossophidine (no formula)	—	—	—	99.0-99.5°	—	(338)
C₇H₁₂N₂O Norloline	15.1°	—	94-95°	226° (dec.)	—	(303)
C₈H₁₄N₂O Loline	18.9°	—	103°/5 mm 229°/731 mm	—	256° (dec.)	(301)
C₁₀H₁₆N₂O₂ Lolinine	36.9°	73°	85-93°	195-196°	—	(301)
Lolinidine	—	—	—	—	215-216°	(301)

Table 9A. The Acids from the Hydrolysis of the Pyrrolizidine Alkaloids.
[Addendum to Table 9 (387).]

[Solvents used in spec. rotation measurements: (c) chloroform; (e) ethanol; (m) methanol and (w) water.]

Acid	Formula	m. p.	$[\alpha]_D$	References		
				Isolation	Structure	Synthesis
Angelic acid	$C_5H_8O_2$	45°	—	—	—	—
Crispatic acid	$C_8H_{14}O_5$	133–134°	0°	(267)	(267)	—
Echidiminic acid [≡ Macrotomic acid]	$C_7H_{14}O_5$	—	16.4°(e)	(276)	(276, 249)	—
Fulvinic acid	$C_8H_{14}O_5$	113–114°	0°	(369, 267)	(267)	—
Grantianic acid	$C_{10}H_{14}O_7$	not crystalline	—	(3)	(202)	—
Heliotrinic acid [≡ Heliotric acid]	$C_8H_{16}O_4$	92–93°	−9.8°(w)	(125, 53)	(133)	—
Hygrophyllenecic acid						
Monolactone	$C_{10}H_{14}O_5$	180–181°	−187.9°(e)	(362)	(362)	—
Dilactone	$C_{10}H_{12}O_4$	103–105°	−97.6°(e)			
Incanic acid	$C_{10}H_{18}O_4$	161–163°	+25°(?)	(306)	(307)	—
Integerrinecic acid	$C_{10}H_{16}O_5$	151°	+159°(e)	(121)	(105)	(257)
Isoincanic acid	$C_{10}H_{16}O_4$	122–123.5°	−25°	(306)	—	—
Isoseneciphyllic acid	$C_{10}H_{14}O_5$	105–108°	−8.6°(e)	(98)	(8, 105, 340)	—
Jacolinecic acid dilactone	$C_{12}H_{16}O_6$	157–58°	+4.6°(?)	(230, 271)	(230, 287)	—
Isojaconecic acid	$C_{10}H_{16}O_6$	113–114°	+74.8°	(232)	(227, 229)	—
Junceic acid	$C_{10}H_{16}O_6$	180–182°		(201)	(201)	—
Latifolic acid	$C_7H_{10}O_5$	165–166°	+94.0°(e)	(248)	(248)	—
Mikanecic acid	$C_{10}H_{12}O_4$	234°	—	(120)	(256)	—
α-Retusanecic acid lactone	$C_8H_{12}O_4$	130–131°	+3.3°(e)	(11, 261)	(261, 389)	(261)
β-Retusenecic acid lactone	$C_8H_{12}O_4$	118°	−60°(e)	(11)	(389)	(261)
Sarracinic acid	$C_5H_8O_3$	57°	—	(272)	(272)	—

	Formula	M.p.	$[\alpha]_D$			
Senecic acid (Senecionic)	$C_{10}H_{16}O_5$	152°	+ 11.8°(e)	(36)	(36, 104)	(257)
[Squalinecic] acid(a)	—			(36)	(105, 293)	—
Thesinic acid	$C_{18}H_{16}O_6$	> 300°	—	(218)	(219, 220)	(220)
Tiglic acid	$C_5H_8O_2$	64°		—	—	—
Trichodesmic(b) acid or Tricho- desminic acid	$C_{10}H_{18}O_6$	209°		(145, 200)	(306, 308)	—
α,β-Dihydroxy-β-methylvaleric acid (ex Strigosine)	$C_6H_{12}O_4$	oil	—17.6°(e)	(343)	(343)	(380)
Viridifloric acid	$C_7H_{14}O_4$	142°	— 1.3°(c)	(136, 246)	(136)	(28, 65)
Acid from alkaloid SC	$C_{10}H_{16}O_7$	180—182°	—9±4°(e) 0°(w)	(318)	—	—

(a) Squalinecic acid, envisaged as a mixture (105) of integerrinecic acid and its lactone, has been confirmed as such (293).

(b) Previously reported (Table 9, 387) as "acid from trichodesmine" $C_7H_{12}O_3$ (129) called trichodesmic acid by ADAMS and GIANTURCO (200) and trichodesminic acid by YUNUSOV and PLEKHANOVA (310).

Table 16. The Structure of the Pyrrolizidine Alkaloids Containing C_{10}-Adipic Acids (R = Retronecine, P = Platynecine, Ros = Rosmarinecine and O = Othonecine).

Alkaloid	*Base*	$CH_3 \cdot CH = C(COOH) \cdot CHX \cdot CH(CH_3) \cdot C(OH)(COOH) \cdot$ $\cdot CH_2 Y$		
Senecionine	R	Senecic	$X = H$	$Y = H$
Renardine	O	Senecic		
Platyphylline	P	Senecic		
Rosmarinine	Ros	Senecic		
Neoplatyphylline	P	Senecic		
Integerrimine	R	Integerrinecic	$X = H$	$Y = H$
Usaramoensine	R	Usaramoensinecic	$X = H$	$Y = H$
Retrorsine	R	Isatinecic	$X = H$	$Y = OH$
Hygrophylline	P	Hygrophyllinecic	$X = OH$	$Y = H$
		$CH_3 \cdot CHX \cdot CY(COOH) \cdot CH_2 \cdot CH(CH_3) \cdot$ $\cdot C(OH)(COOH) \cdot CH_3$		
Jacobine	R	"Jacobinecic"	$-X-Y-$	$= -O-$
Othosenine	O	"Jacobinecic"	$-X-Y-$	$= -O-$
Jaconine	R	"Jaconinecic"	$X = Cl$	$Y = OH$
Jacoline	R	"Jacolinecic"	$X = OH$	$Y = OH$
Onetine	O	"Jacolinecic"	$X = OH$	$Y = OH$
		$CH_3 \cdot CH = C(COOH) \cdot CH_2 \cdot C(=CH_2) \cdot C(OH)(COOH) \cdot$ $\cdot CH_2 X$		
Seneciphylline	R	Seneciphyllic acid	$X = H$	
Spartioidine	R	"Spartioidinecic acid"	$X = H$	
Angularine	Ros	Seneciphyllic acid	$X = H$	
Riddelline	R	Riddellic acid	$X = OH$	
Jacozine	R	Seneciphyllic epoxide		
		$HOOC \cdot C(OH)(CH_3) \cdot CH(CH_3) \cdot CH(CH_3) \cdot$ $\cdot C(OH)(COOH)CH_2 X$		
Sceleratine	R	Sceleranecic acid	$X = OH$	
Sceleratinyl chloride	R	Sceleratinic acid	$X = Cl$	

Table 19. Alkaloids Containing Hydroxylated and Methoxylated 2-Hydroxy-2-isopropyl-butanoic Esters.

$$HOOC \cdot C(OH) \cdot CHY \cdot CH_3$$
$$\mid$$
Monoesters		$CX(CH_3)_2$	X	Y
Trachelanthamine	Trachelanthamidine	(+)Trachelanthic Acid	H	OH
Lindelofine	Lindelofidine	(+)Trachelanthic Acid	H	OH
Supinine	Supinidine	(+)Trachelanthic Acid	H	OH
Viridiflorine	Trachelanthamidine	Viridifloric Acid	H	OH
Echinitine	Heliotridine	Viridifloric Acid	H	OH
Heliotrine	Heliotridine	Heliotrinic Acid	H	OCH_3
Heleurine	Supinidine	Heliotrinic Acid	H	OCH_3
Europine [Base G]	Heliotrine	Lasiocarpic Acid	OH	OCH_3
Macrotomine	Trachelanthamidine	Macrotomic Acid	OH	OH

(Table 19, continued.) $\text{HOOC} \cdot \text{C(OH)} \cdot \text{CH}Y \cdot \text{CH}_3$

7-Angelyl Derivatives		$\text{C}X(\text{CH}_3)_2$	X	Y
Echimidine	Heliotrine	Echimidinic Acid	OH	OH
Heliosupine	Heliotrine	Isoechimidinic Acid	OH	OH
Lasiocarpine	Heliotrine	Lasiocarpic Acid	OH	OCH$_3$
Alkaloid	Heliotrine	Trachelanthic Acid	H	OH
Alkaloid	Heliotrine	Viridifloric Acid	H	OH
Echiumine	Retronecine	Trachelanthic Acid	H	OH

References.

(See also pp. 396 and 406.)

a) References Repeated from the First Paper (387).

3. ADAMS, R., M. CARMACK and E. F. ROGERS: The Alkaloid of *Crotalaria grantiana*. I. Grantianine. J. Amer. Chem. Soc. **64**, 571 (1942).

5. ADAMS, R. and T. R. GOVINDACHARI: *Senecio* Alkaloids: The Isolation of Senecionine from *Senecio cineraria* and Some Observations on the Structure of Senecionine. J. Amer. Chem. Soc. **71**, 1953 (1949).

6. — — *Senecio* Alkaloids: The Alkaloids of *Senecio douglasii, carthamoides, eremophilus, ampullaceus* and *parksii*. J. Amer. Chem. Soc. **71**, 1956 (1949).

8. ADAMS, R., T. R. GOVINDACHARI, J. H. LOOKER and J. D. EDWARDS, Jr.: *Senecio* Alkaloids: alpha-Longilobine; Structure of alpha-Longinecic Acid. J. Amer. Chem. Soc. **74**, 700 (1952).

11. ADAMS, R. and F. B. HAUSERMAN: The Total Structure of Monocrotaline. XIII. Synthesis of Dihydroanhydro-monocrotalic Acid. J. Amer. Chem. Soc. **74**, 694 (1952).

13. ADAMS, R. and N. J. LEONARD: Structure of Monocrotaline. XI. Proof of the Structure of Retronecine. J. Amer. Chem. Soc. **66**, 257 (1944).

15. ADAMS, R. and J. H. LOOKER: The Identity of alpha-Longilobine and Seneci-phylline. J. Amer. Chem. Soc. **73**, 134 (1951).

18. ADAMS, R. and E. F. ROGERS: The Structure of Monocrotaline, the Alkaloid in *Crotalaria spectabilis* and *Crotalaria retusa*. I. J. Amer. Chem. Soc. **61**, 2815 (1939).

19. — — Structure of Monocrotaline. V. Retronecine, a Derivative of 1-Methyl-pyrrolizidine. J. Amer. Chem. Soc. **63**, 228 (1941).

25. ADAMS, R. and B. L. VAN DUUREN: Usaramoensine, the Alkaloid in *Crotalaria usaramoensis* E. G. BAKER. Integerrimine from *Crotalaria incana* LINN. and Senecionine from *Senecio glabellus* D. C. Stereochemical Relationships. J. Amer. Chem. Soc. **75**, 4631 (1953).

26. — — The Structure of Heliotrinic Acid. J. Amer. Chem. Soc. **75**, 4636 (1953).

28. — — Trachelanthic and Viridifloric Acids. J. Amer. Chem. Soc. **74**, 5349 (1952).

36. BARGER, G. and J. J. BLACKIE: Alkaloids of *Senecio*. II. Senecionine and Squalidine. J. Chem. Soc. (London) **1936**, 743.

37. — — Alkaloids of *Senecio*. III. Jacobine, Jacodine, and Jaconine. J. Chem. Soc. (London) **1937**, 584.

40. BLACKIE, J. J.: The Alkaloids of the Genus *Senecio*. Pharmac. J. **138**, 102 (1937).

41. BRADBURY, R. B.: The Relationship between Jacobine and Jaconine and the Structure of Jaconecic Acid. Chem. and Ind. **1954**, 1022.

42. BRADBURY, R. B. and C. C. J. CULVENOR: The Alkaloids of *Senecio jacobaea* L. Chem. and Ind. **1954**, 1021; Austral. J. Chem. **7**, 378 (1954).

43. BRIGGS, L. H., J. L. MANGAN and W. E. RUSSELL: Alkaloids of New Zealand *Senecio* Species. I. The Alkaloid from *Senecio kirkii*. J. Chem. Soc. (London) **1948**, 1891.

49. CHRISTIE, S. M. H., M. KROPMAN, L. NOVELLIE and F. L. WARREN: The *Senecio* Alkaloids. IV. The Structure of Retronecic and Isatinecic Acids. J. Chem. Soc. (London) **1949**, 1703.

50. CLEMO, G. R. and T. A. MELROSE: The Synthesis of 2-Methylpyrrolizidine. J. Chem. Soc. (London) **1942**, 424.

53. CULVENOR, C. C. J.: The Alkaloids of *Heliotropium europaeum* L. II. Isolation and Structures of the Third Major Alkaloid and Two Minor Alkaloids, and Isolation of the Principal N-Oxides. Austral. J. Chem. **7**, 287 (1954).

54. CULVENOR, C. C. J., L. J. DRUMMOND and J. R. PRICE: The Alkaloids of *Heliotropium europaeum* L. I. Heliotrine and Lasiocarpine. Austral. J. Chem. **7**, 277 (1954).

59. DANILOVA, A. V. and R. A. KONOVALOVA: Alkaloids of *Senecio* Species. VII. Alkaloids from *Senecio renardi*. J. Gen. Chem. (USSR) **20**, 1921 (1950) [Chem. Abstr. **45**, 2960 (1951)].

61. DANILOVA, A. V., R. A. KONOVALOVA, P. (S.) MASSAGETOV and M. GARINA: The Alkaloids of *Senecio sarracenius*. C. R. (Doklady) Acad. Sci. (USSR) **89**, 865 (1953) [Chem. Abstr. **48**, 5875 (1954)].

65. DRY, L. J. and F. L. WARREN: The *Senecio* Alkaloids. IX. The Synthesis and Resolution of 3 : 4-Dihydroxy-2-methylpentane-3-carboxylic Acids: Viridifloric and Trachelanthic Acids. J. Chem. Soc. (London) **1952**, 3445.

66. — — The *Senecio* Alkaloids. XII. The Synthesis of (±)-*erythro*-3-Hydroxy-2-methoxy-4-methylpentane-3-carboxylic Acid. J. Chem. Soc. (London) **1955**, 65.

68. ELI LILLY and Co.: *Senecio* and Related Alkaloids. Research Today **5** (1949).

70. FONSECA, E. DE CAMARGO: Chemical and Toxicological Study of *Senecio brasiliensis*. Anais faculd. farm. odontol. Univ. São Paulo **9**, 85 (1951) [Chem. Abstr. **47**, 4888 (1953)].

72. GALINOVSKY, F., H. GOLDBERGER und M. PÖHM: Über das Laburnin, ein Alkaloid aus *Cytisus laburnum*. Monatsh. Chem. **80**, 550 (1949).

76. GRESHOFF, M.: Mitteilungen aus dem chemisch-pharmakologischen Laboratorium des Botanischen Gartens zu Buitenzorg (Java). Ber. dtsch. chem. Ges. **23**, 3537 (1890).

94. KONOVALOV, V. S. and G. P. MEN'SHIKOV: Alkaloids of *Cacalia hastata*. J. Gen. Chem. (USSR) **15**, 328 (1945) [Chem. Abstr. **40**, 3760 (1946)].

98. KONOVALOVA, R. A. and A. (V.) DANILOVA: Alkaloids of *Senecio* Species. VI. Structure of Seneciphylline. J. Gen. Chem. (USSR) **18**, 1198 (1948) [Chem. Abstr. **43**, 1427 (1949)].

101. KONOVALOVA, R. A. et A. P. OREKHOV: Sur les alcaloïdes du seneçon. 5e mémoire. Sur la constitution de la séneciphylline. Bull. soc. chim. France (5) **4**, 2037 (1937).

102. KONOVALOVA, R. A., A. P. OREKHOV and V. TIDEBEL: Alkaloids of *Senecio platyphyllus*. J. Gen. Chem. (USSR) **8**, 273 (1938) [Chem. Abstr. **32**, 5403 (1938)].

104. KROPMAN, M. and F. L. WARREN: The *Senecio* Alkaloids. V. The Structure of Senecic Acid. J. Chem. Soc. (London) **1949**, 2852.

105. — — The *Senecio* Alkaloids. VI. The Isomerisation of Senecic Acid to *trans*-Senecic (Integerrinecic) Acid, and the General Structure of the "Necic" Acids. J. Chem. Soc. (London) **1950**, 700.

106. LABENSKIĬ, A. S. and G. P. MEN'SHIKOV: Alkaloids of *Lindelofia anchusoides*. I. New Alkaloids, Lindelofine and Lindelofamine, and their Structures. J. Gen. Chem. (USSR) **18**, 1836 (1948) [Chem. Abstr. **43**, 3827 (1949)].

110. LEISEGANG, E. C. and F. L. WARREN: The *Senecio* Alkaloids. VII. The Structure of Retrorsine and Isatidine. The Ester Groupings. J. Chem. Soc. (London) **1950**, 702.

116. LEONARD, N. J. and G. L. SHOEMAKER: The Synthesis of Pyrrolizidines. IV. Condensation of Nitroparaffins with Methyl Methacrylate and Subsequent Formation of 2-, 6- and 8-Alkyl-Substituted Pyrrolyzidines. J. Amer. Chem. Soc. **71**, 1762 (1949).

118. MacKAY, A. H.: The Alkaloids of *Senecio jacobaea*. Nature **106**, 503 (1920).

119. MANSKE, R. H. F.: The Alkaloids of *Senecio* Species. I. The Necines and Necic Acids from *S. retrorsus* and *S. jacobaea*. Canad. J. Res. **5**, 651 (1931).

120. — The Alkaloids of *Senecio* Species. II. Some Miscelleneous Observations. Canad. J. Res. **14 B**, 6 (1936).

121. — The Alkaloids of *Senecio* Species. III. *Senecio integerrimus*, *S. longilobus*, *S. spartioides* and *S. riddellii*. Canad. J. Res. **17 B**, 1 (1939).

125. MEN'SHIKOV, G. P.: Über die Alkaloide von *Heliotropium lasiocarpum*. I. Ber. dtsch. chem. Ges. **65**, 974 (1932).

129. — Alkaloids of *Heliotropium lasiocarpum* and *Trichodesma incanum*, fam. Boraginaceae. Bull. Acad. Sci. USSR, Classe sci. math. nat., Sér. chim. **4**, 969 (1936) [Chem. Abstr. **31**, 5365 (1937)].

133. — Alkaloids of *Heliotropium lasiocarpum*. The Structure of Heliotrinic Acid. J. Gen. Chem. (USSR) **9**, 1851 (1939) [Chem. Abstr. **34**, 4071 (1940)].

135. — Alkaloids of *Trachelanthus korolkovi*. IV. Structure of Trachelanthamine. J. Gen. Chem. (USSR) **17**, 343 (1947) [Chem. Abstr. **42**, 556 (1948)].

136. — Alkaloids of *Cynoglossum viridiflorum*. I. New Alkaloid, Viridiflorin, and its Structure. J. Gen. Chem. (USSR) **18**, 1736 (1948) [Chem. Abstr. **43**, 2625 (1949)].

137. MEN'SHIKOV, G. P. and G. M. BORODINA: Alkaloids of *Trachelanthus korolkovi*. J. Gen. Chem. (USSR) **11**, 209 (1941) [Chem. Abstr. **35**, 7111 (1941)].

138. — — Alkaloids of *Trachelanthus korolkovi*. II. J. Gen. Chem. (USSR) **15**, 225 (1945) [Chem. Abstr. **40**, 2141 (1946)].

139. MEN'SHIKOV, G. P., S. O. DENISOVA and P. S. MASSAGETOV: Alkaloids of *Turneforcia sibirica*. I. New Alkaloid, Turneforcine. J. Gen. Chem. (USSR) **22**, 1465 (1952) [Chem. Abstr. **47**, 7512 (1953)].

141. MEN'SHIKOV, G. P. and E. L. GUREVICH: Alkaloids of *Heliotropium supinum*. I. Supinine and its Structure. J. Gen. Chem. (USSR) **19**, 1382 (1949) [Chem. Abstr. **44**, 3486 (1950)].

143. MEN'SHIKOV, G. P. and A. D. KUZOVKOV: Synthesis in the Pseudoheliotridane Series. J. Gen. Chem. (USSR) **21**, 2515 (1951) [Chem. Abstr. **47**, 5947 (1953)].

144. MEN'SHIKOV, G. P. and M. F. PETROVA: Alkaloids of *Makrotomia echoides*. I. New Alkaloid Makrotomine and its Structure. J. Gen. Chem. (USSR) **22**, 1457 (1952) [Chem. Abstr. **47**, 7512 (1953)].

145. MEN'SHIKOV, G. P. und W. RUBINSTEIN: Über ein Alkaloid aus *Trichodesma incanum* D. C. I. Ber. dtsch. chem. Ges. **68**, 2039 (1935).

148. MÜLLER, A.: Zur Kenntnis der *Senecio*-Arten in botanisch-medizinischer und pflanzenchemischer Hinsicht mit besonderer Berücksichtigung der Alkaloide. Heil- und Gewürzpflanzen **1924**, 58 SS [Chem. Zbl. **1925**, II, 1049].

149. NEAL, W. M., L. L. RUSOFF and C. F. AHMANN: The Isolation and Some Properties of an Alkaloid from *Crotalaria spectabilis* ROTH. J. Amer. Chem. Soc. **57**, 2560 (1935).

151. OREKHOV, A. (P.) und R. A. KONOVALOVA: Über *Senecio*-Alkaloide, III.: Abbau des Platynecins zum Heliotridan. Ber. dtsch. chem. Ges. **69**, 1908 (1936).

153. OREKHOV, A. und V. TIDEBEL: Über *Senecio*-Alkaloide. I. Die Alkaloide von *Senecio platyphyllus* D. C. Ber. dtsch. chem. Ges. **68**, 650 (1935).

155. PRELOG, V.: Über die zweifache intramolekulare Alkylierung. Liebigs Ann. Chem. **545**, 229 (1940).

159. RICHARDSON, M. F. and F. L. WARREN: The *Senecio* Alkaloids. I. Rosmarinine. J. Chem. Soc. (London) **1943**, 452.

168. STEYN, D. G.: Recent Investigations into the Toxicity of Known and Unknown Poisonous Plants in the Union of South Africa. Report Dir. Vet. Res. Onderstepoort (1929), (1931); Onderstepoort J. Vet. Sci. (1933), (1934), (1936), (1937).

189. WAAL, H. L. DE and J. TIEDT: The *Senecio* Alkaloids. 4. Platyphylline, the Active Principle of *Senecio adnatus* D. C. Onderstepoort J. Vet. Sci. Animal Ind. **15**, 251 (1940).

191. WARREN, F. L., M. KROPMAN, R. ADAMS, T. R. GOVINDACHARI and J. H. LOOKER: The Identity of beta-Longilobine with Retrorsine. J. Amer. Chem. Soc. **72**, 1421 (1950).

195. ZHDANOVICH, E. S. and G. P. MEN'SHIKOV: The Alkaloids of *Senecio othonnae*. J. Gen. Chem. (USSR) **11**, 835 (1941) [Chem. Abstr. **36**, 4123 (1942)].

b) New References.

196. ADAMS, R., C. C. J. CULVENOR, C. N. ROBINSON and H. A. STINGL: Synthesis of Heliotramide. Austral. J. Chem. **12**, 706 (1959).

197. ADAMS, R. and D. FLEŠ: The Absolute Configuration of the C_1 Atom in Retronecanone (1-Methyl-7-oxopyrrolizidine). J. Amer. Chem. Soc. **81**, 4946 (1959).

198. — — The Absolute Configuration of the C_8 Atom in the Pyrrolizidine Moieties of the Senecio Alkaloids. J. Amer. Chem. Soc. **81**, 5803 (1959).

199. ADAMS, R. and M. GIANTURCO: The Alkaloids of *Crotalaria juncea*. J. Amer. Chem. Soc. **78**, 1919 (1956).

200. — — Senecio Alkaloids: The Structure of Trichodesmine. J. Amer. Chem. Soc. **78**, 1922 (1956).

201. — — Crotalaria Alkaloids: The Structure of Junceine. J. Amer. Chem. Soc. **78**, 1926 (1956).

202. — — The Structures of Grantianine and Sceleratine. A Suggested Biogenesis of the Acids in the Alkaloids from *Senecio* and *Crotalaria* Species. J. Amer. Chem. Soc. **78**, 4458 (1956).

203. — — Senecio Alkaloids: The Alkaloids of *Senecio brasiliensis, fremonti* and *ambrosioides*. J. Amer. Chem. Soc. **78**, 5315 (1956).

204. — — Senecio Alkaloids: Mikanoidine, the Alkaloid from *Senecio mikanoides*. J. Amer. Chem. Soc. **79**, 166 (1957).

205. — — Senecio Alkaloids: Spartiodine, the Alkaloid from *Senecio spartioides*; Stereochemical Relationship to other *Senecio* Alkaloids. J. Amer. Chem. Soc. **79**, 174 (1957).

206. — — Recent Advances in the Chemistry of Pyrrolizidine Alkaloids. Angew. Chem. **69**, 5 (1957); Festschr. Arthur Stoll, p. 72. Basel: Birkhäuser. 1957.

207. ADAMS, R., M. GIANTURCO and B. L. VAN DUUREN: The Alkaloids of *Senecio tomentosus*. Observations on the Alkaloid Jacobine and on the Structure of Jaconecic Acid. J. Amer. Chem. Soc. **78**, 3513 (1956).

208. ADAMS, R., S. MIYANO and D. FLEŠ: 1-Hydroxypyrrolizidine and Related Compounds. J. Amer. Chem. Soc. **82**, 1466 (1960).

209. ADAMS, R., S. MIYANO and M. D. NAIR: Synthesis of Substituted Pyrrolidines and Pyrrolizidines. J. Amer. Chem. Soc. 83, 3323 (1961).

210. ALEKSEEV, V. S.: Alkaloids of 1-Methylpyrrolizidine Series. III. Alkaloids of Senecio palmatus. Zhurn. Obshchei Khimii 30, 3139 (1960) [Chem. Abstr. 55, 19 973 (1961)].

211. — Alkaloid Derivatives of 1-Methylpyrrolizidine from Senecio poludosus. Farmatsevt. Zh. (Kiev) 16, 39 (1961) [Chem. Abstr. 56, 13011 (1962)].

212. — 1-Methylpyrrolizidine Alkaloids. VII. Seneciphylline from Senecio borysthenicus. Med. Prom. (USSR) 15, 27 (1961) [Chem. Abstr. 57, 7384 (1962)].

213. — Chemical and Chromatographic Studies of Alkaloids from the Ukranian and Sakhalin Ragweed. Sb. Nauchn. Tr. Dnepropetr. Gos. Med. Inst. 19 (2), 179 (1961) [Chem. Abstr. 59, 8812 (1963)].

214. ALEKSEEV, V. S., T. G. BILYUGA and O. E. TALDYKIN: 1-Methylpyrrolizidine Alkaloids. V. Alkaloids from Senecio cineraria D. C. — Cineraria maritima. Farmatsevt. Zh. (Kiev) 17, No. 1, 42 (1962) [Chem. Abstr. 57, 7384 (1962)].

215. ALEKSEEV, V. S., T. G. BILYUGA, O. E. TALDYKIN, A. M. OLEKSANDRUK, A. G. TIMOSHENKO, N. N. MALUKHA, A. F. MINKO, V. S. SHABEL'NYUK, P. P. GIRENKO and V. V. MAZENKO: Content in Alkaloids of the Group of 1-Methylpyrrolizidine in Senecio borysthenicus at Different Vegetation Periods and the Effect of Mowing upon Alkaloid Content in Plants Grown after Mowing. Nauchn. Dokl. Vyashei. Shkoly. Biol. Nauki 1962, No. 2, 152 [Chem. Abstr. 57, 10 224 (1962)].

216. ALLEN, J. R., J. J. LALICH and S. C. SCHMITTLE: Crotalaria spectabilis. Induced Cirrhosis in Turkeys. Lab. Invest. 12, 512 (1963) [Chem. Abstr. 59, 4310 (1963)].

217. ALONSO DE LAMA, J. M., A. LÓPEZ-BLANCO and I. RIBAS: Papilionaceae Alkaloids. XXXIII. Alkaloids of Adenocarpus decorticans. The Chemistry of Decorticasine. Anales real soc. españ. fís. quím B 55, 717 (1959).

218. ARENDARUK, A. P., N. F. PROSKURNINA and R. A. KONOVALOVA: Alkaloids of Thesium minkwitzianum. Zhurn. Obshchei Khimii 30, 670 (1960) [Chem. Abstr. 54, 24835 (1960)].

219. ARENDARUK, A. P. and A. P. SKOLDINOV: Cyclobutanedicarboxylic Acids. I. Structure of Thesinic Acid. Zhurn. Obshchei Khimii 30, 484 (1960) [Chem. Abstr. 54, 24835 (1960)].

220. — — Cyclobutanedicarboxylic Acids. II. Structure of Thesinic Acid and Thesine. Zhurn. Obshchei Khimii 30, 489 (1960) [Chem. Abstr. 54, 24835 (1960)].

221. ARESHKINA, L. YA.: Nitrogen Oxides of Alkaloids of Plants, Senecio platyphyllus. Doklady Akad. Nauk. (USSR) 61, 483 (1948) [Chem. Abstr. 43, 280 (1949)].

222. — N-Oxides of Alkaloids of Plants of Senecio platyphyllus. Doklady Akad. Nauk. (USSR) 65, 711 (1949) [Chem. Abstr. 45, 9546 (1951)].

223. BAN'KOVSKAYA, A. N. and A. I. BAN'KOVSKIĬ: Alkaloids of Senecio platyphyllus. Trudy Vsesoyuz. Nauch. — Issledovatel. Inst. Lekarstv. i Aromat. Rast. 1959, No. 11, 46 [Chem. Abstr. 55, 18893 (1961)].

224. BARNES, J. M. and R. SCHOENTAL: Experimental Liver Tumours. Brit. Med. Bull. 14, No. 2, 165 (1958).

225. BOIT, H.-G.: Ergebnisse der Alkaloid-Chemie bis 1960. Berlin: Akademie Verl. 1961.

226. BOTTOMLEY, W. and T. A. GEISSMAN: Pyrrolizidine Alkaloids. The Biosynthesis of Retronecine. Phytochem. 3, 357 (1964).

227. Bradbury, R. B.: The Relationship between Jacobine and Jaconine and the Structure of Jaconecic Acid. Chem. and Ind. **1954**, 1022.
228. — The Alkaloids of *Senecio jacobaea* L. III. The Structure of Jaconecic Acid. Austral. J. Chem. **9**, 521 (1956).
229. — The Structures of Jaconecic and *iso*Jaconecic Acids. Tetrahedron **2**, 363 (1958); erratum ibid. **4**, 204 (1958).
230. Bradbury, R. B. and S. Masamune: The Alkaloids of *Senecio jacobaea* L. IV. The Structure of Jacobine, Jaconine and Jacoline and their Constituent Acids. J. Amer. Chem. Soc. **81**, 5201 (1959).
231. Bradbury, R. B. and S. Mosbauer: The Alkaloids of *Senecio jacobaea*. L. Chem. and Ind. **1956**, 1236.
232. Bradbury, R. B. and J. B. Willis: Alkaloids of *Senecio jacobaea*. II. Structures of the Acids and the Relationship between Jacobine and Jaconine. Austral. J. Chem. **9**, 258 (1956).
233. Briggs, L. H., R. C. Cambie, B. J. Candy, G. M. O'Donovan, R. H. Russell and R. N. Seelye: Alkaloids of New Zealand *Senecio* Species. Part II. Senkirkine. J. Chem. Soc. (London) **1965**, 2492.
234. Brummerhof, S. W. D. and H. L. de Waal: Retrorsine from *Crotalaria spartioides*. J. South African Chem. Inst. **14**, 101 (1961).
235. Bull, L. B., A. T. Dick and J. S. McKenzie: The Acute Toxic Effects of Heliotrine and Lasiocarpine and their N-Oxides on the Rat. J. Pathol. Bacteriol. **75**, 17 (1958).
236. Cahn, R. S., C. K. Ingold and V. Prelog: The Specification of Asymmetric Configuration in Organic Chemistry. Experientia **12**, 81 (1956).
237. Cason, J. and M. J. Kalm: Branched-chain Fatty Acids. XXXI. Assignment of Geometric Configuration in the 2-Methyl-2-alkenoic Acids. J. Organ. Chem. (USA) **19**, 1947 (1954).
238. Červinka, O.: Stereospecific Synthesis of *D,L*-Pseudoheliotridane. Chem. Listy **52**, 307 (1958); Collect. Czech. Chem. Comm. **24**, 1880 (1959) [Chem. Abstr. **52**, 11004 (1958); **53**, 22034 (1959)].
239. Červinka, O., K. Pelz and I. Jirkovský: Synthesis of the Alkaloid Trachelanthamidine. Collect. Czech. Chem. Comm. **26**, 3116 (1961) [Chem. Abstr. **56**, 10204 (1962)].
240. Clark, A. M.: Mutagenic Activity of the Alkaloid Heliotrine in *Drosophila*. Nature **183**, 731 (1959).
241. Clemo, G. R. and G. R. Ramage: New Derivatives of Pyrrole. Part I. The Synthesis of 3-Keto-4 : 5-dihydrodi-(1 : 2)-pyrrole and 8-Keto-5 : 6 : 7 : 8-tetrahydropyrrocoline. J. Chem. Soc. (London) **1931**, 49.
242. Corcilius, F.: Investigations on the Necine and Necinic Acid of the Alkaloid Ester Fuchsisenecionine from *Senecio fuchsii*. Planta Med. (Germany) **3**, 147 (1955) [Chem. Abstr. **50**, 5983 (1956)].
243. Crow, W. D.: Alkaloids of the Gramineae; *Thelepogon elegans*. Austral. J. Chem. **15**, 159 (1962).
244. Crowley, H. C. and C. C. J. Culvenor: Occurrence of Supinine in *Tournefortia sarmentosa*. Austral. J. Chem. **8**, 464 (1955).
245. — — Alkaloid Assays of *Heliotropium europaeum* L. Austral. J. Appl. Sci. **7**, 359 (1956).
246. — — The Alkaloids of *Heliotropium supinum* L. with Observations on Viridifloric Acid. Austral. J. Chem. **12**, 694 (1959).
247. — — The Decomposition of Lasiocarpic and Echimidinic Acids in Hydrochloric Acid. Austral. J. Chem. **13**, 269 (1960).

248. CROWLEY, H. C. and C. C. J. CULVENOR: Alkaloids of *Cynoglossum latifolium* R. BR. Latifoline and 7-Angelylretronecine. Austral. J. Chem. **15**, 139 (1962).

249. CULVENOR, C. C. J.: The Alkaloids of *Echium plantagineum* L. I. Echiumine and Echimidine. Austral. J. Chem. **9**, 512 (1956).

250. — Some Alkaloids of Australian *Crotalaria* Species. In: A. Albert, G. M. Badger and C. W. Shoppee (Edits.), Current Trends in Heterocyclic Chemistry. New York: Academic Press and London: Butterworths. 1958.

251. — *Senecio magnificus* E. MUELL. A Source of Senecionine. Austral. J. Chem. **15**, 158 (1962).

252. — The Structure of Jacozine, an Alkaloid of *Senecio jacobaea* L. Austral. J. Chem. **17**, 233 (1964).

253. CULVENOR, C. C. J. and R. DAL BON: Carbonyl Stretching Frequencies of Pyrrolizidine Alkaloids. Austral. J. Chem. **17**, 1296 (1964).

254. CULVENOR, C. C. J., A. T. DANN and A. T. DICK: Alkylation as the Mechanism by which the Hepatotoxic Alkaloids Act on Cell Nuclei. Nature **195**, 570 (1962).

255. CULVENOR, C. C. J., A. T. DANN and L. W. SMITH: Recombination of Amino Alcohols and Acids Derived from Pyrrolizidine Alkaloids. Chem. and Ind. **1959**, 20.

256. CULVENOR, C. C. J. and T. A. GEISSMAN: Structure of Mikanecic Acid. Chem. and Ind. **1959**, 366.

257. — — The Total Synthesis of Senecic and Integerrinecic Acids. J. Amer. Chem. Soc. **83**, 1647 (1961).

258. — — Alkaloids of *Senecio mikanioides* OTTO. Sarracine and Sarracine N-Oxide. J. Organ. Chem. (USA) **26**, 3045 (1961).

259. CULVENOR, C. C. J., J. D. MORRISON, A. J. C. NICHOLSON and L. W. SMITH: Alkaloids of *Crotalaria trifoliastrum*. II. 1-Methoxymethyl-1,2-epoxypyrrolizidine. Austral. J. Chem. **16**, 131 (1963).

260. CULVENOR, C. C. J. and L. W. SMITH: Alkaloids of *Erechtites quadridenta*. Austral. J. Chem. **8**, 556 (1955).

261. — — Alkaloids of *Crotalaria retusa* L. Austral. J. Chem. **10**, 464 (1957).

262. — — Alkaloids of *Crotalaria spectabilis* ROTH. Austral. J. Chem. **10**, 474 (1957); addendum ibid. **11**, 97 (1958).

263. — — 1-Methylene-pyrrolizidine, the Major Alkaloid of *Crotalaria anagyroides* H. B. & K. Austral. J. Chem. **12**, 255 (1959).

264. — — The Alkaloids of *Crotalaria goreensis* GUILL. et PERR. 7β-Hydroxy-1-methylene-8β- and 7β-Hydroxy-1-methylene-8α-pyrrolizidine. Austral. J. Chem. **14**, 284 (1961).

265. — — The Alkaloids of *Crotalaria trifoliastrum* WILLD. and *C. aridicola* DOMIN. I. Methyl Ethers of Supinidine and Retronecine. Austral. J. Chem. **15**, 121 (1962).

266. — — Identity of the Alkaloid from *Crotalaria damarensis* ENGL. with (—)-1-Methylenepyrrolizidine, now Shown to Occur Partially Racemized in *C. anagyroides* H. B. & K. Austral. J. Chem. **15**, 328 (1962).

267. — — Alkaloids of *Crotalaria crispata*. The Structures of Crispatine and Fulvine. Austral. J. Chem. **16**, 239 (1963).

268. DANILOVA, A. V., R. (A.) KONOVALOVA, P. (S.) MASSAGETOV and M. GARINA: Alkaloids of *Senecio* Species. VIII. Alkaloids from *Senecio sarracenius*. Zhurn. Obshchei Khimii **23**, 1417 (1953) [Chem. Abstr. **47**, 12759 (1953)].

269. DANILOVA, A. V. and N. I. KORETSKAYA: Structure and Properties of Seneciphylline. Zhurn. Obshchei Khimii **35**, 584 (1965) [Chem. Abstr. **63**, 644 (1965)].

270. DANILOVA, A. V., N. I. KORETSKAYA and L. M. UTKIN: Structure of the Alkaloid Renardine. II. Zhurn. Obshchei Khimii **31**, 3815 (1961) [Chem. Abstr. **57**, 9901 (1962)].

271. — — — A New Alkaloid from *Senecio othonnae*. Zhurn. Obshchei Khimii **32**, 647 (1962) [Chem. Abstr. **58**, 2477 (1963)].

272. DANILOVA, A. V. and A. D. KUZOVKOV: Alkaloids from *Senecio* Species. IX. Structure of Alkaloids from *Senecio sarracenius*. Zhurn. Obshchei Khimii **23**, 1597 (1953) [Chem. Abstr. **48**, 11436 (1954)].

273. DANILOVA, A. V. and L. M. UTKIN: Structure of the Alkaloid Macrophylline. Zhurn. Obshchei Khimii **30**, 345 (1960) [Chem. Abstr. **54**, 22698 (1960)].

274. DANILOVA, A. V., L. M. UTKIN, G. V. KOZYREVA and YU. I. SYRNEVA: A New Alkaloid Isomeric with Platyphylline. Zhurn. Obshchei Khimii **29**, 2432 (1959) [Chem. Abstr. **54**, 9980 (1960)].

275. DANILOVA, A. (V.), L. (M.) UTKIN and P. (S.) MASSAGETOV: Alkaloids of *Senecio macrophyllus*. Zhurn. Obshchei Khimii **25**, 831 (1955) [Chem. Abstr. **50**, 2626 (1956)].

276. DENISOVA, S. I., G. P. MEN'SHIKOV and L. M. UTKIN: New Alkaloid from the Plant *Heliotropium supinum*. Doklady Akad. Nauk. (USSR) **93**, 59 (1953) [Chem. Abstr. **49**, 3992 (1955)].

277. DENISOVA, S. I., M. F. PETROVA and G. P. MEN'SHIKOV: Decomposition of Macrotomic Acid and the Acid from Heliosupine in Solutions of Alkalis. Zhurn. Obshchei Khimii **28**, 1882 (1958) [Chem. Abstr. **53**, 1395 (1959)].

278. DICK, A. T., A. T. DANN, L. B. BULL and C. C. J. CULVENOR: Vitamin B_{12} and the Detoxification of Hepatotoxic Pyrrolizidine Alkaloids in Rumen Liquor. Nature **197**, 207 (1963).

279. DOROSH, T. P. and V. S. ALEKSEEV: Electrochemical Isolation of Seneciphylline from an Alkaloid Mixture Obtained from *Senecio paludosus*. Farm. Zhur. (Kiev) **15**, No. 6, 44 (1960) [Chem. Abstr. **55**, 18012 (1961)].

280. EVANS, W. C. and E. T. EVANS: Poisoning of Farm Animals by the Marsh Ragwort (*Senecio aquaticus* HUDS.). Nature **164**, 30 (1949).

281. FODOR, G.: The Stereochemistry of the Pyrrolizidine Alkaloids. Chem. and Ind. **1954**, 1424.

282. FODOR, G., I. SALLAY and F. DUTKA: Stereochemistry of Pyrrolizidine Alkaloids. II. The Configuration of Retronecine and Related Compounds. Acta Univ. Szegediensis, Acta Phys. et Chem. [N. S.] **2**, 80 (1956) [Chem. Abstr. **51**, 16498 (1957)].

283. FRIDRICHSONS, J. and A. McL. MATHIESON: Molecular Structure and Absolute Configuration of Thelepogine Methiodide. Tetrahedron Letters **1960**, No. 26, 18 [Chem. Abstr. **55**, 10488 (1961)].

284. FRIDRICHSONS, J., A. McL. MATHIESON and J. D. SUTOR: The Molecular Structure and Absolute Configuration of Jacobine Bromhydrin. Tetrahedron Letters **1960**, No. 23, 35.

285. GALINOVSKY, F., O. VOGL and H. NESVADBA: A New Alkaloid of *Cytisus laburnum*. Sci. Pharm. **21**, 256 (1953) [Chem. Abstr. **49**, 6977 (1955)].

286. — — — The Structure of Laburnine. Monatsh. Chem. **85**, 913 (1954) [Chem. Abstr. **50**, 1055 (1956)].

287. GEISSMAN, T. A.: The Alkaloids of *Senecio jacobaea* L. The Structures of the Alkaloids and the Necic Acids. Austral. J. Chem. **12**, 247 (1959).

288. GEISSMAN, T. A. and A. C. WAISS, Jr.: The Total Synthesis of (+)-Retronecine. J. Organ. Chem. (USA) **27**, 139 (1962).

289. GELLERT, E. and C. MÁTÉ: The Isolation of Integerrimine from *Senecio magnificus* F. MUELL. Austral. J. Chem. **17**, 158 (1964).

290. GLONTI, SH. I.: The Process of Alkaloid Accumulation in *Senecio grandifolia* Growing in a Wild State in Some Regions of the Georgian S. S. R. Samml. Arbb. Chem-pharm. Forsch-Inst. Tiflis **1956**, No. 8, 31 [Chem. Abstr. **52**, 12322 (1958)].

291. GOLDSCHMIDT, B. M.: Some Substituted Pyrrolizidines. J. Organ. Chem. (USA) **27**, 4057 (1962).

292. GONZALES, A. G. and A. CALERO: Alkaloids of Canary Island Plants. IV. *Senecio kleinia* SCH. BIP. Anales real soc. españ. fís. quím. B **54**, No. 3, 223 (1958) [Chem. Abstr. **54**, 19740 (1960)].

293. — — Squalidine and Integerrimine. Chem. and Ind. **1958**, 126.

294. GORDON-GRAY, C. G.: *Senecio* Alkaloids. The Isolation of Sceleratinyl Chloride from *Senecio sceleratus* SCHWEIK. (In preparation.)

295. GORDON-GRAY, C. G., D. MORGAN, F. D. SCHLOSSER and F. L. WARREN: unpublished.

296. HENNIG, A. J.: *Senecio discolor.* The Isolation and Identification of Senecionine and Retrorsine. Lloydia **24**, 68 (1961) [Chem. Abstr. **57**, 11309 (1962)].

297. HUGHES, C. A., C. G. GORDON-GRAY, F. D. SCHLOSSER and F. L. WARREN: The *Senecio* Alkaloids. Part XVII. A Study of the Formation of the Total Alkaloid in Plant in Relation to the Synthesis of the "Necine" Base and "Necic" Acid. J. Chem. Soc. (London) **1965**, 2370.

298. HUGHES, C. A., R. LETCHER and F. L. WARREN: *Senecio* Alkaloids. Part XVI. Biosynthesis of the "Necine" Bases from Carbon-14 Precursors. J. Chem. Soc. (London) **1964**, 4974 [cf. C. A. HUGHES, Ph. D. Thesis, Univ. Natal, S. Africa].

299. HUGHES, C. A. and F. L. WARREN: The *Senecio* Alkaloids. Part XIV. The Biological Synthesis of the "Necic" Acids using C-14. J. Chem. Soc. (London) **1962**, 34 [cf. C. A. HUGHES, Ph. D. Thesis, Univ. Natal, S. Africa].

300. JEŽO, I. and V. KALÁČ: Synthesis of some Alkaloid Derivatives. XII. (±)-1-Hydroxymethylpyrrolizidines. Chem. zvesti **11**, 696 (1957) [Chem. Abstr. **52**, 10052 (1958)].

301. JUNUSOV (YUNUSOV), S. YU. and S. T. AKRAMOV: Alkaloids of Seeds of *Lolium cuneatum.* Zhurn. Obshchei Khimii **25**, 1813 (1955) [Chem. Abstr. **50**, 7117 (1956)].

302. — — Structure of Norloline, Loline and Lolinine. Doklady Akad. Nauk Uzbek. SSR **1959**, No. 4, 28 [Chem. Abstr. **54**, 11028 (1960)].

303. — — Alkaloids of *Lolium cuneatum.* II. Zhurn. Obshchei Khimii **30**, 677 (1960) [Chem. Abstr. **54**, 24831 (1960)].

304. — — Alkaloids of *Lolium cuneatum.* III. Zhurn. Obshchei Khimii **30**, 683 (1960) [Chem. Abstr. **54**, 24831 (1960)].

305. — — Structure of Norlolin, Lolin and Lolinine. IV. Zhurn. Obshchei Khimii **30**, 3132 (1960) [Chem. Abstr. **55**, 19981 (1961)].

306. JUNUSOV, S. YU. and N. V. PLEKHANOVA: Alkaloids from Seeds of *Trichodesma incanum.* Doklady Akad. Nauk Uzbek. SSR **1953**, No. 4, 28 [Chem. Abstr. **51**, 1539 (1957)].

307. — — The Structure of Incanine. Doklady Akad. Nauk Uzbek. SSR **1957**, No. 5, 13 [Chem. Abstr. **53**, 7137 (1959)].

308. — — The Structure of Trichodesmine. Doklady Akad. Nauk Uzbek. SSR **1957**, No. 6, 19 [Chem. Abstr. **53**, 6276 (1959)].

309. — — Alkaloids of *Trichodesma incanum.* Doklady Akad. Nauk Uzbek. SSR **1957**, No. 4, 31 [Chem. Abstr. **52**, 13017 (1958)].

310. — — Alkaloids of *Trichodesma incanum.* Structure of Incanine and Trichodesmine. Zhurn. Obshchei Khimii **29**, 677 (1959) [Chem. Abstr. **54**, 1580 (1960)].

311. KLYNE, W. and J. A. MILLS: The Correlation of Configurations. Progr. Stereo-
chem. 1, 177 (1954).

312. KOCHETKOV, N. K., A. M. LIKHOSHERSTOV and E. I. BUDOVSKIĬ: Pyrrolizidine
Alkaloids. I. Synthesis of 1-Hydroxymethylpyrrolizidine (dl-Trachelanth-
amidine). Zhurn. Obshchei Khimii 31, 1735 (1961) [Chem. Abstr. 55, 7386,
22 354 (correction) (1961)].

313. KOCHETKOV, N. K., A. M. LIKHOSHERSTOV and A. S. LEBEDEVA: The Pyrrolizidine Alkaloids. II. A Stereospecific Synthesis of dl-Isoretronecanol. Zhurn.
Obshchei Khimii 31, 3461 (1961) [Chem. Abstr. 57, 3490 (1962)].

314. KOCHETKOV, N. K., A. M. LIKHOSHERSTOV and L. M. LIKHOSHERSTOV: New
Methods of Synthesizing Pyrrolizidine and Quinolizidine Amino-alcohols.
Zhurn. Vsesoyuz. Khim. Obshch. im. D. I. Mendeleeva 5, 109 (1960) [Chem.
Abstr. 54, 21 099 (1960)]; LIKHOSHERSTOV, A. M., L. M. LIKHOSHERSTOV and
N. K. KOCHETKOV: Pyrrolizidine Alkaloids. V. General Route for the Synthesis
of Amino Alcohols of the Pyrrolizidine and Quinolizidine Series. Zhurn.
Obshchei Khimii 33, 1801 (1963) [Chem. Abstr. 59, 10 143 (1963)].

315. KOCHETKOV, N. K. and A. E. VASIL'EV: Pyrrolizidine Alkaloids. III. Synthesis
of Some Derivatives of Dihydrosenecinoic Acid (3-Methyl-2-hydroxyheptane-
2,5-dicarboxylic Acid). Zhurn. Obshchei Khimii 32, 1703 (1962) [Chem. Abstr.
58, 6689 (1963)].

316. KOCHETKOV, N. K., A. E. VASIL'EV and S. N. LEVCHENKO: Synthesis of
Dihydrosenecinic Acid. Izv. Akad. Nauk (USSR) Otd. Khim. 1962, 2240
[Chem. Abstr. 58, 12 412 (1963)].

317. — — — Pyrrolizidine Alkaloids. VI. Total Synthesis of (±)-Integerrinecinic
Acid. Zhurn. Obshchei Khimii 34, 2202 (1964) [Chem. Abstr. 61, 9399 (1964)].

318. KOMPIŠ, I. and F. ŠANTAVÝ: Alkaloids of Senecio erucifolius. Collect. Czech.
Chem. Comm. 27, 1413 (1962) [Chem. Abstr. 57, 15 166 (1962)].

319. KOMPIŠ, I., H.-B. SCHRÖTER, H. POTĚŠILOVÁ and F. ŠANTAVÝ: Alkaloids of
Spreading Leafy Groundsel (Senecio erraticus BERT., ssp. barbaraeifolius KROCK).
II. Collect. Czech. Chem. Comm. 25, 2449 (1960) [Chem. Abstr. 55, 1673 (1961)].

320. KONOVALOVA, R. A.: Alkaloids of Senecio platyphyllus. Isolation of the
N-Oxides of Platyphylline and Seneciphylline. Doklady Akad. Nauk. (USSR)
78, 905 (1951) [Chem. Abstr. 46, 2086 (1952)].

321. KORETSKAYA, N. I., A. V. DANILOVA and L. M. UTKIN: Structure and Interrelationship of Senecic and epoxyJaconecic Acids. Zhurn. Obshchei Khimii
32, 3823 (1962) [Chem. Abstr. 58, 12 504 (1963)].

322. — — — Structure of the Alkaloid Renardine. Part 3. Structure of Dihydro-
deoxyothonecine. Khim. prirod. Soedinenii 1965, 22.

323. KUZOVKOV, A. D., M. D. MASHKOVSKIĬ, A. V. DANILOVA and G. P. MEN'SHIKOV:
Syntheses in the Series of Pseudoheliotridane and Heliotridane. Preparation
of Curare-like Substances. Doklady Akad. Nauk. (USSR) 103, 251 (1955)
[Chem. Abstr. 50, 5695 (1956)].

324. KUZOVKOV, A. D. and G. P. MEN'SHIKOV: Syntheses in the Pseudoheliotridane
Series. Zhurn Obshchei Khimii 21, 2245 (1951) [Chem. Abstr. 46, 8130 (1952)].

325. LABENSKIĬ, A. S., N. A. SEROVA and G. P. MEN'SHIKOV: Stereochemical
Transformations in the Heliotridane Series. Doklady Akad. Nauk. (USSR)
88, 467 (1953) [Chem. Abstr. 48, 2721 (1954)].

326. LALICH, J. J. and L. A. EHRHART: Monocrotaline-induced Pulmonary Arteritis
in Rats. J. Atherosclerosis Res. 2, 482 (1962) [Chem. Abstr. 59, 5611 (1963)].

327. LEISEGANG, E. C. and B. O. G. SCHÜLER: A Preliminary Study of the Infrared Absorption Spectra of the Senecio Alkaloids. J. South African Chem. Inst.
10, 1 (1957).

328. LEONARD, N. J.: Absolute Configurations of the Necines. Chem. and Ind. 1957, 1455.

329. — Senecio Alkaloids. In: R. H. F. Manske, The Alkaloids. VI. New York: Academic Press. 1960.

330. LEONARD, N. J., R. C. FOX, M. OKI and S. CHIAVARELLI: Cyclic Aminoacyloins. Ring-Size Limitation of Transannular Interaction between N and C_{CO}. J. Amer. Chem. Soc. **76**, 630 (1954).

331. LEONARD, N. J., M. OKI, J. BRADER and H. BOAZ: Cyclic Aminoacyloins and Aminoketones. V. Detailed Infrared Spectral Studies of Transannular Interaction between N and C_{CO}. J. Amer. Chem. Soc. **77**, 6237 (1955).

332. LIKHOSHERSTOV, A. M., A. M. KRITSYN and N. K. KOCHETKOV: Pyrrolizidine Alkaloids. IV. Total Synthesis of the Alkaloid, 1-Methylene-pyrrolizidine. Zhurn. Obshchei Khimii **32**, 2377 (1962) [Chem. Abstr. **58**, 9154 (1963)]; N. K. KOCHETKOV, A. M. LIKHOSHERSTOV and A. M. KRITSYN: Synthesis of 1-Methylenepyrrolizidine. Tetrahedron Letters **1961**, 92.

333. LIKHOSHERSTOV, A. M., V. N. KULAKOV and N. K. KOCHETKOV: Pyrrolizidine Alkaloids. VII. Stereoisomeric Conversions of Pyrrolizidine-1-carboxylic Acids. Zhurn. Obshchei Khimii **34**, 2798 (1964) [Chem. Abstr. **61**, 14734 (1964)].

334. LOUW, P. G. J.: A Note on the Alkaloid of *Crotalaria damarensis*. Onderstepoort J. Vet. Sci. Animal Ind. **25**, 111 (1952).

335. LUKEŠ, R. and M. JANDA: Cyclization of Aminodicarboxylic Acids. II. Synthesis of 1-Methylpyrrolizidine. Chem. Listy **52**, 450 (1958) [Chem. Abstr. **53**, 4252 (1959)].

336. MCKENZIE, J. S.: Some Pharmacological Properties of Pyrrolizidine Alkaloids and their Relationship to Chemical Structure. Austral. J. Exper. Biol. Med. Sci. **36**, 11 (1958) [Chem. Abstr. **52**, 11265 (1958)].

337. MCMILLAN, J. A. and R. E. DICKERSON: unpublished.

338. MAN'KO, I. V. and YU. G. BORISYUK: Chemical Investigation of *Cynoglossum officinale*. Ukrain. Khim. Zhurn. **23**, 362 (1957) [Chem. Abstr. **52**, 2187 (1958)].

339. MARKSON, L. M.: The Pathogenesis of the Hepatic Lesion in Calves Poisoned Experimentally with *Senecio jacobaea*. Proc. Roy. Soc. Med. (London) **1960**, 283.

340. MASAMUNE, S.: Structure of *iso*Seneciphyllic Acid. Chem. and Ind. **1959**, 21.

341. — Stereochemistry of Jacobine. J. Amer. Chem. Soc. **82**, 5253 (1960).

342. MASSAGETOV, P. S. and A. D. KUZOVKOV: Alkaloids of Plants, *Nardosmia laevigata*. Zhurn. Obshchei Khimii **23**, 158 (1953) [Chem. Abstr. **48**, 697 and 12758 (1954)].

343. MATTOCKS, A. R.: Strigosine, the Major Alkaloid of *Heliotropium strigosum*. J. Chem. Soc. (London) **1964**, 1974.

344. MATTOCKS, A. R., R. SCHOENTAL, H. C. CROWLEY and C. C. J. CULVENOR: Indicine: The Major Alkaloid from *Heliotropium indicum* L. J. Chem. Soc. (London) **1961**, 5400.

345. MEN'SHIKOV, G. P. and S. O. DENISOVA: Alkaloids of *Rindera echinata*. I. New Alkaloid, Echinatine and its Structure. Sbornik Statei Obschei Khim. **2**, 1458 (1953) [Chem. Abstr. **49**, 5496 (1955)].

346. MICHEEL, F. und W. FLITSCH: Eine einfache Synthese des 3,4-Dioxo-pyrrolizidins. (Pyrrolizidinderivate III.) Chem. Ber. **88**, 509 (1955).

347. NAIR, M. D. and R. ADAMS: Synthesis of *dl*-Isoretronecanol. J. Organ. Chem. (USA) **26**, 3059 (1961).

348. NOVELLI, A.: *Senecio* Species of the Argentine Republic. II. *Senecio pampeanus*. Anais farm. quím São Paulo **9**, 38 (1958) [Chem. Abstr. **53**, 3606 (1959)].

349. NOWACKI, E. and R. U. BYERRUM: A Study on the Biosynthesis of the *Crotalaria* Alkaloids. Life Sciences **1**, 157 (1962).

350. OREŠČANIN-MAJHOFER, B. and R. SEIWERTH: 3-Methylpyrrolizidine (2-Methyl-1-aza-bicyclo-[o.3.3]-octane). Monatsh. Chem. 83, 1298 (1952) [Chem. Abstr. 48, 668 (1954)].

351. PETROVA, M. F., S. I. DENISOVA and G. P. MEN'SHIKOV: Alkaloids of *Heliotropium lasiocarpum*. Decomposition of Lasiocarpic Acid and its Esters in Solutions in Alkali. Doklady Akad. Nauk (USSR) 114, 1073 (1957) [Chem. Abstr. 52, 2877 (1958)].

352. PORTER, L. A.: Naturally-occurring Lactones. Part I. Angularine, a new *Senecio* Alkaloid. Dissertation Abstr. 25, 105 (1964).

353. PORTER, L. A. and T. A. GEISSMAN: Angularine, a new Pyrrolizidine Alkaloid from *Senecio angulatus* L. J. Organ. Chem. (USA) 27, 4132 (1962).

354. PRETORIUS, T. P.: The Alkaloids of *Senecio paucicalyculatus* PLATT. Onderstepoort J. Vet. Sci. Animal Ind. 22, 297 (1949).

355. PUKHALSKAYA, E. CH., M. F. PETROVA and I. W. MANKO: The Action of Six Alkaloids (1-Methylpyrrolizidine Derivatives) upon the Growth of Hepatomas and Several Other Mutually Inoculable Animal Tumors. Bull. Exp. Biol. Med. (Moscow) 48, 91 (1959) [Chem. Abstr. 58, 2753 (1963)].

356. RED'KO, A. L.: Alkaloids from *Senecio borysthenicus*. Wiss. Arbb. höh. Lehranstalten Ukraine SSR 1956, 193 [Chem. Abstr. 53, 20695 (1959)].

357. ROBINSON, R.: The Structural Relations of Natural Products, p. 72. London and New York: Oxford Univ. Press. 1955.

358. ŠANTAVÝ, F.: Alkaloids of *Senecio erraticus* Subspecies *barbaraeifolius*. Planta Med. 6, 78 (1958).

359. ŠANTAVÝ, F., H. POTĚŠILOVÁ und R. KUBÍČEK: Isolierung des Rutins aus *Senecio erraticus*, ssp. *barbaraeifolius*. Collect. Czech. Chem. Comm. 24, 646 (1959).

360. ŠANTAVÝ, F., B. ŠULA und V. MANIŠ: Isolierung der Alkaloide aus *Senecio viscosus* L. und *S. rivularis* D. C. Collect. Czech. Chem. Comm. 27, 1666 (1962).

361. SAPIRO, M. L.: The Alkaloids of *Senecio bupleuroides* D. C. Onderstepoort J. Vet. Sci. Animal Ind. 22, 291 (1949).

362. SCHLOSSER, F. D. and F. L. WARREN: The *Senecio* Alkaloids. Part XVIII. The Structure of Hygrophylline, the Alkaloid from *Senecio hygrophylus* DYER and SM. J. Chem. Soc. (London) 1965, 5707.

363. SCHOENTAL, R.: Kwashiorkor-like Syndromes and other Pathological Changes in Rats as a Result of Feeding with *Senecio* Alkaloids (Isatidine). Voeding (The Hague) 16, 268 (1955).

364. — Hepatotoxic Action of Pyrrolizidine *(Senecio)* Alkaloids in Relation to their Structure. Nature 179, 361 (1957).

365. — Liver Lesions in Young Rats Suckled by Mothers Treated with the Pyrrolizidine *(Senecio)* Alkaloids, Lasiocarpine and Retrorsine. J. Pathol. Bacteriol. 77, 485 (1959).

366. — The Chemical Aspects of Seneciosis. Proc. Roy. Soc. Med. 1960, 284.

367. — Alkaloidal Constituents of *Senecio discolor* D. C. Retrorsine and Isatidine. J. Chem. Soc. (London) 1960, 2375.

368. — Pyrrolizidine *(Senecio)* Alkaloids and their Hepatotoxic Action. Biochem. J. 88, 57 P (1963).

369. — Alkaloidal Constituents of *Crotalaria fulva* ROXB. Fulvine and its N-Oxide. Austral. J. Chem. 16, 233 (1963).

370. — Liver Disease and "Natural" Hepatotoxins. Bull. World Health Organiz. 29, 823 (1963).

371. SCHOENTAL, R. and J. P. M. BENSTED: Effects of Whole Body Irradiation and of Partial Hepatectomy on the Liver Lesions Induced in Rats by a Single Dose of Retrorsine, a Pyrrolizidine *(Senecio)* Alkaloid. Brit. J. Cancer **17**, 242 (1963).

372. SCHOENTAL, R. and M. HEAD: Pathological Changes in Rats as a Result of Treatment with Monocrotaline. British J. Cancer **9**, 229 (1955).

373. SCHOENTAL, R. and P. N. MAGEE: Chronic Liver Changes in Rats after a Single Dose of Lasiocarpine, a Pyrrolizidine *(Senecio)* Alkaloid. J. Pathol. Bacteriol. **74**, 305 (1957) [Chem. Abstr. **52**, 2270 (1958)].

374. — — Evolution of Liver Lesions in the Rat after a Single Dose of Pyrrolizidine Alkaloids. Acta Unio Intern. contra Cancrum **15** (1959).

375. SCHOENTAL, R. and A. R. MATTOCKS: Hepatotoxic Activity of Semi-synthetic Analogues of Pyrrolizidine Alkaloids. Nature **185**, 842 (1960).

376. SCHRÖTER, H.-B. und F. ŠANTAVÝ: Isolierung der im gespreiztblättrigen Kreuzkraut *(Senecio erraticus* BERT., ssp. *barbaraeifolius* KROCK) enthaltenen Alkaloide. Collect. Czech. Chem. Comm. **25**, 472 (1960).

377. SEIWERTH, R.: New Synthesis of Pyrrolizidine (1-Azabicyclo[3.3.0]octane). Arhiv Kem. **23**, 77 (1951) [Chem. Abstr. **46**, 10183 (1952)].

378. SEIWERTH, R. and S. DJOKIĆ: Syntheses in the Pyrrolizidine and Indolizidine Series. Croat. Chem. Acta **29**, 403 (1957).

379. SEIWERTH, R. and B. OREŠČANIN-MAJHOFER: The Synthesis of 1-Methylpyrrolizidine (4-Methyl-1-azabicyclo[o.3.3]octane). Arhiv Kem. **24**, 53 (1952) [Chem. Abstr. **49**, 295 (1955)].

380. SJOLANDER, J. R., K. FOLKERS, E. A. ADELBERG and E. L. TATUM: α,β-Dihydroxyisovaleric Acid and α,β-Dihydro-β-methylvaleric Acid, Precursors of Valine and Isoleucine. J. Amer. Chem. Soc. **76**, 1085 (1954).

381. SYKULSKA, Z.: Alkaloids of *Cynoglossum officinale.* Acta Polon. Pharm. **19**, 183 (1962) [Chem. Abstr. **59**, 2876 (1963)].

382. TSUDA, K. and S. SAEKI: Syntheses of Pyrrolizidine, Indolizidine and Related Compounds. J. Organ. Chem. (USA) **23**, 91 (1958).

383. TSYRUL'NIKOVA, L. G., A. S. LABENSKIĬ and L. M. UTKINA: Alkaloids of *Lindelofia macrostyla* Plants. Zhurn. Obshchei Khimii **32**, 2705 (1962) [Chem. Abstr. **59**, 680 (1963)].

384. WAAL, H. L. DE and P. VAN TWISK: Die Chemiese Ondersoek van Vier *Senecio* Spesies van die Nasionale Krugerwildtuin. Koedoe (South Africa) **1964**, Nr. 7, 40.

385. WAAL, H. L. DE, A. WIECHERS and F. L. WARREN: The *Senecio* Alkaloids. Part XV. The Structure of Sceleranecic and Sceleratinic Acids and Sceleratine. J. Chem. Soc. (London) **1963**, 953.

386. WALI, B. K. and K. L. HANDA: Alkaloidal Constituents of *Senecio chrysanthemoides.* Current Sci. (India) **33**, 585 (1965).

387. WARREN, F. L.: The Pyrrolizidine Alkaloids. Fortschr. Chem. organ. Naturstoffe **12**, 198 (1955).

388. WARREN, F. L. and M. E. VON KLEMPERER: The *Senecio* Alkaloids. Part XIII. The Absolute Configuration of Heliotridane (1-Methylpyrrolizidine) and the "Necine" Bases. J. Chem. Soc. (London) **1958**, 4574.

389. WUNDERLICH, J. A.: The Molecular Structure of Retusamine, Otosenine, Renardine and Onetine. Chem. and Ind. **1962**, 2089.

390. YATES, S. G.: Paper Chromatography of Alkaloids of Tall Fescue Hay. J. Chromatogr. **12**, 423 (1963).

391. YATES, S. G. and H. L. TOOKEY: Festucine, an Alkaloid from Tall Fescue *(Festuca arundinacea).* Chemistry of the Functional Groups. Austral. J. Chem. **18**, 53 (1965) [Chem. Abstr. **62**, 11865 (1965)].

c) References Added in Proof.

ATAL, C. K., C. C. J. CULVENOR, R. S. SAWKNEY and C. W. SMITH: Grozemperine, a new Otonecine Ester from *Crotalaria semperflorens* VENT. Austral. J. Chem. (in press).

ATAL, C. K., K. K. KAPUR, C. C. J. CULVENOR and L. W. SMITH: A New Pyrrolizidine Aminoalcohol in Alkaloids form *Crotalaria* Species. Tetrahedron Letters **1966**, 537.

ATAL, C. K., R. K. SHARMA, C. C. J. CULVENOR and C. W. SMITH: Alkaloids of *Crotalaria rubiginoza* WILLD. Trichadesimine and Junceine. Austral. J. Chem. (in press).

BAKER, E. G., C. C. J. CULVENOR and L. W. SMITH: Usaramine, a New Pyrrolizidine Alkaloid from *Crotalaria usaramoensis*. Austral. J. Chem. (in press).

CHALMERS, A. H., C. C. J. CULVENOR and L. W. SMITH: Characterisation of Pyrrolizidine Alkaloids by Gas, Thin-layer and Paper Chromatography. J. Chromatogr. **20**, 270 (1965).

COOKS, R. G., F. L. WARREN and D. H. WILLIAMS: *Rhizophoraceae* Alkaloids. Part III. Cassipourine. J. Chem. Soc. (London) (in press).

CROUT, D. H. G., M. H. BENN, H. IMASEKI and T. A. GEISSMAN: Pyrrolizidine Alkaloids: the Biosynthesis of Seneciphyllic Acid. Phytochem. **5**, 1 (1966).

CULVENOR, C. C. J.: The Conformation of Esters and the Acylation Shift. NMR Evidence from Pyrrolizidine Alkaloids. Tetrahedron Letters **1966**, 1091.

CULVENOR, C. C. J., G. M. O'DONOVAN and L. W. SMITH: Alkaloids of *Crotalaria trifoliastrum* WILLD. and *C. aridicola* DOMIN. III. Additional Pyrrolizidine Derivatives. Austral. J. Chem. (in press).

CULVENOR, C. C. J., G. M. O'DONOVAN and C. W. SMITH: The Identity of the Aminoalcohol of Retusamine with Otonecine. Austral. J. Chem. (in press).

EDWARDS, J. D., Jr., T. HASE, C. HIGNITE and T. MATSUMOTO: *Senecio* Alkaloids: Synthesis of Necic Acids. J. Organ. Chem. (USA) **31**, 2282 (1966).

EDWARDS, J. D., Jr., T. MATSUMOTO and T. HASE: *Senecio* Alkaloids: Synthesis of Sarracinic Acid. J. Organ. Chem. (USA) (in press).

GANDHI, R. N., T. R. RAJAGOPALAN and T. R. SESHADRI: Chemical Components of *Heliotropium eichwaldi*. Current Sci. **35**, 121 (1966).

HART, N. K. and J. A. LAMBERTON: Pyrrolizidine Alkaloids from *Planchonella* Species (Family Sapotaceae). I. Alkaloids of *Planchonella thyrsoidea* C. T. WHITE and *P. anteridifera* (WHITE and FRANCIS). Austral. J. Chem. **19**, 1259 (1966).

KOOLAKOV, B. N., A. M. LIKHOSHERSTOV and N. K. KOCHETKOV: Pyrrolizidine Alkaloids. IX. The Complete Structure of Trachelanthamine. Zhurn. Obshchei Khimii (in press).

LIKHOSHERSTOV, A. M., B. N. KOOLAKOV and N. K. KOCHETKOV: Pyrrolizidine Alkaloids. X. Absolute Configuration of Alkaloid Viridiflorine. Zhurn. Obshchei Khimii (in press).

MATTOCKS, A. R. and F. L. WARREN: The *Senecio* Alkaloids. Part XIX. The Conversion of Retrorsine into Senecionine, and the Preparation of ^{14}C-Senecionine. J. Chem. Soc. (London) (in press).

WRIGHT, W. G. and F. L. WARREN: Rhizophoraceae Alkaloids. Part I. Four Sulphur Containing Bases from *Cassipourea* spp. J. Chem. Soc. (London) (in press).

(Received, November 15, 1965.)

Some Aspects of Virus Chemistry.

By H. FRAENKEL-CONRAT, Berkeley, California.

With 9 Figures.

Contents.

I. Introduction.

Viruses were recognized and identified as biological entities prior to their characterization by chemical and physical methods. In functional terms, viruses are exogenous biological agents which enter living cells, become replicated in them, and usually cause them to become diseased or die. Viruses differ from microorganisms in lacking the metabolic apparatus which enables the latter to replicate in or on nonliving nutrient media.

In physico-chemical terms, viruses are biological particles weighing as much as 10^6–10^9 hydrogen atoms, thus falling between the molecular weights of proteins and the weight of small bacteria. All viruses that have been purified are composed of protein and nucleic acid. However, the functional definition given above would include infectious nucleic acids among the viruses and it appears quite probable that some virus diseases are caused by naturally free-occurring nucleic acids.

Thus, the chemistry of viruses is the chemistry of proteins and nucleic acids. Since proteins are not absolutely obligatory components of viruses (as defined above), and since their chemistry is discussed in other chapters of this Series, we will give main emphasis to the chemistry of the nucleic acid moieties of viruses in this discussion. This is all the more justified since little definite chemical information exists about nucleic acids of molecular weights of the order of 10^6 to 10^8, which are not viral. Thus, the study of the chemistry of viral nucleic acids is the forerunner to the detailed investigation of cellular nucleic acids which is as yet in its infancy.

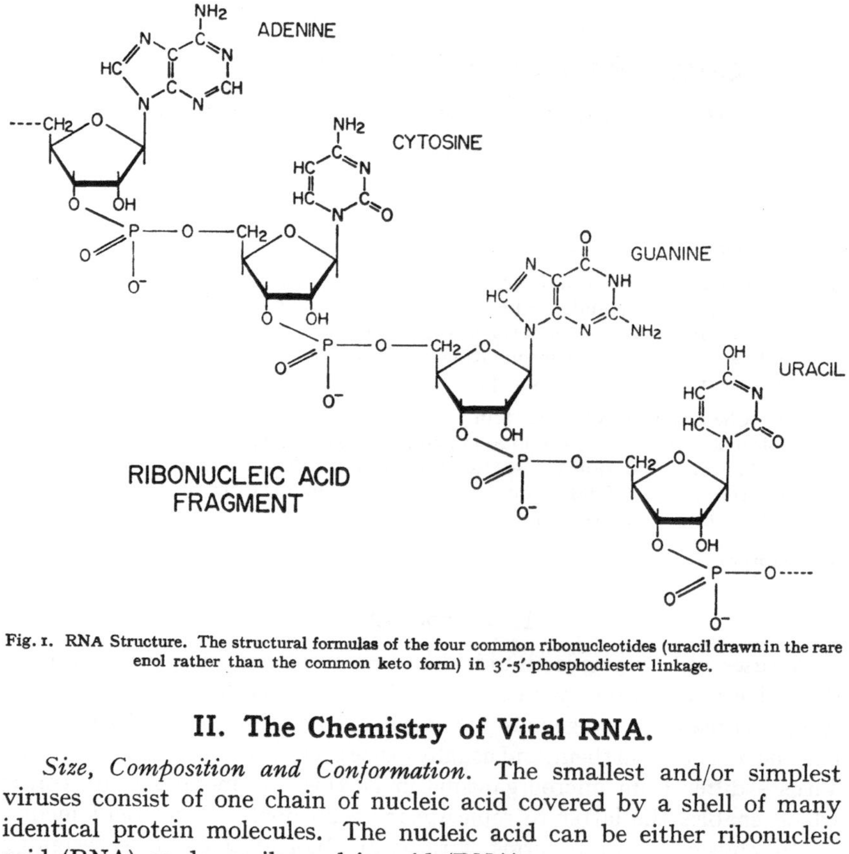

Fig. 1. RNA Structure. The structural formulas of the four common ribonucleotides (uracil drawn in the rare enol rather than the common keto form) in 3′-5′-phosphodiester linkage.

II. The Chemistry of Viral RNA.

Size, Composition and Conformation. The smallest and/or simplest viruses consist of one chain of nucleic acid covered by a shell of many identical protein molecules. The nucleic acid can be either ribonucleic acid (RNA) or deoxyribonucleic acid (DNA).

The RNA occurring in bigger viruses may be either single-stranded, or composed of pairs of complementary strands (p. 410). The DNA of the bigger viruses has only been found to occur in the double-stranded state. There seems to exist a rather narrow limit of about $1-2.5 \times 10^6$ for the molecular weight of complete single-stranded viral RNA or DNA, and this approximate amount suffices viruses of greatly different size and shape as well as host range. Examples are the RNA of mammalian viruses (polio virus, 7×10^6 particle weight; and influenza virus,

100 × 10⁶, both approximately spherical), of plant viruses (tobacco mosaic virus, TMV, 40 × 10⁶, rod shaped; and bushy stunt virus, 9 × 10⁶, spherical), as well as for the RNA and DNA, respectively, in the roughly spherical bacteriophages f 2 and Φ X 174 of about 5 × 10⁶ particle weights. Of special interest is an exceptionally small plant virus (2 × 10⁶), containing a correspondingly small RNA of 0.4 × 10⁶ molecular weight, which seems to be too small to perform all the functions of a virus and is infective only in conjunction with an otherwise unrelated group of viruses, the tobacco necrosis virus (TNV). The incomplete virus has been termed a satellite of TNV (SVTNV) (33).

Viral RNA consists, as far as is known, of only the four common ribonucleotides, adenylic, guanylic, cytidylic and uridylic acid (Ap, Gp, Cp, Up), in unbranched and uninterupted 3'–5'-phosphodiester linkage (Fig. 1). The range of base compositions of viral RNA preparations is not different from that of cellular RNA, although a given virus may show a characteristic pattern, e. g., relatively low C content in TMV and high C content in the turnip yellows mosaic virus (TYMV). Strains or mutants of a virus show the same general pattern, although minor differences may be detectable between groups of not closely related strains (37) (Table 1).

Table 1. Proportions of Nucleotides in RNA of Some Plant Viruses and Strains.

Virus	Adenylic acid	Guanylic acid	Cytidylic acid	Uridylic acid	References
	Moles per 100 moles total nucleotide				
Broad bean mottle	27.3	24.6	19.4	28.7	(79)
Cucumber 4	25.7	25.7	18.7	30.2	(36)
Potato X	32.2	21.8	23.8	22.2	(8)
Tobacco mosaic	28.0	24.0	20.0	28.0	(7)
Holmes ribgrass	28.0	24.0	20.0	28.0	(7)
Tobacco necrosis	28.0	25.5	22.0	25.7	(38)
Tobacco ringspot	23.9	24.7	23.2	28.2	(32)
Tomato bushy stunt	25.7	27.9	20.8	25.7	(21)
Turnip crincle (TC)	26.1	27.8	23.7	22.4	(64)
Turnip yellow mosaic (TYM)......	22.4	17.2	38.2	22.1	(64)
TYM, Rothamsted strain	21.3	16.7	41.7	20.3	(64)
Southern bean mosaic	24.3	25.8	24.0	26.3	(8)
Wild cucumber mosaic (WCM)	17.0	16.4	41.0	25.6	(80)

Among the DNA viruses there are many which contain unusual and characteristic bases. The most notable case is that of the T-even bacteriophages which contain 5-hydroxymethyl cytosine instead of cytosine. These phages further contain typically varying amounts of glucose or diglucose in α or β glucosidic linkage on the hydroxymethyl groups (*55*). Other odd DNA components are 5-hydroxymethyl uracil or deoxyuracil instead of thymine (T) (*31, 65*).

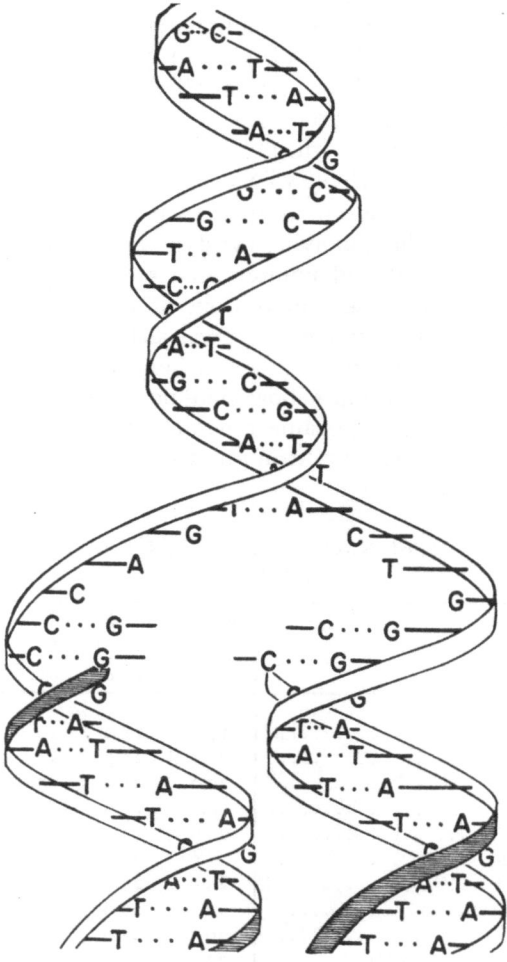

Fig. 2. DNA Structure. Schematic presentation of the double helix structure of DNA, in the state of being replicated by complementary base binding in the lower part.

When the molecular biologist speaks of the structure of RNA or DNA he usually refers to the three-dimensional structure as exemplified by the multiple interactions of two complementary chains or parts of chains. Thus, double-stranded RNA and DNA occur as intertwined helices with complementary nucleotide sequences on the two strands, as illustrated

References, pp. 433—437.

in *Fig. 2*. The main points of evidence for this are: (1) Nucleotide analysis show that A = T (or U) and G = C. (2) X-ray diffraction patterns show marked regularities characteristic for the specific helical structure. (3) The buoyant density of single-stranded nucleic acid is higher than that of double-stranded material. (4) The U. V. absorbance of double-stranded nucleic acid is lower than expected (hypochromicity) and loses this feature upon heating to high temperature with a sharp and characteristic transition ("melting point"). (5) Double-stranded nucleic acids are resistent to attack by certain nucleases.

In the absence of long range complementarity the bases of nucleic acid chains nevertheless tend to interact in media of sufficient ionic strength to dampen the ionic repulsions due to their phosphate charges. However, these random interactions of short chain segments or single bases are even cumulatively weak. Thus, single-stranded RNA usually 'melts' over a wide temperature range (e. g. 25–50° in 0.05 M salt), and is hyperchromed in the absence of divalent metals, or at monovalent cation concentrations below 0.01 M. In contrast to typically double-stranded DNA, typical single-stranded RNA has no definite three-dimensional structure.

Nucleotide Sequences and End Groups. In classical chemical terms all nucleic acids have a definite structure. Though mostly indirect, the evidence is strong that they have a unique sequence of nucleotides (or deoxynucleotides) and that it is this infinite range of possible nucleotide sequences that empowers nucleic acids as the specific agents of genetic information storage and transfer.

The elucidation of the nucleotide sequence of viral nucleic acids is unfortunately an extremely difficult and laborious task, which may require several decades for its accomplishment. The paucity of methodological tools for sequence analysis is even greater for polydeoxyribonucleotides than for polyribonucleotides. The simplicity of composition of viral RNA and the absence of rare base components which could serve as landmarks represent additional disadvantages for structural analysis. The main problem is certainly that of magnitude, since typical viral nucleic acids contain 3500 to 7000 nucleotides, i. e. 1000–2000 of each of its four components.

Before describing the more limited aims and the approaches which are now being tested in the endeavor of determining long nucleotide sequences, I would like to insert a brief discussion of a group of nonviral RNAs, the structure of which is currently being elucidated with great success. There is known to occur in all living cells a group of relatively small RNA molecules, which combine with and thus transport the activated amino acids, aligning them for sequential incorporation into newly synthesized proteins. Several of this group of over 40 different S-RNAs (one or more for each amino acid) have been purified. One of these (an alanine S-RNA from yeast) is composed of 78 nucleotides and contains like all members of this class a number of "odd" bases [these nine are: 1-methylguanine, N-2-dimethylguanine, dihydrouracil (two), hypoxanthine, 1-methylhypoxanthine, pseudouracil (two), thymine]. It appears, in contrast to what was said earlier, that this group of small RNA molecules has a definite secondary structure with groups of 2–4 nucleotides interacting with complementary sequences in the same chain, thus stabilizing nonpaired segments of the chain which can then combine with amino acids, ribosomes, and messenger RNA, respectively, and thus achieve their functional purpose. Thus, this chain of a molecular weight of about 26 000 is believed to maintain a fixed three-dimensional structure. It was a major achievement in the field of nucleic acid chemistry that the nucleotide sequence of this molecule has recently been estabilshed (29). This was achieved by ingenious use of the methods to be described below, and it was feasible because of (1) the above described secondary structure which made it possible to selectively break the molecule by carefully controlled enzyme treatment in one or a few exposed places, (2) the presence of one or two residues of several of the odd bases, and (3) the comparative smallness of this RNA. Thus, rather than encouraging virus nucleic acid chemists, this achievement further confirms their realization of the magnitude of their own problem.

Two main avenues of approach exist for the investigation of nucleic acid structure. One consists in fragmenting the RNA with specific endonucleases (enzymes which attack particular sites along the chain), analysing the nucleotide sequence of the fragments and finally, deriving their relative positions from overlapping sequences, the location of odd bases or exceptional sequences, etc. The other approach consists in systematically degrading the nucleic acid from one or the other end of the chain by chemical or enzymatic methods. The latter techniques are also frequently resorted to in characterizing the fragments obtained in the first mentioned method of approach.

Endoattack. Two enzymes are known which attack RNA at specific sites. Pancreatic RNase attacks the 5'-phosphate ester bond following, or on the right of, pyrimidines (a on *Fig. 3a*), while the T-1 nuclease of takadiastase attacks that bond on the right of guanine residues (b on *Fig. 3a*). Of several pyrimidines in a row all but the first one are released as mononucleotides by the pancreatic enzyme, and the same result is obtained with several guanines in a row upon treatment with the T-1 enzyme. The amounts of mononucleotides released, as well as the relative abundances of di-, tri-, tetra-, etc. nucleotides terminating in pyrimidines or guanines, respectively, is characteristically different for different RNA preparations (9, 44–46). With small RNA molecules an appreciable part of the total structure can be derived from the fragments obtained by these two enzymes. The controlled use of T-1 under very gentle conditions (0°) has been the key to the selective splitting at only one or a few bonds in alanine S-RNA which made the elucidation of the complete structure of this molecule possible (29).

The results of the application of these methods to the RNA of several viruses and of groups of strains more or less closely related to them are summarized in *Table 2*, p. 414.

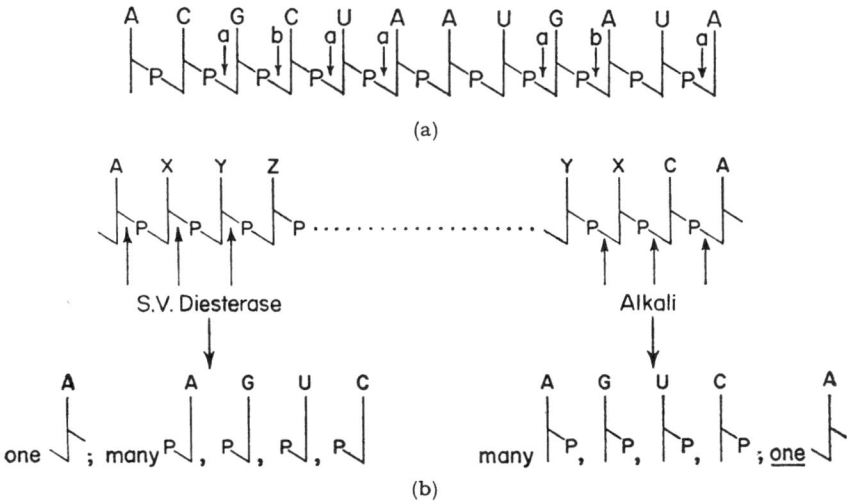

Fig. 3. Degradation of RNA. In these schematic presentations the vertical line represents the ribose, the letters A, G, U, C; X, Y, Z the purine and pyrimidine bases (identified or not) and the P the phosphate groups, in either single or double ester linkage to ribose. (a) Attack of pancreatic ribonuclease *a* or T-1 ribonuclease from takadiastase *b* on oligonucleotide. (b) Degradation by alkali or snake venom diesterase to 2'- or 3'- and 5'-nucleotides, respectively. One terminal residue appears as a nucleoside.

No enzymes are known to attack DNA with a similar degree of specificity.

Exoattack for Identification of End Groups. All methods to search for a single end group or nucleotide sequence in a molecule as big as viral RNA require forbiddingly large amounts of material or preferably the use of nucleic acids labeled with radioactive elements (P^{32} or C^{14}). This is usually done by growing the virus in the presence of labeled metabolites, which may be as simple as inorganic phosphate or carbon dioxide.

Another aspect, important for all structural studies but particularly so when focusing on the end groups of a chain polymer, is the homogeneity of the preparation under study. If an RNA composed of 6000 nucleotides contains only 2% of short fragments (e. g. averaging 120 nucleotides in length), these would double the total number of end groups. Fortunately, most of the end groups of fragments carry a 3′-phosphate group and thus differ characteristically from the bona fide end groups of those viral RNAs that have been studied. It is nevertheless of obvious importance that the large polynucleotide under study, which in the case of TMV-RNA has a sedimentation constant of about 30 S, be as free of fragments or smaller contaminating polynucleotides as possible. Density gradient centrifugation has been the most effective tool to achieve this (*18*).

Table 2. Oligonucleotides in Pancreatic Ribonuclease Digests of Viral RNA [SYMONS et al. (*64*)].

	% of Nucleotides Fractionated			
	TYM, wild type	TYM, Rothamsted	WCMV	TCV
Cp	26.5	29.2	31.0	13.5
Up	14.6	13.4	18.3	12.6
ApCp	10.1	10.0	6.8	6.5
ApUp	5.6	4.9	4.5	5.1
GpCp................	6.3	6.6	7.1	5.4
GpUp	4.1	3.7	4.0	5.4
ApApCp	4.0	4.5	2.1	3.5
ApApUp.............	1.5	2.1	1.3	2.5
GpApCp	6.7	6.4	2.7	2.6
ApGpCp			2.3	3.5
GpApUp	1.8	1.4	1.6	2.6
ApGpUp	1.3	1.1	1.2	2.7
GpGpCp			1.6	2.2
ApApApCp...........			1.0	0.7

(p stands for a phosphate group; when to the right of a nucleoside: 3′-phosphate, when to the left: 5′-phosphate.)

The determination of the end groups of nucleic acids usually starts with a treatment with a phosphomonoesterase (PME), usually the alkaline phosphatase of *Escherichia coli*, which splits monoesterified phosphate, as well as consecutively di- or triphosphates off the 5′-OH group at the left end of the chain or the 3′- (or 2′-) OH group at the right end of the chain. If no phosphate is released, the interpretation that neither end was phosphorylated appears justified. If phosphate is released, the techniques to be described indicate its location. Of those viral RNAs which were tested, TMV-RNA contained no appreciable amounts (i. e. < 1 residue/mole) of PME susceptible phosphate groups (*18*) while this is still uncertain for MS2-RNA.

The simplest method for identification of end groups in RNA is by degradation of the RNA to a mixture of 2′- and 3′-nucleotides by means of 0.1 N alkali at 25—37° *(Fig. 3b)*. If the right end of the chain is not phosphorylated, that residue alone will appear as a nucleoside, and can be identified as such. If no nucleoside is found, this could be due to it being phosphorylated or otherwise blocked. If the left end is phosphory-lated it will appear as the only nucleoside 3', 5'-diphosphate, again clearly different from the nucleotides. Chromatographic or electrophoretic methods can be used to separate the nucleosides and the nucleoside diphosphates from the bulk of the nucleotides. TMV-RNA and MS2-RNA were both found by this technique to carry an A in the right end of the chain (60, 62). DNA is not degraded by alkali.

Snake venoms contain an enzyme which attacks RNA (and DNA) by splitting the 3'-phosphate ester bond, thus degrading the nucleic acids to 5'-nucleotides. This enzyme transforms the left terminal residue, if unphosphorylated, to a nucleoside, and the right terminus, if phosphory-lated, to a 3',5'-diphosphate (Fig. 3b). The results obtained with this method are unfortunately markedly affected by the presence of frag-mented molecules and of contaminating endonucleases which create fragments in the course of the digestion by the snake venom exonuclease. This potential error is serious because most fragments carry no phosphate on the 5'-OH end, and thus yield nucleosides upon snake venom diesterase digestion. With the most highly purified preparation of enzyme and TMV-RNA available the total amount of nucleosides was as low as 2.5 (1.3 adenosine and 0.3–0.5 for each of the other 3 nucleosides) (63), but frequently this background is appreciably higher. Yet the conclusion that adenosine is also the bona fide left end of the TMV-RNA chain appears justified, since it is always one residue higher than the "noise level" due to fragments.

Determination of Terminal Sequences. The methods which have been outlined serve to identify the terminal nucleoside. Information about terminal sequences can be obtained in various ways. If the right end of the chain is unphosphorylated (or has been dephosphorylated by the monoesterase), and if the neighboring base is a pyrimidine or a guanine, then pancreatic ribonuclease or T-1 nuclease, respectively, liberates the terminal residue as the only nucleoside present in the mixture of mono- and oligonucleotides. If the third or fourth residue is a pyrimidine or guanine, then the same methods may yield a single dinucleoside monophosphate or trinucleoside diphosphate — products which can in principle be separated from the bulk of the digestion products, all of which carry a 3' terminal phosphate. This approach has indicated that a pyrimidine is located next to the right terminal A in both TMV-RNA and MS2-RNA, for one mole of adenosine was released from both viral RNAs by pancreatic ribonuclease (61, 74).

Alternatively, the end group can be labeled and the labeled terminal nucleotide or family of nucleotides may be identified. One procedure is to bind C^{14}-aniline by means of dicyclohexyl carbodiimide to the 5'-phos-phate end, if any (42). Introduction of a P^{32} phosphate group on this end by means of the newly discovered enzyme polynucleotide phosphokinase represents another means of identifiying the 3'-linked end, and may actually be combined with the anilid method (43).

Several reagents have been proposed for labeling the right end of the chain. RNA with free terminal 2'- and 3'-OH groups represents a glycol which is susceptible to selective oxidation by periodate. The resulting terminal dialdehyde can be reduced to a di-alcohol by tritium containing sodium borohydride and the end group thus labeled. C^{14} containing isonicotinic acid hydrazide or semicarbazide has also been advocated as blocking agent (30, 58). The nicotinic hydrazone appears more stable than the semicarbazone and this method has successfully been applied to terminal sequence analysis of ribosomal RNA, but not yet to viral RNA.

Stepwise Degradation of RNA. Finally, a systematic method of stepwise degradation, first proposed in 1954 (73), has been adapted in recent years to the requirements of macromolecular RNA. The periodate oxidation reaction mentioned above is the first step. Aldehyde groups are known to weaken ester linkages on the β-carbon atoms, and thus the transformation of the $C_{(3)}$ of the ribose to a carbonyl function weakens the $C_{(5)}$ phosphoester bond. It was found that aniline catalyses this β-elimination reaction optimally at pH 5, in contrast to the pH 7–9 required with aliphatic amines as catalysts, or pH 10–12 required for $OH^{\ominus}$ catalysis (59). Thus, the oxidized terminal nucleoside can be eliminated, and a chain terminating in a 3'-linked phosphate group remains. This can be dephosphorylated by alkaline monoesterase, resulting in a new terminal glycol group, which can again be subjected to the cycle of oxidation, β-elimination and dephosphorylation. The eliminated fragment of the terminal nucleotide is unstable and yields the free base upon chromatography in acidic and basic solvents. Thus, the sequence of bases at the 5'-linked end of an RNA can be systematically elucidated (*Fig. 4*).

The applicability of such a method depends largely on how discriminating the elimination step is in splitting the last and no other of the 12800 diester bonds in the molecule. Aniline catalysis is at least one order of magnitude better in this respect than the aliphatic amines advocated for this purpose in recent years (41). With aniline it was possible to carry the reaction with TMV-RNA through several cycles, and get what seems to be a definite predominance of one base at 3 or 4 steps. These data indicate the sequence —(G)—(C)—C—C—A, with decreasing confidence from right to left, for TMV-RNA. The fact that the level of label in the other three bases increases progressively at each step may represent an indication of gradual fragmentation of the molecule. Sedimentation analyses also indicate that the 30 S material disappears after several cycles. The total amount of all bases found is slightly less than

References, pp. 433—437.

one per mole at each step (59). This could in part be due to incomplete removal of the oxidized ribose from the bases; in any case one would expect the total to increase, as degradation is presumed to contribute additional end groups, and it is not clear why this fails to occur. Improvements on the procedure must clearly be sought through further detailed study.

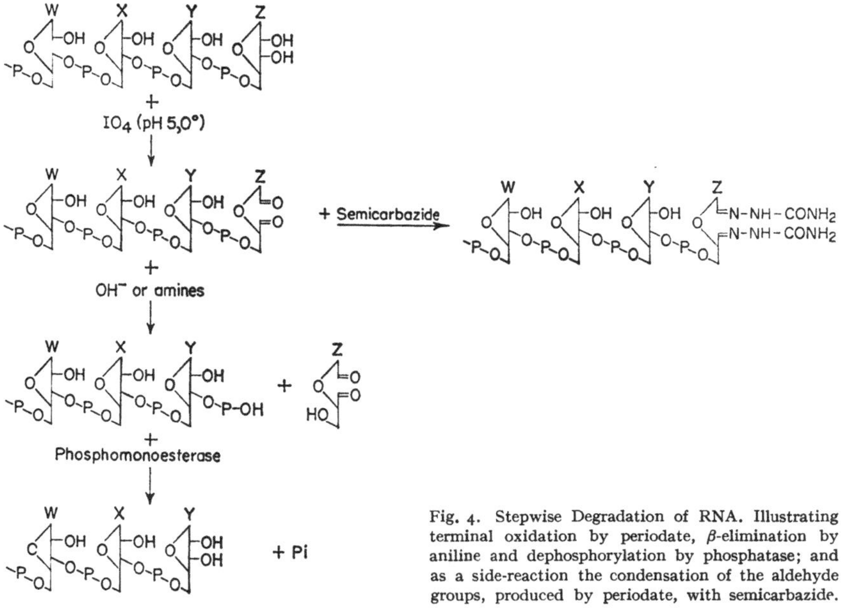

Fig. 4. Stepwise Degradation of RNA. Illustrating terminal oxidation by periodate, β-elimination by aniline and dephosphorylation by phosphatase; and as a side-reaction the condensation of the aldehyde groups, produced by periodate, with semicarbazide.

A less laborious manner for the degradation of nucleic acids from one end or the other would seem to be by means of exonucleases. Snake venom phosphodiesterase, while able to degrade nucleic acids completely, has been shown to start its attack preferentially at the 5'-linked unphosphorylated end. Thus, limited treatment by this enzyme appears to liberate 5'-nucleotides from the right end of the chain. Inversely, spleen diesterase and a recently described enzyme from *Bacillus subtilis* attack with preference from the 3'-linked end of the chain, splitting off 3'-nucleotides consecutively from the left end (34).

Yet the action of such enzymes on macromolecular RNA, particularly when the substrate is not completely homogeneous, appears difficult to interpret. The disappearance of the known —CCA sequence from S-RNA under the influence of the venom diesterase indicates that in this case the reaction proceeds in the expected manner. When spleen diesterase was used with the same substrate, stepwise degradation from the opposite end was assumed to occur but the expected release of pGp was not proven.

When TMV-RNA was treated with the snake venom diesterase at 0° the right terminal pA was readily released. No predominance of adenosine was found upon subsequent alkali degradation (*50*). However, the pattern of nucleotides released upon further treatment with the diesterase (predominantly pA and pU) did not agree with the terminal sequence as later determined by the stepwise method described above and it is now believed that further enzyme degradation of TMV-RNA proceeds very slowly and that the release of predominantly pA and pU is due to pre-ferential attack by the enzyme on these contaminating short fragments of TMV-RNA which are particularly rich in A and U and thus less structured than G and C rich fragments.

Polynucleotide phosphorylase from *E. coli* or *Azotobacter vinelandii* is also believed to attack polynucleotides preferentially from the un-phosphorylated right end, releasing nucleoside 5'-diphosphates. Con-trolled treatment of TMV-RNA with this enzyme released again from the beginning more A, and U, than G and C, but in this case it could be shown by alkali degradation that the terminal A remained intact, when many nucleotides were released. Thus, it was concluded that this enzyme attacked only the smaller fragments rich in A and U (*52*). The fact that TMV-RNA retained most of its infectivity after the treatment with polynucleotide phosphorylase supports this conclusion. Loss of 73% of the infectivity, indicative of an average of one inactivating event per molecule according to the Poisson distribution, or one "lethal hit", was observed only when about 150 nucleotides had been released by the enzyme. With the snake venom diesterase the release of about 4–5 nucleotides corresponds to one lethal hit.

Controlled exonuclease attack of viral RNA from the left end of the chain has been attempted only in preliminary manner. The fact that more adenylic and uridylic acid is released than cytidylic or guanylic may again be due to the degradation of A and U rich fragments, rather than representing a true indication of the nucleotide composition of the left end of the TMV-RNA chain. However, in contrast to attack from the other end of the chain, this enzymatic degradation rapidly inactivated the RNA, with one lethal hit resulting from the release of two or possibly one nucleotide per 6400, while no noticeable internal chain breakage had occurred at this stage as indicated by an unchanged proportion of the U. V. absorbing material of 30 S (*54*). Thus, it appears that the biological activity of this viral RNA is more critically dependent upon the integrity of the 3'-linked than of the 5'-linked end of the chain.

While we know little about the structure of viral RNA, we know less about that of viral DNA. Obviously the smallest representatives of this group would be the most encouraging subjects for study. As stated some of these occur in single-stranded form in small viruses and thus have

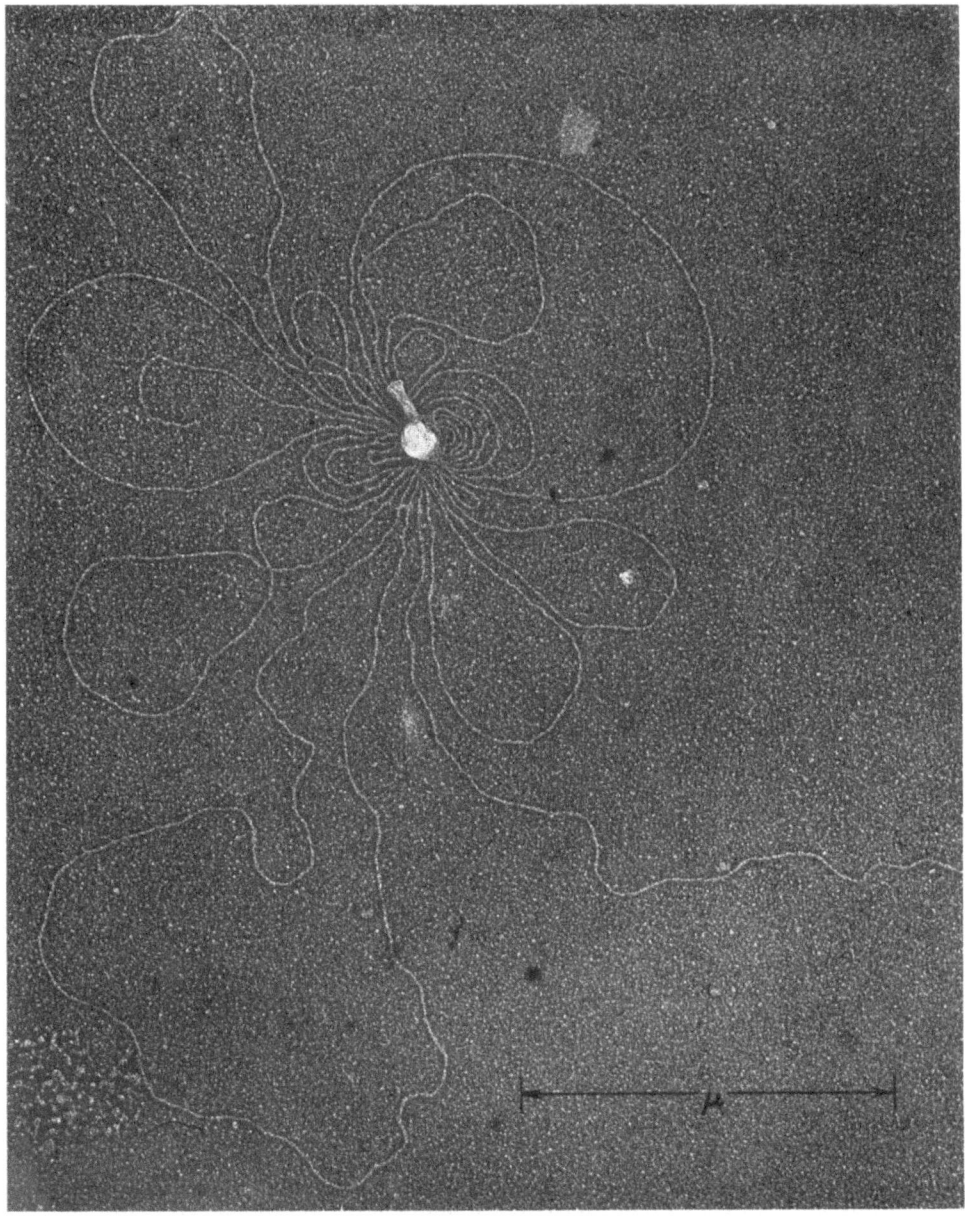

Fig. 5. Phage-DNA. One DNA molecule, a thread about 50 μ long, comprises the entire DNA of the T4 bacteriophage. (Courtesy of Dr. A. KLEINSCHMIDT.)

compositions with unequal amounts of A, T, G and C. The attempts to detect and determine end groups in Φ X 174 DNA by means of exo-nucleases were unsuccessful. The fact that these enzymes caused no loss of infectivity, until degradation of the viral DNA became apparent, combined with data on the physico-chemical properties of this viral DNA, led to the conclusion that it represented a cyclic molecule (10). This was a completely novel concept, for no other natural polymeric macromolecules of such dimensions had been shown to be cyclic in nature. The question whether the ring is closed by a 3′,5′-phosphodiester bond, or whether there is some non-DNA material bridging the ends of the chain remains unresolved.

It appears that the cyclic nature is not confined to the single-stranded DNA of certain small phages, but that also the somewhat bigger double-stranded DNA of polioma virus and others are cyclic. On the other hand the DNA of the T-even bacteriophages consisting of about 200 000 base pairs, definitely does not occur in the phage particle in cyclic form, since electronmicrographs clearly show the two ends of each molecule and the length of the double-stranded chain has been measured with some exactitude (6, 35) (Fig. 5, p. 419).

III. The Chemistry of Viral Protein.

Molecular Size. All viruses appear to consist of one molecule of a large RNA or DNA (molecular weight of the order of 10^6–10^9) and many molecules of relatively small-molecular protein (molecular weight of the order of 10^4–10^5). TMV and other rod-shaped viruses contain only one protein consisting of a single peptide chain; but some of the small spherical viruses seem to contain two or more peptide chains, and the more complex viruses carry different proteins in different parts of each particle. The primary function of virus proteins is the formation of a protective coat which must be readily disassembled prior to replication of the nucleic acid, and quickly assembled during maturation of the virus. This is achieved by means of small proteins with particularly high affinities for specific interaction and aggregation. The fact that some virus proteins are easily obtained in pure form and are comparatively small, has led to their being used for sequence studies. On the other hand, their tendency to aggregate has rendered their physico-chemical characterization a difficult and ever-challenging job.

Because of the tendency of virus proteins to aggregate it was due to analytical data, rather than to physico-chemical parameters, that first indications of their true molecular weights were obtained. Amino acid analyses, which are now quite routinely performed by automatized equip-ment, gave a minimal molecular weight of 10000–20000 for many virus proteins (i. e. one mole of cysteine and two of lysines per 18 000 g in TMV protein; two of tryptophane and three of histine, arginine and tyrosine in 22 000 g TYMV, etc.).

This minimal molecular weight was later identified with the true molecular weight by means of end group analyses. The methodology for the determination of N-terminal amino acids has been well developed over the past 20 years, but it soon became evident that most virus proteins lacked terminal amino groups. This mystery was resolved when it was found that the terminal serine of TMV-protein was acetylated and the same could be proven for the methionine of TYMV and other viral N-termini (26, 39).

The first indication of the true chain-length of a virus protein came from a method which in general is quite unreliable and inadequate for this purpose. The pancreatic enzymes, the carboxypeptidases, degrade proteins by splitting off one amino acid at a time from the carboxyl end. This usually yields progressively more complex mixtures of free amino acids. Yet from the TMV, or its protein, only threonine was split off and the quickly reached maximum corresponded to one threonine per 18 000 g, i. e., the minimal molecular weight as derived from amino acid analyses (27, 28). That this threonine was really C-terminal was confirmed by the chemical method of hydrazinolysis: the protein is heated with dry hydrazine until all its peptide bonds are replaced by hydrazide bonds, only the C-terminal amino acid being free (41). Thus, the first definitive evidence concerning the chain-length of a viral protein came from the C-terminal analysis of TMV-protein by carboxypeptidase, and that finding represented the stimulus for complete sequence analysis of this protein.

Amino Acid Sequence. Studies of the complete amino acid sequence of virus proteins have been performed by what is now regarded as standard procedure, although the methodology was constantly being developed and improved in the course of these studies. Trypsin is the most commonly used primary splitting agent, since it shows great specificity in degrading virus and other proteins exclusively at the lysine and arginine residues. Another agent of good specificity is cyanogen bromide (at about pH 1–2) which splits peptide chains only at methionine residues (25).

Other enzymatic and chemical agents have been found to act with sufficient selectivity for a given peptide under carefully controlled conditions, but none has the general applicability of trypsin and CNBr.

The greatest subsequent problem is that of separating the fragments. Most commonly used are the "finger printing" or mapping techniques which fractionate the mixture of peptide fragments by two-dimensional electrophoresis and chromatography on paper or thin layer plates. The advantage of these procedures lies in the ease with which a characteristic peptide pattern can be obtained on a microscale, and differences between related patterns can be detected. The disadvantage lies in the difficulty of eluting many of the bigger peptides from the paper, and the impossibility of obtaining enough material for further purification, followed by compositional and sequential analysis.

Column chromatographic separation is applicable to both smaller and bigger amounts of material. Fractionation on Dowex-I with the use of volatile buffers of decreasing pH has given reproducible patterns, characteristic of a given peptide mixture — the basic peptides emerging first, followed by neutral and acidic peptides. Besides size and negative charge, the content of aromatic residues in the peptides decreases their rate of movement on this column (22).

Alternatively, Dowex-50 and increasing pH can be used to achieve better resolution of negatively charged peptides. That method has been automatized by use of the amino acid analyzer (69). By means of a stream divider, it is possible to obtain simultaneously an elution pattern, which is characteristic for a given mixture, and sufficient material for further purification and characterization of each peptide. Further purification is often not necessary for the peak tubes, but the shoulders of peaks are frequently contaminated with other peptides. One-dimensional paper chromatography with butanol-acetic acid-water or with pyridine-butanol-acetic acid-water appears the best method for secondary purification of peptides isolated by ion exchange chromatography, but electrophoresis or further ion exchange column fractionation have also been employed.

Peptides, the analysis of which indicates close to integer values for the component amino acids are usually considered pure enough for sequence analysis. The carboxy-terminal amino acid is often defined by the method of degradation of the protein (i. e., lysine or arginine after tryptic digestion). The N-terminal amino acid is determined by dinitrophenylation or the phenylisothiocyanate method (15). The latter can also be used to determine the sequence of several residues from the N-terminus. The enzyme, leucine aminopeptidase also degrades peptides sequentially from the N-terminus and can be used alternatively or better, additionally. On the other hand, a mixture of carboxypeptidases A and B (i. e., commercial carboxypeptidase) degrades peptides from the carboxyl end.

For up to octapeptides these methods may suffice under favorable circumstances. Frequently, however, tryptic degradation yields peptides of 15–40 members and the products of CNBr degradation are also generally quite long. A secondary fragmentation of such peptides after their isolation and purification is then advisable. Any tryptic peptide containing methionine can be split secondarily by CNBr, and any CNBr-peptide containing lysine or arginine by trypsin. Treatment of peptides with weak acid (HCl at pH 1.7–2.0) at 100° leads to a release of aspartic acid residues and more slowly, after loss of the amide groups, of asparagine residues, and the method can be used for the selective fragmentation of peptides containing these residues (69). Finally, enzymes of limited specificity can be used to a desired point of 1–4 bonds split. Under such conditions chymotrypsin which prefers large hydrophobic residues or subtilisin which favors hydrophilic groups are found to preferentially split certain bonds in a given peptide and a limited number of fragments result which can be separated and sequentially analyzed (23a).

Peptide Sequences. The remaining problem lies in determining the sequential order of such subfragments of a bigger peptide, as well as the order of the primary fragments in the original protein. The identification of the N- and C-terminal peptide is usually automatic. This is particularly simple after tryptic digestion, when the C-terminal peptide is the only one lacking arginine or lysine, while the N-terminal one, in the case of plant virus proteins, is usually acetylated. The position of other peptides must be more laboriously determined. Blocking of the lysine amino group by acylation confines the trypsin action to the arginine residues and analysis of these bigger fragments established the sequential order of some peptides. The same result was obtained by comparing strains of TMV which included among their differences from the wild type the replacement of one or several lysine or arginine residues by nonbasic residues (74a). The more generally applicable method is the systematic search for overlap peptides after degradation of the protein by a different agent than that used primarily. Thus, if trypsin was used as the primary agent, then chymotrypsin may be used to obtain overlap peptides. With trypsin as primary splitting agent, overlap peptides in a chymotryptic digest will be any basic peptide with at least one amino acid residue on each side of the lysine or arginine. For proteins of the size of virus proteins, the amino acid sequence of most of such peptides will serve to determine in unique manner the sequential arrangement of two of the originally studied tryptic fragments. The location of disulfide bonds represents a similar and often more difficult problem, but disulfide bonds have not yet been encountered in virus proteins.

The amino acid sequence of common TMV (the geneticists use the term "wild type" to distinguish the common type from the usually less viable strains or mutants) was reported in 1960 from two laboratories, and after a few minor corrections it can now be regarded as definitively established (3, 23a, 68) (*Fig. 6*, p. 425). The structure of many related strains showing only 1–3 amino acid replacements are known with similar certainty (23a, 66, 67). One strain called Dahlemense, which differs extensively, yet shows definite similarities, has also been analyzed in sequential detail (78) (Fig. 6). Other strains differing appreciably from the wild type have only partly been analyzed.

Partial amino acid sequence work has been done on the TYMV (26) and a number of other plant viruses. The small RNA phages which have been discovered in recent years in various laboratories and can be isolated in appreciable amounts are now the subject of intense study in terms of protein structure. It appears that most of these are closely related, differing by only one or a few amino acid replacements (36 a). The complete amino acid sequence of one has recently been reported (72 a).

The various functional parts of the T-even phage proteins have been separated, but little has as yet been reported concerning their chemical fine structure.

Chain Conformation. Peptide chains are transformed into functional proteins through the conformation of their chain(s). It has only been realized in recent years that the unique conformation of a native protein may be the consequence solely of its amino acid sequence. This was observed at about the same time with pancreatic ribonuclease and with the TMV protein. TMV-protein shows the characteristic properties of native proteins. Its sulfhydryl group is not autoxidizable nor does it react with most of the typical —SH reagents (12). Its tyrosine groups are also of variably lowered reactivity (14). In regard to function, the characteristic tendency of the protein to aggregate helically and form rods of about 18 mμ diameter is also a property of the native protein (*Fig.* 7, p. 426). This aggregation in turn renders the protein more native as judged by the nonreactivity of its groups, its heat stability, enzyme resistance and other criteria. All these properties are lost if the protein is denatured by heat, surface forces, etc. The protein then aggregates randomly and precipitates rather than forming soluble rod-shaped particles. However, if the protein is denatured under conditions where it remains soluble (alkali or urea) or if it is redissolved in such media after denaturation and slowly brought back to physiological conditions (e. g., dialysis against dilute neutral phosphate buffer), it then reacquires its typically native properties, including its ability to form rods upon lowering the pH (1). Thus, it appears that transition is freely possible from the random coil state assumed for denatured proteins in dissociating solvents to the specific three-dimensional conformation of the protein required for rod formation. The exact nature of this conformation is not yet known for this as for

---

Fig. 6. Amino Acid Sequence of TMV-Protein, Dahlemense Strain, and other Mutants. The upper line (V) gives the sequence of amino acids in wild-type TMV (23 a), the lower line (D) that of the Dahlemense strain (78) as recently modified (3 a). Amino acids differing in V and D are underlined. The* denotes sites at which changes were observed frequently in mutants resulting from nitrous acid treatment of the RNA of common TMV, as follows: Res. Nr. 5: thr→ileu; Nr. 20: pro → leu; Nr. 25: aspN → ser; Nr. 33: aspN → ser; Nr. 46, 61, 134: arg → gly; Nr. 81: thr → ala; Nr. 97: glu → gly; Nr. 107: thr → met; Nr. 129: ileu → val, thr; Nr. 138 and 148: ser → phe; Nr. 156: pro → leu.

References, pp. 433—437.

Fig. 6 — Amino acid sequence comparison (V and D chains).

Position	V	D
1	Acetyl-Ser	Acetyl-Ser
2	Tyr	Tyr
3	Ser	Ser
4	Ileu	Ileu
5	Thr*	Thr
6	Thr	Ser
7	Pro	Pro
8	Ser	Ser
9	GluN	GluN
10	Phe	Phe
11	Val	Val
12	Phe	Phe
13	Leu	Leu
14	Ser	Ser
15	Ser	Ser
16	Ala	Val
17	Try	Try
18	Ala	Ala
19	Asp	Asp
20	Pro*	Pro
21	Ileu	Ileu
22	Glu	Glu
23	Leu	Leu
24	Ileu	Leu
25	AspN*	AspN
26	Leu	Val
27	Cys	Cys
28	Thr	Thr
29	AspN	Ser
30	Ala	Ser
31	Leu	Leu
32	Gly	Gly
33	AspN*	AspN
34	GluN	GluN
35	Phe	Phe
36	GluN	GluN
37	Thr	Thr
38	GluN	GluN
39	GluN	GluN
40	Ala	Ala
41	Arg	Arg
42	Thr	Thr
43	Val	Thr
44	Val	Val
45	GluN	GluN
46	Arg*	GluN
47	GluN	GluN
48	Phe	Phe
49	Ser	Ser
50	GluN	Glu
51	Val	Val
52	Try	Try
53	Lys	Lys
54	Pro	Pro
55	Ser	Phe
56	Pro	Pro
57	GluN	GluN
58	Val	Ser
59	Thr	Thr
60	Val	Val
61	Arg*	Arg
62	Phe	Phe
63	Pro	Pro
64	Asp	Gly
65	Ser	Asp
66	Asp	Val
67	Phe	Tyr
68	Lys	Lys
69	Val	Val
70	Tyr	Tyr
71	Arg	Arg
72	Tyr	Tyr
73	AspN	AspN
74	Ala	Ala
75	Val	Val
76	Leu	Leu
77	Asp	Asp
78	Pro	Pro
79	Leu	Leu
80	Val	Ileu
81	Thr*	Thr
82	Ala	Ala
83	Leu	Leu
84	Leu	Leu
85	Gly	Gly
86	Ala	Thr
87	Phe	Phe
88	Asp	Asp
89	Thr	Thr
90	Arg	Arg
91	AspN	AspN
92	Arg	Arg
93	Ileu	Ileu
94	Ileu	Ileu
95	Glu	Glu
96	Val	Val
97	Glu*	Glu
98	AspN	AspN
99	GluN	GluN
100	Ala	GluN
101	AspN	Ser
102	Pro	Pro
103	Thr	Thr
104	Thr	Thr
105	Ala	Ala
106	Glu	Thr
107	Thr*	Leu
108	Leu	Leu
109	Asp	Thr
110	Ala	Ala
111	Thr	Thr
112	Arg	Arg
113	Arg	Arg
114	Val	Val
115	Asp	Asp
116	Asp	Asp
117	Ala	Ala
118	Thr	Thr
119	Val	Val
120	Ala	Ala
121	Ileu	Ileu
122	Arg	Arg
123	Ser	Ser
124	Ala	Ala
125	Ileu	Ileu
126	AspN	AspN
127	AspN	AspN
128	Leu	Leu
129	Ileu*	Val
130	Val	AspN
131	Glu	Glu
132	Leu	Leu
133	Ileu	Val
134	Arg*	Arg
135	Gly	Gly
136	Thr	Thr
137	Gly	Gly
138	Ser*	Leu
139	Tyr	Tyr
140	Tyr	Tyr
141	Arg	GluN
142	Ser	AspN
143	Ser	Thr
144	Phe	Phe
145	Glu	Glu
146	Ser	Ser
147	Ser	Met
148	Ser*	Ser
149	Gly	Gly
150	Leu	Leu
151	Val	Val
152	Try	Try
153	Thr	Thr
154	Ser	Ser
155	Gly	Ala
156	Pro*	Pro
157	Ala	Ala
158	Thr	Ser

Fig. 6.

most proteins. The facts that in the virus rod the C-terminus of the peptide chain is very susceptible to carboxypeptidase, and only the tyrosine nearest the C-terminus is reactive towards iodine (*14*) indicates that this part of the molecule is near the surface of the rod, and thus near one end of the oblong protein molecule. Immunological studies with anti-TMV serum confirm this (*2*) (Fig. 7). In contrast the use of

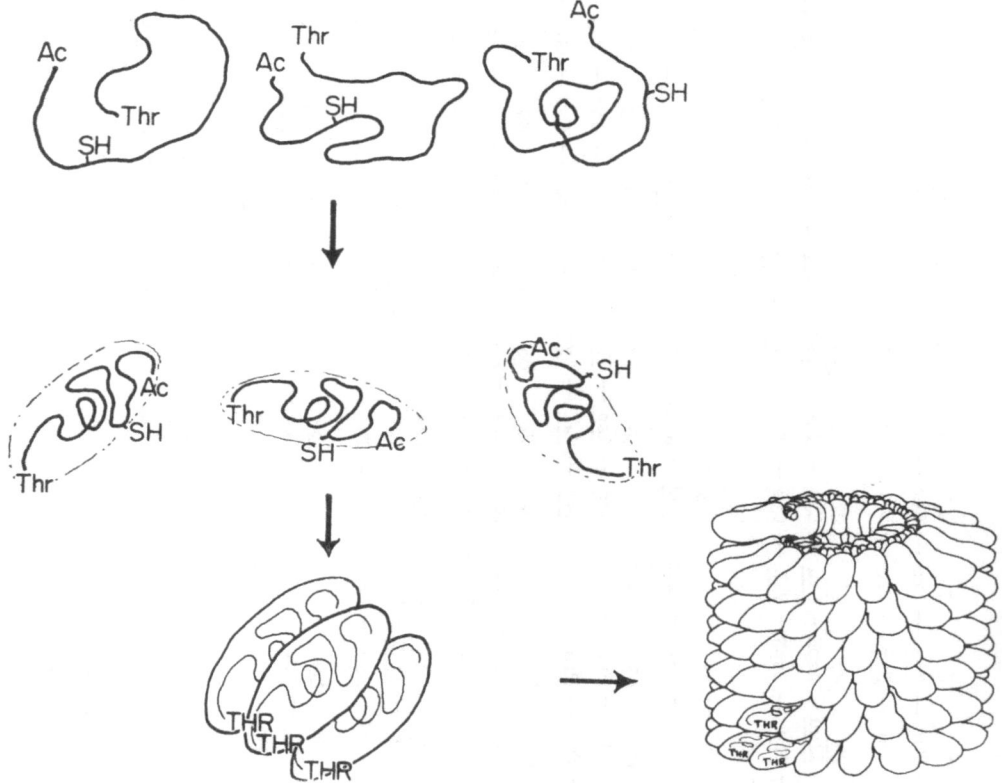

Fig. 7. Reversible Denaturation. Transition from denatured TMV-protein to native protein, native protein aggregates, and virus rods. The C-terminal part of the chain (Thr) is peripheral in the rod.

antiserum to dissociated TMV-protein indicated that a strongly determinant group was located between residues 93 and 112 (peptide 8), which thus must be near an inter-subunit surface (*4*).

Not only the immunological but also the electrical charge properties of viruses are determined by the protein surface rather than the entire composition. Thus, the electrophoretic mobilities of many viruses are unaffected by the presence of the highly charged nucleic acid inside the protein shell. This is true for in vitro aggregated TMV-protein as well as

References, pp. 433—437.

for the viral protein particles which appear in many virus-host systems together with the virus.

Relation of Viral Infectivity to the Structure and Interaction of its Components. As stated, many viruses occur in the host cell together with identical protein shells or rods, as the case may be, but lacking the nucleic acid. In the case of TMV, the protein tends to aggregate in vitro to rods which are not readily distinguishable in electrophoretic, immunological and electron microscopical regards from the complete virus. In structural terms the nucleic acid contributes mainly a stabilizing backbone so that the intact virus rod is stable over a much wider range of pH, temperature and ionic strength than the rod composed of protein alone. On the other hand, the latter can "grow" to any length, while the stabilizing action of the RNA is limited to the length of that molecule when passing through the groove formed by the protein subunits with its characteristic diameter and pitch.

That the 300 mμ length of the TMV particle is determined by the length of its RNA, was recently newly demonstrated when it proved possible to coaggregate TMV-protein with the RNA of the bacteriophage MS2 as the backbone. This RNA has a molecular weight of only about 60% of that of TMV-RNA, and the rods that it formed with TMV protein, while of the same diameter as TMV, were predominantly about 60% as long as complete TMV rods (*61*).

The tendency of TMV-protein to aggregate to virus-like particles, and the ease with which TMV-RNA is incorporated and the actual virus is formed in vitro has presented a strong stimulus to the development of our current concepts of the nature of viruses. Ten years ago it was demonstrated that TMV-RNA could be isolated under gentle conditions avoiding high or low pH and that it then remained largely intact and infective when applied to host leaves (*11, 24*). However, in quantitative terms the infectivity of TMV-RNA is 2–3 orders of magnitude below that of its virus equivalent, a fact which is attributed to the sensitivity of free single-stranded RNA to nucleases, hydroxyl and metal ions and other agents it may encounter upon entering a cell. Yet, when this RNA is allowed to interact with TMV-protein at 25–37° for a few hours in 0.1 M salt at pH 7, virus is formed and most of the infectivity of the original virus from which the RNA was isolated is regenerated (*20*). This reconstitution reaction presented the first and best proof that the RNA, though actually of low infectivity was potentially fully infective. Further, by reconstituting "mixed" virus from RNA of a strain of TMV with wild type protein or vice versa, it was first clearly and incontrovertibly shown that the genetic function of RNA viruses was carried entirely and exclusively by the RNA and was uninfluenced by the nature of the protein (*16, 17, 19, 20*).

The infectivity of viral nucleic acids was subsequently demonstrated with most types of viruses, although very large nucleic acids (such as the DNA of the T-even bacteriophages) or the presence of great amounts of complex proteins (such as in the influenza virus) seem to present major technical obstacles to this demonstration. However, in vitro reconstitution of virus from its two components has not been clearly demonstrated for any virus other than TMV and its related strains. The limiting factor appears to lie in the ease with which the protein tends to aggregate. Thus TMV protein readily forms rods without any RNA and interacts with variable efficiency and reproducibility with a variety of different types of RNA, including synthetic polynucleotides rich in purines (*19*). On the other hand, the proteins of spherical viruses have never been observed to form virus-like particles in vitro.

Modification of Viral RNA and its Consequences. With viral RNA as with other compounds of complex chemical structure, modification can proceed pending elucidation of the details of the structure.

Table 3, p. 430 summarizes the reagents that have been used with viral nucleic acids, the presumed or established reactions involved, and the results in terms of the biological properties of the nucleic acid. All chemical modifications cause loss of infectivity, usually according to exponential kinetics. The number of chemical modifications required for inactivation can be inferred by intrapolation from analytical data obtained on much more intensely treated preparations (*47*), or it can be directly determined with radioactive reagents which become bound during the reaction (*13*). One or two chemical hits appear usually sufficient to inactivate the molecule, unless reversible reactions like that with formaldehyde are being studied. Molecules which have sustained one or a limited number of hits are obviously heterogeneous in terms of both number and location of hits. Thus they cannot be chemically analyzed and identified. However, by applying the reaction mixture to a suitable host, pure cultures of the progeny of those chemically altered molecules which have retained their infectivity can be isolated. The transformation of one base to another would, in principle, be detectable by analysis of the RNA of the progeny of such a molecule if sufficiently accurate methods of analysis were available. Actually, at present, it is not possible to detect such a change of approximately 0.02% in the analytical values and thus the chemistry of mutagenesis would appear to be beyond the present range of methodology.

This is not necessarily the case, however, because of the fortunate circumstance that each RNA molecule occurs in the virus in association with 1000 or more relatively small protein molecules. Any mutagenic event in the part of the RNA which codes for that coat protein is thus actually potentiated by a factor of at least 10^3, and if it results in an

References, pp. 433—437.

amino acid replacement, this is easily within range of present day analysis. Thus, extensive studies of the amino acid composition and sequence of the coat proteins of chemically produced virus mutants have been carried out in our laboratory, as well as that of WITTMANN in Tübingen and will briefly be summarized below (*23, 66, 67, 75–77*).

The most interesting modification is the deamination reaction, for it is the only one able to transform one typical base, cytosine, into another, uracil. If the sequence of bases along the polynucleotide chain represents

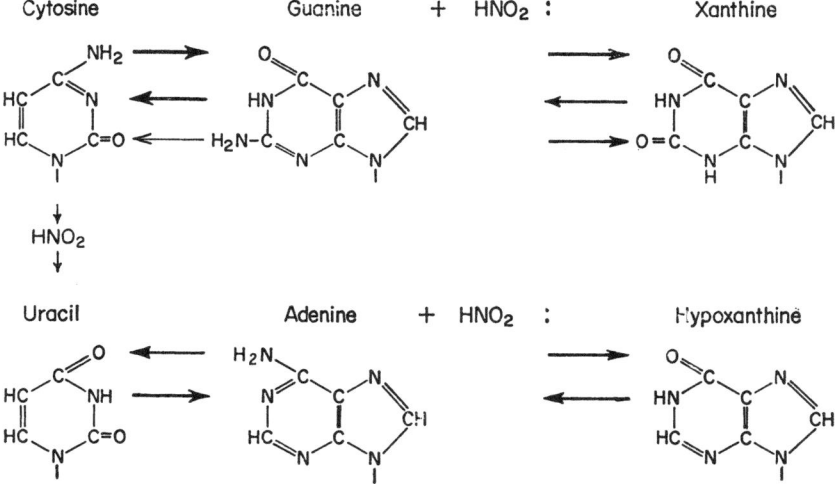

Fig. 8. Effect of Deamination on the Base-binding Ability of Nucleotides. Besides the obvious direct change of cytosine to uracil, the deamination of adenine to hypoxanthine makes it simulate guanine as shown by the arrows indicating the direction of hydrogen bond donation. The deamination of guanine to xanthine gives a product which does not fit to either pyrimidine.

the chemical basis of genetic function, then a change from one base to another would change that part of the genetic information and a mutation would thus have been achieved. This obligatory and direct relationship between chemical modification and mutagenesis obviously holds only for the RNA viruses, and it was definitely shown with these that most of the surviving molecules after deamination were mutants.

Deamination of adenine transforms it to hypoxanthine which resembles guanine in its base pairing properties and thus is also believed to lead to mutations, see *Figure 8* (*71*). The change of C to U in DNA is of similar nature, since U resembles T in its binding properties, yet does not represent a direct transformation of one normal component of that nucleic acid to another such component. Some of the other reaction products listed in Table 3 (p. 430) may also belong in this category.

Table 3. Modification of Viral RNA.

Reagent	pH	Specificity and Reaction Mechanism	Biological Effects		References
			Chemical Events per Lethal Event	Mutagenesis	
Nitrous acid	4–5	C → U; A → hypoxanthin; G → xanthine, etc.	2	++++	(47)
Formaldehyde	7	A > C, G (addition of one or two HCHO, Schiff's base formation)	∼ 20		(56)
Glyoxal, Kethoxal	7	G (substitution)	∼ 20		(57)
Hydroxylamine	6	C ≫ U (substitution)			
	9–10	U > C (ring opening)		+	(48, 70)
Bromine	7	C ≫ G			(5)
	9	G ≫ C		+	
Iodine (N-Iodosuccinimide)	7	C	2–3	—	(5)
Dimethylsulfate, Epoxides, etc.	7	G > A (alkylation at G_7, A_1)	1–3	+	(13)
Iron + light	7	loss of bases			(51)
Dyes + light	7	G	∼ 20		(53)
U. V.-light	7	C, U	(1–3)		(38)

References, pp. 433—437.

Table 4. Amino Acid Exchanges in Chemically Produced Mutants*.

Amino Acid Exchange	HNO₂**	Br₂	CH₃—	Others	Codon Exchange***	
Arg < Gly	4	3	1		A → G	AGpu < GGpu
Arg < Lys	1!	1			G → A	AGpu < AApu
Asp < Ala	(4)!				A → C	GApy < GCpy
Asp < Gly	(2)				A → G	GApy < GGpy
AspNH₂→Ser	4 (2)	3			A → G	AAC→AGC
Glu < Gly	1 (1)				A → G	GApu < GGpu
Glu < Val	(2)!				A → U	GApu < GUpu
GluNH₂ < Arg	1				A → G	CApu < CGpu
GluNH₂ < His (?)				1	pu → py	CApu < CApy
Ileu < Met	(2)				A → G?	A→AUG
Ileu → Thr		1		1	U → C	AUpy → ACpy
Ileu < Val	3 (4)				A → G	AUpy < GUpy
Leu→Phe	(1)				C → U	CUpy→UUpy
Pro < Leu	3 (4)	6	4	(1)	C → U	CCp < CUp
Pro < Ser	(4)				C → U	CCp < UCp
Ser < Gly		1		(1)	A → G	AGC→GGC
Ser → Leu	(2)				C → U	UCp < UUG
Ser < Phe	4 (4)	2	2		C → U	UCp < UUpy
Thr < Ala	2			(1)	A → G	ACp < GCp
Thr → Ileu	(9)				C → U	ACp < AUpy
Thr < Met	(3)				C → U	ACp < AUG
Tyr→Cys	(1)				A → G	UApy→UGpy
Val < Ala				(1)	U → C	GUG < GCp
Val < Met	1!				G → A	GUG < AUG

* Numbers of mutants found to show listed amino acid exchanges after various RNA modification reactions (nitrous acid, bromination, methylation, etc.), and the possible explanations of these exchanges on the basis of exchanges of single nucleotides. Each of the trinucleotides in the right column has been shown to code for the correspondingly positioned amino acid in the left column.

** Data of H. G. WITTMANN are listed in parentheses.

*** pu stands for A or G; py stands for C or U; and p stands for A, G, U or C.

Other chemical modifications which destroy a base completely or alter it in such a manner that it no longer has bonding properties complementary to any of the four nucleotides are believed to be lethal. Among these is the action of nitrous acid on guanine (71), and the majority of the modification reactions listed in Table 3 (Fig. 8, p. 429).

The results of the studies of protein modifications resulting from various chemical modifications of TMV-RNA are summarized in *Table 4*. Also listed are base exchanges which could account for these results on the basis of the now known codons, viz. triplets of nucleotides which code for specific amino acids [data of Nierenberg's, Khorana's and other groups, as summarized by Watson (72)]. By far most of the amino acid exchanges can be accounted for by single base exchanges, most of these being C → U and A → G. Since most of the mutants studied were evoked by nitrous acid treatment, this represents statistical evidence supporting both the mechanism of mutagenesis given above and the correctness of the codon indentifications. Exceptions are two singly observed exchanges from our laboratory and, more disturbingly, two repeatedly observed ones from Tübingen. Also unexplained is the fact that the most frequently recurring exchanges are the same for nitrous acid treatment and for agents which are believed to act in quite different manner on the RNA, such as dimethylsulfate which methylates guanine and to a lesser extent adenine. Also unexplained are the observed frequencies of some replacements and the failure to detect others which should be equally

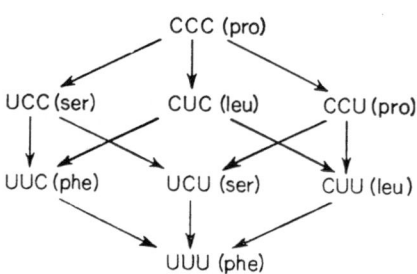

	EXPECTED RATIO	FOUND ※
Pro —➤ Ser	I	4
Pro —➤ Leu	I	7
Ser —➤ Phe	I	8
Leu —➤ Phe	I	I

Fig. 9. Effect of HNO₂ on all Codons Consisting of C and U. The amino acid exchanges expected from the now known codon allocations are the only ones observed (within this group), but the observed frequencies do not correspond to expectation, even if one includes the other known codons for leucine and serine (not consisting of U and C alone) in this consideration.

possible in terms of codon alterations by nitrous acid. It may safely be assumed that not all amino acid replacements yield functional protein, but one would predict that of the four exchanges accounted for by C and U containing codons alone *(Fig. 9)* the most frequent one on the basis of protein chemical considerations would be leucine → phenylalanine and the rarest, serine → phenylalanine, in exactly opposite order to the observed frequencies.

In most instances the location of each exchange in the peptide chain has been determined and it has become evident that these are far from random. Only two of the 16 serines were repeatedly found to be replaced, and the same is true for the prolines, etc. All the frequently changed sites (see Fig. 6, p. 425) are in the terminal segments of the peptide chain, but not necessarily the last residues of each type. It appears very probable

References, pp. 433—437.

that this frequency distribution is due to the fact that the same replacement at other sites renders the protein nonfunctional. Such nonfunctional proteins have actually been detected in a few instances, and one such protein which had lost its ability to form a proper virus coat was found to differ by only two amino acid replacements from common TMV, apparently in functionally critical locations on the peptide chain (49, 81).

One fact remains to be pointed out. Many mutants (over half) produced by nitrous acid or other agents and detected as usual by changes in their symptomatology on one or another host, did not show any amino acid exchanges at all. The interpretation of this lies in the comparative numbers of total codons (nucleotide triplets) in viral RNA (about 1000–2000) and of those required to code for the coat protein (125–200), which is the only protein accessible to us for study. It is apparent from these numbers that many hits do not affect the gene (cystron is the now accepted term for a gene representing a specific single peptide chain) which is responsible for protein structure. Among those which do, some will not be mutagenic because of the redundancy of the code. Thus, not each nucleotide change specifies a different amino acid (e. g. UCC = = UCU = serine, Fig. 9). Other exchanges are lethal such as G → X, or represent a functional codon turning into a nonsense codon, (UCA → UAA), which may signify chain termination (72). Finally, other exchanges lead to production of a protein which is not able to fulfill its function.

Viewed as a whole, the chemical production of mutants by modification of viral RNA and the study of consequent changes in their coat protein have proven to be a valuable experimental tool in support of the nature of the genetic code. Furthermose, the availability of many variants of the coat protein which are able to function, and others which are not, is contributing to our understanding of the chain conformation and location of the binding groups which cooperatively enable virus proteins to function. Thus, it may be hoped that in the not too distant future the three-dimensional structure of such proteins will become established.

References.

1. ANDERER, F. A.: Reversible Denaturierung des Proteins aus Tabakmosaikvirus. Z. Naturforsch. **14b**, 642 (1959).
2. — Versuche zur Bestimmung der serologisch determinanten Gruppen des Tabakmosaikvirus. Z. Naturforsch. **18b**, 1010 (1963).
3. ANDERER, F. A., H. UHLIG, E. WEBER and G. SCHRAMM: Primary Structure of the Protein of Tobacco Mosaic Virus. Nature **186**, 922 (1960).
3a. ANDERER, F. A., B. WITTMANN-LIEBOLD and H. G. WITTMANN: Weitere Untersuchungen zur Aminosäuresequenz des Proteins im Tabakmosaikvirus. Z. Naturforsch. **20b**, 1203 (1965).

4. Benjamini, E., J. D. Young, W. J. Peterson, C. Y. Leung and M. Shimizu: Immunochemical Studies on the Tobacco Mosaic Virus Protein. II. The Specific Binding of a Tryptic Peptide of the Protein with Antibodies to the Whole Protein. Biochemistry 4, 2081 (1965).

5. Brammer, K. W.: Chemical Modification of Viral Ribonucleic Acid. II. Bromination and Iodination. Biochim. Biophys. Acta 72, 217 (1963).

6. Cairns, J.: An Estimate of the Length of the DNA Molecule of T2 Bacteriophage by Autoradiology. J. Mol. Biol. 3, 756 (1961).

7. Cooper, W. D. and H. S. Loring: The Purine and Pyrimidine Composition of the Tobacco Mosaic Virus and the Holmes Masked Strain. J. Biol. Chem. 211, 505 (1954).

8. Dorner, R. W. and C. A. Knight: The Preparation and Properties of Some Plant Virus Nucleic Acids. J. Biol. Chem. 205, 959 (1953).

9. Fiers, W., L. Lepoutre and L. Vandendriessche: Studies on the Bacteriophage MS2. I. Distribution of Purine Sequences in the Viral RNA and in Yeast RNA. J. Mol. Biol. 13, 432 (1965).

10. Fiers, W. and R. L. Sinsheimer: The Structure of the DNA of Bacteriophage ΦX174. I. The Action of Exopolynucleotidases. J. Mol. Biol. 5, 408 (1962).

11. Fraenkel-Conrat, H.: The Role of Nucleic Acid in the Reconstitution of Active Tobacco Mosaic Virus. J. Amer. Chem. Soc. 78, 882 (1956).

12. — The Masked —SH Group in Tobacco Mosaic Virus Protein. In: R. Benesch et al. (Edits.), Sulfur in Proteins (Symposium, 1958), p. 339. New York: Academic Press. 1959.

13. — Chemical Modification of Viral Ribonucleic Acid. I. Alkylating Agents. Biochim. Biophys. Acta 49, 169 (1961).

14. — Iodination of TMV Protein. Abstract, 142nd Amer. Chem. Soc. Meeting, Atlantic City, p. 44C (1962).

15. Fraenkel-Conrat, H., J. I. Harris and A. L. Levy: Recent Developments in Techniques for Terminal and Sequence Studies in Peptides and Proteins. In: D. Glick (Edit.), Methods of Biochemical Analysis, Vol. II, p. 359. New York and London: Interscience. 1955.

16. Fraenkel-Conrat, H. and B. Singer: Virus Reconstitution. II. Combination of Protein and Nucleic Acid from Different Strains. Biochim. Biophys. Acta 24, 540 (1957).

17. — — Reconstitution of Tobacco Mosaic Virus. III. Improved Methods and the Use of Mixed Nucleic Acids. Biochim. Biophys. Acta 33, 359 (1959).

18. — — The Absence of Phosphorylated Chain Ends in TMV-RNA. Biochemistry 1, 120 (1962).

19. — Reconstitution of Tobacco Mosaic Virus. IV. Inhibition of Enzymes and other Proteins, and Use of Polynucleotides. Virology 23, 354 (1964).

20. Fraenkel-Conrat, H. and R. C. Williams: Reconstitution of Active Tobacco Mosaic Virus from its Inactive Protein and Nucleic Acid Components. Proc. Nat. Acad. Sci. (USA) 41, 690 (1955).

21. Fremery, D. de and C. A. Knight: A Chemical Comparison of Three Strains of Tomato Bushy Stunt Virus. J. Biol. Chem. 214, 559 (1955).

22. Funatsu, G.: Separation of Tryptic Peptides of Tobacco Mosaic Virus and Strain Proteins by an Improved Method of Column Chromatography. Biochemistry 3, 1351 (1964).

23. Funatsu, G. and H. Fraenkel-Conrat: Location of Amino Acid Exchanges in Chemically Evoked Mutants of Tobacco Mosaic Virus. Biochemistry 3, 1356 (1964).

23a. FUNATSU, G., A. TSUGITA and H. FRAENKEL-CONRAT: Studies on the Amino Acid Sequence of Tobacco Mosaic Virus Protein. V. Amino Acid Sequences of Two Peptides from Tryptic Digests and Location of Amide Group. Arch. Biochem. Biophys. 105, 25 (1964).

24. GIERER, A. and G. SCHRAMM: Infectivity of Ribonucleic Acid from Tobacco Mosaic Virus. Nature 177, 702 (1956).

25. GROSS, E. and B. WITKOP: Nonenzymatic Cleavage of Peptide Bonds: Methionine Residues in Bovine Pancreatic Ribonuclease. J. Biol. Chem. 237, 1856 (1962).

26. HARRIS, J. I. and J. HINDLEY: The Protein Subunit of Turnip Yellow Mosaic Virus. J. Mol. Biol. 13, 894 (1965).

27. HARRIS, J. I. and C. A. KNIGHT: The Action of Carboxypeptidase on Tobacco Mosaic Virus. Nature 170, 613 (1952).

28. — — The Action of Carboxypeptidase on Strains of Tobacco Mosaic Virus. J. Biol. Chem. 214, 231 (1955).

29. HOLLEY, R. W., J. APGAR, G. A. EVERETT, J. T. MADISON, M. MARQUISEE, S. H. MERRILL, J. R. PENSWICK and A. ZAMIR: Structure of a Ribonucleic Acid. Science 147, 1462 (1965).

30. HUNT, J. A.: Terminal-Sequence Studies of High-Molecular Weight Ribonucleic Acid. The Reaction of Periodate Oxidized Ribonucleosides, 5'-Ribonucleotides and Ribonucleic Acid with Isoniazid. Biochem. J. 95, 541 (1965).

31. KALLEN, R. G., M. SIMON and J. MARMUR: The Occurrence of a new Pyrimidine Base Replacing Thymine in a Bacteriophage DNA: 5-Hydroxymethyl Uracil. J. Mol. Biol. 5, 248 (1962).

32. KAPER, J. M. and R. L. STEERE: Infectivity of Tobacco Ringspot Virus Nucleic Acid Preparations. Virology 7, 127 (1959).

33. KASSANIS, B.: Properties and Behaviour of a Virus Depending for its Multiplication on Another. J. Gen. Microbiol. 27, 477 (1962).

34. KERR, I. M., E. A. PRATT and I. R. LEHMAN: Exonucleolytic Degradation of High-Molecular-Weight DNA and RNA to Nucleoside 3'-Phosphates by a Nuclease from B. subtilis. Biochem. Biophys. Res. Comm. 20, 154 (1965).

35. KLEINSCHMIDT, A. K., D. LANG, D. JACHERTS und R. K. ZAHN: Darstellung und Längenmessungen des gesamten Desoxyribonucleinsäure-Inhaltes von T₂-Bakteriophagen. Biochim. Biophys. Acta 61, 857 (1962).

36. KNIGHT, C. A.: The Chemical Constitution of Viruses. In: K. M. SMITH and M. A. LAUFFER (Edits.), Advances in Virus Research, Vol. II, p. 152. New York: Academic Press. 1954.

36a. LIN, J.-Y., CH. M. TSUNG and H. FRAENKEL-CONRAT: The Coat Protein of the RNA Phage MS-2. J. Mol. Biol. (in press).

37. MACLEOD, R. and R. MARKHAM: Experimental Evidence of a Relationship Between Turnip Yellow Mosaic Virus and Wild Cucumber Mosaic Virus. Virology 19, 190 (1963).

38. MARKHAM, R.: Nucleic Acids in Virus Multiplication. In: P. FILDES and W. E. VAN HEYNINGEN (Edits.). The Nature of Virus Multiplication (Symposium), p. 85. Cambridge: Univ. Press. 1952.

39. NARITA, K.: Isolation of Acetylpeptide from Enzymic Digests of TMV-Protein. Biochim. Biophys. Acta 28, 184 (1958).

40. NEU, H. C. and L. A. HEPPEL: Nucleotide Sequence Analysis of Polyribonucleotides by Means of Periodate Oxidation Followed by Cleavage with an Amine. J. Biol. Chem. 239, 2927 (1964).

41. NIU, C.-I. and H. FRAENKEL-CONRAT: C-Terminal Amino Acid Sequence of Tobacco Mosaic Virus Protein. Biochim. Biophys. Acta 16, 597 (1955).

28*

42. Ralph, R. K., R. J. Young and H. G. Khorana: Studies on Polynucleotides. XXI. Amino Acid Acceptor Ribonucleic Acids (2). The Labeling of Terminal 5'-Phosphomonoester Groups and a Preliminary Investigation of Adjoining Nucleotide Sequences. J. Amer. Chem. Soc. 85, 2002 (1963).

43. Richardson, C. C.: Phosphorylation of Nucleic Acid by an Enzyme from T4 Bacteriophage-Infected *Escherichia coli*. Proc. Nat. Acad. Sci. (USA) 54, 158 (1965).

44. Rushizky, G. W. and C. A. Knight: Products Obtained by Digestion of the Nucleic Acids of Some Strains of Tobacco Mosaic Virus with Pancreatic Ribonuclease. Proc. Nat. Acad. Sci. (USA) 46, 945 (1960).

45. Rushizky, G. W., C. A. Knight and H. A. Sober: Studies on the Preferential Specificity of Pancreatic Ribonuclease as Deduced from Partial Digests. J. Biol. Chem. 236, 2732 (1961).

46. Rushizky, G. W., H. A. Sober and C. A. Knight: Products Obtained by Digestion of the Nucleic Acids of Some Strains of Tobacco Mosaic Virus with Ribonuclease T_1. Biochim. Biophys. Acta 61, 56 (1962).

47. Schuster, H. und G. Schramm: Bestimmung der biologisch wirksamen Einheit in der Ribosenucleinsäure des Tabakmosaikvirus auf chemischem Wege. Z. Naturforsch. 13b, 697 (1958).

48. Schuster, H. and H. G. Wittmann: The Inactivating and Mutagenic Action of Hydroxylamine on Tobacco Mosaic Virus Ribonucleic Acid. Virology 19, 421 (1963).

49. Siegel, A., M. Zaitlin and O. P. Sehgal: The Isolation of Defective Tobacco Mosaic Virus Strains. Proc. Nat. Acad. Sci. (USA) 48, 1845 (1962).

50. Singer, B. and H. Fraenkel-Conrat: Studies of Nucleotide Sequences in TMV-RNA. I. Stepwise Use of Phosphodiesterase. Biochim. Biophys. Acta 72, 534 (1963).

51. — — Effects of Light in the Presence of Iron Salts on Ribonucleic Acid and Model Compounds. Biochemistry 4, 226 (1965).

52. — — Action of Polynucleotide Phosphorylase on TMV-RNA. Federat. Proc. (Amer. Soc. Exp. Biol.) 24, 603 (1965).

53. — — Dye-Catalyzed Photoinactivation of Tobacco Mosaic Virus Ribonucleic Acid. Biochemistry 5, 2446 (1966).

54. Singer, B., M. Sherwood and H. Fraenkel-Conrat: Studies of Nucleotide Sequences in Tobacco Mosaic Virus. II. The Action of Spleen Phosphodiesterase. Biochim. Biophys. Acta 108, 306 (1965).

55. Sinsheimer, R. L.: Nucleotides from T2r+ Bacteriophage. Science 120, 551 (1954).

56. Staehelin, M.: Reaction of Tobacco Mosaic Virus Nucleic Acid with Formaldehyde. Biochim. Biophys. Acta 29, 410 (1958).

57. — Inactivation of Virus Nucleic Acid with Glyoxal Derivatives. Biochim. Biophys. Acta 31, 448 (1959).

58. Steinschneider, A. and H. Fraenkel-Conrat: Studies of Nucleotide Sequences in Tobacco Mosaic Virus Ribonucleic Acid. III. Periodate Oxidation and Semicarbazone Formation. Biochemistry 5, 2729 (1966).

59. — — Studies of Nucleotide Sequences in Tobacco Mosaic Virus Ribonucleic Acid. IV. Use of Aniline in Stepwise Degradation. Biochemistry 5, 2735 (1966).

60. Sugiyama, T.: 5'-Linked End Group of RNA from Bacteriophage MS2. J. Mol. Biol. 11, 856 (1965).

61. — Tobacco Mosaic Viruslike Rods Formed by "Mixed Reconstitution" between MS2 Ribonucleic Acid and Tobacco Mosaic Virus Protein. Virology 28, 488 (1966).

62. SUGIYAMA, T. and H. FRAENKEL-CONRAT: Identification of 5'-Linked Adenosine as End Group of TMV-RNA. Proc. Nat. Acad. Sci. (USA) **47,** 1393 (1961).

63. — — The End-Groups of Tobacco Mosaic Virus RNA. II. Nature of the 3'-Linked Chain End in TMV and of Both Ends in Four Strains. Biochemistry **2,** 332 (1963).

64. SYMONS, R. H., M. W. REES, M. N. SHORT and R. MARKHAM: Relationships Between the Ribonucleic Acid and Protein of Some Plant Viruses. J. Mol. Biol. **6,** 1 (1963).

65. TAKAHASHI, I. and J. MARMUR: Replacement of Thymidylic Acid by Deoxy-uridylic Acid in the Deoxyribonucleic Acid of a Transducing Phage for *Bacillus subtilis.* Nature **197,** 794 (1963).

66. TSUGITA, A. and H. FRAENKEL-CONRAT: The Amino Acid Composition and C-Terminal Sequence of a Chemically Evoked Mutant of TMV. Proc. Nat. Acad. Sci. (USA) **46,** 636 (1960).

67. — — The Composition of Proteins and Chemically Evoked Mutants of TMV-RNA. J. Mol. Biol. **4,** 73 (1962).

68. TSUGITA, A., D. T. GISH, J. YOUNG, H. FRAENKEL-CONRAT, C. A. KNIGHT and W. M. STANLEY: The Complete Amino Acid Sequence of the Protein of Tobacco Mosaic Virus. Proc. Nat. Acad. Sci. (USA) **46,** 1463 (1960).

69. TSUNG, C. M. and H. FRAENKEL-CONRAT: Preferential Release of Aspartic Acid by Dilute Acid Treatment of Tryptic Peptides. Biochemistry **4,** 793 (1965).

70. VERWOERD, D. W., H. KOHLHAGE and W. ZILLIG: Specific Partial Hydrolysis of Nucleic Acids in Nucleotide Sequence Studies. Nature **192,** 1038 (1961).

71. VIELMETTER, W. und H. SCHUSTER: Die Basenspezifität bei der Induktion von Mutationen durch salpetrige Säure im Phagen T 2. Z. Naturforsch. **15b,** 304 (1960).

72. WATSON, J. D.: Molecular Biology of the Gene. New York: Benjamin. 1965.

72a. WEBER, K., G. NOTANI, M. WIKLER and W. KONIGSBERG: Amino Acid Sequence of the f₂ Coat Protein. J. Mol. Biol. **20,** 423 (1966).

73. WHITFELD, P. R.: A Method for the Determination of Nucleotide Sequence in Polyribonucleotides. Biochem. J. **58,** 390 (1954).

74. — Identification of End Groups in Tobacco Mosaic Virus Ribonucleic Acid by Enzymatic Hydrolysis. J. Biol. Chem. **237,** 2865 (1962).

74a. WITTMANN, H. G.: Comparison of the Tryptic Peptides of Wild Strains of Tobacco Mosaic Virus. Virology **12,** 613 (1960).

75. — Proteinuntersuchungen an Mutanten des TMV als Beitrag zum Problem des genetischen Codes. Z. Vererbungsl. **93,** 491 (1962).

76. — Übertragung der genetischen Information. Naturwiss. **50,** 76 (1963).

77. — Proteinanalysen von chemisch induzierten Mutanten des Tabakmosaikvirus. Z. Vererbungsl. **95** 333 (1964).

78. WITTMANN-LIEBOLD, B. und H. G. WITTMANN: Die primäre Proteinstruktur von Stämmen des Tabakmosaikvirus. Aminosäuresequenzen des Proteins des Tabakmosaikvirusstammes *Dahlemense.* Teil III: Diskussion der Ergebnisse. Z. Vererbungsl. **94,** 427 (1963).

79. YAMAZAKI, H., J. BANCROFT and P. KAESBERG: Biophysical Studies of Broad Bean Mottle Virus. Proc. Nat. Acad. Sci. (USA) **47,** 979 (1961).

80. YAMAZAKI, H. and P. KAESBERG: Biophysical and Biochemical Properties of Wild Cucumber Mosaic Virus and of Two Related Virus-Like Particles. Biochim. Biophys. Acta **51,** 9 (1961).

81. ZAITLIN, M. and W. F. MCCAUGHEY: Amino Acid Composition of a Nonfunctional Tobacco Mosaic Virus Protein. Virology **26,** 500 (1965).

(Received, December 20, 1965.)

Namenverzeichnis. Index of Names. Index des Auteurs.

Kursiv gedruckte Seitenzahlen beziehen sich auf Literaturverzeichnisse.

Page numbers printed in *italic* refer to References.

Les chiffres en *italique* indiquent les pages de bibliographie.

Sachverzeichnis. Index of Subjects. Index des Matières.

Fortschritte der Chemie organischer Naturstoffe. Progress in the Chemistry of Organic Natural Products. Progrès dans la chimie des substances organiques naturelles. Herausgegeben von **L. Zechmeister,** California Institute of Technology, Pasadena, California, U. S. A.

Springer-Verlag / Wien · New York

Bisher erschienen:

Erster Band: Mit 41 Abbildungen im Text. VI, 371 Seiten. Gr.-8°. 1938.
Ganzleinen S 348.—, DM 72.25, $ 17.20

Zweiter Band: Mit 24 Abbildungen im Text. VII, 366 Seiten. Gr.-8°. 1939.
Ganzleinen S 348.—, DM 72.25, $ 17.20

Dritter Band: Mit 10 Abbildungen im Text. VI, 252 Seiten. Gr.-8°. 1939.
Ganzleinen S 264.—, DM 55.45, $ 13.20

Vierter Band: Mit 47 Abbildungen im Text. VIII, 499 Seiten. Gr.-8°. 1945.
Ganzleinen S 474.—, DM 99.10, $ 23.60

Fünfter Band: Mit 34 Abbildungen. VIII, 417 Seiten. Gr.-8°. 1948.
Ganzleinen S 305.—, DM 50.40, $ 12.—

Sechster Band: Mit 32 Abbildungen. VIII, 392 Seiten. Gr.-8°. 1950.
Ganzleinen S 338.—, DM 55.80, $ 13.30

Siebenter Band: Mit 12 Abbildungen. VII, 330 Seiten. Gr.-8°. 1950.
Ganzleinen S 325.—, DM 53.70, $ 12.80

Achter Band: Mit 47 Abbildungen. XI, 400 Seiten. Gr.-8°. 1951.
Ganzleinen S 427.—, DM 70.50, $ 16.80

Neunter Band: Mit 20 Abbildungen. XI, 535 Seiten. Gr.-8°. 1952.
Ganzleinen S 498.—, DM 82.50, $ 19.60

Zehnter Band: Mit 19 Abbildungen. IX, 529 Seiten. Gr.-8°. 1953.
Ganzleinen S 498.—, DM 83.—, $ 19.80

Elfter Band: Mit 67 Abbildungen. VIII, 457 Seiten. Gr.-8°. 1954.
Ganzleinen S 448.—, DM 74.80, $ 18.—

Zwölfter Band: Mit 15 Abbildungen. X, 550 Seiten. Gr.-8°. 1955.
Ganzleinen S 497.—, DM 82.80, $ 19.80

Dreizehnter Band: Mit 48 Abbildungen. XII, 624 Seiten. Gr.-8°. 1956.
Ganzleinen S 645.—, DM 107.50, $ 25.60

Vierzehnter Band: Mit 38 Abbildungen. VIII, 377 Seiten. Gr.-8°. 1957.
Ganzleinen S 450.—, DM 75.—, $ 17.85

Weitere Bände siehe nächste Seite!

Zu beziehen durch Ihre Buchhandlung
Auslieferung für die U. S. A. und Canada: Springer-Verlag New York Inc.,
175 Fifth Avenue, New York N. Y. 10010

Springer-Verlag / Wien · New York

Fortsetzung von vorhergehender Seite

Fünfzehnter Band: Mit 81 Abbildungen. VI, 244 Seiten. Gr.-8°. 1958.
Ganzleinen S 246.—, DM 41.—, $ 9.75

Sechzehnter Band: Mit 27 Abbildungen. VI, 226 Seiten. Gr.-8°. 1958.
Ganzleinen S 240.—, DM 40.—, $ 9.50

Siebzehnter Band: Mit 57 Abbildungen. X, 515 Seiten. Gr.-8°. 1959.
Ganzleinen S 498.60, DM 83.10, $ 19.80

Achtzehnter Band: Mit 65 Abbildungen. X, 600 Seiten. Gr.-8°. 1960.
Ganzleinen S 618.—, DM 103.—, $ 24.50

Neunzehnter Band: Mit 16 Abbildungen. VIII, 420 Seiten. Gr.-8°. 1961.
Ganzleinen S 490.—, DM 78.—, $ 19.50

Zwanzigster Band: Mit 33 Abbildungen. XIII, 509 Seiten. Gr.-8°. 1962.
Ganzleinen S 604.—, DM 96.—, $ 24.—

Über den Inhalt der Bände gibt der Verlag bereitwilligst Auskunft

Einundzwanzigster Band: Mit 14 Abbildungen. VII, 362 Seiten. Gr.-8°. 1963.
Ganzleinen S 479.—, DM 76.—, $ 19.—

Inhalt: **Bonner, J.** The Biosynthesis of Rubber. — **Oroshnik, W.,** and **A. D. Mebane.** The Polyene Antifungal Antibiotics. — **Muxfeldt, H.,** und **R. Bangert.** Die Chemie der Tetracycline. — **Brockmann, M.** Anthracyclinone und Anthracycline (Rhodomycinone, Pyrromycinone und ihre Glykoside). — **Jaenicke, L.,** und **C. Kutzbach.** Folsäure und Folat-Enzyme. — **Crombie, L.** Chemistry of the Natural Rotenoids.

Generalregister / Cumulative Index / Index Général I—XX. 1938—1962.
XVI, 369 Seiten. Gr.-8°. 1964. Ganzleinen S 378.—, DM 60.—, $ 15.—

Zweiundzwanzigster Band: Mit 8 Abbildungen. VII, 370 Seiten. Gr.-8°.
1964. Ganzleinen S 554.—, DM 88.—, $ 22.—

Inhalt: **Schaffner, K.** Photochemische Umwandlungen ausgewählter Naturstoffe. — **Billek, G.** Stilbene im Pflanzenreich. — **Halsall, T. G.,** and **R. T. Aplin.** A Pattern of Development in the Chemistry of Pentacyclic Triterpenes. — **Grove, J. F.** Griseofulvin and Some Analogues. — **Scheuer, P. J.** The Chemistry of Toxins Isolated from Some Marine Organisms. — **Keller-Schierlein, W.,** **V. Prelog** und **H. Zähner.** Siderochrome.

Dreiundzwanzigster Band: Mit 58 Abbildungen. VIII, 397 Seiten. Gr.-8°.
1965. Ganzleinen S 590.—, DM 93.60, $ 23.40

Inhalt: **Peat, S.,** and **J. R. Turvey.** Polysaccharides of Marine Algae. — **Schlubach, H. H.** Der Kohlen-hydratstoffwechsel in Gerste, Hafer und Rispenhirse. — **Schlenk, F.** The Chemistry of Biological Sulfonium Compounds. — **Schroeder, W. A.,** and **R. T. Jones.** Some Aspects of the Chemistry and Function of Human and Animal Hemoglobins. — **Grassmann, W.** Kollagen. — **Jackman, L. M.** Some Applications of Nuclear Magnetic Resonance Spectroscopy in Natural Products Chemistry.

Subskribenten auf die „Fortschritte der Chemie organischer Naturstoffe" er-halten die Bände zu einem um 10% ermäßigten Vorzugspreis.

Bei Bezug der Serie Band 1—20 mit Generalregister 20% Nachlaß.

Zu beziehen durch Ihre Buchhandlung
Auslieferung für die U. S. A. und Canada: Springer-Verlag New York Inc.,
175 Fifth Avenue, New York N. Y. 10010